Quantum 20/20

Quantum 20/20

Fundamentals, Entanglement, Gauge Fields, Condensates and Topology

Ian R. Kenyon

OXFORD
UNIVERSITY PRESS

OXFORD
UNIVERSITY PRESS

Great Clarendon Street, Oxford, OX2 6DP,
United Kingdom

Oxford University Press is a department of the University of Oxford.
It furthers the University's objective of excellence in research, scholarship,
and education by publishing worldwide. Oxford is a registered trade mark of
Oxford University Press in the UK and in certain other countries

© Ian R. Kenyon 2020

The moral rights of the author have been asserted

First Edition published in 2020

Impression: 1

Published in the United States of America by Oxford University Press
198 Madison Avenue, New York, NY 10016, United States of America

British Library Cataloguing in Publication Data

Data available

Library of Congress Control Number: 2019945245

ISBN 978–0–19–880835–0 (hbk.)
ISBN 978–0–19–880836–7 (pbk.)

DOI: 10.1093/oso/9780198808350.001.0001

Printed and bound by
CPI Group (UK) Ltd, Croydon, CR0 4YY

Preface

In writing this book the intention has been to provide support for lecture courses on general quantum physics for university undergraduates in the final year(s) of a physics degree programme. The audience would be expected to have taken the courses in introductory quantum physics and in basic quantum mechanics that are normally met in the first or second years of a physics degree programme. The first chapter is a review of the basic quantum mechanics needed for getting the best out of the text. Instructors are then free to concentrate on a group of chapters, or select components from all chapters, whichever suits their needs.

Thanks to the sheer variety and rapid advance of research across the discipline, the average course in the later years of a physics degree programme is designed to address one area of physics. Instead, this text covers key themes of quantum physics, taking the perspective achieved after more than a century of research, and emphasizing the effectiveness and the subtlety of quantum concepts in explaining diverse physical phenomena. The book is used to bring out these unifying ideas and illustrate them with important examples from modern experiments and applications.

The themes developed in the text, and listed next, are the essence of quantum physics.

One theme contrasts boson condensation and fermion exclusivity. Bose–Einstein condensation is basic to superconductivity, superfluidity and gaseous BEC. Fermion exclusivity leads to compact stars and to atomic structure, and thence to the band structure of metals and semiconductors with applications in material science, modern optics and electronics.

A second theme is that a wavefunction at a point, and in particular its phase is unique (ignoring a global phase change). If there are symmetries, conservation laws follow and quantum states which are eigenfunctions of the conserved quantities. By contrast with no particular symmetry topological effects occur such as the Bohm–Aharonov effect: also stable vortex formation in superfluids, superconductors and BEC, all these having quantized circulation of some sort. The quantum Hall effect and quantum spin Hall effect are *ab initio* topological.

A third theme is entanglement: a feature that distinguishes the quantum world from the classical world. This property led Einstein, Podolsky and Rosen to the view that quantum mechanics is an incomplete physical theory. Bell proposed the way that any underlying local hidden variable theory could be, and was experimentally rejected. Powerful tools in quantum optics, including near-term secure communications, rely on entanglement. It was exploited in the the measurement of CP violation in the decay of beauty mesons.

A fourth theme is the limitations on measurement precision set by quantum mechanics. These can be circumvented by quantum non-demolition techniques and by squeezing phase space so that the uncertainty is moved to a variable conjugate to that being measured. The boundaries of precision are explored in the measurement of g-2 for the electron, and in the detection of gravitational waves by LIGO; the latter achievement has opened a new window on the universe.

The fifth and last theme is quantum field theory. This is based on local conservation of charges. It reaches its most impressive form in the quantum gauge theories of the strong, electromagnetic and weak interactions, culminating in the discovery of the Higgs. Where particle physics has particles condensed matter has a galaxy of pseudoparticles that exist only in matter and are always in some sense special to particular states of matter. Emergent phenomena in matter are successfully modelled and analysed using quasi-particles and quantum theory. Lessons learned in that way on spontaneous symmetry breaking in superconductivity were the key to constructing a consistent quantum gauge theory of electroweak processes in particle physics.

Care has been taken to maintain a level of presentation accessible to undergraduates reading physics, and to provide exercises and solutions to reinforce the learning process.

Solutions to the exercises are accessible via the OUP webpage link for this text.

Acknowledgements

I thank two Heads of the School of Physics and Astronomy at Birmingham University, Professors Andy Schofield and Martin Freer, and also Professor Paul Newman, Head of the Elementary Particle Physics Group at Birmingham, for their support and encouragement during the lengthy preparation of this textbook. Dr Sonke Adlung, the senior science editor at Oxford University Press, has always been unfailingly helpful and courteous in dealing with the many aspects of the preparation, and my thanks go to him for making my path easier. I am also grateful to his colleagues at Oxford University Press and SPi Global for the smooth management of copy editing, layout, production, and publicity.

Many colleagues have been more than generous in finding time in busy lives to read and comment on material for which they have a particular interest and expertise. My thanks go to Professor Ted Forgan, who was kind enough to read the chapters dealing with metals and semiconductors: his guidance and suggestions on particular points were very helpful. Thanks too, to Professor Peter Jones for reading the chapter on Transitions and his helpful comments on suitability. I am indebted to both Dr Rob Smith and Dr Martin Long, who generously found time on numerous occasions for patient explanations on condensed matter physics: their insights clarified many difficult points. In addition, Rob went the extra mile in reading the chapter on superconductivity, unearthing an embarassing number of subtle misunderstandings. I am under a further obligation to him for reading and commenting on the chapter on symmetry and topology. Thanks to Dr Elizabeth Blackburn for reading and making useful comments on the chapter on photon physics. I am very much indebted to Professor Vinen, who read and cross-questioned me on the liquid helium and superfluidity chapter, subjects about which he is a world authority. Dr Giovanni Barontini indoctrinated me in the subtleties of atomic Bose–Einstein condensates, and afterwards reviewed the chapter on BEC with great care: his input was indispensable. I much appreciate the interaction with Dr Hannah Price, who considerably sharpened my understanding of the quantum Hall effects, in particular, the role of topological quantization. Professor Alberto Vecchio took valuable time off from discovering gravitational waves to look over the chapter on quantum measurement, for which I am most grateful. I much appreciate Dr Alastair Rae offering to read three chapters where he has particular expertise: those on entanglement, the EPR controversy, and quantum measurement. He helped me appreciate a number of important points that had escaped my attention. My colleagues in the particle physics group have helped enormously in many ways. Kostas Nikolopoulos and Miriam Watson provided access to AT-LAS publications and event displays. In addition, Miriam and Chris Hawkes took on the task of reading and commenting on the chapters on particle physics. Their valuable critique is highly appreciated. Thanks too, to David Charlton and Mark Colclough who looked over sections where their expertise was particularly useful. Ian Styles of the Computer Science Department scrutinized the chapter relating to the potential for quantum cryptography and computing, for which I am most grateful. I am equally grateful to Dr Brooker who carefully searched the whole text for errors on behalf of Oxford University Press. Last, and definitely not least, I thank Mark Slater warmly for rescuing the text from a problem with missing fonts in the figures, which could have required the redrawing of 120 figures: his prompt and expert help was invaluable.

The input from all these colleagues removed misunderstandings on my part, helped me to clarify arguments, and brought points to my attention that I would otherwise have missed. The responsibility for any remaining errors should be laid at my door.

My thanks also go to the authors and publishers who have allowed me to use published figures, or adaptations of figures, or tables, each acknowledged individually in the text. I am grateful on this account to the American Physical Society, the American Society for the Advancement of Science, the American Chemical Society, the Royal Society, Elsevier publishers, Springer Nature, and Springer Verlag.

In producing some 290 diagrams I made almost exclusive use of the ROOT package developed by Dr Rene Brun and Dr Fons Rademakers and described in *ROOT – An Object Oriented Data Analysis Framework*, which appeared in the Proceedings of AIHENP '96 Workshop, Lausanne, Nuclear Instruments and Methods in Physics Research A389(1997)81-6. ROOT can be accessed at http://root.cern.ch. I also thank Dr Brun for help while learning to use this sophisticated tool.

To Valerie

Contents

The cover picture shows one end of a double dipole magnet from the CERN LHC. Two counter-rotating beams are steered around a 27 km closed path inside an evacuated tube by 1232 such magnets. For around 90 per cent of the year the particles in the beams are protons; for the remainder of the year heavy ions (such as lead). The beams are brought into collision at four locations on this circuit: around three of these interaction points the ATLAS, CMS, and LHCb detectors were built to study the products of proton-proton collisions at 13 TeV energy; around the fourth the ALICE detector was built, explicitly to study heavy ion collisions. In operation, the dipole current is carried by a *superconducting condensate* of Cooper pairs. Liquid helium flowing through the magnet coils provides the necessary cooling to 1.9 K. The flow is dissipation-free thanks to the *superfluid condensate* component of ultra-cold liquid helium. A final condensate is the *Higgs condensate* in the vacuum, whose particle excitations were discovered by the ATLAS and CMS experiments. See Chapters 14, 15 and 19 respectively.

Review of basic quantum physics

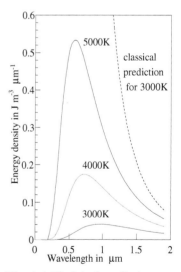
1

1.1   Introduction

This chapter reviews the evidence for quantization, and basic quantum mechanics, including Schrödinger's equation. The material should provide a refresher and a reference for students who have had exposure to a first course on quantum physics. The uncertainty principle, wave packets, wavefunction collapse and the no-cloning theorem are discussed. State vectors in Hilbert space are introduced. A summary of useful formulae from special relativity and electromagnetism closes the chapter.

1.2   The fundamental evidence

The three pieces of compelling evidence for the quantum nature of electromagnetic radiation were provided by the black body radiation spectrum, the photoelectric effect and Compton scattering. They are reviewed briefly here. The black body radiation spectrum, that is, the radiation in thermal equilibrium within a closed volume, whose walls are at a constant temperature, is shown in Figure 1.1 for three temperatures. The ingredients for calculating this spectrum are the count of distinct modes of oscillation of the electromagnetic field, and the energy per mode in thermal equilibrium. Of these, the mode density per unit frequency per unit volume $\rho(f)$ is given by[1]

$$\rho(f)\mathrm{d}f = [8\pi f^2/c^3]\mathrm{d}f, \tag{1.1}$$

for frequency f, *including both* polarizations. Classical and quantum predictions differ on how the energy per mode is assigned: in classical thermodynamics all modes have energy $k_{\mathrm{B}}T$, where T K is the temperature and k_{B} is Boltzmann's constant, $1.381\,10^{-28}$ J K^{-1}. The predicted energy spectrum therefore diverges at high frequencies and short wavelenths, as shown in Figure 1.1. This was the *ultraviolet catastrophe*. Planck, in a step that signalled the birth of quantum physics, proposed that in each mode electromagnetic radiation is emitted and absorbed in

Fig. 1.1 Black body radiation spectra at 3000 K, 4000 K and 5000 K. Planck's quantum predictions which are indicated with full lines fit the data. The classical prediction for 3000 K is shown with a broken line.

---

[1]This is derived in Appendix B, eqn. B.12, together with similar mode calculations.

Quantum 20/20: Fundamentals, Entanglement, Gauge Fields, Condensates and Topology.
Ian R. Kenyon. © 2020. Published in 2020 by Oxford University Press.
DOI: 10.1093/oso/9780198808350.001.0001

energy packets, the quanta, of magnitude

$$E = hf = \hbar\omega, \tag{1.2}$$

where ω is the angular frequency of the radiation and h is known as Planck's constant; $\hbar$ is simply $h/2\pi$. The rest of the quantum calculation proceeds as follows. If there are ℓ quanta in the mode its energy is ℓhf. The probability of ℓ quanta in any mode within the enclosure is then given by the Boltzmann distribution $\exp[-\ell hf/(k_{\mathrm{B}}T)]$. Evaluating the mean number gives

$$\bar{\ell} = 1/\left[\exp\left(hf/k_{\mathrm{B}}T\right) - 1\right]. \tag{1.3}$$

Multiplying this expression by the density of modes given in eqn. 1.1, the energy spectrum of black body radiation is predicted to be[2]

$$W(f)\mathrm{d}f = (8\pi hf^3/c^3)\mathrm{d}f/[\exp\left(hf/k_{\mathrm{B}}T\right) - 1], \tag{1.4}$$

[2]In a material of refractive index n there would be an additional factor n^3 in the numerator.

with $W(f)$ in $\mathrm{J\,m^{-3}\,Hz^{-1}}$. This expression fits the the observed black body spectra at all temperatures with one value of h, namely $6.626\,10^{-34}\,\mathrm{J\,s}$. When $hf/k_{\mathrm{B}}T$ is very small, the quantum formula reduces to the classical expression.

The *photoelectric effect* occurs when visible or ultraviolet light falls on a clean metal or an alkali metal surface causing electrons to be emitted. By 1902, Lenard had shown that for each such metal there exists a threshold frequency below which no photoelectrons are produced, however high the intensity of the incoming radiation. Classically, the electrons would continuously acquire energy from incident radiation and eventually break free from the surface, at any radiation frequency.

In 1905, Einstein proposed that in the photoelectric effect a single electron absorbs one quantum of energy from the radiation and escapes from the surface. Then electrons located at the surface will emerge with kinetic energy equal to

$$\mathrm{KE_{max}} = hf - \phi, \tag{1.5}$$

where ϕ is the *work function* of the metal surface. Electrons originating deeper in the metal lose further energy through collisions on their way out. Einstein's proposal provides the required cut-off, with no photoelectron emission for radiation of frequencies below $f_{\mathrm{co}} = \phi/h$.

Millikan showed that the maximum kinetic energy of the *photoelectrons* fitted Einstein's predicted linear relation. He extracted a value of Planck's constant from the data, which agreed with that found from the fit to the black body spectrum. Forrester, Gudmundson and Johnson in 1955 studied the photoelectric effect using light modulated at high frequencies and observed that the detector signal followed the modulation faithfully, showing that any delay was much less than the modulation period of 10^{-10} s. According to the classical view, there would be a delay before photoemission because the wave energy is supposed to be spread uniformly over the whole surface, rather than being concentrated in quanta that interact with individual atoms.

Compton followed through the consequences of the quantization of radiation, realizing that when X-rays scatter from matter the underlying process is the scattering of one quantum of electromagnetic radiation off one electron that is initially at rest. Then treating the photon-electron collision as an elastic two body collision, Compton arrived at an expression for the wavelength change of the scattered radiation

$$\lambda - \lambda_0 = (h/mc)(1 - \cos\theta), \tag{1.6}$$

where m is the electron mass and θ the angle through which the photon scatters. Compton found that the shift in wavelength fitted his prediction precisely in magnitude, in its dependence on the scattering angle and in its independence of the target material.

1.3   De Broglie's hypothesis

In 1924, de Broglie realized that if electromagnetic waves possess particle properties, then all material particles such as electrons must possess wave properties. Equally, the relations connecting the wave and particle properties of electromagnetic radiation apply also to material particles. Thus the frequency of the wave f associated with a particle of total energy E would be given by Planck's relation

$$E = hf. \tag{1.7}$$

The parallel relation for the momentum of a photon, p, also extends to material particles

$$p = h/\lambda. \tag{1.8}$$

This is called the *de Broglie* relation, and λ is known as the *de Broglie wavelength* of material particles. De Broglie's ideas were confirmed in 1926 when electron diffraction was observed from crystals.

Wave particle duality extends to gravitational effects. Einstein's general theory of relativity assigns an inertial mass of E/c^2 to a photon of energy E.

1.4   The Bohr model of the atom

In 1911, Rutherford's colleagues used α-particles (bare ^{4}He nuclei) to bombard thin metal foils and observed that substantial numbers were

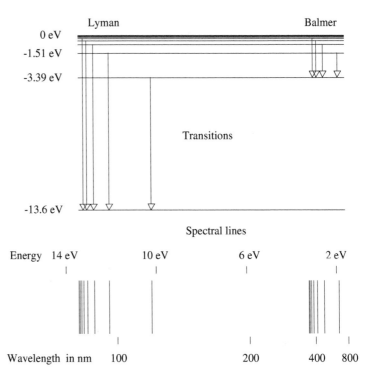

Fig. 1.2 The energy levels in the hydrogen atom are displayed together with transitions producing the Lyman and Balmer series. For clarity only the first few transitions are drawn. The spectral lines are shown in the lower panel. Each series converges to a limit: in the case of upward transitions the limit is reached when the electron just escapes from the atom with zero kinetic energy.

[3]The fine and hyperfine structure of spectral lines are discussed in the next chapter.

scattered into the backward hemisphere and some almost straight backward. Rutherford showed that these observations could only be consistently explained if the object within the atom that scatters the α-particles carries most of the atomic mass, has positive charge and is very much smaller than the atom. This scatterer is the nucleus, which is typically 10^{-15}m across. The electrons circulate around it in orbits extending to 10^{-10}m from the nucleus.

Classically the electrons in such an atom would accelerate into the nucleus and radiate over a broad range of wavelengths. In fact, isolated atoms radiate at discrete wavelengths, the spectral lines. Hydrogen has a particularly simple atomic structure, with one electron orbiting a single proton nucleus. Its spectrum consists of spectral lines whose wavelengths are given by a single formula[3]

$$1/\lambda = R_{\text{H}}(1/n^2 - 1/p^2), \tag{1.9}$$

where n and p are positive integers with the restriction that $p > n$. R_{H} is a constant known as the *Rydberg constant*, with value $1.09678\,10^7\text{m}^{-1}$. Part of the spectrum is shown in Figure 1.2

In Bohr's model of 1913, the electrons are pictured as travelling in stable circular orbits around the nucleus. Here, his model is applied to the hydrogen atom, for which it works best. The transition from an orbit of higher energy to one of lower energy is accompanied by a photon being emitted. Equally, an electron in a lower energy orbit can absorb a photon and jump to a higher energy orbit. In both cases, the photon energy exactly matches the energy difference between the electron states ΔE: the photon frequency, f, is thus given by

$$hf = \Delta E. \tag{1.10}$$

Bohr discovered the quantization condition that one complete electron orbit contains an integral number n of de Broglie wavelengths; any other orbit would interfere destructively with itself. This is a condition on the phase of a stable quantum state: in a closed loop the phase must change by an integral number of multiples of 2π. This simple, but powerful, phase condition has applicability beyond the Bohr model and will be met repeatedly. Bohr's only other requirement was that the Coulomb attraction of the nucleus provides the centripetal force to maintain the electron in its stable orbit. This leads to an expression for the energy of the nth orbit as

$$E_n = -(e^2/4\pi\varepsilon_0)^2(m/2n^2\hbar^2), \tag{1.11}$$

where $-e$ is the electron charge, m its mass and ε_0 the permittivity of free space. From this result the photon energy emitted/absorbed in a transition between the pth and nth orbit is

$$\Delta E = Rhc[1/n^2 - 1/p^2], \tag{1.12}$$

where $p > n$ is required for this to be positive. R is $me^4/[4\pi c(4\pi\varepsilon_0)^2\hbar^3]$ so this reproduces the main features of the hydrogen spectrum with R being equal to the experimental Rydberg constant. The radius of the innermost orbit, called the *Bohr radius*, is

$$a_0 = 4\pi\varepsilon_0\hbar^2/(me^2). \tag{1.13}$$

Despite its undoubted success in the case of hydrogen, Bohr's model fails to explain the spectrum of neutral helium. Its hybrid nature is also very unsatisfying and it lacks any means for calculating transition rates. Further progress in understanding required the quantum mechanics developed by Born, Heisenberg and Schrödinger.

1.5   Wave–particle duality

The connection between the particle and wave properties for electromagnetic radiation, and for material particles, is statistical. Thus the probability of finding a photon in a given volume dV is determined by the instantaneous energy density, I, of the electromagnetic wave over the same volume

$$P dV = I dV / \int I dV, \tag{1.14}$$

where P is called the *probability density*. The integral is taken over the whole of space to ensure that the total probability of the photon being

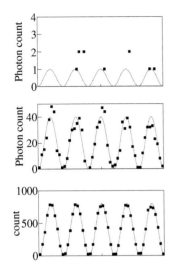

found somewhere is unity.

Young's two slit experiment will be used to illustrate the statistical interpretation. Suppose the observation screen is a pixelated detector with granularity much finer than the fringe widths, and that all the pixels are equally efficient in detecting photons. Further, suppose that an extremely low intensity source is used, so that at any given moment there is only ever a single photon within the volume between source and screen. Figure 1.3 shows typical histograms of the photon distribution across the detector after 10, 1000 and 20 000 photons have been detected. For comparison the calculated wave intensity is superposed. An individual photon may hit anywhere across the screen, other than locations where the wave intensity is precisely zero. Only the probability for arriving at each pixel is known, and the probabilities of reaching a given pixel are identical for each and every photon emerging from the source slit. The distribution with few photons is extremely ragged, but as the number increases the resemblance becomes ever closer to the wave intensity. If the number of photons expected to strike a pixel is n, then the statistical uncertainty in this number is $\sqrt{n}$. Thus the *fractional uncertainty* is $1/\sqrt{n}$, which falls with increasing n.

Fig. 1.3 Distribution of photons in the detection plane in Young's two slit experiment for 10, 1000 and 20 000 photons. The broken curves indicate the classical interference pattern.

The laws of classical physics are deterministic: statistical analysis is used only in dealing with systems containing very large numbers of particles, as in the kinetic theory of gases. By contrast, statistical behaviour is fundamental to quantum systems.

1.6   The uncertainty principle

Heisenberg considered the uncertainty when simultaneous measurements are made of a position component y and corresponnding momentum p_y and arrived at the limit

$$\Delta p_y \Delta y \geq \hbar/2, \tag{1.15}$$

known as Heisenberg's *uncertainty principle*. The equality would only be achieved if the distributions of momentum and position were Gaussian, and there were no instrumental errors. The uncertainty principle applies to each dimension separately, from which it follows that simultaneous measurements of the vector position **r** and vector momentum **p** have uncertainties that satisfy

$$\Delta p_x \Delta p_y \Delta p_z \Delta x \Delta y \Delta z \geq \hbar^3/8. \tag{1.16}$$

The product on the left-hand side of this equation can be pictured as a volume element in a six-dimensional space-momentum *phase space*.

Spatial coordinates and time are treated in a unified way within the special theory of relativity: they define a single space-time location (ct, x, y, z). Similarly, energy and momentum form an energy–

momentum four-vector $(E/c, p_x, p_y, p_z)$, which transforms exactly like a space-time vector under a Lorentz boost from one inertial frame to another. This implies that there must exist an energy–time uncertainty relation

$$\Delta E \Delta t \geq \hbar/2. \tag{1.17}$$

The interpretation differs subtly from that for space-momentum because Δt is the time taken to measure the energy, and ΔE is the resulting uncertainty in this measurement. To be precise, a quantity such as $\Delta \chi$ is the standard deviation of the measurements of χ on a set of similarly prepared states, that is, on an *ensemble*.

1.7   Outline of quantum mechanics

Experiment shows that the laws of conservation of energy, momentum and angular momentum are universal.

The behaviour of electrons or other particles is described by a complex wavefunction that contains all possible information that exists about the system. This wavefunction, $\Psi(q_n, t)$, is a function of time and all the independent variables, written as a set $\{q_n\}$. These variables include the spatial coordinates and its spin (intrinsic angular momentum) state. The interpretation of the wavefunction will always be that the probability for finding a system with variables in a range $dV = dq_1 dq_2 \cdots$ around q_1, $q_2, \cdots$ is

$$P(q_1, q_2, \cdots) \, dV = \Psi^* \Psi dV. \tag{1.18}$$

The wavefunction used is normalized, meaning that a numerical factor is inserted so that integrating $P dV$ over the full range of the independent variables gives unity.

Quantities that are measurable for a particle or a system of particles are known as *observables*. Position, momentum, orbital angular momentum, polarization and energy are all observables. These are therefore *real*, rather than complex, quantities. Each observable has a corresponding operator that acts on the wavefunction describing the system considered. The operators for momentum and the total energy are indicated by placing hats over the respective symbols for the observables

$$\hat{p} = -i\hbar \frac{\partial}{\partial x}, \quad \hat{H} = +i\hbar \frac{\partial}{\partial t}. \tag{1.19}$$

Their action is most easily demonstrated by considering a free electron moving in the x-direction. This has a plane wavefunction.

$$\Phi_k = (1/\sqrt{L}) \exp\left[i(kx - \omega t)\right], \tag{1.20}$$

where L is the range in x to which the electron is restricted and can be increased to infinity as required. Then

$$\hat{p}\Phi_k = \hbar k \Phi_k = p\Phi_k, \quad \hat{H}\Phi_k = \hbar\omega\Phi_k = E\Phi_k. \tag{1.21}$$

Plane waves provide a simple example for discussion and finite realistic wavepackets are all linear sums of plane waves. Plane sinusoidal waves extend to infinity and the range L is needed to give a normalizable wavefunction. The values of measurable quantities are correctly predicted when the limit $L \to \infty$ is taken. On occasion care is needed when taking the limit.

The quantities p and E appearing on the right-hand side, without hats, are the values that can be obtained in measurements of the momentum and kinetic energy respectively. This prediction of an exact value for the momentum of a plane wave does not violate the uncertainty principle because the location is undetermined.

Operators are complex and the quantities measured are real, so it follows that the waves for electrons and other material particles must themselves be complex, unlike magnetic and electric fields, which are always real. We now come to the equation that is the equivalent for non-relativistic electrons of the wave equation for electromagnetic waves.

1.8   Schrödinger's equation

Starting from the equation for conservation of energy for non-relativistic motion in a potential V,

$$E = V + p^2/2m,$$

Schrödinger constructed an operator equation acting on a wavefunction. The total energy and the momentum are replaced by the operator forms from eqn. 1.19, giving

$$i\hbar\partial\Psi(\mathbf{r},t)/\partial t = V(\mathbf{r},\mathbf{t})\Psi(\mathbf{r},t) - (\hbar^2/2m)\nabla^2\Psi(\mathbf{r},t)., \qquad (1.22)$$

This is *Schrödinger's time-dependent equation* for non-relativistic motion in a potential $V(\mathbf{r},\mathbf{t})$. Solutions for the motion in square, harmonic and Coulomb potentials are calculated in Chapter 2.

In the case of a static potential the solution factorizes to give

$$\Psi(\mathbf{r},t) = \psi(\mathbf{r})\exp\left(-iEt/\hbar\right), \qquad (1.23)$$

which, when substituted in Schrödinger's equation, gives its *time independent* form

$$E\psi(\mathbf{r}) = V(\mathbf{r})\psi(\mathbf{r}) - (\hbar^2/2m)\nabla^2\psi(\mathbf{r}), \qquad (1.24)$$

where E is the electron total energy, kinetic plus potential.

Schrödinger's equation is linear in ψ so that the *superposition principle* applies to wavefunctions that satisfy the equation: adding these wavefunctions with constant coefficients produces another valid wavefunction. There are several crucial differences between, on the one hand, Maxwell's equations for electromagnetic waves, and on the other, Schrödinger's equation for electron waves. Schrödinger's equation is non-relativistic, and complex, so the electron waves are complex and not directly measurable: Maxwell's equations are relativistic, and the electromagnetic fields are directly measurable. A basic understanding of atomic states and their radiation is achieved by applying Schrödinger's equation.

Any solution of Schrödinger's equation must satisfy several simple requirements. Firstly, the wavefunction must be finite and continuous everywhere. If instead the wavefunction jumped discontinuously, the derivative, and hence the momentum, would become infinite. Similarly, the first derivative must be continuous everywhere to avoid an infinite term in the energy. These requirements on continuity are essential tools when joining up solutions of Schrödinger's equation at boundaries where the potential changes.

The non-relativistic density of particles is

$$\rho = \Psi^*\Psi, \tag{1.25}$$

so that

$$\partial\rho/\partial t = \Psi^*\partial\Psi/\partial t + [\partial\Psi^*/\partial t]\Psi. \tag{1.26}$$

There is a corresponding vector flux, or current, $\mathbf{j}$, defined to be the number of particles crossing unit area in unit time. That is the real part of $\Psi^*[\mathbf{p}/m]\Psi$:

$$\mathbf{j} = (i\hbar/2m)(\Psi\nabla\Psi^* - \Psi^*\nabla\Psi). \tag{1.27}$$

Flux and density obey the standard continuity equation

$$\partial\rho/\partial t + \nabla \cdot \mathbf{j} = 0, \tag{1.28}$$

which ensures that the number of particles is conserved.

1.9 Eigenstates

The wavefunctions that are solutions of Schrödinger's equation are known as energy *eigenfunctions*. The corresponding energies are called energy *eigenvalues* and the electron is said to be in an *eigenstate* of energy. An eigenstate may be an eigenstate of several observables with each taking unique values for a given eigenstate. These are then known as *compatible* or *simultaneous* observables: examples are the energy, the angular momentum and a component of the angular momentum of an electron in a hydrogen atom. The measurement of an observable leaves the electron in an eigenstate of that variable and its compatible variables. As explained later, the eigenvalues of energy are discrete when the potential localizes the electron in a potential well, but continuous from zero up to any conceivable positive value when an electron is free.

The existence and the properties of eigenstates generalize to systems of electrons and other material particles. Such a system has a set of eigenstates $\{\phi_i\}$ of observables, such as A, with eigenvalues $\{a_i\}$, respectively. With the standard notation the operator corresponding to A is $\hat{A}$, and this acts on the wavefunction ϕ_i in the following way:

$$\hat{A}\phi_i = a_i\phi_i, \tag{1.29}$$

meaning that any measurement of the observable A on the eigenstate ϕ_i always gives the eigenvalue a_i.

An intrinsic property of eigenstates is their *orthogonality* in the sense that the overlap integrals between the wavefunctions of any pair of them over all the free variables vanish. Suppose ϕ_i and ϕ_j are two such eigenfunctions of an electron; then

$$\int \phi_j^* \phi_i \, \mathrm{d}V = 0, \ \ \text{if } j \neq i. \tag{1.30}$$

Eigenfunctions are usually normalized for convenience so that

$$\int \phi_j^* \phi_i \mathrm{d}V = \delta_{ji}, \tag{1.31}$$

where δ_{ji} is the Kronecker δ defined by

$$\delta_{ji} = 0 \text{ for } j \neq i; \ \ \delta_{ji} = 1 \text{ for } j = i. \tag{1.32}$$

In the case of a free particle $\phi(x) = \exp(ikx)$, where k is a continuous variable rather than an integer label, then

$$\frac{1}{2\pi} \int_{-\infty}^{\infty} \phi^*(x')\phi(x)\mathrm{d}k = \frac{1}{2\pi} \int_{-\infty}^{\infty} \exp[i(kx - kx')]\mathrm{d}k = \delta(x - x'). \tag{1.33}$$

where δ is the *Dirac delta function* This has the property that for any function $f(x)$

$$\int f(x)\delta(x - x')\mathrm{d}x = f(x'), \tag{1.34}$$

provided the range of integration includes the point $x = x'$; otherwise it vanishes. The Dirac δ-function $\delta(x - x')$ is effectively an infinitely narrow and tall spike at $x = x'$ such that the area under this spike is exactly unity.

1.10   Observables and expectation values

In the more general case that a system is not in an eigenstate of an observable, the value that is obtained by measuring the observable can only be predicted statistically. Quantum mechanics predicts the *expectation value* of an observable A, which is written $\langle \hat{A} \rangle$, is obtained using the equation

$$\langle \hat{A} \rangle = \int \psi^* \hat{A} \psi \, \mathrm{d}V, \tag{1.35}$$

where ψ is normalized. The equation is to be interpreted in this way. This is the average value found if A is measured on a large number of systems which have been prepared in exactly the same way so that they have identical wavefunctions ψ – such a hypothetical collection of systems is called an *ensemble*. In the case of an eigenstate of the observable A the expectation value is simply the eigenvalue of A for that eigenstate.

Any wavefunction ψ of a system which has an observable A can always be expanded as a linear superposition of the normalized eigenfunctions $\{\phi_i\}$ of A. Assuming for simplicity that the eigenvalues are discrete,

$$\psi = \sum_i c_i \phi_i. \tag{1.36}$$

Then the expectation value of A in a state with wavefunction ψ is

$$\langle \hat{A} \rangle = \sum_j c_j^* c_j a_j. \tag{1.37}$$

where $c_j^* c_j$ is the probability that the system is found to be in the eigenstate ϕ_j with eigenvalue a_j of A. When the eigenvalues are continuous, as for example the momentum $\hbar k$ of a free electron described by eqn. 1.20,

$$\psi = \int c(k) \phi_k \, \mathrm{d}k, \tag{1.38}$$

and

$$\langle \hat{A} \rangle = \int c^*(k) c(k) a(k) \, \mathrm{d}k, \tag{1.39}$$

where $a(k)$ is the value obtained when A is measured on the eigenstate with momentum k. In particular $c^*(k)c(k) \, \mathrm{d}k$ is the probability that the measurement gives a momentum eigenvalue lying between k and $k + \mathrm{d}k$.

All measurements yield real values so that the expectation value of an observable, A, is always real; hence it equals its complex conjugate so that

$$\int \psi^* \hat{A} \psi \mathrm{d}V = \int (\hat{A}\psi)^* \psi \, \mathrm{d}V. \tag{1.40}$$

Operators with this mathematical property are called *hermitean*.

Eigenstates of a system are usually eigenstates of several observables, and it requires knowledge of all of these to completely specify an eigenstate. Here we consider the case where there are just two of these *compatible* observables, A and B. There is a set of eigenfunctions $\{\phi\}$ for which

$$\hat{A}\phi_j = a_j \phi_j; \quad \hat{B}\phi_j = b_j \phi_j.$$

Suppose some eigenfunctions are *degenerate*, that is to say they share the same eigenvalue for A, while each has an eigenvalue of B different from that of all the others. A measurement of A may then leave the system in a state described by a superposition of the degenerate eigenfunctions. A subsequent measurement of B will result in the wavefunction collapsing into a single eigenfunction from this superposition, for example ϕ_k. Further measurements thereafter of A and B yield a_k and b_k respectively. The expectation value for the product of compatible observables

$$\int \psi^* \hat{A}\hat{B}\psi \, \mathrm{d}V = \sum_j c_j^* c_j a_j b_j = \int \psi^* \hat{B}\hat{A}\psi \, \mathrm{d}V,$$

holds true whatever arbitrary state ψ of the system is being considered. Consequently the expectation value of $\hat{A}\hat{B} - \hat{B}\hat{A}$ always vanishes. This operator is called the *commutator* of A and B and is written with square brackets $[\hat{A}, \hat{B}]$. In the case of compatible observables

$$[\hat{A}, \hat{B}] = 0 : \tag{1.41}$$

$\hat{A}$ and $\hat{B}$ are said to commute.

Pairs of conjugate variables like $\hat{p}_x$ and $\hat{x}$ are not compatible observables. It follows from eqn. 1.19 that

$$
\begin{aligned}
[x, p_x]\psi &= x(-i\hbar\partial/\partial x)\psi + i\hbar\partial(x\psi)/\partial x \\
&= -ix\hbar(\partial\psi/\partial x) + i\hbar(\partial x/\partial x)\psi + i\hbar x(\partial\psi/\partial x) \\
&= i\hbar\psi.
\end{aligned}
\tag{1.42}
$$

There are corresponding relations for the commutators of other pairs of conjugate variables.

1.11   Collapse of the wavefunction

A most surprising feature of quantum mechanics has been left for discussion here. Any measurement of the observable A on a system in a superposition of eigenstates of A with wavefunction $\psi = \sum_i \phi_i$ gives some eigenvalue a_j. The system is thereafter in the eigenstate with wavefunction ϕ_j, and no longer in the state with wavefunction ψ. A second measurement of the observable A will again give a_j, and so would further measurements. The result of the first measurement is profoundly different from anything met in classical mechanics. There is a discontinuity: up to the exact moment of the measurement the system is evolving according to the wavefunction ψ and immediately afterwards its wavefunction has become ϕ_j. This step is known as the *collapse of the wavefunction*. It is clearly a very drastic step because, for example when a photon is absorbed on an atom in the photoelectric effect, the photon wavefunction over all space collapses simultaneously.

1.12   Schrödinger's cat

Schrödinger highlighted a logical difficulty arising out of the collapse of the wave function. A cat is locked in a box together with a mechanism which will release a lethal gas if and when a single radioactive nucleus decays. It is then argued that the wavefunction of the contents of the box should contain two terms: the first term describing an undisturbed mechanism and a live cat; the second describing an activated mechanism and a dead cat. Later Schrödinger opens the box and observes the contents. At this instant the wavefunction of the contents collapses to either one that contains a live cat, or to another that contains a dead cat. A popular resolution of this paradox of having a cat simultaneously alive and dead is through what is called *decoherence*. Broadly speaking any interaction of a quantum system with its surroundings, for example gas molecules striking the cat, is supposed to cause the collapse its wavefunction.[4] A counter view is that the interactions only cause loss of phase coherence between the macroscopic states and leave them superposed.[5]

1.13 No-cloning theorem

Quantum mechanics forbids the creation of exact replicas of arbitrary quantum states. If it were possible to do so then one observable could be measured on the original and the conjugate observable measured on the clone, both with high precision. As a result, both observables would be precisely known for the parent quantum state, which would violate the uncertainty principle. Suppose we attempt to clone a state labelled '1', which is an arbitrary superposition of two pure states

$$\psi(1) = \alpha\psi_a(1) + \beta\psi_b(1), \qquad (1.43)$$

where α and β are some unknown constants. Then making a copy labelled '2' gives

$$
\begin{aligned}
\mathrm{Copy}[\psi(1)] &= \alpha\,\mathrm{Copy}[\psi_a(1)] + \beta\,\mathrm{Copy}[\psi_b(1)] \\
&= \alpha\psi_a(1)\psi_a(2) + \beta\psi_b(1)\psi_b(2) \qquad (1.44)
\end{aligned}
$$

whereas true cloning[6] would produce $\psi(1)\psi(2)$.

1.14 Wavepackets

Wavepackets considered here are finite wavetrains of electromagnetic waves emitted by a source, and contain a number of photons determined by the physical situation. Wavepackets exist equally for any particle species. In free space wavepackets travel without change of shape because all the frequency components travel at the same velocity, c. However, in dispersive media the velocity depends on the frequency. For example the electric field

$$E(z,t) = \int \epsilon(\omega)\exp[i(kz - \omega t)]\,\mathrm{d}\omega, \qquad (1.45)$$

k being the wavenumber and $\epsilon(\omega)$ is the wave distribution in angular frequency centred on ω_0. This is portrayed at a given location in the upper plot of Figure 1.4. Rewriting this waveform to first order in $(\omega - \omega_0)$

$$E(z,t) = \exp[i(k_0 z - \omega_0 t)]\int \epsilon(\omega)\exp[i(\omega - \omega_0)(z\,\mathrm{d}k/\mathrm{d}\omega - t)]\,\mathrm{d}\omega, \quad (1.46)$$

[4]See, for example, 'Decoherence and the the Transition from Quantum to Classical' by W. H. Zurek, *Physics Today*, October 1991.

[5]See A. Bassi, K. Lochan, S. Satin, T. P. Singh, and H. Ulbricht, *Reviews of Modern Physics* 85, 471 (2013).

[6]Obviously a single pure state can be copied but we have to know that it is a pure state in the first place. It is cloning of an arbitrary and therefore unknown quantum state that is forbidden. See W. K. Wootters and W. H. Zurek, *Nature* 299, 802 (1982).

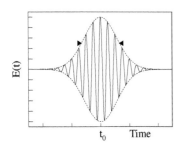

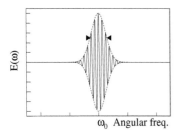

Fig. 1.4 The time and angular frequency distributions for a Gaussian wavepacket. In this case the full width at half maximum is 2.36 times the standard deviation.

In the case of microwaves of much lower frequency, it is the electric field that is detected. Very large numbers of photons make up the signal detected, so that a classical analysis is generally adequate.

where ω_0 and k_0 are the central values. The first term describes the rapidly oscillating waves within the envelope that have the *wave velocity* $v_w = \omega_0/k_0$. The integral describes the envelope. Its maximum is located where all the waves are in phase, that is, where $z\,dk/d\omega = t$. Thus the envelope has the *group velocity*, $v_g = d\omega/dk$. Photons, and energy and information, all travel at the group velocity. Suppose the wave velocity is less than the group velocity, then to an observer moving at the group velocity the waves would appear to travel backward within the envelope in the upper plot.

Measurements on photons from identically prepared wavepackets must satisfy time/angular frequency and position/wave-vector uncertainty relations:

$$\Delta t\Delta\omega \geq 1/2, \quad \Delta x\Delta k \geq 1/2, \tag{1.47}$$

where Δx, etc. are standard deviations in measurements. The most convenient wavepackets for use in analysis are Gaussian in shape, and approximations to this shape are met often enough to be of practical interest. An example is drawn in Figure 1.4, first as a function of time at a fixed location and then as a function of angular frequency: these are both amplitude plots. The corresponding intensity plots, proportional to the probability of finding a photon, are shown in Figure 1.5. These are also Gaussians, narrower than the amplitude envelopes by a factor $\sqrt{2}$. Only the envelope is shown because detectors of light, which typically have nanosecond resolution, cannot follow the wave oscillations under the envelope at 10^{14} Hz. Explicitly the intensity distributions are

$$I(t - t_0) = \frac{1}{\sqrt{2\pi}\sigma_t} \exp[-(t - t_0)^2/(2\sigma_t^2)], \tag{1.48}$$

$$I(\omega - \omega_0) = \frac{1}{\sqrt{2\pi}\sigma_\omega} \exp[-(\omega - \omega_0)^2/(2\sigma_\omega^2)]. \tag{1.49}$$

These are related through Fourier transforms

$$I(\omega) = \int_{-\infty}^{\infty} I(t) \exp(i\omega t)\,dt,$$

$$I(t) = \int_{-\infty}^{\infty} I(\omega) \exp(-i\omega t)\,d\omega. \tag{1.50}$$

These distributions have standard deviations σ_ω and σ_t related by $\sigma_\omega\sigma_t = 1/2$. Projections of the wavepacket in position and wave-vectors are also Gaussian and these too are Fourier transforms of one another.

When measurements are made on photons from a Gaussian wavepacket, taking the errors in the detectors to be negligible, the distributions would have standard deviations σ_t and σ_ω. In this special case of Gaussian wavepackets with negligible intrinsic measurement errors, we have uniquely an equality

$$\Delta\omega\Delta t = \sigma_\omega\sigma_t = 1/2. \tag{1.51}$$

In any other conditions, with intrinsic errors and/or a non-Gaussian wavepacket, $\Delta\omega\Delta t$ is larger.

1.15   State vectors

States of systems have so far been described by wavefunctions. A more flexible description is provided by *state vectors*, which are indispensable when dealing with quantum fields. State vectors are presented here using the notation introduced by Dirac. For a system with a set of orthonormal wavefunctions $\{\phi_i\}$, any normalized wavefunction can be expanded as

$$\psi = \sum_i c_i \phi_i, \tag{1.52}$$

in which the c_is are complex coefficients. Then the integral

$$\int \psi^* \psi \, dV = \sum_{ij} c_i^* c_j \int \phi_i^* \phi_j \, dV = \sum_i c_i^* c_i. \tag{1.53}$$

The right-hand side of this equation can be expressed in matrix notation as

$$\begin{pmatrix} c_1^* & c_2^* & c_3^* & \cdots \end{pmatrix} \begin{pmatrix} c_1 \\ c_2 \\ c_3 \\ \cdot \\ \cdot \end{pmatrix}, \tag{1.54}$$

which is identical to the scalar product of two vectors with coordinate lengths referred to the same set of orthogonal axes $(c_1^*, c_2^*, c_3^*, \cdots)$ and $(c_1, c_2, c_3, \cdots)$. We define the unit vector along the ith axis to be $|\phi_i\rangle$. This space is known as a *Hilbert space* and all the vectors, including $|\phi_i\rangle$, are called *kets*. The state vector for the column matrix above is then a ket,

$$|\psi\rangle = \sum_i c_i |\phi_i\rangle, \tag{1.55}$$

an equation equivalent to eqn. 1.52. Another type of state vector is needed to correspond to the row matrix in eqn. 1.54. These are called *bra vectors*, and are written $\langle \psi |$ and $\langle \phi_i |$. Note that the vectors $|\psi\rangle$ and $\langle \psi |$ describe *exactly* the same state.[7] For the bra vectors

$$\langle \psi | = \sum_i c_i^* \langle \phi_i |. \tag{1.56}$$

The scalar product $\langle \phi_i | \phi_j \rangle$ is the overlap of these states given by

$$\langle \phi_i | \phi_j \rangle = \int \phi_i^* \phi_j \, dV = \delta_{ij}. \tag{1.57}$$

Correspondingly,

$$\langle \psi | \phi_i \rangle = c_i^*, \quad \text{and} \quad \langle \phi_i | \psi \rangle = c_i. \tag{1.58}$$

Then the state vectors $|\psi\rangle$ and $\langle \psi |$ can be expanded in this way

$$|\psi\rangle = \sum_i \langle \phi_i | \psi \rangle |\phi_i\rangle, \quad \text{and} \quad \langle \psi | = \sum_i \langle \psi | \phi_i \rangle \langle \phi_i |. \tag{1.59}$$

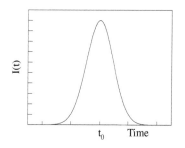

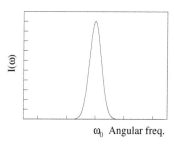

Fig. 1.5 The time and angular frequency intensity distributions for the Gaussian wavepacket amplitudes shown in 1.4.

[7]The Hilbert spaces containing the bra and ket vectors are actually separate vector spaces known as dual spaces. If wavefunctions were real then only one vector space would suffice.

We can rewrite the first equation as

$$|\psi\rangle = \sum_i |\phi_i\rangle\langle\phi_i|\psi\rangle, \tag{1.60}$$

which demonstrates that

$$\sum_i |\phi_i\rangle\langle\phi_i| = I = \begin{pmatrix} 1 & 0 & 0 & . & . \\ 0 & 1 & 0 & . & . \\ 0 & 0 & 1 & . & . \\ . & . & . & . & . \\ . & . & . & . & . \end{pmatrix}, \tag{1.61}$$

where I is the identity matrix. This useful result is called the *closure relation*. The expectation value of an observable A in the state described by $|\psi\rangle$ is given by

$$\langle \hat{A}\rangle = \int \psi^* \hat{A}\psi \, \mathrm{d}V = \sum_{ij} c_i^* c_j \langle\phi_i|\hat{A}|\phi_j\rangle, \tag{1.62}$$

which contains matrix elements $A_{ij} = \langle\phi_i|\hat{A}|\phi_j\rangle$. Observables are real and A_{ij} is hermitian: $A_{ij}^* = A_{ji}$. The complex conjugate transpose of a matrix A is written $A^\dagger$ so that for observables $A^\dagger = A$.

The state vector $|\mathbf{r}\rangle$ describes a state whose wavefunction is a delta function at the point $\mathbf{r}$ in space. In this case the connection between the spatial wavefunction and the state vector is

$$\psi(\mathbf{r}) = \langle\mathbf{r}|\psi\rangle. \tag{1.63}$$

In momentum space the wavefunction would be

$$\psi(\mathbf{p}) = \langle\mathbf{p}|\psi\rangle, \tag{1.64}$$

where $|\mathbf{p}\rangle$ is a delta function at momentum $\mathbf{p}$. In the case of free particles the Hilbert space with unit vectors $|\mathbf{r}\rangle$ has infinite dimensions. Rephrasing eqn. 1.63 is useful: when a state vector is projected onto eigenstates of, for example, $\mathbf{r}$ the outcome is a wavefunction, in this case $\psi(\mathbf{r})$.

Rotations can be made in Hilbert space which leave any product of state vectors $\langle\psi|\xi\rangle$ unaffected. These transformations differ from rigid rotations in cordinate space because the length of a vector in Hilbert space is a complex number. Suppose U is such a *unitary transformation*,

$$\langle\psi|\xi\rangle = \langle U\psi|U\xi\rangle. \tag{1.65}$$

From this it follows that

$$U^\dagger U = I \ \text{ and } \ U^\dagger = U^{-1}. \tag{1.66}$$

Energy being an observable and real, the corresponding operator, the Hamiltonian operator $\hat{H}$ is Hermitian. Hence the time evolution operator $\exp(-i\hat{H}t)$ is unitary:

$$[\exp(-i\hat{H}t)]^{\dagger} = \exp(i\hat{H}^{\dagger}t) = \exp i\hat{H}t, \qquad (1.67)$$

so that states originally orthogonal remain orthogonal and retain their separate identities.

1.16   Special relativity and electromagnetism

The space-time four-vector for a point at $\mathbf{r}$ at time t is $(ct,\mathbf{r})$ or X with components

$$X_0 = ct, \quad X_1 = x, \quad X_2 = y, \quad X_3 = z. \qquad (1.68)$$

The energy-momentum four-vector for a system of energy E and momentum $\mathbf{p}$ is $P = (E/c,\mathbf{p})$. Under a Lorentz transformation from one inertial frame to another (primed) frame with relative velocity $v = \beta c$ parallel to the $x(x')$-axis four-vectors all transform in the same way:

$$
\begin{aligned}
E' &= \gamma(E - p_x v); & t' &= \gamma(t - xv/c^2); \\
p'_x &= \gamma(p_x - Ev/c^2); & x' &= \gamma(x - vt); \\
p'_y &= p_y; & y' &= y; \\
p'_z &= p_z; & z' &= z; & (1.69)
\end{aligned}
$$

where $\gamma = \sqrt{1/(1-\beta^2)}$. Quantities like X and P are four-vector representations of the Lorentz group. For four-vectors we introduce the subscript labelling 0, 1, 2, 3 for the time component, x-component, y-component, z-component respectively. We also introduce the *Einstein convention* that if a subscript is repeated it should be summed over, $a_\mu b_\mu$ is to be read $a_0 b_0 - a_1 b_1 - a_2 b_2 - a_3 b_3$. The scalar product of four-vectors is defined by

$$A \cdot B = g_{\mu\nu} A_\mu B_\nu, \qquad (1.70)$$

The *metric tensor* appearing here is

$$g_{\mu\nu} = \begin{bmatrix} +1 & 0 & 0 & 0 \\ 0 & -1 & 0 & 0 \\ 0 & 0 & -1 & 0 \\ 0 & 0 & 0 & -1 \end{bmatrix}, \qquad (1.71)$$

where the indices run from 0 to 3. These products are scalar representations of the Lorentz group and are invariant under Lorentz transformations. The scalar product of $(ct,\mathbf{r})$ with itself is:

$$X^2 = c^2 t^2 - x^2 - y^2 - z^2 = c^2 t^2 - r^2. \qquad (1.72)$$

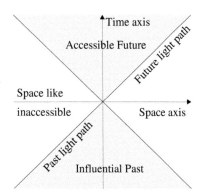

Accessible Future

Future light path

Space like inaccessible

Space axis

Past light path

Influential Past

Fig. 1.6 Section through the light cones at one point in space-time showing one spatial dimension. A representation for two spatial dimensions can be obtained by rotating the image around the time axis.

This is the new form of Pythagoras' theorem after taking special relativity into account. If X^2 is positive the separation in space-time is *time-like*, if it is negative the separation is *space-like* and if it is zero the separation is *light-like*. No signals can travel over space-like separations, since signals would then be travelling faster than light. Figure 1.6 illustrates by the gray shading the regions of the past that can influence an event (at the centre), and the regions of the future that the event can influence.

The scalar product of an energy-momentum four-vector with itself is:

$$s/c^2 = E^2/c^2 - p_x^2 - p_y^2 - p_z^2 = E^2/c^2 - p^2. \tag{1.73}$$

In the centre of mass frame of a system the vector sum of the momenta is zero, so s is the centre of mass energy squared. In the case of an isolated particle

$$m^2c^4 = E^2 - p^2c^2, \tag{1.74}$$

where m is its rest mass. At low energies such that $cp \ll mc^2$ we can expand this as

$$E = mc^2 + p^2/(2m) + \cdots, \tag{1.75}$$

and at high energies such that $cp \gg mc^2$, as met for neutrinos,

$$E = cp + m^2c^3/(2p) + \cdots, \tag{1.76}$$

and in the limit for the massless photons $E = cp$. The scalar product of space-time and energy-momentum four-vectors is equally invariant under Lorentz transformations $P \cdot X = Et - \mathbf{p} \cdot \mathbf{r}$. This appears in the wave function of free particles:

$$\Psi(\mathrm{r},t) = \exp[-i(Et - \mathbf{p} \cdot \mathbf{r})] = \exp[i(\mathbf{k} \cdot \mathbf{r} - \omega t)\hbar], \tag{1.77}$$

where $\mathbf{k} = \mathbf{p}/\hbar$ is the wave-vector and $\omega = E/\hbar$ the angular frequency.

Any quantities that are representations of the Lorentz group are known as *Lorentz covariant*. Relationships between covariant quantities do not change their form under Lorentz transformations and are also Lorentz covariant. Maxwell's equations are a fundamental example.

In the presence of an electromagnetic field the energy and momentum of a particle need to be specified more carefully. Suppose the Lorentz four-vector potential describing the field is $(\phi/c, \mathbf{A})$ with electric and magnetic fields $-\nabla\phi - \partial\mathbf{A}/\partial t$ and $\nabla \wedge \mathbf{A}$. Then the components of a particle's energy-momentum vector become:

$$E/c + q\phi/c \text{ and } m\mathbf{v} = \mathbf{p} - q\mathbf{A}, \tag{1.78}$$

with q being the particle's charge. The symbol e is only used in this textbook in the electron's charge $-e$. The quantity $m\mathbf{v}$ is the usual *physical momentum* of a particle of mass m with velocity $\mathbf{v}$. The quantity

p is the *canonical momentum* conjugate to the coordinate: in operator form $[x, p_x] = i\hbar$. For completeness note that the corresponding non-relativistic Lagrangean is

$$L = mv^2/2 - q\phi + q\mathbf{A} \cdot \mathbf{v}, \tag{1.79}$$
$$\text{and } \mathbf{p} = \partial L/\partial \mathbf{v} = m\mathbf{v} + q\mathbf{A}. \tag{1.80}$$

The Hamiltonian (energy) is

$$H = [\mathbf{p} - q\mathbf{A}]^2/(2m) + q\phi. \tag{1.81}$$

In addition, the current carried by particles of charge q becomes, in the presence of an electromagnetic field, the real part of $(q/m)[\Psi^*(\mathbf{p}-q\mathbf{A})\Psi]$, that is

$$\mathbf{j} = \frac{q}{2m}[\Psi(i\hbar\nabla - q\mathbf{A})\Psi^* - \Psi^*(i\hbar\nabla + q\mathbf{A})\Psi]. \tag{1.82}$$

This reduces to

$$\mathbf{j} = \frac{iq\hbar}{2m}[\Psi\nabla\Psi^* - \Psi^*\nabla\Psi] - \frac{q^2}{m}\mathbf{A}|\Psi|^2. \tag{1.83}$$

The four-vector potential is not unique: under what is called a *local gauge transformation* the physical electric and magnetic fields are unchanged. With $\alpha(t,\mathbf{r})$ being any smoothly varying scalar function of position and time, the corresponding local gauge transformation is

$$\mathbf{A} \rightarrow \mathbf{A} - \nabla\alpha; \quad \phi \rightarrow \phi + \partial\alpha/\partial t. \tag{1.84}$$

The four-vector form of, for example, the momentum operator is

$$\hat{P} = (\hat{E}/c, \hat{\mathbf{p}}) = i\hbar\left[\frac{\partial}{c\partial t}, -\frac{\partial}{\partial x}, -\frac{\partial}{\partial y}, -\frac{\partial}{\partial z}\right], \tag{1.85}$$

or in shorthand notation: $i\hbar\frac{\partial}{\partial x_\mu}$ and $i\hbar\partial_\mu$.

The wave equation of electromagnetism in the presence of a current **j**, obtained from Maxwell's equations, will also be needed:[8]

$$\nabla^2\mathbf{A} - \frac{1}{c^2}\frac{\partial^2\mathbf{A}}{\partial t^2} = -\mu_0\mathbf{j}. \tag{1.86}$$

This requires a choice of gauge fields such that

$$\frac{1}{c}\frac{\partial\phi}{\partial t} + \nabla \cdot \mathbf{A} = 0, \tag{1.87}$$

known as the *Lorentz gauge.*

Finally a relativistic equivalent of Schrödinger's equation is obtained, starting with eqn. 1.74. The kinematic quantities are replaced by operators and the result applied to a wavefunction ϕ representing a massive material particle. This gives the *Klein–Gordon* equation,

$$m^2c^4\phi = -\hbar^2\partial^2\phi/\partial t^2 + \hbar^2c^2\nabla^2\phi. \tag{1.88}$$

[8] Page 410 in *Classical Electrodynamics* by W. Greiner, published by Springer (1996), but using S.I. units.

All particles must satisfy this equation. However, on its own, because it lacks any reference to spin, it is inadequate to describe relativistic electrons. Dirac's relativistic description of electrons is covered in Appendix G. This takes account of both spin and, equally important, the existence of antiparticles. Paralleling this approach the Klein–Gordan equation provides an adequate description of spinless (scalar) particles.

Note the standard use of the symbol ϕ as the wavefunction of a scalar particle and as the electric potential.

1.17   Further reading

There are numerous good introductions to quantum mechanics. A well-tried example is *Quantum Mechanics*, 5th edition, by A. I. M. Rae, published by Taylor and Francis, London (2007). More sophisticated is *Lectures on Quantum Mechanics*, by the Nobel Laureate Steven Weinberg, published by Cambridge University Press (2012).

Exercises

(1.1) In the case of a particle having high energy compared to its rest mass energy, show that the appropriate expansion of the energy equation is $E = pc + m^2c^3/(2p) + ...$

(1.2) Where ψ is some scalar that varies smoothly with position, show that the simultaneous transformations of the electromagnetic vector field $(\mathbf{A}, \phi/c)$

$$\mathbf{A} \rightarrow \mathbf{A} + \nabla\psi$$
$$\phi \rightarrow \phi - \partial\psi/\partial t$$

do not affect measureable quantities.

(1.3) Calculate the particle density and flux in a plane wave $\exp[i(\mathbf{k}\cdot\mathbf{r} - \omega t)]/\sqrt{V}$ and check that the continuity equation holds.

(1.4) Evaluate the commutator of the energy and time operators $[E, t]$.

(1.5) Time dilation has $\Delta t' = \gamma\Delta t$ and the Lorentz contraction has $\Delta x' = \Delta x/\gamma$. In each case spell out what is being measured in which frame. Then deduce these results.

(1.6) Show that in a rigid rotation in normal space the product of the transformation matrix and its transpose equals the identity matrix.

(1.7) $|\xi_i\rangle$ is one of a set of orthonormal eigenstates. What is the operator that projects out the contribution of this eigenstate from a normalized state $|\phi\rangle$?

(1.8) Two lasers illuminate one each of the pair of slits in a Young's two slit interference experiment. They are tuned to the same frequency. Would interference fringes be seen? If so, explain how it is possible when any given photon originates from only one laser?

(1.9) A light source at a frequency f with a spectral width Δf has lateral dimensions r. What are the dimensions of the volume over which the light is coherent at a distance L from the source? $L \gg r$. Light at one point in the coherence volume has a fixed phase with respect to light at another point in the same coherence volume. Light from two such points can be superposed by means of mirrors, lenses, etc. and will show interference. Light from two points not within the same coherence volume have time varying relative phases and intereference fringes are not seen. Take L to be 1 m, Δf to be 10^{10} Hz, r to be 1 μm, and the wavelength to be 0.5 μm.

(1.10) A vertically polarized beam of light is incident in sequence on a polaroid that transmits light with its polarization at 45 degrees to the vertical and a polaroid that transmits light that is horizontally polarized. Using quantum mechanics explain how light initially vertically polarized can pass through the second horizontal polarizer.

Solutions to Schrödinger's equation

2

2.1 Introduction

Solutions to Schrödinger's equation are presented for square, harmonic and Coulomb wells. Barrier penetration in nuclear α-decay is used to illustrate the use of the solution for the square well potential. The solution for the harmonic well potential provides a first step toward the quantum theory of electromagnetic fields. Finally, the solution for the Coulomb potential provides the basis for explaining the states of the hydrogen atom. This is supplemented by a discussion of fine and hyperfine splitting of atomic energy levels.

2.2 The square potential well

The potential is drawn in Figure 2.1, it has a value $-V_0$ over the region $-a/2 < x < a/2$ and is zero elsewhere. Within the attractive well Schrödinger's equation is

$$(-\hbar^2/2m)\mathrm{d}^2\psi/\mathrm{d}x^2 = (E + V_0)\psi \quad \text{(internal)}, \qquad (2.1)$$

while outside the potential it becomes

$$(-\hbar^2/2m)\mathrm{d}^2\psi/\mathrm{d}x^2 = E\psi \quad \text{(external)}. \qquad (2.2)$$

Bound states of the particle, say an electron, for which E is negative and the kinetic energy, $(E + V_0)$, is positive are considered first. A solution inside the well, which is symmetric about the origin is

$$\psi_i = A_i \cos(k_i x), \qquad (2.3)$$

where $k_i = \sqrt{2m(E + V_0)}/\hbar$ and A_i is some constant. Externally

$$\psi_e = A_e \exp(\mp k_e x) \qquad (2.4)$$

where $k_e = \sqrt{-2mE}/\hbar$ and A_e is another constant. The upper sign in the exponent is taken for $x > a/2$ and the lower sign for $x < -a/2$. The opposite choices of sign for the exponentials would give wavefunctions growing exponentially with the distance from the well. These can be rejected because they grow infinitely.

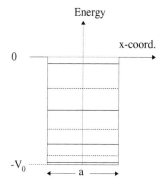

Fig. 2.1 The energy levels of eigenstates in the square potential well.

Quantum 20/20: Fundamentals, Entanglement, Gauge Fields, Condensates and Topology.
Ian R. Kenyon. © 2020. Published in 2020 by Oxford University Press.
DOI: 10.1093/oso/9780198808350.001.0001

Applying the requirements that the wavefunction and its first derivative are continuous at the wall at $x = a/2$ gives

$$A_i \cos(k_i a/2) = A_e \exp(-k_e a/2) \text{ and}$$
$$A_i k_i \sin(k_i a/2) = k_e A_e \exp(-k_e a/2).$$

Dividing one equation by the other gives

$$k_e = k_i \tan(k_i a/2). \tag{2.5}$$

From the definitions of k_i and k_e we also have

$$(k_i a/2)^2 + (k_e a/2)^2 = ma^2 V_0/(2\hbar^2). \tag{2.6}$$

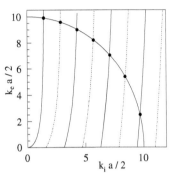

The last two equations can be solved simultaneously either by computer or graphically as exhibited in Figure 2.2 where $(k_e a/2)$ is plotted as a function of $(k_i a/2)$ for a given potential V_0. The relation found in eqn. 2.5 is represented by the full lines, while the quarter circle represents eqn. 2.6 with $ma^2 V_0/(2\hbar^2)$ taken to be 100. Simultaneous solutions to eqns. 2.5 and 2.6 lie at the points where these curves intersect.

Fig. 2.2 Graphical method of solving Schrödinger's equation for the square well potential.

A second set of wavefunctions that are antisymmetric about the origin also satisfy Schrödinger's equation for the square well. The waves inside the well have the form

$$\psi_i = B_i \sin(k_i x), \tag{2.7}$$

where B_i is some constant. Outside the well

$$\psi_e = B_e \exp(\mp k_e x) \tag{2.8}$$

where B_e is another constant. For these wavefunctions the continuity conditions lead to a different transcendental equation

$$k_e = -k_i \cot(k_i a/2). \tag{2.9}$$

This equation is plotted with broken lines in Figure 2.2. On this plot the simultaneous solutions to eqns. 2.9 and 2.6 lie at the intersections of the broken lines and the quarter circle. Then on Figure 2.1 the energy levels of all seven solutions are shown using full and broken lines for the states with even and odd wavefunctions respectively. Finally the wavefunctions of the four lowest energy (most tightly bound) states are plotted in Figure 2.3.

The preceding analysis shows that bound states are restricted to discrete energies. Only then can the sinusoidal waves inside the well join smoothly onto a wave that decays exponentially outside the well. At other energies the requirement of continuity at the boundary makes it necessary to have a sum of a decaying and an *increasing* exponential outside the well. No matter how little the electron's energy differs from the discrete value picked out by the solution of Schrödinger's equation in Figure 2.2, the exponentially increasing component of the wave outside the well will tend to infinity at an infinite distance. Such a wavefunction cannot describe electron states localized in the well. This restriction, in some situations, to states with *discrete* energies is a feature specific to quantum mechanics. Discrete energy states are met in atoms, molecules, nuclei and elementary particles. Particles with positive energies have wavefunctions that are oscillatory both inside and outside the potential well. The continuity conditions at the boundary can now be satisfied at any positive energy and so there is a continuum of allowed states extending from zero energy upwards.

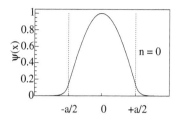

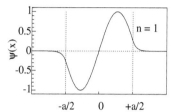

2.2.1   Barrier penetration

In contrast to the classical prediction, electrons and other particles, can penetrate regions where their kinetic energy is negative. The particle wavefunction decays exponentially as it penetrates such a region. When, as in Figure 2.4, the potential barrier is of finite width, the particle's wavefunction penetrates the potential barrier and emerges as an oscillatory wave that travels away from the boundary. Particles can therefore travel through a region where their kinetic energy is negative and penetrate to the far side. This purely quantum process is called *barrier penetration* or *tunnelling*.

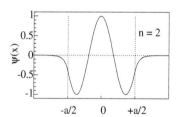

The penetration of evanescent electromagnetic waves through interfaces at which total internal reflection is expected is also due to barrier penetration. Its exploitation in monomode optical fibre underpins the telecomms industry: near-infrared radiation is guided along the glass core by total internal reflection between the core and the glass cladding. Nonetheless, around half the energy is carried by the evanescent wave travelling in the cladding. This parallel between the behaviour of electrons and photons can be better appreciated when their wave equations are compared. The electromagnetic wave equation is

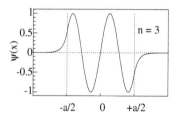

$$\mathrm{d}^2\psi/\mathrm{d}x^2 = -[\omega\mu(x)/c]^2\psi, \qquad (2.10)$$

Fig. 2.3 The four wavefunctions of lowest energy satisfying the square well boundary conditions. The node count is given. Broken lines mark limits of classical motion.

with $\mu(x)$ being the refractive index of the material at x and ω the wave's angular frequency. Writing Schrödinger's equation again

$$\mathrm{d}^2\psi/\mathrm{d}x^2 = -(2m/\hbar^2)[E - V(x)]\psi, \qquad (2.11)$$

we can recognize the equivalence

$$\mu(x) = \sqrt{2m[E - V(x)]}/(\hbar k), \qquad (2.12)$$

where $k = \omega/c$ is the wave number in vacuum. Barrier penetration corresponds to an imaginary refractive index and hence to decay of the electromagnetic wave. What is expressed here is of general significance: namely, that the electromagnetic wave equation has the same relation to photons that Schrödinger's equation has to non-relativistic electrons.

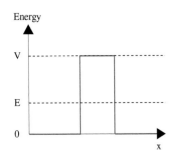

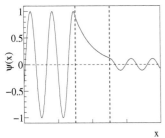

Fig. 2.4 Potential barrier and wavefunction penetration.

Nuclear species emitting α-particles have a huge range in lifetimes but surprisingly small differences in the energy released: for example ^{238}U has a half life of $4.47\,10^9$yr for an energy release of 4.27 MeV, while ^{226}U has a half life of 0.35s for an energy release of 7.70 MeV. This striking variation in lifetimes can only be explained by quantum barrier penetration. The α-particle is pictured as bouncing to and fro inside the attractive nuclear potential well shown in Figure 2.5. Outside this nuclear potential there is a Coulomb barrier due to the charge on the nucleus. A dotted line indicates the energy of the α-particle which has to penetrate this barrier between radii r_1 and r_2. The solution to eqn. 2.11 for barrier penetration where the kinetic energy is negative is

$$\psi(r + \mathrm{d}r) = \exp\left(-g\mathrm{d}r\right)\psi(r), \tag{2.13}$$

where $g = \sqrt{(2M/\hbar^2)(V(r) - E)}$ over a region of length $\mathrm{d}r$ where the potential is $V(r)$, M is the α-particle mass and E is the energy of the α-particle. Integrating across the barrier

$$\psi(r_2) = \psi(r_1)\exp[-\int_{r_1}^{r_2} g(r)\,\mathrm{d}r] = \psi(r_1)\exp[-G], \tag{2.14}$$

and the *probability* of transmission through the barrier is

$$T = \exp\left(-2G\right). \tag{2.15}$$

If the α-particle has a velocity v_i inside the nucleus the rate of collisions with the barrier is $[v_i/(2r_1)]$ and the decay probability is

$$P = [v_i/(2r_1)]T = [v_i/(2r_1)]\exp\left(-2G\right). \tag{2.16}$$

The Coulomb potential energy of the α-particle is

$$V(r) = 2Ze^2/(4\pi\varepsilon_0 r), \tag{2.17}$$

where the atomic number of the daughter nucleus is Z. The energy E equals $V(r_2)$, so we can replace both E and $V(r)$ in eqn. 2.14 to give

$$G = \sqrt{[MZe^2/(\pi\varepsilon_0\hbar^2)]}\int_{r_1}^{r_2} \sqrt{(1/r - 1/r_2)}\,\mathrm{d}r. \tag{2.18}$$

Making the approximation that r_2 is much larger than r_1 this yields

$$G = \sqrt{[MZe^2/(\pi\varepsilon_0\hbar^2)]}\,[\pi\sqrt{r_2}/2 - \sqrt{r_1}]. \tag{2.19}$$

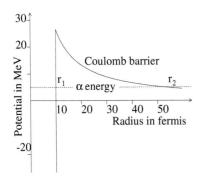

Fig. 2.5 The potential seen by the α particle. Its kinetic energy is negative between r_1 and r_2.

Applying eqn. 2.17 and using the equality of E and $V(r_2)$ again, we have

$$r_2 = 2Ze^2/(4\pi\varepsilon_0 E). \tag{2.20}$$

The radius r_1 varies little between the nuclei considered and this fact allows some simplification from here onward. Then using eqn. 2.16 and substituting for r_2 in G gives

$$P = [v_i/(2r_1)]\exp\left\{-\sqrt{(M/2E)}\,[Ze^2/\varepsilon_0\hbar)] + c_1\right\}, \tag{2.21}$$

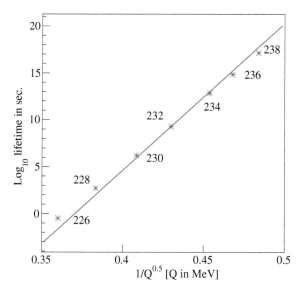

Fig. 2.6 Geiger–Nuttall plot of half-lives versus energy release for α decays of uranium isotopes with even atomic weight. The fit is explained in the text.

where c_1, and later c_2 and c_3 are near constant quantities related to r_1. Taking natural logs of this equation gives

$$\ln P = c_2 - aZ/\sqrt{E} \tag{2.22}$$

where $a = \sqrt{(M/2)}[e^2/(\varepsilon_0 \hbar)]$. Measuring the energy in MeV a takes the value $4.0\sqrt{\text{MeV}}$. Hence in terms of the half-life, $\tau_{1/2}$, in seconds

$$\log \tau_{1/2} = c_3 + \log(e)aZ/\sqrt{E} = c_3 + 1.74Z/\sqrt{Q}. \tag{2.23}$$

We have replaced the α-particle energy by energy released Q, because the α-particle, being very much lighter than the parent nucleus, receives almost all the energy. This expression for the half-life has the same form as an empirical fit to nuclear α decay data first made by Geiger and Nuttall. In Figure 2.6 the data for uranium α decays of even mass nuclei is compared with the above prediction for $Z = 90$ with an arbitrary choice of c_3.[1]

Despite the simplicity of the analysis, the prediction using quantum barrier penetration successfully accounts for the variation of over seventeen orders of magnitude in the lifetime of the uranium nuclei. Predictions made for other sequences of α decays show similarly impressive agreement.

The substitution $r = r_2 \cos^2 \theta$ in the integral of eqn. 2.18 gives

$$-2\sqrt{r_2} \int \sin^2 \theta \mathrm{d}\theta$$
$$= \sqrt{r_2}[\sin 2\theta/2 - \theta]_{r_1}^{r_2}$$
$$= \sqrt{r_2}[\sqrt{r/r_2 - r^2/r_2^2}$$
$$- \arccos \sqrt{r/r_2}]_{r_1}^{r_2}.$$

Now r_1 is much smaller than r_2 so this reduces to $\sqrt{r_2}[\pi/2 - \sqrt{r_1/r_2}]$.

[1] Ze is the charge on the daughter nucleus. It is also possible to extract the nuclear radius from the value of c_3.

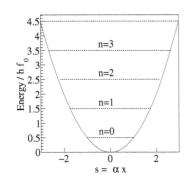

Fig. 2.7 Energy levels in the harmonic potential.

2.3   The harmonic oscillator potential

A frequently met dynamical system is that of a mass m undergoing simple harmonic motion in the x-dimension under a restoring force linear in the displacement from the origin, kx. The eigenstates are of particular interest because they are exact parallels to the eigenstates of electromagnetic radiation. The potential energy is $V = kx^2/2$, which is displayed in Figure 2.7. The classical motion is sinusoidal with angular frequency $\omega_0 = \sqrt{k/m}$ so that we can rewrite the potential energy as

$$V = m\omega_0^2 x^2/2. \tag{2.24}$$

Schrödinger's time independent equation for this potential is

$$-(\hbar^2/2m)\,\mathrm{d}^2\psi/\mathrm{d}x^2 + (kx^2/2)\psi = E\psi.$$

Setting $\alpha = \sqrt{m\omega_0/\hbar}$ and $s = \alpha x$, the equation can be rewritten

$$\mathrm{d}^2\psi/\mathrm{d}s^2 + (\lambda - s^2)\psi = 0, \tag{2.25}$$

where $\lambda = 2E/\hbar\omega_0$. This equation has analytic solutions of the form

$$\psi(s) = H(s)\exp\left(-s^2/2\right). \tag{2.26}$$

Solutions with a term $\exp\left(s^2/2\right)$ are excluded because they diverge at infinity and are not confined to the potential well. $H(s)$ is a polynomial satisfying

$$[\mathrm{d}^2 H(s)/\mathrm{d}s^2] - 2s\,[\mathrm{d}H(s)/\mathrm{d}s] + (\lambda - 1)\,H(s) = 0. \tag{2.27}$$

Finite polynomials can be obtained provided that λ takes the discrete values[2]

$$\lambda = 2n + 1, \tag{2.28}$$

where n is any non-negative integer. When λ is replaced by its defined value, $2E/(\hbar\omega_0)$, it immediately follows that the energy of the nth eigenstate is

$$E_n = \hbar\omega_0(n + 1/2). \tag{2.29}$$

The lowest energy solutions, labelled with the value of n as a subscript, are

$$H_0(s) = 1; \ H_1(s) = 2s; \ H_2(s) = 4s^2 - 2; \ H_3(s) = 8s^3 - 12s. \tag{2.30}$$

These functions are called *Hermite* polynomials. A simple recurrrence relation derived from eqn. 2.27 can be used to generate further members of the sequence

$$H_{n+1}(s) = 2sH_n(s) - 2nH_{n-1}(s).$$

[2] See Chapter 4 of the third edition of *Quantum Mechanics* by L. I. Schiff, published by McGraw-Hill Kogakusha Ltd., Tokyo (1968).

The resulting *Gauss–Hermite* solutions to the Schrödinger equation for the harmonic well, $H_n(s) \exp\left(-s^2/2\right)$, are shown in Figure 2.8. Vertical dotted lines indicate where the classical motion terminates. The lowest energy solution has the familiar Gaussian shape. Corresponding energy levels are indicated by the horizontal lines in Figure 2.7.

In the lowest energy classical state the mass would be at rest at $x = 0$. By contrast the lowest energy quantum state with the Gaussian wavefunction has an energy $\hbar\omega_0/2$ called the *zero point energy*. This is a characteristic of quantum systems: in the previous example with a square well potential the lowest energy state is not at rest, but displaced upward from $-V_0$ by the zero point energy.

Now, the root mean square excursion x_{zpf} in the ground state is the quantum zero point fluctuation in position. Equating kx_{zpf}^2 to the zero point energy gives

$$
\begin{aligned}
\hbar\omega_0/2 &= kx_{\mathrm{zpf}}^2 \text{ giving} \\
x_{\mathrm{zpf}}^2 &= \hbar/(2m\omega_0).
\end{aligned}
\tag{2.31}
$$

This provides a convenient length scale used later

$$
a_{\mathrm{zpf}} = \sqrt{\hbar/(2m\omega_0)}.
\tag{2.32}
$$

There is a corresponding velocity scale

$$
v_{\mathrm{zpf}} = \sqrt{\hbar\omega_0/2m}.
\tag{2.33}
$$

2.4   The hydrogen atom

Schrödinger's analysis of the electron motion within the Coulomb potential due to the nuclear charge provided a detailed and precise description of atomic structure, which incorporated all the successes of the Bohr model. In a hydrogen-like atom with a nucleus carrying a charge Ze and having mass M, the Coulomb potential felt by the single electron of mass m is $-Ze^2/(4\pi\varepsilon_0 r)$ at a distance r from the nucleus. This potential is drawn for the hydrogen atom in Figure 2.9. Then the Schrödinger time independent equation is

$$
-(\hbar^2/2\mu)\nabla^2\psi - Ze^2\psi/(4\pi\varepsilon_0 r) = E\psi,
\tag{2.34}
$$

where $\mu = mM/(M+m)$ is the reduced mass of the electron and E is the total energy of the atom. The same approach is taken as that applied in seeking solutions to the square well potential. Although the analysis is more complicated the solutions in this case are analytic. Here only the results are discussed, but full details of the solution can be found in

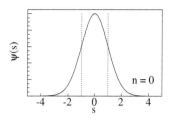

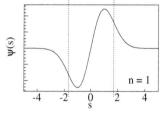

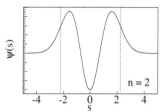

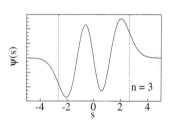

Fig. 2.8 The four wavefunctions of lowest energy in a harmonic well, labelled with the quantum number n.

Table 2.1 Radial eigenfunctions $R_{nl}(r)$.

n	l	$R_{nl}(r)$
1	0	$2(Z/a_0)^{3/2} \exp(-Zr/a_0)$
2	0	$(1/\sqrt{8})(Z/a_0)^{3/2} [2 - Zr/a_0] \exp(-Zr/2a_0)$
2	1	$(1/\sqrt{24})(Z/a_0)^{3/2} [Zr/a_0] \exp(-Zr/2a_0)$
3	0	$(\sqrt{4/3}/81)(Z/a_0)^{3/2} [27 - 18Zr/a_0 + 2(Zr/a_0)^2] \exp(-Zr/3a_0)$

many standard texts on atomic physics or quantum mechanics.[3]

The eigenfunctions separate into radial and angular components, which are best written in spherical polar coordinates

$$\psi_{nlm_l}(r,\theta,\phi) = R_{nl}(r)\, Y_{lm_l}(\theta,\phi). \qquad (2.35)$$

Each solution is identified by three integral *quantum numbers* n, l, m_l. The function R_{nl} contains an associated Laguerre polynomial $F(r)$, and Y_{lm} is a spherical harmonic function. In general, these have the forms

$$R_{nl}(r) = \exp(Cr/n)\, r^l\, F(r), \qquad (2.36)$$
$$Y_{lm_l}(\theta,\phi) = \sin^{|m_l|}\theta \, \exp(im_l\phi)\, G(\theta), \qquad (2.37)$$

where C is a constant and $F(r)$ and $G(\theta)$ are polynomials in r and $\cos\theta$ respectively. The exact forms of the eigenfunctions are given in Tables 2.1 and 2.2 for a few of the lowest values of n, l and m_l. These wavefunctions are *orthonormal* in the usual sense that the volume integrals over all space are

$$\int \psi^*_{nlm_l}\, \psi_{n'l'm'_l}\, \mathrm{d}V = \delta_{n,n'}\, \delta_{l,l'}\, \delta_{m_l,m'_l}. \qquad (2.38)$$

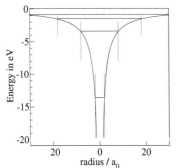

Fig. 2.9 Diametral slice through the Coulomb potential due to the hydrogen nucleus. The first few electron energy levels are shown. a_0 is the Bohr radius.

Valid combinations of the quantum numbers are restricted to the following values:

$$n = 1, 2, 3, \cdots;$$
$$l = 0, \cdots, n-2, n-1;$$
$$m_l = -l, -l+1, -l+2, \cdots, l-2, l-1, l. \qquad (2.39)$$

The energy of an electron, its orbital angular momentum and a component of its angular momentum, which can be chosen to be the z-component, are the three compatible observables and their eigenvalues are specified by the quantum numbers n, l and m_l respectively. The predicted energy eigenvalues duplicate those found with the Bohr model

$$E_n = -\mu Z^2 e^4 / [(4\pi\varepsilon_0)^2 2\hbar^2 n^2], \qquad (2.40)$$

---

[3]For example the third edition of *Quantum Mechanics* by L. I. Schiff, published by McGraw-Hill Kogukusha, Tokyo (1968).

Table 2.2 Angular eigenfunctions $Y_{lm}(\theta, \phi)$.

l	m	$Y_{lm}(\theta, \phi)$
0	0	$1/\sqrt{4\pi}$
1	0	$\sqrt{(3/4\pi)}\cos\theta$
1	± 1	$\mp\sqrt{(3/8\pi)}\sin\theta\exp(\pm i\phi)$
2	0	$\sqrt{(5/16\pi)}[3\cos^2\theta - 1]$
2	± 1	$\mp\sqrt{(15/8\pi)}\sin\theta\cos\theta\exp(\pm i\phi)$
2	± 2	$\sqrt{(15/32\pi)}\sin^2\theta\exp(\pm 2i\phi)$

so that n is called the *principal* or energy quantum number.

Angular momentum operators can be constructed from the momentum operators as follows. In classical mechanics the z-component of the vector angular momentum, **L**, is

$$L_z = xp_y - yp_x,$$

from which the quantum mechanical operator can be obtained by replacing the position and momentum by their operator equivalents

$$\hat{L}_z = -i\hbar\left[x\frac{\partial}{\partial y} - y\frac{\partial}{\partial x}\right]. \tag{2.41}$$

The total orbital angular momentum operator, **L**, is given by

$$\hat{\mathbf{L}}^2 = \hat{L}_x^2 + \hat{L}_y^2 + \hat{L}_z^2. \tag{2.42}$$

$\hat{L}^2$ commutes with all the components L_x, L_y and L_z, but these components do not commute with each other. For example

$$[L_x, L_y] = i\hbar L_z. \tag{2.43}$$

This explains why the total orbital angular momentum and only one of its components can be compatible observables. For completeness the forms of these operators in spherical polar coordinates are

$$\hat{\mathbf{L}}^2 = -\hbar^2\left[\frac{1}{\sin\theta}\frac{\partial}{\partial\theta}\left(\sin\theta\frac{\partial}{\partial\theta}\right) + \frac{1}{\sin^2\theta}\frac{\partial^2}{\partial\phi^2}\right] \tag{2.44}$$

$$\text{and } \hat{L}_z = -i\hbar\frac{\partial}{\partial\phi}. \tag{2.45}$$

When these operate on the wavefunctions they give

$$\hat{\mathbf{L}}^2\psi_{nlm_l} = l(l+1)\hbar^2\psi_{nlm_l} \tag{2.46}$$

$$\text{and } \hat{L}_z\psi_{nlm_l} = m_l\hbar\psi_{nlm_l}. \tag{2.47}$$

It follows that l specifies the magnitude, $\sqrt{l(l+1)}\hbar$, of the orbital angular momentum, while $m_l\hbar$ specifies its component in the z-direction.

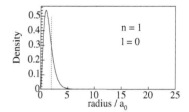

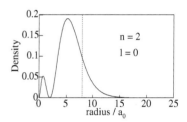

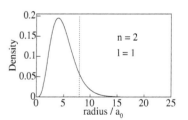

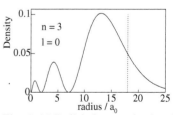

Fig. 2.10 Radial electron density distributions in the hydrogen atom for the lowest energy eigenstates. The dotted lines indicate where the kinetic energy changes sign.

The historical spectral notation is used in labelling the eigenstates of the electron: an electron with *orbital angular momentum* quantum number $l = 0, 1, 2, 3, 4, 5 \cdots$ is said to be in an s, p, d, f, g, h,$\cdots$ state. The radial distribution of the electron probability density $r^2 R_{nl}^2$ is shown in Figure 2.10 for a few values of n and l, where the factor r^2 is included to take account of the growth in the volume element as the radius increases, $dV = r^2 \sin\theta \, d\theta \, d\phi \, dr$. The dotted lines in this diagram indicate where the kinetic energy changes sign. Each wavefunction has $(n - l)$ radial nodes. Figure 2.11 shows diametral sections containing the z-axis through the $l = 1$ electron probability distributions. The three-dimensional distributions are obtained by rotating these planar distributions around the axis indicated by the broken line in Figure 2.11. Note that the three 2p electronic wavefunctions give a combined electron density distribution that is spherically symmetric. Such spherically symmetric distributions always result when there is an electron in each of the $2l + 1$ substates for a given l value; that is to say, when the electron *subshell* is full.

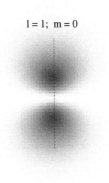

l = 1; m = 0

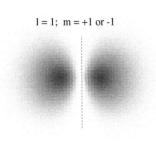

l = 1; m = +1 or -1

The eigenstates have a definite *parity*, that is to say the result of reflecting the coordinates in the origin causes the wavefunction to change by a factor ±1. A wavefunction with orbital angular momentum ℓ has a parity of $(-1)^\ell$. Eigenstates sharing the same value of n but different values of l and m_l have identical energies in the Schrödinger model: such eigenstates are termed *degenerate*. This degeneracy is lifted by the spin-orbit coupling and relativistic effects discussed later.

The most intense lines in the hydrogen spectrum involve transitions in which l changes by unity

$$\Delta l = \pm 1, \tag{2.48}$$

Fig. 2.11 Diametral sections through the probability distributions in the hydrogen atom for electrons in the 2p shell. The lower distribution is toroidal in three dimensions.

which is one example of a *selection rule* in optical spectra. The corresponding rule for the associated changes in the magnetic quantum number is

$$\Delta m = \pm 1, 0. \tag{2.49}$$

Transitions between the states of a hydrogen atom in which the change in ℓ is different from those specified in eqn. 2.48 occur at far lower rates and are known collectively as *forbidden* transitions.

2.5 Intrinsic angular momentum

Electrons and other elementary particles carry intrinsic angular momentum or *spin* in addition to any orbital angular momentum due to their motion. Elementary particles are, to the best of our knowledge, point-like: no structure has been detected in electrons and quarks probed to 10^{-18}m in electron-proton collisions at the HERA collider in Hamburg. Hence the spin is not associated with any internal motion, it is intrinsic.

The initial experiment that identified electron spin was performed in 1922 by Stern and Gerlach. They fired silver atoms through an inhomogeneous magnetic field which produced a deflection proportional to the magnetic moment of the atom. They observed two discrete beams emerging, indicating that the magnetic moment of the silver atom is quantized. Now the magnetic moment expected for an electron with orbital angular momentum $l\hbar$ was[4]

$$\mu = -\mu_B l,   (2.51)$$

where μ_B is the *Bohr magneton* equal to $e\hbar/2m$ and m is the electron mass. A component of this magnetic moment along the quantization axis

$$\mu = \mu_B m_l.   (2.52)$$

In a varying magnetic field the moment feels a force $-\mu_B m_l \left(\partial B/\partial z\right)$.

If an electron has orbital angular momentum quantum number l, then clearly the number of magnetic substates expected is odd, $2l+1$. Hence an odd number of discrete beams are expected to emerge from the region of varying field, not two. Having a pair, or other even number, of magnetic substates is explained by Pauli's proposal that electrons possess an intrinsic angular momentum (spin) that is half integral in units of $\hbar$. Explicitly, the spin and its component along a quantization axis are

$$S = \sqrt{(1/2)(3/2)}\,\hbar \ , m_s = \pm/2.   (2.53)$$

A full description of an electron in a hydrogen atom therefore requires the specification of the quantum numbers n, l, m_l and m_s. Its total angular momentum is the vector sum of orbital and intrinsic angular momentum

$$\mathbf{J} = \mathbf{L} + \mathbf{S}.   (2.54)$$

The total angular momentum has integer eigenvalues, j, with $J = \sqrt{j(j+1)}\hbar$ and $2j+1$ magnetic substates with

$$m_j = -j, -j+1, \cdots, j-1, j.   (2.55)$$

A further unexpected property of the electron is that the intrinsic magnetic moment of the electron is

$$\mu = g\mu_B m_s   (2.56)$$

where g would be exactly 2.0 in Dirac's relativistic model of the electron. This is in contrast to the case of the magnetic moment due to orbital angular momentum for which g is exactly unity. However the electron g departs slightly from two, a departure explained in the quantum field theory of electrons, photons and their interactions.

In summary, the magnetism of atoms originates in the magnetic moment of the closed stable orbits of electrons (and nucleons) and from

[4]An electron travelling in an orbit of radius r with velocity v produces a current $i = -ev/(2\pi r)$. This gives a magnetic moment

$$\mu = i(\text{orbit area}) = -evr/2.   (2.50)$$

This is $-(e/2m)$ times the orbital angular momentum.

their intrinsic magnetic moments. Looking back, there were no closed stable orbits and no intrinsic magnetic moments in classical physics, so the very existence of atomic magnetism should have been a puzzle.[5]

Turning now to the photon, this has unit spin: $S = \sqrt{(1)(2)}\hbar$. A beam of left (right) circularly polarized light consists entirely of photons with spin component $+\hbar$ ($-\hbar$) along the direction of travel. These properties were established in 1926 by Beth: he passed circularly polarized light through a half-wave plate which has the property that it reverses the sense of circular polarization of light, and which therefore reverses the spin component of each photon. The half-wave plate was suspended by a fine wire so that it could swing freely. An equal and opposite torque is developed by the suspension of the half-wave plate to balance the rate of change in angular momentum of the photons passing through it. Beth measured this torque Γ and extracted the individual photon angular momentum component, $\Gamma/2N_\phi$, where N_ϕ is the number of photons passing through the half-wave plate per second. He confirmed that this was consistent with $\hbar$.

There are only two polarization states of electromagnetic radiation, left circular and right circular polarization: the state corresponding to spin component $0\hbar$ is missing. This striking absence is connected with the fact that photons are massless, and it is connected to the way spontaneous symmetry breaking occurs in the standard model of particle physics. The topic is covered in Chapter 19.

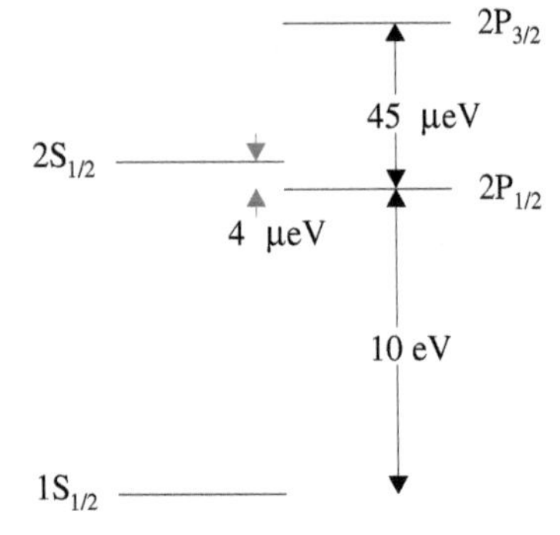

2.6 Fine structure

The single levels of the Bohr model become multiple levels due to the *spin-orbit* interaction, which leads to a difference between the energies of states with different alignments of the spin and orbital angular momentum. Figure 2.12 shows the *fine structure* of the lowest energy levels of the hydrogen atom with the $2P_{1/2}$ state split from the $2P_{3/2}$ state. In this spectroscopic notation S and P refer to zero and unit orbital angular momentum respectively, in units $\hbar$; the prefix is the principal quantum number and the subscript is the total angular momentum quantum number, in units $\hbar$. The relativistic equivalent of Schrödinger's equation was discovered by Dirac, and the resulting expression for the energy of the hydrogen atom is[6]

$$E = -\frac{me^4}{2(4\pi\varepsilon_0\hbar n)^2}\left(1 + \left[\frac{\alpha^2}{n}\right]\left[\frac{1}{j+1/2} - \frac{3}{4n}\right]\right). \tag{2.57}$$

Fig. 2.12 The spin-orbit splitting of the lowest lying hydrogen atomic levels.

[6]To be compared with eqn. 1.11.

[5]In 1911 Bohr proved formally that magnetism is forbidden in classical, isolated non-rotating systems: this may well have influenced his thinking about atomic structure that led to the Bohr model in 1913.

The relativistic and spin-orbit induced splittings are around 10^{-4} of the level spacing in hydrogen; the spin-orbit splitting increases with the atomic number while the relativistic effect does not change appreciably.

Spin-orbit coupling is a magnetic effect that was first encountered in the interaction between the orbital motion and the spin of an electron in an atom, but it is met in many other situations. The interaction felt by an electron in an atom is most easily visualized and analysed in the rest frame of the electron. In this frame the moving nucleus generates a magnetic field which acts on the electron's magnetic moment. If $V(\mathbf{r})$ is the electrostatic potential felt by the electron,

$$\Delta E = \frac{1}{r}\frac{\partial V}{\partial r}\frac{\mathbf{L}\cdot\mathbf{S}}{2m^2c^2} \qquad (2.58)$$

is the spin-orbit interaction energy.

However, the tiny separation between the $2S_{1/2}$ and the $2P_{1/2}$ states in Figure 2.12 is not accounted for. This is known as the *Lamb shift* and is only 1057.864MHz($\times$ h) or 4μeV. A complete consistent explanation of all the features of the hydrogen spectrum including the Lamb shift requires the *quantum electrodynamics* developed Feymann, Tomonaga and Schwinger. Simple features of this theory are presented in Chapters 8, 13, and 18.

2.6.1   Hyperfine splitting

Nuclei are built from neutrons and protons whose spins are $1/2$, the same as that of the electron, and these constituents have orbital motion within a nucleus. Therefore, a nucleus has quantum states with a total angular momentum $I = \sqrt{i(i+1)}\hbar$ where i is integral or half integral for even and odd numbers of nucleons, respectively. Nuclear magnetic moments have the nuclear magneton, $e\hbar/2M$, as a natural unit rather than the Bohr magneton, $e\hbar/2m$, where the nucleon mass M replaces the electron mass m. The nuclear magnetic moment is acted on by the magnetic field due to the electron orbital motion, which serves to couple the electronic and nuclear angular momenta. As a result, an atom has an overall angular momentum $\sqrt{F(F+1)}\hbar$ and magnetic quantum number:

$$\mathbf{F} = \mathbf{I} + \mathbf{J}, \quad m_F = -F, -F+1, \cdots + F. \qquad (2.59)$$

The magnetic interaction causes a further *hyperfine splitting* of the atomic levels which is smaller than the fine structure by a factor approximately $\mu_N/\mu_B \approx 10^{-3}$. Figure 2.13 shows the hyperfine structure of sodium which has a single electron outside closed shells.

An important example is the splitting of the $1S_{1/2}$ level of hydrogen into $F = 0$ and $F = 1$ levels separated by $5.9\,10^{-6}$eV, equivalent to a thermal excitation of only 0.7K. Radiation emitted in the transition has wavelength 21 cm and its presence is a valuable indicator of hydrogen

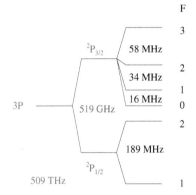

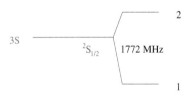

Fig. 2.13 Fine and hyperfine structure of the 3P and 3S levels of sodium. Note the scale change over the columns from THz for 3S-3P splitting, to GHz for spin-orbit splitting, to MHz for hyperfine splitting.

in cold regions of the universe. By measuring the doppler shift of the 21 cm line astromomers have been able to estimate the velocities of the gas making up the outer regions of galaxies. The velocity profile indicates there is much more matter present in galaxies than that visible through its electromagnetic radiation: this is the mysterious *dark matter* discussed in Chapter 19.

2.6.2 Landé g-factor

It has just been noted that the principal contributions to an atom's magnetic moment come from the electrons. The orbital motion gives a term $\mu_B m_\ell$, while the spin gives a term $g\mu_B m_s$ where g is close to 2. Here we take g to be exactly 2 and ignore the thousand times smaller nuclear contribution to the magnetic moment. It follows that if the total angular momentum has a component m_J then the magnetic moment of the atom is $g_J\mu_B m_J$ where g_J will depend on the relative magnitude of the orbital and intrinsic angular momenta. The result is that

$$g_J = 3/2 + [S(S+1) - L(L+1)]/[2J(J+1)]. \qquad (2.60)$$

When the nucleus has angular momentum this is modified to take account of the different combinations of the electron spin and orbital angular momentum that become possible. Now, we have a magnetic moment $g_F\mu_B m_F$ where

$$g_F = g_J[F(F+1) - I(I+1) + J(J+1)]/[2F(F+1)], \qquad (2.61)$$

where the nuclear magnetic moment is neglected because it is smaller than the electron magnetic moment by the ratio of the electron mass to the nucleon mass. Evidently the values of these *Landé g-factors* are of order unity.

2.7 Further reading

Quantum Physics by R. Eisberg and R. Resnick, published by John Wiley (1974) contains a wealth of material on spectral structure in atoms and molecules.

Exercises

(2.1) Using Figure 1.2 calculate the relative proportion of hydrogen atoms in their ground and first excited states in gas at thermal equilibrium at 100 000 K and 1000 K.

The absorption lines due to remote clouds of hydrogen gas are seen in the spectra of powerful sources whose light they intercept on its way to the earth. Clouds at different distances have lines red-shifted differently so that a spectrum of a quasar will exhibit large numbers of hydrogen gas absorption lines known as a Lyman forest. Why is the assignment to the Lyman absorption lines so sure?

(2.2) Show that the sum of the electron distributions for the three 2p wavefunctions in hydrogen atoms is spherically symmetric.

(2.3) Show that $[L_z, L_x] = i\hbar L_y$.

(2.4) An electron drops from the $n = 4$ excited state to the ground state in a hydrogen atom. Calculate the energy release. How is this shared between the photon and the hydrogen atom? Does this mean that the energy of a photon emitted by one atom would have too small an energy to excite another hydrogen atom from the ground state to the same excited state?

(2.5) Positronium is a bound state of an electron and its antiparticle the positron, which has the same mass as the electron but a positive charge equal in magnitude to that of the electron. Calculate the wavelength of the $n = 2$ to $n = 1$ transition.

(2.6) A micromechanical oscillator has frequency of 100 MHz. Calculate the energy of its zero point motion. What temperature is required to hold the number of phonons of excitation to around 10?

(2.7) Electrons of energy E are incident on a surface at which the potential changes from zero to $V < E$. Calculate the transmission and reflection coefficients and determine the value of E that leads to half the electrons being reflected.

(2.8) Show that an operator whose eigenvalues for a set of orthonormal eigenstates are all real is itself hermitian.

(2.9) Check that the current and density given in Section 1.8 satisfy eqn. 1.28.

(2.10) Show that for a scalar b, $\delta(bx) = \delta(x)/|b|$. Also show that

$$\int_{-\infty}^{\infty} \exp[ik(x-y)] \, \mathrm{d}k/(2\pi) = \delta(x-y).$$

Quantum statistics

3.1 Introduction

A key distinction between electrons and photons lies in their different statistics. Electrons belong to one class of particle, the *fermions*, all of which possess half integral spins ($\hbar/2$, $(3/2)\hbar, \cdots$). Photons belong to the other class, the *bosons*, with integral spins ($0\hbar$, $\hbar$, $2\hbar, \cdots$). Here $n\hbar$ is shorthand for $\sqrt{n(n+1)}\hbar$. An eigenstate of a specific fermion species can only contain one such fermion, while an eigenstate of a specific boson species can contain any number of such bosons. Atomic structure and the resulting chemical properties of matter are consequences of fermion statistics, while lasers and Bose–Einstein condensation, are consequences of boson statistics. Sections of this chapter are used to examine examples for which fermion statistics lead to significant physical effects: atomic structure, the origin of ferromagnetism and compact star histories. Then the conditions for bosons to condense in the lowest energy state are calculated and examples introduced that are covered fully later in Chapters 14, 15 and 16. Lasers are treated in Chapter 8.

3.2 Symmetries of fermion and boson states

The statistical behaviour of electrons and photons has a common origin in a uniquely quantum mechanical property: that is *the indistinguishability of any particle of one species from another of the same species.* Particle 1 being in quantum state A and particle 2 in quantum state B is then indistinguishable from 2 being in A and 1 being in B. Consequently, the expectation values for the two alternatives are identical:

$$\psi^*(1_A, 2_B)\psi(1_A, 2_B) = \psi^*(2_A, 1_B)\psi(2_A, 1_B). \qquad (3.1)$$

The only simple possibilities are that either ψ is symmetric under the interchange of 1 and 2, or antisymmetric under the interchange:

$$\psi_s(1_A, 2_B) = \psi_s(2_A, 1_B), \ \text{ or } \ \psi_a(1_A, 2_B) = -\psi_a(2_A, 1_B). \qquad (3.2)$$

Of these possibilities the symmetric choice applies to particle species with integral spin, the *bosons*, and the antisymmetric choice to particle species with half integral spin, the *fermions*. Taking the antisymmetric

Quantum 20/20: Fundamentals, Entanglement, Gauge Fields, Condensates and Topology.
Ian R. Kenyon. © 2020. Published in 2020 by Oxford University Press.
DOI: 10.1093/oso/9780198808350.001.0001

choice we see immediately that if both particles are in the same quantum state

$$\psi_a(1_A, 2_A) = -\psi_a(2_A, 1_A), \tag{3.3}$$

so that wavefunction must vanish. Hence two fermions of the same species cannot occupy the same eigenstate. This is the *Pauli exclusion principle*, first recognized for electrons. More generally, when there are many electrons, the overall wavefunction must be antisymmetric under interchange of any pair of electrons. The appropriate expression is a Slater determinant, which for three electrons is

$$\psi_{A,B,C}(1,2,3) = (1/\sqrt{3!}) \begin{vmatrix} \psi_A(1) & \psi_B(1) & \psi_C(1) \\ \psi_A(2) & \psi_B(2) & \psi_C(2) \\ \psi_A(3) & \psi_B(3) & \psi_C(3) \end{vmatrix}. \tag{3.4}$$

In the alternative case for bosons there is symmetry under the interchange of like particles. When two bosons share the same eigenstate

$$\psi_s(1_A, 2_A) = \psi_s(2_A, 1_A), \tag{3.5}$$

which is unexceptional. It follows that any number of bosons, for example photons, can occupy a given quantum state. The product wavefunction for bosons corresponding to eqn. 3.4 is the *permanent* of the same square matrix; this is like a determinant with all the coefficients now positive.

3.3   Multi-electron atoms

Atoms of elements beyond hydrogen in the periodic table contain many electrons, and the Pauli principle guarantees that each additional electron enters the empty eigenstate with the lowest energy. Bosons would pile up in the lowest eigenstate. Again, nuclear structure is the consequence of the constituent protons and neutrons being fermions.

The periodicity in the chemical behaviour of the elements is reflected in their ionization energies, shown here in Figure 3.1 as a function of atomic number. Ionization energy is the energy required to detach the least well-bound electron from the atom. Clear peaks are evident at the atomic numbers of the chemically inert noble gases which have the electron configurations given in Table 3.1. The inference is that a configuration is extremely well bound and stable in which all the $2(2l + 1)$ available eigenstates for a given l-value contain electrons. These configurations have spherically symmetric wavefunctions with zero total angular momentum.

The alkali metals follow the noble gases in the periodic table, having one extra electron in a previously empty $l = 0$ orbit. For example sodium's electron configuration is $(1s^2\, 2s^2\, 2p^6)3s^1$. The lone 3s electron

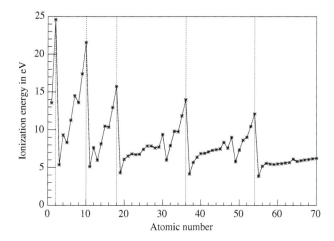

Fig. 3.1 The ionization energies required to remove the least well bound electron from the atomic elements are plotted against the atomic number. The inert noble gases with their completed electron shells are indicated by the vertical lines.

Table 3.1 Noble gas electron configurations. The initial number is the principal quantum number. The letter signifies the orbital angular momentum: s stands for $l = 0$, p for $l = 1$ and so on through d, f, g,.. The superscript is the number of electrons sharing those quantum numbers.

Element	Atomic number	Configuration
Helium	2	$1s^2$
Neon	10	$1s^2\ 2s^2\ 2p^6$
Argon	18	$1s^2\ 2s^2\ 2p^6\ 3s^2\ 3p^6$
Krypton	36	$1s^2\ 2s^2\ 2p^6\ 3s^2\ 3p^6\ 4s^2\ 3d^{10}\ 4p^6$

experiences an electric field due to the nuclear charge, Ze, surrounded by the $(Z-1)$ electrons in the closed shells. This resembles the electric field due to a single positive charge, so that the energy levels and the spectral lines produced when the 3s electron in a sodium atom is excited are similar to those of hydrogen. Similarly, the other alkali metals show spectra with some resemblance to that of hydrogen.

3.3.1 Hartree–Fock calculations

Hartree introduced an iterative method of calculating eigenfunctions of electrons in multi-electron atoms. A first estimate of the potential as a function of radius $V_0(r)$ is chosen, varying smoothly from $-Ze^2/(4\pi\varepsilon_0 r)$ at the nucleus to $-e^2/(4\pi\varepsilon_0 r)$ at large distances. Schrödinger's equation is solved for each orbital (n, l) and electrons are placed in the lowest Z energy states. Then the resulting potential of the nucleus plus electrons, $V_1(r)$, is calculated. The steps in this sequence are repeated until the potential after n steps, $V_n(r)$ is sufficiently close to that after $n-1$ steps $V_{n-1}(r)$. Fock included overall antisymmetrization of the waveform using the Slater determinant of eqn. 3.4 in order to satisfy the Pauli exclusion principle. Predictions from Hartree–Fock wavefunctions of spectra and other properties of atoms agree well with data.

3.3.2   Quantum numbers and selection rules

Note that the mutual Coulomb interaction of the electrons in the un-filled shells cannot alter their total orbital angular momentum, nor their total spin, nor their total angular momentum. Thus the corresponding angular momentum observables for the light elements are the vector sum of the orbital angular momenta, the vector sum of the spins, the total angular momentum and finally, the magnetic component of this total angular momentum:

$$\mathbf{L} = \sum_i \mathbf{l}_i, \ \ \mathbf{S} = \sum_i \mathbf{s}_i, \ \ \mathbf{J} = \mathbf{L} + \mathbf{S}, \ \ M = \sum_i (m_{l_i} + m_{s_i}), \qquad (3.6)$$

where the sums run over all the electrons. The atomic state of any light element atom can be characterized fully by the set of quantum numbers (L, S, J, M).[1] The possible values of J lie at integral steps within the range

$$|L - S| \leq J \leq L + S. \qquad (3.7)$$

The labelling of the eigenstates of an atom is usually in the form of a *term symbol* $^{2S+1}X_J$, where X is the upper case spectroscopic label corresponding to the value of L; that is S, P, D, F, ... for $L = 0, 1, 2, 3$, etc. When there is a single active electron it can be useful to add the principal quantum number n thus, $n^{2S+1}X_J$. The superscript is called the *multiplicity*. It gives the number of eigenstates with different values of J for that combination of S and L provided $L \geq S$. For convenience, and where there is no ambiguity the principal quantum number or the multiplicity are often omitted.

There are more complex selection rules for the dominant transitions in multi-electron atoms

$$\Delta S = 0, \ \ \Delta L = 0, \pm 1, \ \ \Delta J = 0, \pm 1, \ \ \Delta M = 0, \pm 1; \qquad (3.8)$$

but of these the transitions $J = 0 \rightarrow J = 0$ is excluded, and so is $M = 0 \rightarrow M = 0$ for $\Delta J = 0$. Underlying these rules is the requirement that in an allowed transition a single electron emits or absorbs a single photon in an electric dipole transition.

When hyperfine structure is resolved the total angular momentum of the atom including the nucleus, $\mathbf{F}$, becomes relevant. This is the vector sum of the angular momentum of the electrons, $\mathbf{J}$, and that of the nucleus, $\mathbf{I}$:

$$\mathbf{F} = \mathbf{J} + \mathbf{I}, \ \ m_F = m_j + m_i. \qquad (3.9)$$

The atomic state then requires two additional labels (F, m_F). The possible values of the total angular momentum, F, lie at integral steps within the range

$$|J - I| \leq F \leq J + I. \qquad (3.10)$$

[1] This is known as *LS* or *Russell–Saunders* coupling of angular momenta. In heavy elements the coupling between the spin and orbital angular momenta of individual electrons is stronger than the coupling between different electron's spins or between different electron's angular momenta. This produces a different coupling scheme, called *j–j* coupling. See, for example, *Quantum Physics of Atoms, Molecules, Solids, Nuclei and Particles* by R. Eisberg and R. Resnick, published by John Wiley and Sons (1974).

The selection rules become

$$\Delta S = 0, \ \ \Delta L = 0, \pm 1, \ \ \Delta F = 0, \pm 1, \ \ \Delta M_f = 0, \pm 1; \qquad (3.11)$$

but of these the transitions $F = 0 \rightarrow F = 0$, and $M_f = 0 \rightarrow M_F = 0$ for $\Delta F = 0$ are excluded. The underlying allowed process remains the single electron dipole transition.

3.4   White dwarfs, neutron stars and black holes

Within stars the Pauli exclusion principle can act on the grand scale. Stars, after consuming their nuclear fuel, begin to contract under their gravitational self-attraction: gravitational energy is converted into kinetic energy of internal motion. As the volume in space occupied by an electron or nucleon falls the volume of its eigenstate in momentum space increases to satisfy the uncertainty principle. Now, nucleons and electrons are fermions and each fermion occupies a unique volume of momentum-position space. This means that average fermion momenta and energies must grow as the star shrinks. It can happen that at some moment the loss of gravitational energy in further contraction becomes less than the increase in kinetic energy necessary to keep the fermions in separate eigenstates. Then the contraction stalls: the fermions are said to be degenerate and the pressure resisting contraction is called the *degeneracy pressure*.

We must distinguish between two star types. With lighter stars nuclear burning would cease at the stage of converting the core to intermediate nuclei such as carbon: with heavier stars burning only terminates on reaching iron, the nucleus with the tightest binding per nucleon. In the former class of stars the electrons are the fermions, which supply the degeneracy pressure. In heavier stars the electron energy rises to around 1 MeV so that processes become possible in which electrons convert protons to neutrons:

$$e^- + p \rightarrow n + \nu_e$$

where ν_e is an electron neutrino. The process of neutronization of all the nuclei is completed in around one second. Such a star is then almost entirely made of neutrons, and it is the neutron degeneracy pressure that opposes the gravitational self attraction. The two cases of electron and neutron degeneracy pressure can be treated in parallel.

Suppose the star has radius R, that it contains N nucleons of mass M_N and that it contains an equal number of electrons of mass m. The the average spacing between adjacent fermions is therefore

$$\Delta x \approx R/N^{1/3},$$

and the associated uncertainty in the fermion momentum is at least

$$\hbar/\Delta x \approx N^{1/3}\hbar/R. \tag{3.12}$$

We can take this as an estimate for the magnitude, p, of the fermion momentum. Another approximation is to take the fermions to become relativistic as the contraction continues,[2] so that the energy per fermion is

$$E_d = pc \approx N^{1/3}\hbar c/R. \tag{3.13}$$

The corresponding outward pressure for the adiabatic conditions is obtained using the first law of thermodynamics

$$P_d = -\mathrm{d}(NE)/\mathrm{d}V \approx -N(\mathrm{d}E/\mathrm{d}R)/R^2. \tag{3.14}$$

Then taking the fermion energy given in eqn. 3.13

$$P_d \approx (N/R^4)N^{1/3}\hbar c. \tag{3.15}$$

Now because the lower limit from Heisenberg's uncertainty principle was used in eqn. 3.12 the fermion density in position–momentum space is maximized. It follows that P_d is the maximum degeneracy pressure that the compressed fermions can exert to resist further contraction.

The gravitational energy is almost entirely contributed by the nucleons, which are ~2000 times heavier than the electrons. The gravitational energy per nucleon, and hence, that per electron, is

$$E_g \approx -GNM_N^2/R.$$

Using eqn. 3.14 again, the corresponding gravitational pressure is

$$P_g \approx -(N/R^4)GNM_N^2.$$

Thus the overall pressure is

$$P = P_d + P_g \approx (N/R^4)(N^{1/3}\hbar c - GNM_N^2). \tag{3.16}$$

We can see that if the stellar mass, and hence N, is sufficiently large the pressure is negative (inward) and the star shrinks indefinitely. Alternatively, if the pressure is positive this fate is avoided. At the boundary of stability against indefinite contraction $P = 0$ and we have

$$N^{1/3}\hbar c = GNM_N^2. \tag{3.17}$$

This determines the maximum number of nucleons a star can contain while avoiding indefinite contraction to be

$$N_{\mathrm{max}} = (\hbar c/GM_N^2)^{3/2} = 4\,10^{57}. \tag{3.18}$$

This is the *Chandrasekhar limit*, and after more careful evaluation of N_{max} (1.67 10^{57}) the corresponding stellar mass

$$M(\mathrm{Ch}) = 1.4M_\odot. \tag{3.19}$$

[2]This assumption is safe for the electrons which reach MeV energies in white dwarf stars. It is less sound for neutrons whose energies only reach a hundred MeV in neutron stars.

The gravitational pressure overwhelms the degeneracy pressure when the star's mass exceeds 1.4 times the solar mass $M_\odot$. Heavier stars will collapse indefinitely and at some point the radius becomes so small that even light cannot escape from the surface. This radius is equal to $2GM_{\text{star}}/c^2$ and is known as the *Schwarzschild radius*. A star that shrinks within its Schwarzschild radius has become an entirely invisible *black hole*. Several have been detected because they form part of a binary pair. The presence of the black hole Cygnus X-1 was betrayed by several odd features exhibited by its companion HDE 226868. This is a blue supergiant located 7.5 light-years from earth. Firstly, its spectral lines shift regularly every 5.6 days, and secondly it appears to lie almost coincident with a source of X-rays, something not expected from such a cold star. The accepted economic explanation is that HDE 226866 forms a binary pair with an invisible black hole and that they orbit one another every 5.6 days. Material pulled from the blue supergiant radiates the X-rays observed as it accelerates into the black hole.

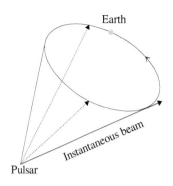

Fig. 3.2 Pulsar illuminating the earth.

Stars whose contraction is halted by the electron degeneracy pressure are *white dwarf stars*, those stabilized by neutron degeneracy pressure are *neutron stars*. Crude estimates of the radius of such stars at the Chandrasekhar limit can be made by taking eqn. 3.12 as the fermion relativistic momentum

$$m_f c = N_{\text{max}}^{1/3}\hbar/R,$$

where m_f is the relevant fermion's mass. Thus

$$R = \hbar N_{\text{max}}^{1/3}/(m_f c). \tag{3.20}$$

This yields $\sim 7000\,\text{km}$ for a white dwarf, that has a degenerate electron core, and $\sim 4\,\text{km}$ for a star containing degenerate neutrons. More satisfactory estimates obtained using appropriate equations of state for the fermions only improve these estimates of radius by small factors.[3] That for a neutron star radius increases to $\sim 10\text{km}$.

[3]See *Black holes, White Dwarfs and Neutron Stars* by S. L. Shapiro and S. A. Teukolsky, published by Wiley Interscience, 1983.

An example of a white dwarf is Sirius B, the companion of Sirius. It has a mass of $0.98\,M_\odot$ and a radius $0.0084\,R_\odot$. It is predominantly made of carbon, at which stage the nuclear burning halted. The mean electron spacing is around $10^{-3}\,\text{nm}$ and the mean electron velocity is around 0.25c. White dwarfs range in mass from 0.2 to $1.4\,M_\odot$. They have a high surface temperature due to the intense radiation and hence a white appearance. A typical density is $10^9\,\text{kg m}^{-3}$ or a million times that of water.

Most neutron stars have been detected as *pulsars*.[4] Such a star spins rapidly and emits a narrow beam of electromagnetic radiation that rotates with the star. This beam marks out the surface of a fixed cone in space. If, as shown in Figure 3.2, the earth lies on this cone then the pulsar is detected through the regular arrival of short radiofrequency

[4]Pulsars were discovered by Jocelyn Bell and Anthony Hewish, the latter receiving the Nobel prize in Physics in 1974.

pulses, one pulse each time the beam illuminates the earth. The pulse rate can be astonishingly stable, to parts in 10^{15}, rivalling that of the optical clocks discussed in Section 12.6. The pulsar masses so far measured cluster around $1.4\,M_\odot$. We can therefore safely state that these pulsars contain $\sim 10^{57}$ neutrons packed to nuclear density at $10^{17}\,\mathrm{kg\,m^{-3}}$.

3.5   The exchange force

In a few elements and in many compounds, the electrons in incomplete shells can have much lower energies when their spins and hence magnetic moments are aligned with those of the corresponding electrons in neighbouring atoms. In the ferromagnetic materials like iron the magnetic moments of the atoms align over macroscopic volumes, and produce strong external magnetic fields. Paradoxically, the magnetic interaction between atoms is far too small to account for the alignment. An electron's magnetic field a distance r along its spin axis is $B = \mu_0\mu_B/(2\pi r^3)$. The energy of a second electron placed there would fall by $\mu_B B$ on aligning its spin. For a typical atomic separation of $0.2\,\mathrm{nm}$ this change is $10\,\mu\mathrm{ev}$. Now, the temperature at which the thermal excitation becomes large enough to fully disrupt the alignment, the Curie temperature (T_c), is in the case of iron $1043\,\mathrm{K}$. The thermal excitation required to disrupt alignment in iron, kT_c, is thus $9\,\mathrm{meV}$, a thousand times the magnetic energy.

The mechanism giving the observed stability of ferromagnetism is not a magnetic interaction but a quantum exchange force. It results from the requirement that under the interchange of a pair of fermions, in this case electrons, the overall wavefunction must change sign. This purely Coulombic force is now examined in detail.

We consider a pair of outer shell electrons in neighbouring atoms in a crystal lattice, close enough that their mutual Coulomb interaction is important. Suppose their spatial wavefunctions would be $\phi_1(\mathbf{r}_1)$ and $\phi_2(\mathbf{r}_2)$ if the atoms were isolated. In the lattice the two possible spatial wavefunctions are symmetric or antisymmetric

$$\psi_{s/a} = [\psi_1(\mathbf{r}_1)\psi_2(\mathbf{r}_2) \pm \psi_1(\mathbf{r}_2)\psi_2(\mathbf{r}_1)]/\sqrt{2},$$

where $\psi_1 \approx \phi_1$ and $\psi_2 \approx \phi_2$. Notice that each electron is located partially around atom 1 and partially around atom 2, but always in the opposite atom to its electron partner. The total spin/spatial wavefunction must be overall antisymmetric under the interchange of electrons, for which the alternatives are

$$\psi_s\chi_a \quad \text{and} \quad \psi_a\chi_s$$

with $\chi_{a/s}$ being antisymmetric and symmetric wavefunctions describing the electrons' spin state. With obvious notation, the *antisymmetric spin state* has a wavefunction with zero total spin

$$\chi_a(s = 0, m_s = 0) = (\uparrow_1\downarrow_2 - \downarrow_1\uparrow_2)/\sqrt{2}. \qquad (3.21)$$

The alternative *symmetric spin states* form a triplet of eigenstates with unit total spin

$$
\begin{aligned}
\chi_s(s = 1, m_s = +1) &= \uparrow_1\uparrow_2, \\
\chi_s(s = 1, m_s = 0) &= (\uparrow_1\downarrow_2 + \downarrow_1\uparrow_2)/\sqrt{(2)}, \\
\chi_s(s = 1, m_s = -1) &= \downarrow_1\downarrow_2 . \qquad (3.22)
\end{aligned}
$$

The antisymmetric spin state has the electron spins antiparallel while in the symmetric spin states they can be parallel. The electrons' energy includes the expectation value of their mutual Coulomb interaction,

$$E = E_0 + \int \psi_{s/a}^* \mathbf{V}_c \psi_{s/a} \mathrm{d}V_1 \mathrm{d}V_2, \qquad (3.23)$$

where $\mathbf{V}_c = e^2/[4\pi\varepsilon_0 \mid \mathbf{r}_1 - \mathbf{r}_2 \mid]$ and E_0 is the energy of the atoms when isolated. The integral is made up of two terms:

$$
\begin{aligned}
E_c &= \int \psi_1^*(\mathbf{r}_1)\psi_2^*(\mathbf{r}_2)\mathbf{V}_c\psi_1(\mathbf{r}_1)\psi_2(\mathbf{r}_2)\, \mathrm{d}V_1\, \mathrm{d}V_2 \ \ \text{and} \qquad (3.24) \\
\pm J &= \pm \int \psi_1^*(\mathbf{r}_1)\psi_2^*(\mathbf{r}_2)\mathbf{V}_c\psi_1(\mathbf{r}_2)\psi_2(\mathbf{r}_1)\mathrm{d}V_1\, \mathrm{d}V_2, \qquad (3.25)
\end{aligned}
$$

with E_c being the equivalent of the classical Coulomb energy. J is the *quantum exchange energy* with the positive (negative) sign referring to the spatially symmetric (anti-symmetric) wavefunction. If J is sufficiently large and positive the spatially antisymmetric state of the two electrons has the lower energy. In the associated symmetric spin state the spins are parallel, as illustrated in the top panel of Figure 3.3, so that the material is ferromagnetic. On the other hand, when the exchange integral, J, is large and negative the electrons' spins are aligned antiparallel in the lower energy state. In this case, electrons in adjacent atoms have antiparallel spins and magnetic moments, while electrons in alternate atoms have parallel spins and magnetic moments. This case is shown in the centre panel of Figure 3.3. There is therefore alignment but no net macroscopic magnetic moment: this makes the material *anti-ferromagnetic*. Another possibility is that shown in the bottom panel of Figure 3.3, called *ferrimagnetic*.

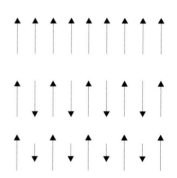

Fig. 3.3 Alignment of atomic magnets for the cases of ferromagnetism, anti-ferromagnetism and ferrimagnetism.

Being Coulombic the quantum exchange force can lead to a large reduction in the energy of the state in which electron magnetic moments are aligned and results in ferromagnetism. In the Heisenberg model the exchange energy is approximated by

$$H_{\mathrm{ex}} = -\sum_{i,j\neq i} J_{ij}\mathbf{S}_i \cdot \mathbf{S}_j, \qquad (3.26)$$

where $\mathbf{S}_i$ is the total angular momentum of the ith atom in a crystal and the jth atom is any one of its nearest neighbours. Conventionally,

Table 3.2 Electronic configurations of transition elements.

shell	Mn	Fe	Co	Ni	Cu
3d	5	6	7	8	10
4s	2	2	2	2	1

Ar core $1s^2\ 2s^2\ 2p^6\ 3s^2\ 3p^6$

the total angular momentum is called *spin* in this context.

Only unpaired electrons not committed to chemical bonds are available for alignment by the exchange force. These electrons lie in partially filled shells that are shielded beneath a closed shell. The only ferromagnetic elements are therefore the transition elements iron, cobalt and nickel, and the rare earths gadolinium and dysprosium. In the transition sequence the 3d shell is being filled inside the closed 4s shell, as shown in Table 3.2. These elements are all good conductors in which individual atoms' energy levels merge into a band shared across the whole body of the conductor. Figure 3.4 shows the Curie temperatures for the transition elements plotted against ρ, the nuclear separation, normalized by dividing it by the 3d-shell diameter. In the wings, and not ferromagnetic, are manganese and copper. Figure 3.5 shows the energy distribution of the electron bands in iron, with the density of states plotted separately for the two relative alignments of spins in adjacent atoms. For iron the displacement between the spin states due to the exchange interaction is comparable to the width of the 3d band. The 3d and 4s bands in iron overlap and *hybridize* so that there effectively 7.4 3d and 0.6 4s electrons. All available states are filled up to a common energy level, the Fermi energy. As a result there is a net $4.8 - 2.6 = 2.2$ aligned spins per iron atom producing an atomic magnetic moment of $2.2\mu_{\mathrm{B}}$. Correspondingly, the measured atomic magnetic moments of cobalt and nickel are respectively $1.7\mu_{\mathrm{B}}$ and $0.6\mu_{\mathrm{B}}$. The only anti-ferromagnetic element ^{24}Cr precedes manganese in the transition elements and for this element the exchange integral has, as one would expect, changed sign.

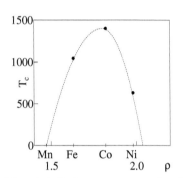

Fig. 3.4 Curie temperatures for the transition elements plotted against the nuclear spacing divided by the 3d shell diameter.

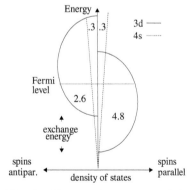

Fig. 3.5 Energy-spin structure of the 4s and 3d bands in iron.

Measurements of the Einstein–de Haas effect confirm that it is the electron spins that are responsible for the magnetization. A thin specimen is hung from a thin thread inside a solenoid. A current pulse aligns the constituent magnets so that the magnetic moment μ and the angular momentum $L\hbar$ of the specimen are related by

$$\mu = g\mu_{\mathrm{B}}L. \tag{3.27}$$

The suspension responds to the torque due to the change in angular momentum and the sample is set in oscillation. Measurements of the motion and the induced magnetic moment lead to determination of g.

This Landé factor would be unity if the electron orbital motion alone contributes to the magnetization, and 2.0 if only the electron spin contributes. Experiment yields values close to 2.0 for iron, cobalt and nickel, showing that the magnetization is indeed mainly due to the alignment of electron spins.

3.6   Fermi–Dirac and Bose–Einstein statistics

The approach is to consider a system consisting of the particles of interest in a uniform condition in good thermal and diffusive contact with a much larger reservoir at temperature T. Both heat and particles can flow between the system and reservoir. We take the total energy of the system plus the reservoir to be E, and the total number of particles to be N. The total number of particles is assumed to be conserved which appears to exclude photons from consideration. However, it will emerge that the formalism can be adapted to cover photons. The probability for the system being in a state having energy τ and containing n particles is

$$P(\tau, n) = g_R(E - \tau, N - n) = \exp[\sigma_R(E - \tau, N - n)], \qquad (3.28)$$

which employs the standard thermodynamic relation between g_R, the number of equivalent configurations of the reservoir, and σ_R its entropy.[5] The reservoir is the dominant thermodynamic partner so we can ignore the rearrangements of the system itself in comparison to those of the reservoir. Similarly, because $\tau \ll E$ and $n \ll N$

$$\begin{aligned} \sigma_R(E - \tau, N - n) &= \sigma_R(E, N) - \tau(\partial \sigma_R / \partial E) - n(\partial \sigma_R / \partial N) \\ &= \sigma_R(E, N) - \tau\beta + n\mu\beta. \end{aligned} \qquad (3.29)$$

Standard thermodynamic relations have been used in writing the second equality. These are $\partial \sigma_R / \partial E = \beta$, and $\partial \sigma_R / \partial N = \beta\mu$; where μ is the energy required to carry a single particle from the reservoir to the system and known as the *chemical potential*, while $\beta = (k_B T)^{-1}$ with k_B being Boltzmann's constant. Substituting this result into eqn. 3.28 gives

$$P(\tau, n) = \exp[(\mu n - \tau)\beta]/Z, \qquad (3.30)$$

where Z depends only on the reservoir entropy.[6] Z can be calculated easily using the property that the sum of the probabilities, $P(\tau, n)$, over all values of τ and n should be unity, consequently

$$Z = \sum_{n,\tau} \exp[(\mu n - \tau)\beta]. \qquad (3.31)$$

Now, we ask the probability for there being n particles in the same eigenstate, each with energy ϵ. This comes to

$$P(\epsilon, n) = \exp[n(\mu - \epsilon)\beta]/Z = \exp(-nx)/Z, \qquad (3.32)$$

[5] In thermodynamics the statistical treatment where the system and reservoir can exchange particles as well as energy is called taking a *grand canonical ensemble*. If only energy is exchanged then it becomes a *canonical ensemble*.

[6] By convention the symbol Z is used for the *partition function*, that is, the sum of weights over all states, or *Zustandssumme* in German.

where $x = (\varepsilon - \mu)\beta$ and

$$Z = \sum_n \exp[n(\mu - \epsilon)\beta] = \sum_n \exp(-nx). \qquad (3.33)$$

At this point the calculations of $\langle n \rangle$ for fermions and for bosons separate because the number of fermions per eigenstate is restricted. First considering the case for fermions, the occupancy of any state is restricted to either 0 or 1 only. Thus

$$Z = 1 + \exp(-x), \qquad (3.34)$$

and for n is 0 or 1

$$P(\varepsilon, n) = \exp(-nx)/[1 + \exp(-x)]. \qquad (3.35)$$

Hence the mean number of fermions in an eigenstate is

$$\begin{aligned}\langle n \rangle &= \sum_{n=0,1} n P(\epsilon, n) = \exp(-x)/[1 + \exp(-x)] \\ &= 1/[\exp(x) + 1] = \frac{1}{\exp[(\epsilon - \mu)/k_{\mathrm B}T] + 1} \qquad (3.36)\end{aligned}$$

which is called *Fermi–Dirac statistics*. It follows that

$$\langle n^2 \rangle = \langle n \rangle. \qquad (3.37)$$

In the case of bosons n can take any value up to infinity, so that

$$Z = \sum_n \exp(-nx) = \frac{1}{1 - \exp(-x)}. \qquad (3.38)$$

Then using eqn. 3.32 again

$$P(\varepsilon, n) = \exp(-nx)[1 - \exp(-x)], \qquad (3.39)$$

and the mean number of bosons in an eigenstate of energy ϵ is

$$\begin{aligned}\langle n \rangle &= \sum_n n P(\epsilon, n) = -1[1 - \exp(-x)]\frac{\mathrm d \sum_n \exp(-nx)}{\mathrm dx} \\ &= \frac{[1 - \exp(-x)]\exp(-x)}{[1 - \exp(-x)]^2} \\ &= 1/[\exp(x) - 1] \\ &= \frac{1}{\exp[(\epsilon - \mu)/k_{\mathrm B}T] - 1} \qquad (3.40)\end{aligned}$$

which is called *Bose–Einstein statistics*. Revisiting the case of photons, the chemical potential is zero, so that

$$\langle n \rangle = \frac{1}{\exp(\epsilon/k_{\mathrm B}T) - 1}. \qquad (3.41)$$

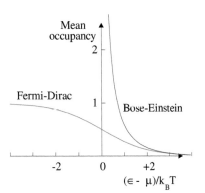

Mean occupancy
2

Fermi-Dirac   1
           Bose-Einstein

-2   0   +2
$(\epsilon - \mu)/k_{\mathrm B}T$

Fig. 3.6 Bose–Einstein and Fermi–Dirac mean occupancy. ϵ is the energy and μ the chemical potential.

Figure 3.6 contrasts the Fermi–Dirac and Bose–Einstein distributions, both plotted against $(\epsilon - \mu)/k_B T$ at some finite temperature. At absolute zero the Fermi–Dirac distribution is unity up to $\epsilon = \mu$ and zero above. At other temperatures μ remains the median value of energy and μ is therefore also called the Fermi energy.

Whether the particles are bosons or fermions the particle distribution in energy is given by the product of the density of available states, $g(\epsilon)\mathrm{d}\epsilon$ from Appendix B, and the mean occupation number $\langle n \rangle$:

$$N(\epsilon)\mathrm{d}\epsilon = g(\epsilon)\langle n(\epsilon, \mu)\rangle \mathrm{d}\epsilon. \tag{3.42}$$

The quantity μ is well defined but nothing so far could be said about its value. This value can always be deduced from the requirement that the integral $\int N(\varepsilon)\mathrm{d}\varepsilon$ is the number of particles present.

3.7   Bose–Einstein condensation

When certain materials with boson constituents are cooled to low temperatures a substantial fraction of these bosons enter a common ground state. They constitute a *Bose–Einstein condensate*, whose overall wavefunction Ψ is the product of identical ground state wavefunctions ψ for each boson:

$$\Psi(\mathrm{r}_1, \mathrm{r}_2, \cdots, \mathrm{r}_n, t) = \prod_{i=1}^{n} \psi(\mathrm{r}_i, t), \tag{3.43}$$

with t being the time. This is a *macroscopic* quantum state, in which the state of bosons at one place is strongly correlated with the state of bosons at another distant point in the condensate. The system then possesses what is called *off diagonal long range order* (ODLRO).

Condensation was anticipated by Einstein in 1924, well before it was recognized as the underlying cause of the superfluidity of liquid helium below 2.17 K and of superconductivity. The former example is worked through to bring out the dramatic departure from classical physics, and to show that Bose–Einstein condensation is a *statistics driven process*.[7] To begin with, we note that the number of atoms in the ground state given by eqn. 3.40 is

$$N_0 = 1/[\exp(-\mu/k_B T) - 1], \tag{3.44}$$

and hence

$$\exp(-\mu/k_B T) = 1 + 1/N_0 \tag{3.45}$$

which is always greater than unity. This requires that the chemical potential, μ, is negative and that it is very close to zero for large values of N_0.

As a numerical example consider a volume of helium gas confined to a cube of side length, L, equal to $10\,\mu m$. At 300 K there are $2.71\,10^{10}$

[7] Einstein was led to deduce the existence of such a condensation mechanism by Bose's statistical derivation of the black body spectrum.

atoms in the cube and the energy between the ground and first excited eigenstates of motion in the box, $\Delta\epsilon$, is $1.56\,10^{-12}\,\mathrm{eV}$. Now consider cooling the sample to $1\,\mathrm{mK}$, making $k_\mathrm{B}T$ equal to $8.6\,10^{-8}\,\mathrm{eV}$. The classical prediction for the population of the first excited state would be

$$N_\mathrm{CL} \;=\; N_0 \exp(-\Delta\epsilon/k_\mathrm{B}T), \tag{3.46}$$

which is close to N_0. Excited states are therefore well populated because $\Delta\epsilon$ is small compared to the thermal excitation.[8] By contrast the Bose–Einstein prediction for the population of the first excited state is

$$
\begin{aligned}
N_\mathrm{BE} &= 1/\{\exp[(\Delta\epsilon - \mu)/k_\mathrm{B}T] - 1\} \\
&= 5.5\,10^4,
\end{aligned} \tag{3.47}
$$

taking $\exp(-\mu/k_\mathrm{B}T)$ equal to unity. Only one in a million atoms is in the first excited state. Evidently almost all the atoms have condensed into the ground state. This example makes it clear that Bose–Einstein condensation is purely a quantum statistical process.

The physical conditions for Bose–Einstein condensation are these. As the bosons cool their de Broglie wavelengths expand and at some point become as large as their mutual separation. At this point the indistinguishability of bosons means that they must all share a single eigenstate and wavefunction. This sequence is beautifully displayed on Bose-Einstein_Condensation.ogv. The de Broglie wavelength for non-relativistic particles is conventionally[9] defined as

$$\lambda_{\mathrm{deB}} = (2\pi\hbar^2/mk_\mathrm{B}T)^{1/2}. \tag{3.48}$$

Condensation requires that the atoms are to overlap. This means that the *phase space density*, the number of atoms within the de Broglie volume, should become of order unity. Rigorous analysis gives condensation when

$$n\lambda_{\mathrm{deB}}^3 = 2.612, \tag{3.49}$$

where $n = N/V$ is the number density. Then using the last two equations the transition to a condensate is predicted to occur at a temperature:

$$T_c = \frac{2\pi\hbar^2}{mk_\mathrm{B}} \left[\frac{n}{2.612}\right]^{2/3}. \tag{3.50}$$

At a lower temperature, T, the same analysis shows that the content of the continuum above the ground state,

$$n_x = 2.612 \left(\frac{mk_\mathrm{B}T}{2\pi\hbar^2}\right)^{3/2}, \tag{3.51}$$

with n_x becoming the total number of atoms, n, at the critical temperature. From this we get the temperature dependence of the condensate fraction

$$n_0/n = (n - n_x)/n = 1 - (T/T_c)^{3/2}. \tag{3.52}$$

[8] The density is $0.179\,\mathrm{kg\,m^{-3}}$, giving $2.71\,10^{10}$ atoms in the cube. ^{4}He atoms have mass, M, of $6.6\,10^{-27}\,\mathrm{kg}$. Permissible wavelengths have nodes at walls, so $\lambda = 2L/n$, giving a wavenumber $k = n\pi/L$, a momentum $p = \hbar k = n\hbar\pi/L$, and a kinetic energy $\epsilon = p^2/(2M) = (n\hbar\pi/L)^2/(2M)$. The energy of standing waves in the box is

$$\epsilon_n = (n^2\hbar^2/2M)(\pi/L)^2,$$

with n a positive integer giving the number of antinodes across the box. Putting in the values for L and M the separation in energy between the lowest two states, $\Delta\epsilon = \epsilon_2 - \epsilon_1$ is $1.56\,10^{-12}\,\mathrm{eV}$.

[9] This definition differs from h/p, p being the momentum, by a factor $\sqrt{\pi}$. It is standard in condensation studies.

Once the temperature falls a little below T_c the number n_0 of condensate atoms becomes numerically very large. Then eqn. 3.45 requires that the chemical potential is zero below T_c. Figure 3.7 shows the behaviour of μ and the condensate fraction as a function of temperature.

Bose–Einstein condensates produced in low density gases of alkali atoms have been studied intensively since their creation in the laboratory in 1995. The atoms are confined in millimetre sized three-dimensional harmonic potential wells by magnetic fields. Confinement modifies the expression for the critical temperature[10]

$$T_c = \frac{\hbar\omega}{k_B}\left[\frac{N}{1.204}\right]^{1/3}, \tag{3.53}$$

where N is the *number of atoms in the potential well* and ω is the angular frequency of oscillation of an atom in such a well. Equivalently

$$k_B T_c = 0.94\hbar\omega[N]^{1/3}. \tag{3.54}$$

Then, in turn the dependence of the condensate fraction on temperature changes

$$N_0/N = 1 - (T/T_c)^3. \tag{3.55}$$

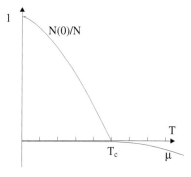

Fig. 3.7 The condensate fraction and the chemical potential plotted against the absolute temperature.

3.8 Condensates

In Figure 3.8 the region where Bose–Einstein condensation is expected for an ideal gas of bosons with vanishing interactions is to the right of the sloping straight line. Regions where matter is solid, liquid, or a gas are indicated. The shaded region indicates conditions not usually realized in nature. Suppose an attempt is made to move from the gas phase toward the BEC boundary by adding additional gas. On entering the shaded region the gas will condense into a liquid with the release of latent heat; and exits this region instantaneously. Helium alone forms a condensate when cooled: other elements are solids below 20 K. Condensation endows ^{4}He with superfluidity, as discussed in Chapter 14.

The BEC region in Figure 3.8 is entered by trapping and cooling mm sized clouds of bosonic atoms to sub-microkelvin temperatures using laser beams and magnetic fields. These condensates only endure for seconds or minutes, before collisions destroy them by forming complexes of multiple atoms. As an example, a rubidium cloud at a density 10^{13} cm^{-3}, corresponding to an average spacing between atoms of $\sim$500 nm, would be cooled to 100 nK in order to produce the condensation predicted by eqn. 3.50: this is over a million times colder than the temperature found

[10]See the second edition of *Bose-Einstein Condensation in Dilute Gases* by C. J. Pethick and H. Smith, published by Cambridge University Press (2008).

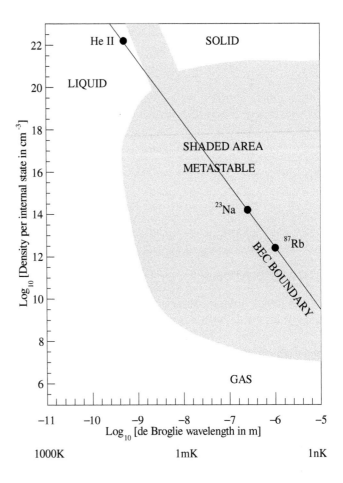

Fig. 3.8 State density plotted against de Broglie wavelength for an ideal gas. The region to the right of the straight line is where Bose–Einstein condensates would exist for such a gas. The grey region is one of meta-stability for real materials. The temperature scale is appropriate for sodium atoms. Typical conditions for forming Bose–Einstein condensates of ^{23}Na and ^{87}Rb atoms are indicated.

in intergalactic space. These conditions are indicated on Figure 3.8 by a point labelled Rb. Properties of these condensates are described in Chapter 16.

Many metals become superconducting below a critical temperature, 7.2 K in the case of lead. All magnetic field is expelled from within the superconductor and if a voltage is applied across the superconductor a surface electric current flows without resistance. These properties are due to the formation of a wildly different BEC, the bosons involved are *Cooper pairs of electrons*, weakly bound by interactions involving the lattice ions. Superconductivity is discussed in Chapter 15.

Is condensation by cooling an option for photons? In fact, no: they are simply absorbed by the container when the temperature falls. However the beam from a single mode He:Ne laser is coherent over a distance of order 300 m. There *is* long range order but it is not a consequence of Bose–Einstein condensation, as will emerge in Chapter 8.[11]

[11] See however J. Klaers, J. Schmitt, F. Vewinger and M. Weitz: Nature 468, 545 (2010).

3.9 Further reading

Introduction to the Theory of Ferromagnetism by A. Aharoni, published by Oxford Science Publications (2001), is a lecture-based text that is very approachable.

Black Holes, White Dwarfs and Neutron Stars by S. L. Shapiro and A. Teukolsky, published by Wiley Interscience (1983), gives a thorough clear account of these compact objects. More recent is *Compact Objects in Astrophysics* by M. Camenzind, published by Springer (2007).

Exercises

(3.1) For a two-dimensional boson gas trapped on some surface show that the content of the excited states is

$$N(>0) = \frac{mk_{\mathrm{B}}T}{2\pi\hbar^2} \int_0^\infty \frac{\lambda \exp(-x)\,\mathrm{d}E}{1 - \lambda \exp(-x)}$$

where $\lambda = \exp(-\mu/k_{\mathrm{B}}T)$ and $x = E/k_{\mathrm{B}}T$. Use the density of states given in Appendix B. Then make a substitution $z = 1 - \lambda \exp(-x)$ to help perform the integration.

(3.2) For a two-dimensional boson gas show that the content of the ground state grows exponentially as the temperature falls, so that there is no divergence at a critical temperature and hence no condensation.

(3.3) A fixed magnetic field **B** is applied to a spherical ferromagnetic specimen with magnetic moment **M** with **B** perpendicular to **M**. What are the torque on the specimen and the rate of change of the magnetic moment? By resolving the rate of change in two directions at right angles, show that the magnetic moment rotates at an angular frequency $\omega_0 = \gamma B$, where γ is the gyromagnetic ratio $g\mu_{\mathrm{B}}/\hbar$.

(3.4) Both material particles and photons follow paths determined by the curvature of space-time. The effective mass of a photon of energy E in a gravitational field is E/c^2. Show that the wavelength shift of a photon moving from the surface of a white dwarf star of mass M, radius R to the earth is

$$\Delta\lambda = \lambda GM/(Rc^2),$$

where G is the gravitational constant $6.67\,10^{-11}\,\mathrm{kg}^{-1}\mathrm{m}^3\mathrm{s}^{-2}$. Calculate the wavelength shift of light from a white dwarf with the maximum possible mass, and also the observed red shift $\Delta\lambda/\lambda$.

(3.5) Free electrons make a contribution to the heat capacity of metals, which is a significant fraction of the total heat capacity at very low temperatures. Show that the energy required to raise the temperature of a free electron gas from $0\,\mathrm{K}$ to $T\,\mathrm{K}$ is

$$U = \int_{E_F}^{\infty} g(E)(E - E_F)f(E)\,\mathrm{d}E$$
$$+ \int_0^{E_F} g(E)(E_F - E)[1 - f(E)]\,\mathrm{d}E$$

where f is the Fermi–Dirac function and g is the three-dimensional density of states. It is helpful to consider separately the energy required to move electrons up to the Fermi level, E_F, and then to their final energy. Hence show that the electronic contribution to the heat capacity is

$$C_{\mathrm{el}} = \int_0^{\infty} g(E)(E - E_F)[\mathrm{d}f/\mathrm{d}T]\,\mathrm{d}E.$$

You can safely neglect changes of the chemical potential with temperature in this and the next exercise.

(3.6) Continuing from the previous question, at low enough temperatures the active electrons have energies close to E_F so that $g(E) \approx g(E_F)$. Making

this approximation establish that at sufficiently low temperatures

$$C_{\mathrm{el}} = \pi^2 N k_B^2 T / 2E_F,$$

where N is the number of electrons per unit volume. Use the result

$$\int_{-\infty}^{\infty} \frac{\exp(x)x^2\,\mathrm{d}x}{(\exp(x) + 1)^2} = \pi^3/3.$$

(3.7) At temperatures below $0.6\,\mathrm{K}$ the total heat capacity of potassium is measured to be

$$C/T = 2.08 + 2.57\,T^2$$

in $\mathrm{mJ\,mol}^{-1}\mathrm{K}^{-3}$. Estimate the Fermi energy and Debye temperature for potassium.

(3.8) The radiowave pulses from pulsars are sharp bursts and the interval between the bursts ranges down to milliseconds. Show that this limits the possible size of the source. Show that if pulsars were as large as white dwarfs the gravitation force would tear them apart.

(3.9) Show that the Fermi–Dirac distribution of occupation numbers satisfies the relation

$$n(E_F + \delta) = 1 - n(E_F - \delta).$$

How far must the particle energy be above E_F for the Fermi–Dirac and Bose–Einstein occupation distributions to differ by less than one per cent.

(3.10) Use the Fermi–Dirac distribution to show that the average energy of free electrons in a metal is $3E_F/5$.

Phonons

4.1 Introduction

This chapter and the two following present an overview of the quantum theory of crystalline solids, with this chapter covering those with few free electrons, that is, the dielectrics. The understanding of the thermal properties of crystalline dielectrics due to Debye was an early success of quantum theory. He developed Einstein's idea that lattice vibrations should be quantized in much the same way as electromagnetic radiation. The corresponding quanta are known as *phonons*. Unlike photons, phonons do not exist outside matter and are therefore known as *quasi-particles*. They owe their existence to the mutual electromagnetic interaction between the ions making up the material. Phonons share many properties of photons: they are neutral bosons and their number is not a conserved quantity, though restricted by the material and its physical condition. Later, we shall meet several other types of quasi-particles that are similarly confined to some material environment.

The kinematic properties of phonons, or any particle species, are contained in their dispersion relations linking momentum and energy, or equivalently between wave-vector (k) and angular frequency (ω). After treating the Debye model, features of more realistic dispersion relations are deduced by considering the dynamics of a linear chain of atoms. Experimental measurements of phonon dispersion relations in crystals are described next. It emerges that the phonon energy spectrum in crystals consists of energy bands over which free propagation occurs separated by gaps in energy within which the propagation is absorptive. Surprisingly, if lattice vibrations were exactly harmonic the thermal conduction in dielectrics would be instantaneous. How the anharmonic component leads to a finite thermal conductivity is described next.[1] The subsequent section covers Raman and Brillouin scattering of light, in which a phonon is emitted or absorbed. After this we examine what happens when the dispersion relations for photons and phonons cross one another, so that they produce identical lattice excitations: a hybrid quasi-particle called a *polariton* is formed. Finally, the Mössbauer effect is introduced, a process in which the *absence* of phonon excitation is the defining property.

[1] As we shall see in the following chapters, phonons being neutral, cannot give any electrical conductivity, but electron scattering off phonons can impede electrical conductivity.

Quantum 20/20: Fundamentals, Entanglement, Gauge Fields, Condensates and Topology.
Ian R. Kenyon. © 2020. Published in 2020 by Oxford University Press.
DOI: 10.1093/oso/9780198808350.001.0001

4.2   Debye model of heat capacity

Treating vibrations of the atomic lattice of a crystal as quantum harmonic oscillators (Section 2.3) leads to an understanding of the heat capacity of a broad range of solids. In particular, the thermal energy of non-magnetic insulators resides solely in the vibrations of the atoms in the crystal lattice. Conduction electrons contribute significantly to heat capacity in metals only at low temperatures, while their magnetic effects contribute in the magnetic solids. Taking the analysis of black body radiation as a template the essential ingredients needed to calculate the thermal energy of lattice vibrations are the mean energy per mode, the mode density and the wave velocity. The energy is quantized in packets of $\hbar\omega$, hence using eqn. 3.41 the mean energy per mode including the zero point energy $\hbar\omega/2$ is

$$\bar{\epsilon} = \hbar\omega/[\exp(\hbar\omega/k_{\mathrm{B}}T) - 1] + \hbar\omega/2. \tag{4.1}$$

At high temperatures this reduces to the classical prediction:

$$
\begin{aligned}
\bar{\epsilon} &= \hbar\omega/\left[(\hbar\omega/k_{\mathrm{B}}T) + (\hbar\omega/k_{\mathrm{B}}T)^2/2 + \cdots\right] + \hbar\omega/2 \\
&= k_{\mathrm{B}}T[1 - (\hbar\omega/k_{\mathrm{B}}T)/2 + \cdots] + \hbar\omega/2 \to k_{\mathrm{B}}T. \tag{4.2}
\end{aligned}
$$

The quanta of lattice vibration are called *phonons*. Unlike photons they have no independent existence outside the material considered: there are no phonons in free space. In addition, phonons are only detectable indirectly via their effect on photons or electrons that absorb, emit, or scatter off phonons. Phonons are one example of what are called *quasi-particles*.

If there are N atoms in the crystal the number of mechanical degrees of freedom of the crystal is $3N$ and there are $3N$ modes of vibration of the lattice. All three polarization possibilities are physically realized, in the two transverse modes and the one longitudinal mode. Low-frequency longitudinal waves are simply *sound waves*. The heat capacity (*i.e.* specific heat) of a mode is

$$C_v = \mathrm{d}\bar{\epsilon}/\mathrm{d}T = k_{\mathrm{B}}(\hbar\omega/k_{\mathrm{B}}T)^2 \frac{\exp(\hbar\omega/k_{\mathrm{B}}T)}{[\exp(\hbar\omega/k_{\mathrm{B}}T) - 1]^2}. \tag{4.3}$$

Einstein made the simplifying assumption that all modes had the same frequency, so that the thermal energy of a solid would be $N\bar{\epsilon}$ and the heat capacity NC_v. The temperature variations of both quantities are displayed in Figure 4.1, which exhibits the *universal* dependence of any quantum harmonic oscillator's behaviour on temperature. At the absolute zero of temperature the zero point energy remains, while at high enough temperatures the behaviour merges with the classical prediction.

At very low temperatures the heat capacity of solids was found to vary as the cube of the temperature rather than exponentially, as predicted by eqn. 4.3. The failure of Einstein's prediction in this temperature regime has its origin in assigning the same frequency to every mode.

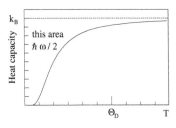

Fig. 4.1 Thermal energy and heat capacity of an harmonic vibrational mode according to Einstein's model, drawn as solid lines. The broken lines indicate the classical predictions.

The zero point energy common to all physical oscillation was discussed in Section 2.3.

Debye instead took the phonon mode distribution to be the same as that for photons, using the box normalization given in eqn. B.11,

$$g(\omega)\mathrm{d}\omega = V\omega^2\mathrm{d}\omega/(2\pi^2 v_s^3). \qquad (4.4)$$

where ω is the mode angular frequency, and v_s is the, as yet, unspecified wave velocity. Debye assumed that *waves of any frequency have the same velocity*, and that this velocity depends only on the material involved. For simplicity we shall also take velocity to be independent of polarization, which presumes an isotropic material. The dispersion relation is thus

$$v_s = \omega/k. \qquad (4.5)$$

In general, at low enough energies a linear dispersion relation holds for thermal excitations not only in solids but also in, for example, Bose–Einstein condensates: the excitations are then said to have *phonon-like* dispersion relations in that energy regime.

Debye's analysis closely resembles the analysis already described for the black body radiation spectrum, with two important differences. First, the wave velocity is very much lower; secondly, the number of vibrational modes is restricted by the *finite* number of degrees of freedom of the lattice, and this imposes an upper limit to the possible mode frequencies. We can gain useful insight into this restriction by considering vibrational waves on a one dimensional linear lattice of atoms.

Let us take the spacing between adjacent atoms to be a. Evidently modes of wavelength shorter than a can have many peaks and troughs between each pair of atoms, which elicit no response from the atoms. We can establish whether such modes are distinct as follows. A wave whose half wavelength matches the atomic spacing has wave number $\pm\pi/a$. Any wavenumber of larger modulus can be written

$$k = 2mk_\mathrm{D} + k' \qquad (4.6)$$

where m is an integer, $k_\mathrm{D} = \pi/a$ and $|k'| < \pi/a$. The displacement of the nth atom in the lattice due to such a wave is

$$\begin{aligned}
\exp[i(\omega t - n\,a\,k)] &= \exp[i(\omega t - 2nm\,a\,k_\mathrm{D} - n\,a\,k')]\\
&= \exp[i(\omega t - 2nm\,\pi - n\,a\,k')]\\
&= \exp[i(\omega t - n\,a\,k')].
\end{aligned}$$

This demonstrates what is called *aliasing*: a mode with wave number greater than k_D mimics one with wave number inside this range, producing an identical displacement of the lattice. An example is given in Figure 4.2. Hence the modes with wavenumber k, such that $|k| \le k_\mathrm{D}$ are sufficient to describe the physical lattice vibrations. A similar analysis in three dimensions converts eqn. 4.6 into a vector equation

$$\mathbf{k} = \mathbf{G} + \mathbf{k}', \qquad (4.7)$$

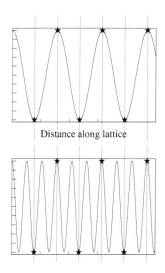

Distance along lattice

Fig. 4.2 An example of aliasing. The stars indicate the location and displacement of the ions a distance a apart. The lower panel shows an aliasing wave with wave number greater than π/a. In the upper panel the equivalent wave with wave number π/a is shown.

where **G** is any *reciprocal lattice vector*. These vectors are defined as follows: if under a displacement **R** the lattice would appear identical, ignoring effects from the finite extent of the lattice, then

$$\mathbf{G} \cdot \mathbf{R} = 2n\pi, \tag{4.8}$$

where n is any integer. Returning to the example of the linear chain, the corresponding limiting angular frequency, following Debye's analysis, is $\omega_D = k_D v_s$. The count of these modes must equal the number of degrees of freedom, N, where there are N atoms in the chain. Thus

$$\int_0^{\omega_D} g(\omega)\mathrm{d}\omega = N. \tag{4.9}$$

In three dimensions, with N being the number of atoms, this result holds for *each polarization*. Using eqn. 4.4 to replace $g(\omega)$, and integrating gives

$$V\omega_D^3/(6\pi^2 v_s^3) = N, \tag{4.10}$$

whence

$$\omega_D = v_s(6\pi^2 N/V)^{1/3}. \tag{4.11}$$

A corresponding *Debye temperature* is defined by

$$\Theta_D = \hbar\omega_D/k_B. \tag{4.12}$$

For simplicity, the three polarizations are regarded as equivalent: the total thermal energy of the solid is then

$$
\begin{aligned}
U &= \int_0^{\omega_D} 3\,g(\omega)\,\bar{\epsilon}\,\mathrm{d}\omega \\
&= \frac{3V\hbar}{2\pi^2 v_s^3}\int_0^{\omega_D}\left[\frac{1}{[\exp(\hbar\omega/kT)-1]}+\frac{1}{2}\right]\omega^3\,\mathrm{d}\omega\,,
\end{aligned}
\tag{4.13}
$$

where we have used eqns. 4.4 and 4.1. Differentiating eqn. 4.13 with respect to temperature gives the heat capacity

$$C_v = \frac{3V\hbar^2}{2\pi^2 v_s^3 k_B T^2}\int_0^{\omega_D}\frac{\omega^4\,\exp(\hbar\omega/k_B T)\,\mathrm{d}\omega}{[\exp(\hbar\omega/k_B T)-1]^2}.$$

Putting $\hbar\omega/k_B T$ equal to x and simplifying, this becomes

$$C_v = 9Nk_B\left[\frac{T}{\Theta_D}\right]^3\int_0^{x_D}\frac{x^4\,\exp(x)\,\mathrm{d}x}{[\exp(x)-1]^2}. \tag{4.14}$$

The contribution to C_v per mole is plotted as a function of (T/Θ_D) in Figure 4.3. This predicted universal curve adequately describes the data for many crystalline materials, with a value of Θ_D obtained by a fit in each case. Values of Θ_D are generally a few hundred degrees Kelvin for everyday solids at room temperature. Notice that the broken line in Figure 4.3 is the classical prediction, $3Nk_B$, to which both the data and the quantum prediction converge at temperatures well above Θ_D.

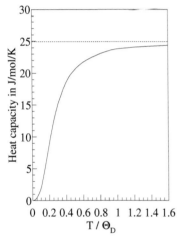

Fig. 4.3 Heat capacity of a solid according to the Debye approximation. The broken line indicates the classical Dulong–Petit law, which only becomes an acceptable approximation at temperatures very much greater than Θ_D.

At very low temperatures eqn. 4.14 can be approximated by setting the upper limit of the integral to infinity. The resulting definite integral has value $4\pi^4/15$, and inserting this in eqn. 4.14 gives

$$C_v \approx 234 N k_{\mathrm{B}} (T/\Theta_{\mathrm{D}})^3. \tag{4.15}$$

The heat capacity of solids at temperatures below $\Theta_{\mathrm{D}}/50$ does show this cubic dependence on temperature. However, the values extracted for Θ_{D} from fits to Debye's prediction in the low and room temperature regimes can differ significantly: for magnesium the values are respectively 403 K and 330 K. In metals at low temperatures the additional linear contribution of electron heat capacity becomes significant.

The measurements used to determine the dispersion relations and mode densities have been made by scattering neutrons or X-rays from materials of interest. X-ray measurements are described below. These measurements show that the quadratic dependence of the mode density assumed by Debye is an oversimplification. Figure 4.4 compares this quadratic dependence with the actual mode density for a face centred cubic crystal.

The prediction is safest at very low and high temperatures. At low temperatures the excitations are of low energy and long wavelength so that they are less sensitive to the detailed crystal structure. At high temperatures all modes are excited and hence it is mainly their number that is important. Debye's quadratic approximation for the mode density provides a useful interpolation over intermediate temperature for dielectrics. Note that the stronger the crystal bonding, the higher the phonon frequencies will range. Hence the Debye frequency, ω_{D}, and temperature, $\hbar\omega_{\mathrm{D}}/k_{\mathrm{B}}$, are measures of material strength and hardness. Thus Θ_{D} is 38 K for wax-like cesium and 1550 K for diamond.

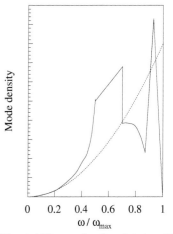

Fig. 4.4 Phonon density of states. The curve is the quadratic distribution used by Debye. The solid line shows the distribution calculated for an fcc lattice.

4.3   Linear chains of atoms

The analysis here of a linear chain of atoms provides an introduction to more detailed models of excitations of lattices of atoms. This takes us beyond Debye's choice of a linear dispersion relation: $k v_s = \omega$. Along the chain the forces between atoms fall off rapidly with distance so that to a first approximation only the forces due to nearest neighbours need to be considered. Again the amplitude of vibration is taken to be small compared to the equilibrium separations so that the force is at least approximately proportional to the displacement from equilibrium. The chain, shown in Figure 4.5, is chosen to contain alternate atoms of masses m and M at locations x_n and X_n respectively for the nth pair, with an equilibrium spacing a between adjacent *like* masses. The restoring force is assumed to be linear, $C\Delta y$ for a displacement Δy from equilibrium. This is the condition for pure harmonic motion. Then the equations of

Fig. 4.5 Linear chain of atoms with masses m and M alternating.

motion are:

$$m\,d^2x_n/dt^2 = C(X_n + X_{n-1} - 2x_n),$$
$$M\,d^2X_n/dt^2 = C(x_{n+1} + x_n - 2X_n),$$

Travelling wave solutions have the form

$$x_n = x\,\exp(inka)\exp(-i\omega t),\quad X_n = X\,\exp(inka)\exp(-i\omega t). \quad (4.16)$$

Substituting these expressions in the equations of motion gives

$$-\omega^2 mx = CX[1 + \exp(-ika)] - 2Cx, \quad (4.17)$$
$$-\omega^2 MX = Cx[1 + \exp(+ika)] - 2CX. \quad (4.18)$$

Eliminating x from these equations gives

$$X[mM\omega^4 - 2C(m+M)\omega^2 + 2C^2(1 - \cos(ka)] = 0.$$

The trivial solution $X = 0$ can be ignored, so we have

$$\omega^2 = \frac{2C(M+m) \pm \sqrt{4C^2(M+m)^2 - 8C^2mM[1 - \cos(ka)]}}{2mM}. \quad (4.19)$$

In the limit of long wavelengths close to $k = 0$, $\cos(ka) \approx 1 - k^2a^2/2$, and the two roots of the last equation are

$$\omega_O^2 \approx 2C(1/M + 1/m), \quad (4.20)$$
$$\omega_A^2 \approx Ck^2a^2/[2(m+M)]. \quad (4.21)$$

Earlier in this chapter it was seen that if the separation between adjacent like atoms in a crystal is a, the wave numbers that give physically distinct lattice oscillations extend up to $|k| = \pi/a$. At this other extreme, when $k = \pi/a$, these roots simplify to

$$\omega_O^2 = 2C/m, \quad (4.22)$$
$$\omega_A^2 = 2C/M. \quad (4.23)$$

These solutions are displayed in Figure 4.6. The significant separation in energy between the two solutions comes about because in the lower *acoustic* branch neighbouring atoms move in phase, while in the *optical* branch they move in antiphase. For clarity, these modes are presented in Figure 4.7, as lateral rather than longitudinal displacements. The higher energy optical branch can be excited when an optical photon is absorbed by the lattice, hence its name. The waves in the acoustic branch at frequencies within the range of hearing are sound waves. Both longitudinal and transverse oscillations occur in three-dimensional crystals, and a simple example of their dispersion relations is shown in Figure 4.8. Notice that the range of wave numbers extends from $-\pi/a$ to $+\pi/a$ to include motion in both directions. The density of states along each segment of a branch is given in Appendix B. Projecting this density onto the ω-axis gives the phonon distribution in angular frequency,

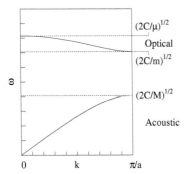

Fig. 4.6 Dispersion plot for vibrations of a linear chain of atoms, with alternately, masses m and $M = 3m$. $\mu = Mm/(m+M)$, and C is the spring constant.

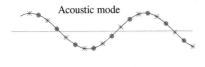
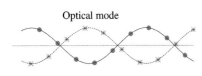

Fig. 4.7 Motion of linear array of atoms when excited in optical and acoustic modes. The displacements of the two species of atoms are made equal for this plot. In practice they will be in inverse relation to their masses.

and hence the energy distribution. Immediately we see that there are *allowed energy bands* with gaps between. Inside these gaps solutions can be obtained to the wave equations only if the wave-vector is complex, meaning that the wave does not propagate freely, but is absorbed. In a three-dimensional crystal the equivalent of Figure 4.8 will depend on the phonon direction with respect to the crystal axes because the spacing and nature of the atoms will in general be different for each axis. The region of inequivalent wave numbers between zero and $\pm\pi/a$ is known as the *first Brillouin zone*. In non-cubic lattices this zone becomes more complex: see Figures 4.13 and 4.14.

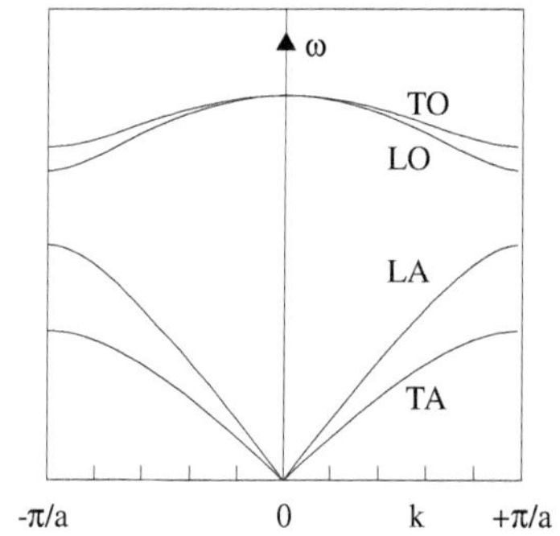

The group velocity of the phonons $d\omega/dk$ is approximately constant at low frequencies, which agrees with observation, and reproduces the classical prediction for waves in a continuous medium. Close to the edge of the Brillouin zone the group velocity of all modes smoothly tends to zero. When the condition $k = \pm\pi/a$ holds true, the waves travelling in opposite directions are equivalent and the result is a stationary standing wave.

Fig. 4.8 Dispersion relations in a cubic lattice for longitudinal (L) and transverse phonons (T), in the optical (O) and acoustic (A) modes.

4.3.1   X-ray scattering studies

Phonon dispersion relations, and hence the energy band structure, have been mapped in experiments in which neutrons or X-ray beams are Bragg-scattered off sample crystals. The incoming and outgoing particle vector momenta, and hence energies, are measured. The difference between them is the vector momentum and energy of the phonon exchanged with the crystal.

The techniques also work for phonon excitations of liquids or amorphous solids.

X-rays scatter off electrons attached to nuclei while neutrons scatter off the nuclei themselves. The techniques for producing and detecting high fluxes of X-rays are simpler than those for neutrons. Nonetheless experimentalists using X-rays have had to overcome a severe problem.

Bragg scattering requires that the X-ray wavelength is similar to the atomic spacing. X-rays of such wavelengths around 0.1 nm have energies of order 10 keV, while the energies of phonons in crystals are typically tens of meV. There is a spectacularly bad mismatch in energy: in order to determine the energy of a phonon using X-rays the incident and scattered photon energies must be measured to precisions of one part in 10^6. Today this precision has been surpassed and X-ray measurements have overtaken neutron measurements in importance, despite the latter's better energy match. The greater availability of photons from synchrotrons has facilitated this change.

X-ray beams are generated as synchrotron radiation from high-energy electron beams produced at accelerators. An electron guided in a circular arc by a magnetic field accelerates radially. As a result, it emits a fan of radiation in the plane of the orbit, centred on the tangent forward to

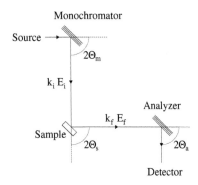

Fig. 4.9 Triple axis spectrometer. Θ_m, Θ_s and Θ_a are the Bragg-scattering angles. The incident and outgoing four momenta of the selected beam are also indicated.

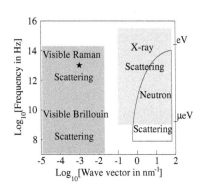

Fig. 4.10 Regions of the frequency/wave-vector space for phonons that are accessible using various scattering techniques. The star locates the experiment on GaN polaritons.

its path. The half-angle of the fan is $\phi \sim 1/\gamma$ where $\gamma = 1/\sqrt{1 - v^2/c^2}$ for an electron velocity v. An electron beam of a few GeV is required to give intense synchrotron radiation over the range of tens of keV.

Measurements of phonon dispersion curves with X-rays use triple axis spectrometers sharing the same overall outline appearance shown in Figure 4.9. The beam from the source is first Bragg-scattered from a chemically pure and structurally uniform crystal, often a silicon crystal. Narrow aperture lead collimators define the incident beam, and the resultant nearly monochromatic photons are Bragg-scattered from the target studied. The outgoing beam is collimated and analysed by a further Bragg-scattering off a second high quality reference crystal. The X-ray intensities at this stage are measured with silicon diodes.

Beryllium lenses are used to focus the X-ray beams and hence shorten the time needed to take data. X-rays have a refractive index slightly less than unity in matter: $[1 - 4.1 \, 10^{-6}]$ in beryllium. Focussing before and after scattering relies on a series of *bi-concave* beryllium blocks, whose faces have radii of curvature $\sim$3mm.

Applying the law of conservation of momentum to the scattering of an X-ray gives

$$\mathbf{k} \pm (\mathbf{k}_\phi + \mathbf{G}) = \mathbf{k}', \qquad (4.24)$$

where $\mathbf{k}(\mathbf{k}')$ is the wave-vector of the incident (outgoing) X-ray, $\mathbf{k}_\phi$ is the phonon wave-vector. Here, $\mathbf{G}$ is any reciprocal lattice vector (all of which are usually well known to the experimenter) accounting for the momentum absorbed or emitted *by the lattice as a whole*. The kinetic energy acquired or lost by a crystal of mass M in this momentum exchange, $\hbar^2 G^2/(2M)$, is negligible. Then, applying the law of conservation of energy to the scatter gives

$$E = E_\phi + E', \qquad (4.25)$$

where $E(E')$ is the incident (outgoing) energy of the beam particle and E_ϕ that of the phonon.

Bragg-scattering at the monochromator and analyser crystals is used to determine the wavelength:

$$\lambda = 2d \sin\theta \qquad (4.26)$$

where d is the crystal layer spacing and θ the Bragg angle. It follows that the precision with which the energy can be determined is

$$\Delta E/E = \Delta\lambda/\lambda = \cot\theta\Delta\theta, \qquad (4.27)$$

where $\Delta\theta$ is the angle uncertainty. In order to attain the necessary precision of one part in 10^6 with 10 keV X-rays the Bragg angle has to be set close to $90°$. The precision necessary in the measured energies requires that the silicon crystals are each placed in their own constant temperature enclosures. Silicon has a coefficient of linear expansion $2.6\,10^{-6}\mathrm{K}^{-1}$, which implies that the relative temperature between the enclosures has be tuned to millikelvin accuracy. Figure 4.10 shows the regions over which phonon kinematics are accessible with scattering techniques, from far infrared to optical frequencies. The lower limits in wave-vector and in angular frequency are where the precision of measurement becomes comparable to the quantity measured. An example of a Raman measurement is described below.

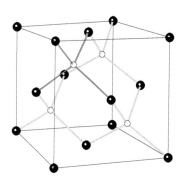

4.3.2 Lattice structure

Three-dimensional crystals are not isotropic with the result that the dispersion relations depend in general on the direction of the wave-vector relative to the crystal axes. As an example the crystal form of germanium is shown in Figure 4.11, made up of two interpenetrating face centred cubic (*fcc*) lattices offset from one another by one quarter the length of a long diagonal axis. Germanium, silicon and diamond share this structure, all semiconductors of great technological importance. The cube side length is 5.64, 5.43 and 3.56 Å respectively in Ge, Si and diamond. An atom in one lattice is covalently bound to its four nearest neighbours in the other lattice, forming a tetrahedron: one set of links is shown in red in Figure 4.11. The electrons involved lie in clouds between the ions rather than lying along the lines in the Figure. In the case of diamond this interlocked structure makes it the hardest natural material, about four times harder than the next hardest, corundum.

Fig. 4.11 Germanium crystal lattice: two interpenetrating fcc lattice displaced by one quarter of the cube diagonal. Atoms in one fcc lattice are shown as solid circles; those in the other as hollow circles.

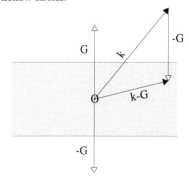

In general, a *unit* cell is defined as one that when repeated, produces the whole crystal. For germanium this is a cube containing the tetrahedron shown with the red bonds. Its side length is half the cube side length. The whole crystal can be reproduced from the unit cell by placing copies at locations relative to the original cell given by

Fig. 4.12 First Brillouin zone in one dimension with origin O.

$$\mathbf{R} = \Sigma_i n_i \mathbf{a_i} \qquad (4.28)$$

where n_1, n_2 and n_3 are any integers and $\mathbf{a}_1$, $\mathbf{a}_2$ and $\mathbf{a}_3$ three displacement vectors. These displacement vectors depend on the crystalline form and are not necessarily orthogonal. In Figure 4.11, the three basic displacements would move an open circle onto the three adjacent open circles. There are correspondingly three basic *reciprocal lattice vectors* with the property that

$$\mathbf{G}_i = 2\pi \frac{\mathbf{a}_j \wedge \mathbf{a}_k}{\mathbf{a}_1 \wedge \mathbf{a}_2 \cdot \mathbf{a}_3}, \qquad (4.29)$$

where (i, j, k) are taken cyclically: (1,2,3) or (2,3,1) or (3,1,2). A general reciprocal lattice vector has the form

$$\mathbf{G} = \Sigma_j [m_j \mathbf{G}_j], \qquad (4.30)$$

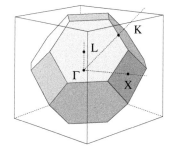

Fig. 4.13 Reciprocal crystal lattice for an fcc crystal. L lies at the centre of a hexagonal face.

where m_1, m_2 and m_3 are all integers. Then for any reciprocal lattice vector **G** and any lattice vector **R**

$$\mathbf{G} \cdot \mathbf{R} = 2p\pi \qquad (4.31)$$

where p is an integer. The wave-vectors whose projections along the directions (plus or minus) of the reciprocal lattice vectors are all smaller than π define a region called the first Brillouin zone for that crystal. Any

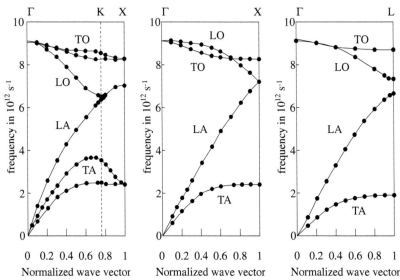

Fig. 4.14 Measured dispersion relations for Germanium using the data in Table I from G. Nilson and G. Nelin, Physical Review B3, 364 (1971) published by the American Physical Society.

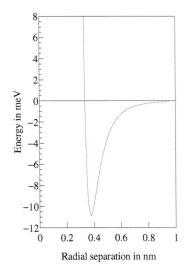

Fig. 4.15 Inter-atom potential (for argon).

phonon whose wave-vector ends outside this zone is equivalent, so far as lattice oscillations are concerned, to a phonon with a wave-vector ending inside. An example in a two-dimensional lattice is shown in Figure 4.12 where subtracting a reciprocal lattice vector **G** brings the wave-vector into the first Brillouin zone. It is easy to construct the first Brillouin zone in three dimensions according to the Wigner–Seitz prescription: planes are drawn in wave-vector space (reciprocal space) perpendicular to each reciprocal lattice vector so as to bisect that vector. The first Brillouin zone for a fcc crystal is drawn in Figure 4.13.

Figure 4.14 displays the phonon dispersion relations obtained by neutron scattering from germanium. The labels above the plots in Figure 4.14 indicate the directions of the wave-vectors in the reciprocal lattice of Figure 4.13: Γ is the origin at the centre of the first Brillouin zone. The locations of K, X, etc. are shown in Figure 4.13. Projecting these distributions onto the frequency axis produces a mode density like that shown in Figure 4.4. The mode density is inversely proportional to the slope of the dispersion curve and is low if one of the modes has a bandgap.

4.4   Anharmonic effects

In general, interatomic potentials resemble those shown in Figure 4.15 and do not have the precise parabolic shape of an harmonic well. The restoring force is larger for compression than for extension. Consequently, the actual modes of oscillation of a crystal lattice are not precisely harmonic. They contain a range of frequencies/energies that broaden the dispersion curve. A phonon (a quantum of a single harmonic mode) is then a linear superposition of actual modes. The components of this superposition travel at varying speeds and their phases change as this superposition propagates. In effect the phonon exchanges lower energy phonons with the lattice. One such process in which the phonon absorbs another is drawn in Figure 4.16. Both energy and *crystal momentum* must be conserved, where the latter includes any momentum given to or absorbed by the crystal as a whole. Then

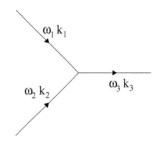

Fig. 4.16 Phonon interaction caused by anharmonic response of the crystal lattice.

$$\omega_3 = \omega_1 + \omega_2; \quad \mathbf{k}_3 = \mathbf{k}_1 + \mathbf{k}_2 \pm \mathbf{G}, \tag{4.32}$$

where the energies are $\hbar\omega$, the momenta $\hbar\mathbf{k}$, and $\mathbf{G}$ is a reciprocal lattice vector so that $\hbar\mathbf{G}$ is the momentum taken by the lattice as a whole. Figure 4.16 is drawn for the case that the resultant momentum, k_3, lies within the first Brillouin zone and $\mathbf{G}$ is zero. In Figure 4.17, by contrast, an example is shown in which the sum of $\mathbf{k}_1$ and $\mathbf{k}_2$ lies outside the physical range and must be displaced by $\mathbf{G}$ to lie within the first Brillouin zone. Such a scattering is called an *Umklapp process*. Normal processes have $\mathbf{G} = 0$, Umklapp processes have $\mathbf{G} \neq 0$. If all scatters were normal processes then thermal conduction by phonons across a crystal would be instantaneous! However, in Umklapp scatters the phonon momentum is reversed in direction, and these collisions hold down the thermal conductivity of dielectrics. The thermal conductivity in conductors and semiconductors (which is discussed in Chapters 5 and 6) is typically around ten to a hundred times larger than in dielectrics, thanks to the electron transport of heat.

The German verb umklappen means to fold back.

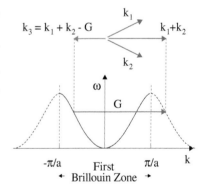

4.4.1   Thermal conductivity of dielectrics

The success of Debye's model shows that to a fair approximation the phonons in a crystal can be regarded as a gas of particles, thus making it reasonable to apply an expression for thermal conductivity taken from the kinetic theory of gases

Fig. 4.17 Umklapp process bringing the phonon wave-vector into the first Brillouin zone. The wave-vectors are drawn in the upper half of the diagram and the dispersion relation in the lower half.

$$K = \frac{1}{3}v\lambda C_v, \tag{4.33}$$

where C_v is the heat capacity per unit volume, v the mean phonon velocity, and λ the mean free path between collisions.[2] The behaviour

[2]See page 402, second edition of *Thermal Physics* by C. Kittel and H. Kroemer, published by W. H. Freeman and Co.

of the conductivity varies enough with temperature that three distinct regimes can be distinguished. Well above the Debye temperature, Θ_D, the phonon energies pile up around $k_B\Theta_D$ while the crystal's thermal energy is proportional to T. Thus the number of phonons becomes proportional to T and so does the collision rate. This makes the mean free path proportional to $1/T$ and in turn $K \propto 1/T$. As the temperature falls below Θ_D a second regime is entered: the number of phonons capable of Umklapp scattering falls and the mean free path also rises. Together these effects cause the conductivity to increase steeply as the temperature falls. Eventually, the mean free path exceeds the size of the sample, and the sample's dimensions restrict the mean free path. In this third low temperature regime the heat capacity has been shown to vary as T^3 (see eqn. 4.15) and therefore $K \propto T^3$, as well as depending on the specimen's dimensions. Figure 4.18 shows the overall generic behaviour of the thermal conductivity for dielectric crystals. The dotted line at low temperatures shows the trend when the crystal dimensions are reduced. In imperfect crystals scattering of phonons from the defects reduces the mean free path and the conductivity is reduced as shown by the broken curve. The T^3 dependence at the lowest temperatures is robust because the phonon wavelengths are very long and do not sense the defects. In very chemically pure and highly regular crystals the peak thermal conductivity can be of order $10^3\,\mathrm{Wm^{-1}K^{-1}}$, compared to $\sim$300 $\mathrm{Wm^{-1}K^{-1}}$ for commercial copper.

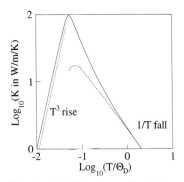

Fig. 4.18 Dielectric conductivity versus temperature. The broken curve indicates the effect of defects and impurities. The dotted curve indicates the change if the specimen size is reduced.

4.5 Phonons and photons

Rayleigh scattering is an elastic process in which a photon is absorbed and re-emitted by a molecule or atom or larger structure. Rayleigh scattering is ubiquitous. Sunlight, Rayleigh scattered from the upper atmosphere, gives the blue of the sky. Rayleigh scattering is the residual unavoidable cause of attenuation and degradation of signals carried on monomode optical fibre, so that signals need refreshing after a few hundred kilometres.

Raman and Brillouin scattering are very much weaker processes that always accompany Rayleigh scattering. However, in these scattering processes a phonon is exchanged between the incident photon and the scattering material. In Raman scattering the phonon is a high-energy optical phonon and in Brillouin scattering the phonon is an acoustic phonon. Figure 4.19 compares the Raman and Rayleigh scattering processes. If the final state has a lower energy than the initial state this is called a Stokes transition, and if a higher energy it is called an anti-Stokes transition. Specializing to Raman scattering from nitrogen gas molecules the energy change of the photon is around 0.3 eV for a vibrational excitation, and around 0.001 eV for a rotational excitation of the molecule. The corresponding intensities are only 0.07 per cent and 2.3 per cent of the Rayleigh intensity. Figure 4.20 shows the spectrum of

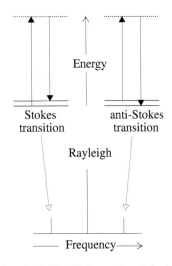

Fig. 4.19 The Stokes and anti-Stokes transitions are shown in the upper panel. In the lower panel the frequencies of these transitions are compared with the Rayleigh transition frequency.

light scattered out from the core of optical fibres, revealing the strong Rayleigh line being accompanied by the weaker Stokes and anti-Stokes lines. Raman scattering is widely used to excite those states of atoms and molecules that cannot be excited directly because of selection rules given in Section 3.3. It is employed, for example, in sensing undesirable chemicals in industry and elsewhere. The acoustic phonons involved in Brillouin scattering have much lower energies so that the scattered photons would be buried within the shoulders of the Rayleigh peak in the figure.

4.6   Polaritons

In some crystals absorbing a phonon produces an oscillating relative displacement between ions of opposite charges, that is, an oscillating electric dipole moment. Such an excitation is indistinguishable from that produced by absorbing a photon having the same direction and polarization. If such a phonon and photon also have identical phase velocities their continuing overlap will lead to strong coupling between them through the excitation of the crystal. Whenever another quantum species couples with photons in this way they form a hybrid quantum state called a *polariton*, in this case, a phonon-polariton.

Figure 4.21 shows the effect of the coupling on the dispersion relations in an ionic crystal, in this case GaN. The blue lines indicate the dispersion relations expected for photons and for optical transverse polarized phonons in GaN if the other species could be excluded. The black lines are the actual dispersion relations. Labels indicate whether the corresponding branch is behaving like a phonon or a photon in the limit of very small and very large wave numbers. Note that there is an energy gap between the two branches. Within this gap the wave number is pure imaginary $k = iK$, signifying that the polariton wave is attenuated as it travels. We now deduce the analytic form of these polariton dispersion relations.

From Figure 4.21 it is apparent that the phase velocities, ω/k, of the phonon and photon only match at a low wave number. In this region of low wave number the transverse optical phonon has an almost constant angular frequency ω_T. Let E be the electric field and P the polarization of the crystal. The polarization is carried by pairs of oppositely charged ions: suppose that the charges are $\pm q$, the displacement from the equilibrium separation u , their reduced mass m and that the number of dipoles per unit volume is N. Then $P = Nqu$. Next the linear restoring force acting on the charges is $m\omega_T^2 u$. Hence the equation of relative motion of the charges is

$$m\partial^2 u/\partial t^2 = qE - m\omega_T^2 u. \qquad (4.34)$$

In the case of an excitation with angular frequency ω this equation

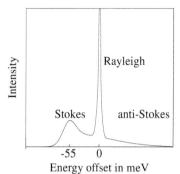

Fig. 4.20 The spectrum of near infrared laser light after travelling over a long stretch of telecom fibre.

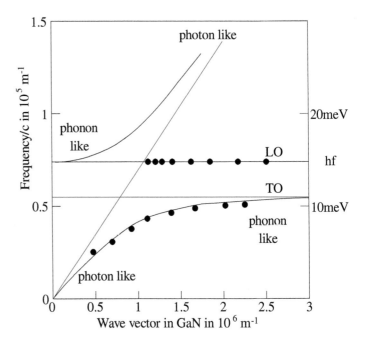

Fig. 4.21 Dispersion relations for GaN where the photon-phonon overlap results in a polariton. Adapted from Figure 7, G. Irmer, C. Röder, C. Himcinschi and J. Kortus, *Physical Review* B88, 104303 (2013). Courtesy Dr. Irmer and the American Physical Society.

reduces to

$$-m\omega^2 u = qE - m\omega_T^2 u. \tag{4.35}$$

Multiplying throughout by Nq/m and rearranging gives

$$\omega_T^2 - \omega^2 = (Nq^2/m)E/P. \tag{4.36}$$

The dielectric constant (relative permittivity) is given by

$$\varepsilon_r = 1 + P/(\varepsilon_0 E) = 1 + \frac{Nq^2/(m\varepsilon_0)}{\omega_T^2 - \omega^2}. \tag{4.37}$$

Taking account of any contributions due to the intrinsic polarization of the ions themselves, which persists up to optical frequencies, this equation is replaced by

$$\varepsilon_r(\omega) = \varepsilon_{\text{opt}} + \frac{Nq^2/(m\varepsilon_0)}{\omega_T^2 - \omega^2}, \tag{4.38}$$

where ε_{opt} is the square of the usual optical refractive index. By introducing $\varepsilon_r(0)$ we can eliminate $Nq^2/m\varepsilon_0$,

$$\varepsilon_r(\omega) = \frac{\omega_T^2 \varepsilon_r(0) - \omega^2 \varepsilon_{\text{opt}}}{\omega_T^2 - \omega^2}. \tag{4.39}$$

Whenever $\varepsilon_r(\omega)$ is negative the refractive index is imaginary, so that the wave is attenuated as it travels. Therefore there is attenuation at frequencies between $\omega = \omega_T$ and $\omega_L = \omega_T \sqrt{\varepsilon_r(0)/\varepsilon_{\mathrm{opt}}}$, where ω_L is the angular frequency of the LO mode. This equation linking the LO and TO modes is called the Lyddane–Sachs–Teller relationship. It is quite general for ionic crystals at low wavenumbers. The attenuation length is short immediately above the angular frequency ω_T and then falls off as ω_L is approached. The crossover occurs in general in ionic crystals at angular frequencies $\omega_T \sim 10^{14}\,\mathrm{s}^{-1}$ or wavelengths $\sim$ μm.

4.6.1   Raman measurements

In the case of a typical crystal of interest, GaN, the polariton region lies in the infrared at a wavelength 10μm, and wave number $10^6\,\mathrm{m}^{-1}$. There is no suitable infrared laser to excite the polariton directly, so Raman scattering is used instead. From Figure 4.10 we can see that this region is inaccessible when using either neutron or X-ray scattering to determine the dispersion relation. Figure 4.22 shows the setup used

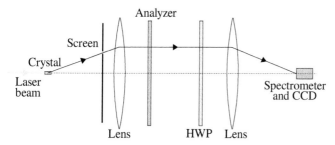

Fig. 4.22 Raman scattering apparatus to measure the polariton dispersion relation in GaN.

to produce the data points shown in Figure 4.21. Light of wavelength 514.5 nm from an argon ion laser is incident on a GaN single crystal. The scattered light travels forward to a screen placed 80 mm from the crystal. This screen is perforated by a single small hole to capture light scattered within a narrow range around a particular small angle. This light includes both the dominant Rayleigh scattered light as well as the component of interest, the Raman scattering accompanying the excitation of a polariton. A lens focusses the captured light into a parallel beam. It then travels in turn through a polarization analyzer and a half wave plate (HWP), and is finally re-focused onto a spectrometer. In this sequence the analyser is used to select light polarized in one of two orthogonal directions that correspond to the axes of symmetry of the GaN crystal. The HWP is used to rotate the electric field axis so that it is always perpendicular to the grating lines in the spectrometer: this maintains uniform grating response. Raman scattering has an intensity only one millionth of that of Rayleigh scattering for GaN, hence, the backgrounds from direct Rayleigh light and Rayleigh light rescattered by the optical components are large. The spectrometer is therefore a three-stage device, each stage comprising an input slit, a grating and an output slit, with refocussing between stages. Of these stages the sec-

In for example a fcc cubic crystal the dispersion relations may be different for light polarized in directions perpendicular to a face and along a diagonal of the basic cube.

ond stage is designed to reverse the dispersion of the first so that its output is a very narrow spectral slice determined by the slit openings. With this light as input the final third stage produces a spectrum that fans out across a silicon CCD (*charge coupled device*) detector, similar to those used in digital cameras. Unfortunately, any electrons thermally excited across the bandgap in silicon during the long exposures required are indistinguishable from those excited across the gap by absorbing a Raman scattered photon. It is therefore essential to cool the CCD with liquid nitrogen, suppressing the thermal counts in pixels to negligible levels (of order one per hour).

4.7 Mössbauer effect

In 1958 Mössbauer, then a graduate student, discovered that in a significant proportion of the γ-ray decays of $^{191}\mathrm{Ir}^*$ nuclei bound in a crystal, the crystal as a whole recoiled to balance the photon momentum, rather than the daughter nucleus alone recoiling. In this case no lattice oscillations are excited, that is to say, zero phonons enter the lattice modes of vibration. This property of *recoilless* emission is shared by around forty other nuclear species undergoing γ-decay. Recoilless emission exhibits unequivocally the transfer of momentum to the whole lattice: something which we have presumed could take place, in particular, in the Umklapp process.

The reaction most commonly used to observe the Mössbauer effect is

$$^{57}\mathrm{Fe}^* \rightarrow {}^{57}\mathrm{Fe} + \gamma: \quad Q = 14.4\,\mathrm{keV}, \tag{4.40}$$

with a lifetime of $10^{-7}\,\mathrm{s}$ and a corresponding natural linewidth, Γ, of $4.6\,10^{-9}\,\mathrm{eV}$. In all decays the photon takes almost all the energy release, Q, and has momentum close to Q/c. Then the recoil, whether nucleus or crystal, has kinetic energy $E_\mathrm{r} = (Q/c)^2/(2M)$, where M is the recoil mass. If the nucleus recoils $E_\mathrm{r} = 1.95\,\mathrm{meV}$ and the photon has energy $Q - E_\mathrm{r}$; if the crystal itself recoils its energy is negligible in comparison. In the reverse process of absorption the γ-ray energy required is correspondingly larger, $Q + E_\mathrm{r}$. Evidently a photon from a recoiling $^{57}\mathrm{Fe}^*$ nucleus cannot excite a $^{57}\mathrm{Fe}$ nucleus because its energy is twice 1.95 meV too low. On the other hand, a photon emitted in a recoilless decay can undergo a recoilless absorption.

Nuclei in crystals have thermal energies of about 25 meV at 300 K, enough to make up the energy deficit just noted. However, in order to make the absorption possible, the difference between the speeds of the two nuclei and their directions need to be correlated precisely. This double requirement drastically reduces the rate of absorption of γ-rays emitted by a $^{57}\mathrm{Fe}^*$ source and then incident on a target foil containing $^{57}\mathrm{Fe}$. Before Mössbauer's discovery it was supposed that this *thermal compensation* explained the observed limited absorption. Mössbauer killed this idea when, studying $^{191}\text{Ir}^*$, he observed that on reducing the temperature of source and absorber from $300\,\text{K}$ to $80\,\text{K}$ the absorption increased, while if thermal compensation were at work the absorption should decrease.

Mössbauer turned to a quantum mechanical explanation, based on re-coilless emission. He interpreted the nuclear recoil energy, E_r calculated above as the *expectation value* of the energy in the vibrational modes of the parent crystal. Suppose the phonon mode active in the transition has angular frequency ω. Consider first the simpler situation at absolute zero. Initially, the nucleus has the ground state harmonic wavefunction given in Section 2.3:

$$\psi_0(x) = \exp\left[\frac{-M\omega x^2}{2\hbar}\right], \tag{4.41}$$

where M is from here onward the mass of a single nucleus. In a *re-coilless* process this will also be the nucleus's final state wavefunction. Normalization of this wavefunction can be safely ignored in the calculation below because the exponent factor alone determines the physical result of interest. The crystal as a whole absorbs the momentum and has a corresponding wave-vector k: hence its wavefunction is $\exp(-ikx)$. The amplitude for the recoilless process is simply the overlap integral between initial and final state wavefunctions:

$$f = \int_{-\infty}^{\infty} \psi_0^*(x)\,\exp(-ikx)\,\psi_0(x)dx, \tag{4.42}$$

and the probability of the recoilless process is $P = f^*f$. We have

$$\begin{aligned} f &= \int \exp[-M\omega x^2/\hbar - ikx]\,\mathrm{d}x \\ &= \exp\left[\frac{-\hbar k^2}{4M\omega}\right] \int \exp\left[-\left(\sqrt{\frac{M\omega}{\hbar}}\,x + i\sqrt{\frac{\hbar}{4M\omega}}\,k\right)^2\right]\mathrm{d}x. \end{aligned}$$

The integral here of the Gaussian factor doesn't give an exponential term, so the important component of the probability is

$$P = f^*f = \exp\left[\frac{-\hbar k^2}{2\omega M}\right] = \exp\left[\frac{-E_r}{\hbar\omega}\right]. \tag{4.43}$$

Consequently, the probability for recoilless emission will be significant if E_r is less than the upper limit for $\hbar\omega$, namely $\hbar\omega_D$. As the temperature rises the phonon count n in the modes rises; stimulated emission by the nucleus increases correspondingly[3] and recoilless emission becomes less likely. This completes Mössbauer's explanation. The restriction on E_r means that recoilless transitions are limited to nuclear γ decays with energy release smaller than

$$Q = \sqrt{2ME_r^{\max}c^2} = \sqrt{2M\hbar\omega_D c^2} = \sqrt{2(Mc^2)(k_B\theta_D)}, \tag{4.44}$$

[3]See Section 8.3 where this process is detailed for photons.

which is below $\sim$150 keV.

In 1960, Rebka and Pound exploited Mössbauer's discovery to measure the gravitational spectral shift of light predicted by Einstein's strong equivalence principle. This principle assigns to a photon of frequency f and energy hf an equivalent inertial mass[4] m_I of hf/c^2. Thus light emitted downward and falling a distance Δr on earth should gain an energy $m_I g \Delta r$, where g is the local gravitational acceleration. The consequent fractional shift in its frequency is

[4]This was discussed earlier in Section 1.3

$$\Delta f/f = m_I g \Delta r/(m_I c^2) = g \Delta r/c^2. \qquad (4.45)$$

Rebka and Pound placed a $^{57}\text{Fe}^*$ source at the top of a tall tower and at its base a thin ^{57}Fe absorber, which shielded a scintillator viewed by a photomultiplier. The source was driven at a low velocity upward, with the object of compensating the gravitational spectral shift of γ-rays falling down the tower by a Doppler redshift. With perfect compensation the absorption by the ^{57}Fe absorber would be maximized and the photomultiplier current minimized. The gravitational fractional spectral shift predicted for the 22.6 m fall is only $2.46\,10^{-15}$, about 500 times smaller than the $^{57}\text{Fe}^*$ γ-decay line width. Rebka and Pound therefore scanned across the line width by varying the drive velocity in small steps and in this way achieved the necessary sensitivity in locating the line centre. Their measured fractional spectral shift was $2.57\pm0.26\,10^{-15}$, agreeing within error with Einstein's prediction. Experiments comparing the frequencies of masers, one on earth and the other carried to 10,000 km on a rocket, have improved the agreement to better than one part in 10^4.

4.8 Further reading

Introduction to Solid State Physics, 5th or later editions by C. Kittel, published by John Wiley, has a detailed account of crystal structure, and is a reliable source on solid state physics.

'Phonon Spectroscopy by Inelastic X-ray Scattering' by E. Burkel in *Reports on Progress in Physics* 63, 171 (2000), published by The Institute of Physics, gives a thorough account of this technique.

Exercises

(4.1) Prove the result given in eqn. 4.43.

(4.2) Calculate the precision with which a 12 keV X-ray's energy can be measured using Bragg-scattering from a silicon crystal at an angle of incidence 89.9°

if the angle can be determined to 0.1 mrad. Calculate the change in energy selected by Bragg-scattering when the crystal temperature changes by 1 mK. Silicon has a linear thermal expansion coefficient of $2.6\,10^{-6}\,\mathrm{K}^{-1}$.

(4.3) Show that in a linear chain of ions with equal spacing r, alternately carrying charges $+e$ and $-e$, the electrostatic energy of an ion is $\alpha e^2/(4\pi\varepsilon_0 r)$, where α, known as the *Madelung constant*, is $2\sqrt{2}$.

(4.4) The frequency of longitudinal optical phonons in KBr crystals is $4.4\,10^{12}$ Hz, where the wave number $k = \pi/a$, a being the repeat distance along the lattice. Calculate the restoring force between neighbouring ions and also the frequency of the longitudinal acoustic (LA) phonons. The atomic weights of potassium and bromine are respectively 39.1 and 79.9 amu, where an amu is $1.6605\,10^{-27}$kg. Treat this as a one-dimensional crystal.

(4.5) Show that r_m is the equilibrium separation of atoms using the Lennard–Jones potential, eqn. 14.1. Calculate the binding energy for argon atoms with $r_m = 0.381$ nm and $\varepsilon = 125k_{\mathrm{B}}$. The measured binding energy in a three-dimensional crystal is 0.08 eV. Can you explain the discrepancy? Make an estimate of the zero-point energy for modes in solid argon from the Debye temperature 92 K.

(4.6) Using the density ρ, the atomic weight A in atomic mass units and the velocity of sound v_s, calculate the Debye temperature Θ_{D} for silicon, diamond, germanium, copper and gold.
ρ [Si,C,Ge,Cu,Au] = 2300, 3510, 5320, 8960, 19300 kgm^{-3},
A [Si,C,Ge,Cu,Au] = 28, 12, 72.6, 63.5, 197 amu,
v_s [Si,C,Ge,Cu,Au] = 9620, 18000, 5400, 3810,

2030 ms^{-1}.
Comment on any trend that you observe given that the values measured around room temperature are Θ_{D} [Si,C,Ge,Cu,Au] = 645, 2230, 374, 343, 165 K.

(4.7) What is the ratio between the thermal conductivities of diamond at 4 K and 60 K, assuming that the specimen is small enough that the phonons scatter almost entirely from the crystal boundaries?

(4.8) A linear chain of identical atoms has an equilibrium spacing a. The restoring force, due to nearest neighbours, is linear in the displacement with force constant K. The atomic mass is M. Show that the angular frequency of oscillations of the chain, ω is given by

$$\omega^2 = (2K/M)[1 - \cos(ka)].$$

Forces between atoms that are not nearest neighbours can be ignored.

(4.9) In a Mössbauer experiment with ^{57}Fe the magnetic splitting of the ground state (m =-1/2 and m=+1/2) corresponds to a velocity shift of 3.92 mms^{-1}. The splitting between adjacent magnetic substates of the first excited state (m$_F$=-3/2, -1/2, 1/2, 3/2) is 2.24 mms^{-1}. Calculate the corresponding energy splitting between the levels in both cases. The nuclear g-factors are -0.181 and 0.100 for the ground and excited states, respectively. Make estimates of the magnetic field at the ^{57}Fe nucleus. Take the nuclear magneton to be $3.15\,10^{-8}$ eVT^{-1}.

Electrons in solids

5

5.1 Introduction

Quantum properties of electrons in matter are easiest to analyse and
best understood for structured materials. This understanding has un-
derpinned much of the use of metals and semiconductors in modern
technology. Attention here centres on electrons within crystalline metals, where the inadequacy of classical physics is immediately apparent.
Classically conduction electrons regarded as particles would have a mean
free path of order the crystal cell size or $\sim 1\,$nm: by contrast quantum
electron waves are Bragg scattered by the lattice, raising the mean free
path to $\sim 100\,$nm, the observed value in good conductors. Then again
the electron energy expressed as $k_{\mathrm{B}}T$ corresponds to several thousands
of degrees in metals at room temperature, a paradox resolved by quantum mechanics.

In condensed matter the outer atomic electron are affected by the electric fields of the neighbouring atoms: the discrete and coincident energy
levels of isolated atoms are spread out by these interactions into a band
of energy levels shared across the crystal. Figure 5.1 shows the effect as
the separation between atoms is reduced. In a micron sized crystal the
$\sim 10^{12}$ valence electron eigenstates, taking one for each atom present,
form an effectively continuous band of width typically 1 eV. All the electrons populating such bands are no longer bound to a particular atom
but have wavefunctions extending across the crystal. Their fermion nature forces the electrons to fill the lowest energy eigenstates, one per
eigenstate, up to the *Fermi level*. At room temperature the boundary
is blurred by thermal excitations, $k_{\mathrm{B}}T \sim 0.025\,$eV. Figure 5.2 contrasts
the band structure in metals, semiconductors and insulators, with the
energy levels filled at absolute zero shaded. In a metal the Fermi level
lies within a band: any small excitation can promote an electron to an
empty region of the band and it can travel across the metal and contribute to conduction. On the other hand an electron in an insulator
or semiconductor must acquire an energy equal to the band separation
before it can travel across the material. In an insulator the bands are
well separated while in a semiconductor thermal excitation at room temperature gives weak conduction. Not only is the electron excited into an
unoccupied energy level in the conduction band free to travel but so too
is the *hole* which it vacated in the valence band.

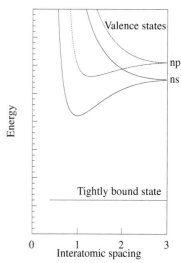

Fig. 5.1 Broadening of atomic states
into continuous energy bands as the
interatomic spacing is reduced. Unit
spacing is the equilibrium spacing. At
2 units separation the *np* and *ns* bands
already overlap forming a continuous
energy band.

Quantum 20/20: Fundamentals, Entanglement, Gauge Fields, Condensates and Topology.
Ian R. Kenyon. © 2020. Published in 2020 by Oxford University Press.
DOI: 10.1093/oso/9780198808350.001.0001

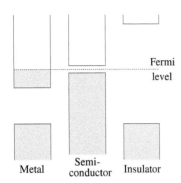

Fig. 5.2 Band structure of metals, semiconductors and insulators compared.

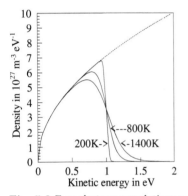

Fig. 5.3 Free electron populations per unit energy interval, at three temperatures for a Fermi energy of 1 eV. The broken line follows the density of states.

Some authors prefer to restrict the term Fermi energy to describe the chemical potential at zero degrees Kelvin.

We infer that elements with odd numbers of valence electrons will have half empty bands and must therefore be metallic conductors. Such are the alkali metals, sodium, potassium and rubidium. In fact three quarters of the elements are metallic including the transition elements (scandium, titanium.....copper and zinc). Among the semiconductors are the elements silicon and germanium, and compounds such as GaAs, GaP and AlGaAs, all of which have immense technological importance, being the building materials of transistors, lasers, LEDs and modern camera sensors.

The model of metallic behaviour considered in the first section below is that of a gas of free electrons obeying Fermi–Dirac statistics in a uniform potential well representing the electrostatic attraction of the ions. Examples of the successes in explaining electrical, thermal and magnetic behaviour are described. Electron screening and electron plasma formation, that are the response of electrons to very low and very high frequency radiation respectively, are both discussed.

Attention then turns to more realistic models using a periodic electrostatic potential to match the crystal structure. The electron dispersion relations become functions of the crystal structure and the number of available valence electrons. These and the associated band structure are discussed for representative metals. The modern technique for extracting electron dispersion relations, *angle resolved photoelectron spectroscopy (ARPES)*, is described.

Techniques for calculating band structure called *Density function theory* will be outlined briefly.

5.2   The free electron gas

The outer shell electrons in a metal are so weakly bound that to a first approximation they can be regarded as free electrons permeating the metal. Then the available modes have the energy distribution appropriate to non-relativistic motion given by eqn. B.15. For each spin state the density of states is

$$g_E(E)\mathrm{d}E = (2m/\hbar^2)^{3/2}\sqrt{E}\,\mathrm{d}E/(4\pi^2) \qquad (5.1)$$

per unit volume, and the mean occupation number of each state is given by eqn. 3.36

$$\langle n \rangle = \frac{1}{\exp[(E-\mu)/k_B T] + 1}. \qquad (5.2)$$

The quantity μ, the chemical potential, is the normalizing factor required to give the correct total number of electrons. It equals the energy at which $\langle n \rangle$ is 0.5 and is also called the Fermi level or Fermi energy. Overall the energy distribution, $f(E)$, counting both spin states is given by

$$df(E) = 2\langle n\rangle g_E(E)dE. \tag{5.3}$$

This function is drawn for temperatures of 200 K, 800 K and 1400 K in Figure 5.3 for an electron gas with E_F equal to 1 eV. The broken line curve is the envelope $2g_E(E)$. At absolute zero all energy states up to the Fermi energy are full, and those above empty. Using eqn. 5.1 the total number of electrons per unit volume is then

$$\begin{aligned} n &= \int_0^{E_F} 2g_E(E)dE \\ &= (2mE_F/\hbar^2)^{3/2}/(3\pi^2), \end{aligned}$$

so that

$$E_F = (\hbar^2/2m)(3\pi^2 n)^{2/3}. \tag{5.4}$$

The corresponding Fermi wave-vector is

$$k_F = (3\pi^2 n)^{1/3}. \tag{5.5}$$

The Fermi velocity, v_F, is $\hbar k_F/m$ and the Fermi temperature, T_F, is E_F/k_B. The mean electron energy can be shown to be $3E_F/5$, and the proof is given in one of the exercises. In the case of the metal copper[1] n is $8.5\,10^{28}\,\mathrm{m^{-3}}$, E_F is 7.0 eV, k_F is $1.36\,10^{10}\,\mathrm{m^{-1}}$, T_F is 81,720 K and v_F is $1.57\,10^6\,\mathrm{ms^{-1}}$. Thus the electron energy far exceeds the classical expectation of the thermal energy, $\sim k_B T$, even if copper were on the point of melting. This is true for solids in general. At finite temperature the Fermi energy expressed in terms of that at zero temperature is

$$E_F(T) = E_F[1 - (\pi k_B T)^2/12E_F^2], \tag{5.6}$$

such a small change that E_F is usually an adequate approximation for $E_F(T)$.

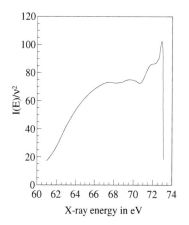

Fig. 5.4 Soft X-ray energy spectrum emitted by aluminium following ejection of a core electron. $I(E)$ is the X-ray intensity and ν the X-ray frequency. Adapted from Figure 2 from H. Neddermeyer and G. Wiech, Physics Letters 31A, 17 (1970).

The energy distribution of the free electrons can be inferred from measurements of the energy distribution of soft X-rays emitted following the removal of an inner shell electron. Figure 5.4 shows data for aluminium. The spread of energies is 11.7 eV, which gives a measurement of the Fermi energy. The sharp cut-off at the upper edge is consistent with the very limited smearing expected from the thermal excitation at room temperature, namely $k_B T \sim 0.025$ eV.

A fundamental question is how to take account of the Coulomb interactions between the electrons in a metal: this will depend on whether the Coulomb energy dominates the kinetic energy or not. Now, the Coulomb energy is of order $e^2/(4\pi\varepsilon_0 r_0)$ where r_0 is the mean spacing of the free electrons. Taking the kinetic energy from eqn. 5.4, setting

[1]Taking the electron effective mass to equal to the free electron mass.

$(3\pi^2)^{2/3}$ equal to 10, and using $n = r_0^{-3}$, the ratio of Coulomb to kinetic energy

$$r_s = \frac{e^2/(4\pi\varepsilon_0 r_0)}{5\hbar^2/mr_0^2} = 0.2r_0/a_0 \qquad (5.7)$$

where a_0 is the Bohr radius 0.0529 nm. Continuing with copper as our example r_0 is $1/n^{1/3}$ or 0.22 nm making r_s around 0.8; but r_s will be smaller for metals in general because the electron spacing is larger. Landau argued that in such a case the actual quantum states could be obtained from the non-interacting states by turning the interaction on adiabatically. This implies that the valence electrons in a metal can be treated as a quantum sea of electron-like quasi-particles.

5.2.1   Electrical conductivity

Using the electron gas model the equation of motion of an individual electron in an electric field E is

$$m\mathrm{d}v/\mathrm{d}t = -eE - mv/\tau, \qquad (5.8)$$

where v is the drift velocity produced by the action of the electric field. The last term represents the viscous drag exerted by lattice vibrations and impurities/defects, with τ being called the *relaxation time*. Classically τ would be interpreted as the average interval between collisions in which the electron is brought to rest. In equilibrium the drift velocity becomes constant with value $-eE\tau/m$. If there are n electrons per cubic metre the current density is

$$J = ne^2 E\tau/m, \qquad (5.9)$$

so that the conductivity is

$$\sigma_0 = ne^2\tau/m. \qquad (5.10)$$

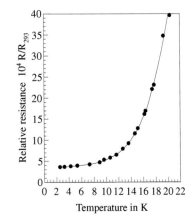

Fig. 5.5 Relative resistance of a pure sodium crystal. Adapted from Figure 4, D. K. C. MacDonald and K. Mendelssohn, Proceedings of the Royal Society A202, 103 (1950).

This expression for the DC conductivity survives in more refined analyses. In copper at room temperature σ_0 is $6\,10^7\,\mathrm{S\,m^{-1}}$, making τ equal to $2.5\,10^{-14}\,\mathrm{s}$. This gives a drift velocity of 44 mm s^{-1} in a field of 10 V m^{-1}. With a Fermi velocity in copper of $1.57\,10^6\,\mathrm{m\,s^{-1}}$ the electron travels 39 nm in the relaxation time of $2.5\,10^{-14}$s, which corresponds to about 180 atomic spacings in the lattice. A classical electron viewed only as a particle would be expected to have a mean free path comparable to the lattice spacing. In the quantum view the electrons interact with the positive ions, with the result that the electron waves are continuously Bragg scattered by the lattice. The effect is to produce stable electron waveforms that are in the form of modified plane waves. Electron-electron scattering is suppressed by the Pauli principle: because both the electrons involved must, after scattering, enter unoccupied states. It follows that electrons in metals principally scatter off phonons, that is to say off lattice excitations. In addition there is scattering from lattice defects, that is from crystal defects and impurity atoms. Figure 5.5 shows how the resistance of a pure metal changes with temperature. At high temperatures the phonon energies pile up around $k_\mathrm{B}\Theta_\mathrm{D}$ and their number increases with temperature. As a result the phonon wavelength and the electron Fermi wavelength, $2\pi/k_\mathrm{F}$, are of similar size: in copper λ_D is ~ 0.68 nm and λ_F is ~ 0.46 nm. This means that the momenta of the electrons active in conduction are similar to those of the phonons, and it follows that absorption and emission of phonons is effective in deflecting and stopping the conduction electrons. Consequently, at high temperatures, the dependence of the resistance on temperature is

$$R \propto n(\text{phonon}) \propto T. \qquad (5.11)$$

By contrast at very low temperatures the phonon energies are $\sim k_\mathrm{B}T \ll k_\mathrm{B}\Theta_\mathrm{D}$. Thus the phonons become less effective in scattering the electrons. The resistance of metals falls to a constant value at temperatures below ~ 20 K: the residual resistance is due scattering from impurities and lattice imperfections rather than phonons. The *residual resistance ratio* between resistance at room temperature and at liquid helium temperature ranges from 50 for everyday copper wire to around 10^5 for the purest defect-free copper crystals. In the latter case the mean free path at low temperatures is an astonishing 10^7 lattice spacings.

5.2.2   Thermal conductivity

The contribution of free electrons to thermal conductivity can be deduced by taking over an expression for thermal conductivity from the kinetic theory of gases:

$$K = v^2 \tau C_\mathrm{el}/3. \qquad (5.12)$$

Here C_el is the heat capacity of the electron gas per unit volume, v the electron velocity and τ the mean time between collisions. Thermal excitation leads to partially filled electron energy levels extending over a range $\sim k_\mathrm{B}T$ around the Fermi energy. Only the electrons in these levels carry heat and current so we take v to be v_F. The Pauli exclusion principle tethers the rest of the electrons in place. C_el was calculated as an exercise in Chapter 3, giving

$$C_\mathrm{el} = \pi^2 k_\mathrm{B}^2 nT/2E_\mathrm{F}, \qquad (5.13)$$

where n is the the electron number density. Then

$$K = \pi^2 k_\mathrm{B}^2 nT\tau/3m. \qquad (5.14)$$

Using eqn. 5.10, we obtain the *Wiedermann–Franz* law:

$$K/(\sigma_0 T) = \pi^2 k_\mathrm{B}^2/(3e^2) = 2.45\,10^{-8} \mathrm{W\,\Omega\,K}^{-2} \qquad (5.15)$$

where the constant is known as the *Lorenz number*. This relationship works quite well, at and above room temperatures for common metals, with the constant lying in the range 2.2 to $2.6\,10^{-8}\mathrm{W\,\Omega\,K}^{-2}$. At these temperatures electrons scatter from phonons with wave numbers around the lattice spacing and thus comparable to the electron wave numbers: scattering through large angles dominates and impedes thermal and electrical conduction equally. At lower temperatures collisions with the lower energy phonons only produce small angle deflection, which reduces the electron energy but not the charge carried. At very low temperatures only scattering from defects and impurities remains, which is elastic and affects the heat and charge transport equally: thus the Wiedermann-Franz law again holds.

5.2.3   Hall effect

In 1879, Hall discovered that when a magnetic field, B, is applied perpendicular to the current flowing in a conductor a voltage is induced across the conductor that is orthogonal to both the current and magnetic field directions. Figure 5.6 illustrates this with the broken arrowed lines showing how the electrons are steered by the magnetic field. They accumulate on one side face of the conductor until the electric field they exert cancels the lateral force on the current due to the magnetic field. In this equilibrium state the transverse electric field, E, is given by

$$vB = E, \tag{5.16}$$

where v is the velocity of the charges carrying the current. The current density

$$J = -nev \tag{5.17}$$

where there are n electrons per unit volume. Then

$$JB = -neE. \tag{5.18}$$

The Hall resistance is defined as

$$R_{\mathrm{H}} = E/J = -B/ne. \tag{5.19}$$

Now if the number density of atoms is N we have

$$-B/(R_{\mathrm{H}}Ne) = n/N, \tag{5.20}$$

which is the number of electrons per atom available to carry the current. The values obtained by measuring the Hall coefficient R_{H}/B are listed in Table 5.1 for various solids. The simplest cases to interpret are those of the alkali metals, sodium and potassium, each atom contributing a single electron from the valence shell. Negative values for solid beryllium and zinc and for many other elements defy any classical explanation. The quantum explanation is that the carriers are holes at the top of the valence band. We shall see that these holes behave like positively charged particles with positive mass. In the liquid state the energy bands become continuous, thus eliminating holes, and as a result liquid beryllium and zinc have positive values of R_{H}.

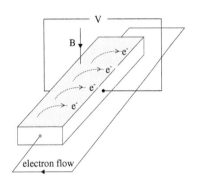

Fig. 5.6 Hall effect with electrons carrying the current.

Table 5.1 Hall valencies.

Element	Equivalent electrons/atom
Na	+0.9
K	+1.1
Cu	+1.3
Ag	+1.3
Au	+1.5
Be	-2.2
Zn	-2.9

5.3 Plasma oscillations

The response of the electron sea in a metal to electromagnetic radiation depends very much on the frequency. At frequencies comparable to, or greater than E_F/h the motion of the electrons is nearly free and results in regions of compression and rarefraction in the electron density. These regions exert restoring electric forces so that oscillations in charge density may occur. The free electrons and stationary ions form an overall neutral system known as a *plasma* and the oscillations are called *plasma oscillations*. By contrast at low frequencies, approaching the static limit, the electrons move to compensate the field changes more rapidly than these changes can develop. This effectively screens the individual electrons. Both these extreme cases will be analysed, starting with the response to high frequency radiation.

In an oscillating electric field $E = E_0 \exp(-i\omega t)$ the velocity of electrons is $v = v_0 \exp(-i\omega t)$. Inserting these quantities into eqn. 5.8 gives

$$v = -eE\tau/[m(1 - i\omega\tau)]$$

(5.21)

so that the conductance, using eqn. 5.10, is

$$\sigma = -nev/E = \sigma_0/(1 - i\omega\tau).$$

(5.22)

The imaginary component parametrizes the absorption of electromagnetic waves. To proceed further we use an expression for the displacement current

$$\partial D/\partial t = \varepsilon_0 \partial E/\partial t + \sigma E,$$

(5.23)

where D is $\varepsilon_0 \varepsilon_r E$ and ε_r is the dielectric constant (relative permittivity). Inserting the oscillating electric field in this expression gives

$$-i\varepsilon_0 \varepsilon_r \omega = -i\varepsilon_0 \omega + \sigma,$$

(5.24)

or

$$\varepsilon_r = 1 + \frac{i\sigma}{\varepsilon_0 \omega}.$$

(5.25)

An imaginary part of the relative permittivity means that the refractive index is complex: $n = \sqrt{\varepsilon_r} = n_r + in_i$. Thus $\varepsilon_r = (n_r^2 - n_i^2) + 2in_i n_r$. A plane wave travelling in the medium is

$$\exp[i(kx - \omega t)]$$
$$= \exp\{i[(\omega/c)(n_r + in_i)x - \omega t]\}$$
$$= \exp\{i[\omega n_r x/c - \omega t]\} \times$$
$$\exp[-w n_i x/c].$$

so that the intensity falls off with depth. Energy is absorbed through Joulean heating of the material.

Substituting for σ from eqn. 5.22 gives

$$\varepsilon_r = 1 + \frac{ine^2\tau}{\varepsilon_0\omega m(1 - i\omega\tau)} = 1 - \frac{\omega_p^2}{\omega^2 + i\omega/\tau}, \tag{5.26}$$

where

$$\omega_p = \sqrt{ne^2/\varepsilon_0 m}. \tag{5.27}$$

In metals at frequencies beyond the infrared $\omega\tau \gg 1$, and the equation reduces to

$$\varepsilon_r = 1 - \omega_p^2/\omega^2 = (\omega^2 - \omega_p^2)/\omega^2. \tag{5.28}$$

Evidently there is a striking change in response to electromagnetic waves as the angular frequency passes through ω_p. Below this frequency ε_r is negative and the refractive index is purely imaginary, indicating strong absorption: above this frequency ε_r is positive so that the refractive index is real. Figure 5.7 shows just this behaviour of the real and imaginary components of the refractive index for sodium around ω_p: absorption at longer *wavelengths* and refraction at shorter *wavelengths*. In the case of sodium, assuming one free electron per atom, the predicted wavelength corresponding to ω_p is 209 nm close to the observed value of 210 nm.

The reflection coefficient of a material at normal incidence is related to the real and imaginary parts of the refractive index, n_r and n_i, through the relation

$$R = \frac{(n_r - 1)^2 + n_i^2}{(n_r + 1)^2 + n_i^2}. \tag{5.29}$$

Below the plasma frequency a large value of n_i requires that R too is large and that most of the incident radiation is reflected. As a result the colours of metals depend sensitively on the plasma frequency. For example the plasma frequency of the electrons in silver lies in the ultraviolet, at 326 nm wavelength: hence silver absorbs and reflects strongly across the whole visible spectrum and looks white. With gold and copper the plasma wavelengths lie at 496 nm and 620 nm respectively so they only reflect the longer wavelengths, which gives them their characteristic gold and copper sheens. Plasma oscillations also occur in the earth's ionosphere. Taking an electron density $10^{16}\,\mathrm{m}^{-3}$, the plasma frequency is 9 MHz. As a result we can see the stars through the ionosphere, while we can also receive short wave radio transmissions from the other side of the planet because they are of low enough frequency to reflect off the ionosphere.

We can show that *longitudinal* oscillations of the electromagnetic field are possible in a plasma, with the electric field along the wave's direction of travel. Consider free oscillations without any external charges so that $\nabla \cdot \mathbf{D} = 0$, and take $\mathbf{D}$ to be zero. The electrons in the plasma are displaced mechanically a distance ξ from their equilibrium location. Then the local polarization per unit volume is $P = -ne\xi$, and the resulting electric field is

$$E = -P/\varepsilon_0 = ne\xi/\varepsilon_0. \tag{5.30}$$

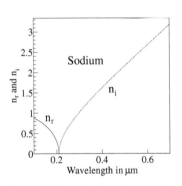

Fig. 5.7 The variation with wavelength of the real and imaginary parts of the refractive index of sodium.

Ignoring damping, the electrons' equation of motion is

$$md^2\xi/dt^2 = -eE = -ne^2\xi/\varepsilon_0. \tag{5.31}$$

This has an oscillatory solution

$$\xi = \xi_0 \exp(-i\omega_p t). \tag{5.32}$$

Because ω_p is the angular frequency at which the electrons oscillate whenever they are disturbed from equilibrium, it is called the *plasma angular frequency*. The quanta of plasma oscillations are known as *plasmons*. Plasmons can be excited by light or by the impact of electrons of high enough energy. Figure 5.8 shows the distribution in energy loss of electrons with initial energy 2.02 keV after scattering through 90° off a nanometre thick magnesium foil. The principal peaks lie at multiples of 10.6 eV. This is quite close to the plasma energy, $\hbar\omega_p$, of 10.9 eV calculated for magnesium assuming two free electrons per atom.

The electric D-field will only penetrate a short distance into a plasma at frequencies below the plasma frequency. A parallel effect occurs in superconductors, into which the magnetic B-field can only penetrate a short distance. Such effects are reinterpreted in Chapters 15 and 19 in the contexts of superconductivity and the the standard model of particle physics. It will become apparent that photons in these conditions behave as if they had become massive, very different from their behaviour in free space.

5.4 Electron screening

Free electrons in metals are only a distance of about 0.3 nm apart so that the Coulomb potential energy of a neighbouring pair is around 3 eV. This is comparable to the Fermi energy. The body of electrons responds to *slow changes* in the electric field so as to cancel these changes and this screens any individual charge. This illustrates one aspect of why the model of electrons in a metal as a non-interacting Fermi sea is successful. We now show that this compensation causes each mobile electron to become a quasi-particle accompanied by a *cloud* with net positive charge.

Any change in the electron density, δn, from the equilibrium value of n_0 involves a change in E_F given by differentiating eqn. 5.4

$$\delta E_F/E_F = (2/3)\delta n/n_0, \tag{5.33}$$

so that

$$\delta n = \frac{3n_0}{2E_F}\delta E_F. \tag{5.34}$$

If the electrostatic potential is ϕ, the energy required to add one electron is

$$E_F - e\phi. \tag{5.35}$$

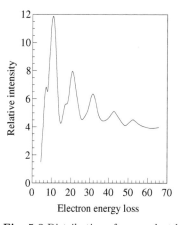

Fig. 5.8 Distribution of energy lost by electrons scattered through 90° from a magnesium foil. The incident electron energy was 2020 eV. Adapted from C. J. Powell and J. B. Swan, *Physical Review* 116, 81 (1959).

This is maintained constant by the good electrical contact with the environment and hence $\delta E_F = e\delta\phi$. Substituting this value of δE_F in the previous equation gives

$$\delta n = \frac{3en_0}{2E_F}\delta\phi. \tag{5.36}$$

Applying Gauss' law of electrostatics for the charge density $-e\delta n$,

$$\nabla^2\delta\phi = e\delta n/\varepsilon_0 = \frac{3e^2 n_0}{2E_F\varepsilon_0}\delta\phi. \tag{5.37}$$

With an isotropic medium, and using eqn. A.3, this becomes

$$\left[\frac{\mathrm{d}^2}{\mathrm{d}r^2} + \frac{2}{r}\frac{\mathrm{d}}{\mathrm{d}r}\right]\delta\phi = \delta\phi/r_0^2, \tag{5.38}$$

where r_0 is the *Fermi–Thomas screening length*

$$r_0 = \sqrt{2\varepsilon_0 E_F/(3e^2 n_0)}. \tag{5.39}$$

Hence the disturbance in ϕ falls off with distance r from a point charge like

$$\delta\phi = \frac{-e}{4\pi\varepsilon_0}\frac{\exp[-r/r_0]}{r}. \tag{5.40}$$

The Fermi–Thomas length changes only slowly with n_0, varying like $n_0^{-1/6}$. Now eqn. 5.36 becomes

$$\delta n = \frac{\varepsilon_0}{er_0^2}\delta\phi = \frac{-1}{4\pi r_0^2}\frac{\exp(-r/r_0)}{r}. \tag{5.41}$$

In copper r_0 has the value 0.055 nm, much smaller than the lattice spacing of 0.36 nm. It follows that the total charge displaced from around a fixed electron is

$$-e\int 4\pi r^2 \delta n \,\mathrm{d}r = +e, \tag{5.42}$$

which exactly compensates the electron charge. Each electron is no longer a single point charge as it is in free space, but rather an carries its own screening cloud. We have here another example of a *quasiparticle*: one that behaves like an electron but with an *effective mass* different from that of an electron in free space.

The screening length becomes the effective size of an electron as seen by other electrons. Therefore, the cross-section, σ_{ee}, for electron-electron scattering should be about $2\pi r_0^2$, that is $2\,10^{-2}\,\mathrm{nm}^2$. This would produce a mean free path

$$\lambda_{ee} = 1/(n\sigma_{ee}) \tag{5.43}$$

amounting to ~ 0.6 nm in copper, much shorter than the actual mean free path of 39 nm noted earlier. However most of the final states accessible energetically to the electrons after scattering are already occupied, and not available. The available empty states of importance are limited to a range in energy of $\sim k_B T$ below the Fermi level. Thus only a fraction $k_B T / E_F$ of the states are available to each of the two electrons after scattering. These fractions are around 0.0036 at 300K so that the mean free path for electron-electron scattering is increased by a factor $7\,10^4$. Scattering from phonons is only suppressed by one factor of $k_B T / E_F$, and this scattering process dominates, producing a shorter observed mean free path of 39 nm.

5.5   Electrons in crystalline materials

The free electron model provides a simple framework for explaining thermal and electrical properties of metals. However it cannot explain the energy bands shown in Figure 5.2, nor the distinctive properties of metals and semiconductors. In order to understand these features we have to take into account the interaction of the valence electrons with the array of positive ions. In essence the electron waves are Bragg scattered from the ions in the crystal lattice forming what are called *Bloch waves*. The particular crystal structure is crucial in determining the range of frequencies of Bloch waves that propagate freely across the crystal. From this comes the band structure and the variety of metallic and semiconductor behaviour. The analysis of band structure commences here with the introduction and analysis of Bloch waves.

5.6   Bloch waves

Our starting point is Bloch's theorem, which states that the solution to Schrödinger's equation for electrons moving in a regular lattice of ions has a simple general form that reflects the lattice structure. The wavefunctions of freely propagating electrons with angular frequency ω and wave-vector $\mathbf{k}$

$$\psi(\mathbf{r}, t) = \exp[i(\mathbf{k} \cdot \mathbf{r} - \omega t)], \tag{5.44}$$

are replaced by *Bloch waves*,

$$\psi_{\mathbf{k}}(\mathbf{r}, t) = u_{\mathbf{k}}(\mathbf{r}) \exp[i(\mathbf{k} \cdot \mathbf{r} - \omega t)]. \tag{5.45}$$

The function $u_{\mathbf{k}}(\mathbf{r})$ has the property that for *all* lattice vectors, $\mathbf{R}_j$, joining equivalent points in the lattice

$$u_{\mathbf{k}}(\mathbf{r} + \mathbf{R}_j) = u_{\mathbf{k}}(\mathbf{r}). \tag{5.46}$$

It follows that the waveforms $u_{\mathbf{k}}(\mathbf{r})$ can be expressed in terms of the reciprocal lattice vectors, $\mathbf{G}_i$ given in eqn. 4.29:

$$u_{\mathbf{k}}(\mathbf{r}) = \sum_i c_i(\mathbf{k}) \exp[i\mathbf{G}_i \cdot \mathbf{r}]. \tag{5.47}$$

Under a displacement of $\mathbf{R}_j$ the phase of this waveform changes by $\mathbf{R}_j \cdot \mathbf{G}_i$ which is always an integral multiple of 2π. These *Bloch waves*

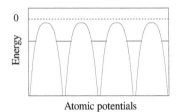

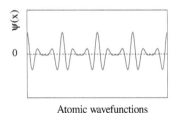

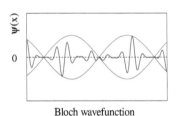

Fig. 5.9 Upper panel: potential wells
of atoms (blue) and electron bound
states (red). Mid panel: atomic wavefunctions. Lower panel: Bloch wavefunctions where the blue curve shows
the wave envelope.

are the stable equilibrium outcome of the Bragg scattering of electrons
off a uniform static lattice of ions (such perfect lattices would offer no
resistance to electron flow). In actual crystals above absolute zero the
electrons (Bloch waves) are scattered by the excitations of the lattice
(phonons), and also by any lattice defects and impurities.

Bloch wavefunctions which have the same energy, but whose wave-
vectors $\mathbf{k}$ and $\mathbf{k}'$ differ by some reciprocal lattice vector $\mathbf{G}$, are equivalent.
To show this we expand as follows

$$
\begin{aligned}
\psi_{\mathbf{k}}(\mathbf{r}) &= u_{\mathbf{k}}(\mathbf{r}) \exp[i\mathbf{k} \cdot \mathbf{r}] \\
&= u_{\mathbf{k}}(\mathbf{r}) \exp[i\mathbf{k}' \cdot \mathbf{r}] \exp(-\mathbf{G} \cdot \mathbf{r}) \\
&= \exp[i\mathbf{k}' \cdot \mathbf{r}] \sum_i c_i(\mathbf{k}) \exp[i(\mathbf{G}_i - \mathbf{G}) \cdot \mathbf{r}] \\
&= \exp[i\mathbf{k}' \cdot \mathbf{r}] \sum_i c_i(\mathbf{k}) \exp[i\mathbf{G}'_i \cdot \mathbf{r}],
\end{aligned}
\tag{5.48}
$$

where $\mathbf{G}'_i$ is a reciprocal lattice vector. Writing

$$
u_{\mathbf{k}'}(\mathbf{r}) = \sum c_i(\mathbf{k}) \exp[i\mathbf{G}'_i \cdot \mathbf{r}],
\tag{5.49}
$$

we arrive at the postulated equivalence

$$
\psi_{\mathbf{k}}(\mathbf{r}) = \psi_{\mathbf{k}'}(\mathbf{r})
\tag{5.50}
$$

with the periodicity preserved. The equivalence of two Bloch waves with
wave-vectors differing by a reciprocal lattice vector $\mathbf{G}$ can be ascribed
to the massive lattice absorbing the momentum $\hbar\mathbf{G}$ with a negligible
change in energy. Therefore, any Bloch wave with wave-vector and energy ending outside the first Brillouin zone is equivalent to some Bloch
wave with wave-vector lying inside this zone, as indicated in Figure
4.12 for phonons. In contrast to phonons in crystals, with energies constrained to be less than the Debye energy $k_B\theta_D$, electron band structure
extends upward to well beyond the Fermi level.

5.7 Tight and weak binding

The next step toward understanding the band structure is to infer how
bands and bandgaps arise in the dispersion relations for electrons in
crystals. The initial approximation made when calculating the mechanical and electromagnetic properties of metal and semiconductor crystals
is to assume that the electrons in the inner closed shells are unaffected
by the presence of other atoms, while the outer electrons move non-
relativistically in the electrostatic potential due to all the ions. In crystals of group I elements the single weakly bound s-shell valence electrons
spend little time close to an ion and are nearly free: this is a case of *weak
binding*. By contrast in crystals having strong covalent or ionic bonds
the valence electrons spend most of their time close to an ion. This *tight*

binding is considered now, using a one dimensional linear crystal as an example that brings out the salient features.

Figure 5.9 shows how the potential wells, single atomic orbits and Bloch wavefunction appear in one dimension. The spatial part of the Bloch wavefunction of an electron can be written

$$\psi(x) = \exp(ikx) \sum_j \exp[-ik(x - x_j)]\phi(x - x_j)/\sqrt{N}, \qquad (5.51)$$

where x_j is the coordinate of the jth ion and the wavefunction ϕ is very similar to the wavefunction of a bound outer shell electron. N is the number of atoms. Any separation $x_i - x_j$ is an integral multiple of the ion spacing a. The electrons' wavefunction has to be antisymmetric under interchange of a pair of electrons, which leads to an exchange term in the energy. The exchange energy is

$$E(k) = \frac{1}{N} \sum_{i,j} \exp[ik(x_i - x_j)] \int \phi^*(x - x_j)\hat{H}\phi(x - x_i)\mathrm{d}x, \qquad (5.52)$$

where $\hat{H}$ is the energy operator. There are N equivalent terms, and picking out that for one value of j and setting $x_j = 0$ gives

$$\begin{aligned}
E(k) &= \sum_i \exp[ikx_i] \int \phi^*(x)\hat{H}\phi(x - x_i)\mathrm{d}x \\
&= \int \phi^*(x)\hat{H}\phi(x)\mathrm{d}x \\
&\quad + \sum_{i \neq j} \exp[ikx_i] \int \phi^*(x)\hat{H}\phi(x - x_i)\mathrm{d}x. \qquad (5.53)
\end{aligned}$$

The first term in this expression is the energy of an outer electron in orbit around an ion and the second term is the *hopping energy* associated with the electron transfering from one ion to another. Only transfers between neighbouring ions are important because the wavefunctions fall off rapidly with distance. The electron energy operator is

$$\hat{H} = -\frac{\hbar^2}{2m}\frac{\mathrm{d}^2}{\mathrm{d}x^2} + v(x) + V(x), \qquad (5.54)$$

where $V(x)$ is the potential due to the nearest ion and $v(x)$ a smoother potential due to all the other ions. Then the energy associated with orbiting the parent ion is

$$\int \phi^*\hat{H}\phi\,\mathrm{d}x = E_0 + E_v, \qquad (5.55)$$

where E_0 is the energy in an isolated atom and E_v that due to distant ions. The remaining contribution to the energy is the hopping energy which involves an electron transfering from one ion to another. The dominant contribution involving nearest neighbours is

$$\exp[ika] \int \phi^*(x) V(x-a) \phi(x-a) \, \mathrm{d}x$$

$$+ \quad \exp[-ika] \int \phi^*(x) V(x+a) \phi(x+a) \, \mathrm{d}x$$

$$= \quad 2E_H \cos[ka], \tag{5.56}$$

where $E_H = \int \phi^*(x) V(x \pm a) \phi(x \pm a) \, \mathrm{d}x$. This integral depends strongly on the overlap of the wavefunctions of electrons orbiting neighbouring ions. Thus

$$E = E_0 + E_v + 2E_H \cos[ka], \tag{5.57}$$

which is plotted in Figure 5.10. Evidently the larger the hopping energy will be, the wider the energy band will be.

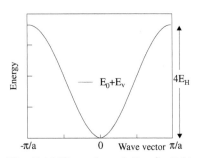

Fig. 5.10 Dispersion relation for tight binding of a linear chain of atoms. The energies appearing are discussed in the text.

5.7.1   Bands and kinematics

We now develop this model of bands and explore the resulting electron dispersion relations. There are important features common to weak and tight binding which are discussed here. First we examine the behaviour at the zone boundaries. Superposition of the Bloch waves which only differ through travelling in opposite directions gives a standing wave thus:

$$\psi_\pm = [\exp(ikx) \pm \exp(-ikx)] u_k(x) = 2[\cos(kx) \text{ or } i\sin(kx)] u_k(x). \tag{5.58}$$

At the zone boundary the sinusoids become $\cos[\pi x/a]$ or $\sin[\pi x/a]$. These functions are shown in Figure 5.11, with the locations of the ion cores indicated by dots. The antinodes of the wavefunction ψ_+ coincide with the ion cores and this state will have the minimum energy, while ψ_- has the least overlap with the cores and hence the maximum energy. The difference is

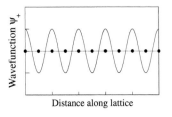

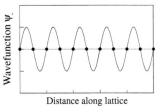

Fig. 5.11 Electron standing waves on the lattice.

$$\Delta E = \frac{\int_{-a/2}^{+a/2} V(x)[\cos^2(\pi x/a) - \sin^2(\pi x/a)] \, \mathrm{d}x}{\int_{-a/2}^{+a/2} \cos^2(\pi x/a) \, \mathrm{d}x}. \tag{5.59}$$

There is therefore an *energy bandgap* which is larger the tighter the binding.

For example consider the group IV elements, C(diamond), Si, Ge and Sn, whose crystals have in common the tightly bound interpenetrating fcc lattice drawn in Figure 4.11. Their band gaps and lattice spacings are shown in Table 5.2. Reading downward in the table, the strength of the binding steadily weakens as the lattice spacing increases, and the bandgap shrinks correspondingly. In tin the bands overlap making it a metal. Silicon and germanium are semiconductors, while diamond, with its much larger band gap, is an insulator.

Table 5.2 Group IV gaps and lattice spacings.

Element	gap in eV	a in nm
C(diamond)	5.0	0.36
Si	1.1	0.54
Ge	1.0	0.57
Sn	metallic	0.65

The dispersion relation of weakly coupled outer electrons is shown by the solid line in the upper panel of Figure 5.12. Away from the bandgap these follow the broken line showing the parabolic dispersion relation for free electrons, $E = \hbar^2 k^2/(2m)$. Whenever the wave-vector k equals the value of any reciprocal lattice vector there can be a bandgap. It was proved in Section 5.6 that Bloch waves differing by a reciprocal lattice wave-vector are equivalent; therefore the dispersion relations can be re-drawn with each segment displaced by an integral number of reciprocal lattice vectors so that it ends up in the first Brillouin zone. The resulting, fully equivalent *reduced zone scheme*, is shown in the lower panel of Figure 5.12. Projecting the curves in this plot (the dispersion relations) onto the energy axis exhibits the band structure and the density of states within the bands. It follows that the density of states, in this one dimensional example, rises sharply as the edge of any band gap is approached: the states lost from the bandgap pile up at its edges.

The group velocity of electrons is given in terms of wave parameters by

$$v = \mathrm{d}\omega/\mathrm{d}k = \mathrm{d}E/\mathrm{d}[\hbar k]. \tag{5.60}$$

It can be seen from Figure 5.12 that as k tends to zero the velocity and momentum also tend to zero. However at the zone boundary, while the electron velocity and momentum drop to zero, as expected for standing waves, k does not vanish. Thanks to an electron's interactions with the lattice ions $\hbar k$ is in general no longer the same as its momentum. Differentiating the above expression with respect to time gives the acceleration:

$$\dot{v} = \frac{\mathrm{d}}{\mathrm{d}t}\left[\frac{\mathrm{d}(E/\hbar)}{\mathrm{d}k}\right] = \frac{\mathrm{d}k}{\mathrm{d}t}\left[\frac{\mathrm{d}^2(E/\hbar)}{\mathrm{d}k^2}\right]. \tag{5.61}$$

Rearranging this equation gives

$$\frac{\hbar^2 \dot{v}}{\mathrm{d}^2 E/\mathrm{d}k^2} = \hbar\frac{\mathrm{d}k}{\mathrm{d}t}. \tag{5.62}$$

Fortunately we can show that $\hbar \mathrm{d}k/\mathrm{d}t$ is still the response of an electron to the force due to an applied electromagnetic field. Taking an applied electric field E_{em} the rate of change of an electron's energy is

$$\mathrm{d}E/\mathrm{d}t = -eE_{em}v. \tag{5.63}$$

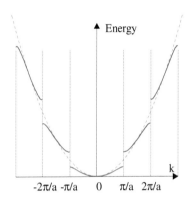

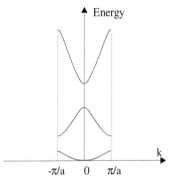

Fig. 5.12 Comparison of the weak binding dispersion relation with that for free electrons in the upper panel. In the lower panel the dispersion relation is shown in the reduced zone scheme.

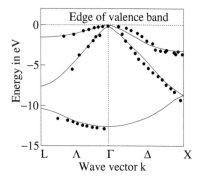

Edge of valence band

Fig. 5.13 Valence band dispersion relations for germanium. Adapted from Figure 2, A. L. Wachs, T. Miller, T. C. Hsieh, A. P. Shapiro and T.-C. Chiang, *Physical Review* B32, 2326 (1985). Courtesy Professor Chiang.

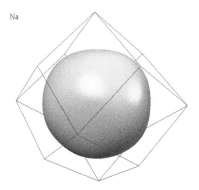

Fig. 5.14 Fermi surface in sodium. Courtesy T.-S. Choy, J. Naset, J. Chen, S. Hershfield and C. J. Stanton: Physics Department, University of Florida. Their collection of Fermi surfaces is accessible at www.phys.ufl.edu/fermisurface/.

Now from above

$$dE = [dE/dk]\,dk = \hbar v dk. \tag{5.64}$$

Comparing the last two equations

$$\hbar v dk = -eE_{em}v dt, \tag{5.65}$$

whence we get

$$\hbar dk/dt = -eE_{em}, \tag{5.66}$$

as anticipated. $\hbar k$ is therefore called the *crystal momentum*. Now looking back at eqn 5.62 the coefficient of $\dot{v}$ must be the electron's *effective mass* m^*, so that

$$\frac{1}{m^*} = \frac{1}{\hbar^2}\left[\frac{d^2E}{dk^2}\right]. \tag{5.67}$$

If m^* is large then the electron is more sluggish to move than a free electron. Referring again to the reduced zone scheme shown in Figure 5.12 we see that d^2E/dk^2 is negative near the zone boundary for the first and third bands and over much of the second band. In these kinematic regions the electron has *negative effective mass*. We shall see shortly that this leads to holes in the valence band having a conventional positive effective mass. Near $k = 0$ the curvature of the bands can be sharp so that the effective mass is then very much smaller than the free electron mass. This results in high electron mobility, which, together with small device size, makes possible *the fast response of the semiconductor devices used in modern electronics*.

Crystal structure in three dimensions is such that the electrostatic force acting on an outer electron is not isotropic. Hence the dispersion relations depend in general on the direction of the three-dimensional wave-vector relative to the crystal axes. In addition with tight binding the dispersion relations become markedly different from free electron parabolae. The results of calculations of dispersion relations for electrons in germanium by Chelikowsy and Cohen are compared in Figure 5.13 with measurements made by Wachs *et al.* using the technique of angle resolved photoelectron spectroscopy, described in the following section. The dispersion relations shown correspond to directions in the fcc lattice labelled according to the notation used in Figure 4.13. In general the degree of complexity of dispersion relations in three-dimensional crystals for electrons is similar to that seen with phonons in Figure 4.14.

The anisotropy of crystal lattices implies that even at absolute zero the boundary of the wave-vector states filled by electrons will not be exactly spherical. Examples of the shape of the Fermi surface[2] are given for sodium and copper in Figures 5.14 and 5.15. In sodium the Fermi

[2] A collection of Fermi surfaces made by T.-S. Choy, J. Naset, J. Chen, S. Hershfield and C. J. Stanton, Physics Department, University of Florida is accessible at www.phys.ufl.edu/fermisurface/. See also *Bulletin of the American Physical Society* 45(1), L3642 (2000).

surface is nearly spherical indicating that the electrons are nearly free, while in copper the effects of binding are more important. In the case of copper the occupied states reach the first zone boundary over a limited angular range. The Fermi surface reaches the boundary of the first zone at regions with circular shape. Copper is by no means the element with the most complex Fermi surface. If it happens that the Fermi surface coincides with the boundary of the first Brillouin zone over a fraction of its area then conduction is suppressed over this fraction by the bandgap. Conductivity in the pentavalent elements arsenic, antimony and bismuth is reduced in this way by a factor 20 compared to copper. These elements are therefore called *semi-metals*.

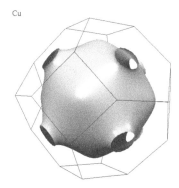

Cu

Fig. 5.15 Fermi surface in copper. Courtesy T.-S. Choy, J.Naset, J. Chen, S. Hershfield and C.Stanton: Physics Department, University of Florida. Their collection of Fermi surfaces is accessible at www.phys.ufl.edu/fermisurface/.

5.8   ARPES

The widely used method for measuring the dispersion relations of electrons in crystalline solids is *angle resolved photoelectron spectroscopy, ARPES*. A collimated monoenergetic beam of X-ray photons is directed at one or other crystal planes of the sample and the energies of the photoelectrons are measured over a wide range of emission angles. Making measurements for the various crystal planes of the sample provides complementary information.

Usually the incident X-ray photon energy, $\hbar\omega$, is in the range of 5 eV to a few hundred eV. The transition generated by the X-rays is shown in Figure 5.16. Electron energies inside the crystal are referred to the lowest point of the band containing the electron initially, and the reference energy in free space is that of an electron at rest: the difference between these reference energies is the sum of the work function W and the Fermi energy E_F. Thus the electron's initial energy, expressed in terms of the energy E, with which it emerges, is

$$E_i = E - \hbar\omega + E_F + W. \tag{5.68}$$

The electron's wavelength is comparable to the lattice spacing $\sim 10^{-10}$ m while the photon wavelength is $\sim 10^{-6}$ m. Hence the absorption of the photon does not alter the electron momentum appreciably. Let the components of the initial electron momentum resolved parallel and perpendicular to the surface be $\hbar k_\parallel$ and $\hbar k_\perp$ respectively. Then

$$E = [\hbar^2/(2m)][k_\parallel^2 + (k_\perp)^2] - V_0, \tag{5.69}$$

where $V_0 = E_F + W - E_0$. E_0 is the unknown energy at the base of the upper band. If the electron emerges at an angle θ to the surface normal

$$\hbar k_\parallel = \sqrt{2mE}\sin\theta, \tag{5.70}$$

and inserting this value into eqn. 5.69 we get

$$\hbar k_\perp = \sqrt{2m(E\cos^2\theta + V_0)}. \tag{5.71}$$

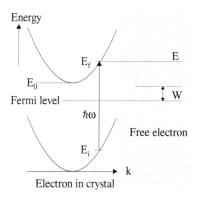

Fig. 5.16 Energetics of ARPES. Energies are measured from the bottom of the lower band in the crystal, and from rest in free space.

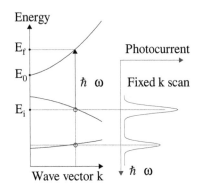

Fig. 5.17 ARPES measurements. Energy levels in the crystal are shown to the left and to the right the resulting photocurrent as a function of photon energy in a scan at fixed wave-vector.

The quantities θ and E are those measured for each electron. The other unknown E_0 can be inferred by making a measurements with electrons emerging along the surface normal so that $k_\parallel$ vanishes. As shown in Figure 5.17 an energy scan at a fixed value of the wave-vector intersects the dispersion relations at the points circled, and produces the electron count distribution shown to the right. A picture of the band structure in one crystal direction is obtained by repeating the scan for a range of wave-vectors.

Figure 5.18 is a sketch of a representative ARPES experiment. The specimen under study has to be freshly prepared and kept under high vacuum so that surface contamination is avoided. Synchrotron radia-

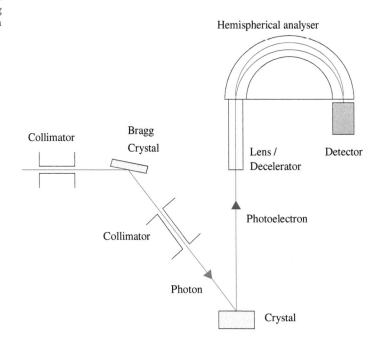

Fig. 5.18 ARPES apparatus set to observe electrons emitted normal to surface.

tion from electrons in an accelerator is the standard X-ray source. A tight beam of monoenergetic photons is directed at the specimen. The emerging photoelectrons are focussed electrostatically onto the slit of the hemispherical analyser. This analyser consists of two concentric metal hemispheres with radii R_1 and R_2 across which a potential V is applied, giving a radial electric field. Electrons of energy $eV[R_1R_2/(R_1^2-R_2^2)]$ follow a circular path between the hemispheres and only they pass through the exit slit diametrically opposite.

The electrons passing through the exit slit of the analyser are first multiplied in a microchannel plate by a factor 10^4, then impinge on a phospor whose light is detected by a CCD similar to those in digital cameras. A scan of the full range of photoelectron energies is obtained by taking data with successive settings of the deceleration before the analyser. The resolution obtained in $k_\parallel$ and E_i is around 10^{-3}.

Figure 5.19 shows the dependence of the mean free path with electron energy averaged for a typical material. Experiments at energies away from the minimum in Figure 5.19 can probe the bulk electron states. The properties of nanometre thick layers of semiconductors are of considerable interest to industry and in research, and for these ARPES is ideally suited. The motion of the electrons within the sample is two dimensional and the electron struck can easily escape without scattering.

5.9   Outline of calculation of band structure

The Hartree–Fock technique introduced in Chapter 3 for calculating wavefunctions of electron orbitals in multi-electron atoms is adapted for the purpose of calculating electron dispersion relations in crystals. The need to solve the $3N$-dimensional Schrödinger equation in condensed matter is approximated by what is called *density functional theory*, *DFT*. DFT is based on general theorems proved by Hohenberg and Kohn.[3] These theorems make it feasible to use electron density distributions rather than wavefunctions in calculating band structure. A $3N$-dimensional problem is then reduced to a three-dimensional problem. Kohn and Sham simplified matters by introducing so-called non-interacting electron orbitals ϕ_i, where the interaction appears as a potential $V_{\mathrm{exc}}[n]$ dependent on the electron density distribution $n(\mathbf{r})$. This potential contains exchange and correlation effects. The resulting equations for the orbitals

$$\left[-\frac{\hbar^2}{2m}\nabla^2 + V + \frac{e^2}{4\pi\varepsilon_0}\int \frac{n(\mathbf{r}')\mathrm{d}\mathbf{r}'}{|\mathbf{r}-\mathbf{r}'|} + V_{\mathrm{exc}}[n]\right]\phi_i(\mathbf{r}) = E_i\phi_i(\mathbf{r}), \quad (5.72)$$

resemble the Hartree–Fock equations. V is the potential due to the ion cores. DFT has had qualitative success reproducing the general features of the band structure. Bond lengths between ions, bulk moduli and the phonon frequency distribution are well determined.

5.10   Further reading

Band Theory and Electronic Properties of Solids by J. Singleton, published by Oxford University Press (2001) pursues these topics at greater depth in a helpful style.

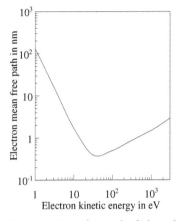

Fig. 5.19 Mean free path of photoelectrons versus their energy. This shows the trend followed by a variety of materials.

[3]For more details see W. Kohn, *Reviews of Modern Physics* 71,1253 (1998).

Exercises

(5.1) Show that the mean kinetic energy of an electron in a free electron gas at low temperatures is $\frac{3}{5}E_F$. Use this expression to calculate the bulk modulus, $-V(\partial p/\partial V)$ of conductors at low temperatures. Compare the value this gives in the case of sodium to the measured value of 6.3 GPa. The electron density is $2.65\,10^{28}\,\mathrm{m}^{-3}$ and E_F is 3.23 eV.

(5.2) Taking the plasma frequency of the electrons in the upper atmosphere to be 9 MHz, what is the electron density?

(5.3) Derive an expression for the minimum energy required to excite an electron out of its ground state in a one-dimensional infinite square well potential of width d. Compare this with the Fermi energy of a two-dimensional free electron gas and decide whether the electron motion in copper sheets of thickness 0.25 nm and 2.0 nm can be considered two-dimensional. The free electron density in copper is $8.45\,10^{28}\,\mathrm{m}^{-3}$.

(5.4) ^{3}He has a density of 81 kg m^{-3} near 0 K. Determine the Fermi energy, wavelength and temperature. ^{3}He atoms are fermions with spin 1/2 like electrons.

(5.5) An electron with wave-vector k_F equal to $1.36\,10^{10}\,\mathrm{m}^{-1}$ absorbs a phonon with the Debye momentum. The Debye temperature is 1000 K and the phonon velocity $10^4\,\mathrm{ms}^{-1}$. What is the angle the electron scatters through?

(5.6) Prove that the energies accepted by the hemispherical analyser in ARPES are centred on $eVR_1R_2/(R_2^2 - R_1^2)$. Assume that the potentials on the hemispheres are the same as those on spheres with equal and opposite charges at their centres.

(5.7) A magnetic field B is applied along the z-direction and an electric field E along the x-direction to a metal with a damping force $-mv/\tau$, v being the electron velocity. What is the motion when these forces are in balance?

(5.8) Consider free electrons travelling in paths perpendicular to a magnetic field B T. Show that the angular frequency of their circular orbits $\omega_c = eB/m$. Show that the areas of the orbits in wave number space S and physical space A are related by

$$A = [\hbar/eB]^2 S.$$

Electrons in a uniform magnetic field follow quantized *Landau orbits* with the magnetic flux through the orbit quantized in units of h/e. The nth Landau orbit containing this flux has an energy $(n + 1/2)\hbar\omega_c$, where ω_c is the cyclotron angular frequency. Show that for the nth Landau orbit

$$1/B = 2\pi(n + 1/2)e/(S\hbar).$$

(5.9) The magnetic moment of a conductor at absolute zero is simply $M = -\partial E/\partial B$, where E is the energy and B the applied magnetic field. Show that M changes cyclically whenever B is increased so that the change in $1/B$ is $\Delta = 2\pi e/(S\hbar)$.

(5.10) A uniform magnetic field is applied to a copper crystal perpendicular to one of the extremal orbits where the Fermi surface meets the zone boundary in Figure 5.15. Oscillations of the magnetic moment are observed with each step Δ in $1/B$ corresponding to this orbit. Can you explain this effect, known as the *de Haas–van Alphen* effect. Why are the orbits away from the Fermi surface not involved, and why are no other orbits on the Fermi surface involved?

(5.11) The Fermi energy of an electron in copper is 7 eV and the equivalent temperature is 81000 K. How is it that one is not burnt by copper? Are there phonons in molten copper?

(5.12) In a semiconductor the current density J where there is a number density gradient in the electron or hole population is

$$J = -De[dn/dx],$$

where D is the *diffusion coefficient*. Write an equation for the current flow when there is an applied electric field taking into account the diffusion. Thermal equilibrium should be assumed. Thence show that

$$D/\mu = k_B T/e,$$

where μ is the electron mobility.

Semiconductors

6

6.1 Introduction

The importance of silicon and other semiconductors such as gallium arsenide in electronics and optoelectronics has resulted in the industrial scale production of crystals of these semiconductors with negligible defects and precisely controlled levels of impurities. Material is laid down from the vapour or liquid phase, or using molecular beams *epitaxially*, that is to say atomic layer by atomic layer onto the substrate with near perfect registration of the crystalline form. For example, the commercial fabrication of silicon crystals with only one displaced bond in 10^{12} is commonplace. The bandgaps of silicon, germanium and gallium arsenide are $1.1\,\mathrm{eV}$, $0.67\,\mathrm{eV}$ and $1.43\,\mathrm{eV}$ respectively, moving the threshold wavelengths for promoting electrons into the conduction band into the near infrared. This makes these semiconductors valuable in producing and detecting light and near infrared radiation, the latter crucial for the optical fibre communication underpinning the internet.

Holes and doping with impurities are of importance in the operation of semiconductor devices, and will considered in first. Semiconductor diodes, photodiodes, and light emitting diodes, LEDs, are introduced next. The following section starts with an account of the *MOSFET* and *HEMT* transistors whose operation requires the provision and control of a two-dimensional electron gas, a *2DEG*. MOSFET indicates a Metal–Oxide–Semiconductor Field Effect Transistor; HEMT indicates a High Electron Mobility Transistor.

The layers of semiconductor in devices such as microprocessor chips can be sculpted using photolithography or electron beams to give features as small as $10\,\mathrm{nm}$. Devices on the scale of $10\,\mathrm{nm}$ to $1\,\mu\mathrm{m}$, neither microscopic nor macroscopic, are called *mesoscopic*. A description of mesoscopic devices and their novel quantum properties rounds out the chapter.

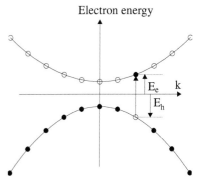

Fig. 6.1 Electron valence and conduction band energy states. A photon has been absorbed by an electron that enters the conduction band.

6.2 Electrons, holes and doping

Close to the lowest point on the conduction band the electron dispersion relation is similar to that of a free electron:

$$E_e = \hbar^2 k^2/(2m_c^*) + E_c, \qquad (6.1)$$

Quantum 20/20: Fundamentals, Entanglement, Gauge Fields, Condensates and Topology.
Ian R. Kenyon. © 2020. Published in 2020 by Oxford University Press.
DOI: 10.1093/oso/9780198808350.001.0001

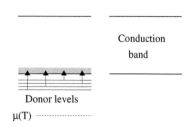

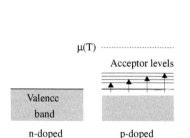

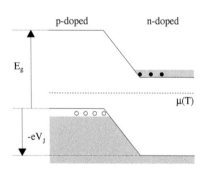

Fig. 6.2 Energy levels produced by n- and p-doping. The arrows indicate thermal excitation of electrons.

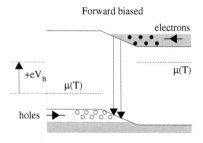

Fig. 6.3 Energy diagrams for a junction diode; with no bias and with forward bias V_b.

where m_c^* is the electron effective mass and E_c the energy at the bottom of the conduction band. When an electron is excited into the conduction band from the valence band a hole appears in the continuum of occupied electron states as shown in Figure 6.1. The hole carries a momentum $-\hbar\mathbf{k}$ because adding an electron of momentum $\hbar\mathbf{k}$ would fill the valence band which when full has *no net momentum*. For the transition drawn the energy required is $E_e + E_h$, both energies being positive. The hole has a positive energy, E_h, and this increases the lower the hole lies in the valence band. Hence the sense of the energy axis for holes is the reverse of that for electrons. The upper end of the valence band is approximately quadratic, so that the hole energy is given by

$$E_h = \hbar^2 k^2/(2m_h^*) + E_v, \tag{6.2}$$

where E_v is the energy of a hole at the top of the valence band, and m_h^* is its effective mass. m_h^* has to be positive because the hole's energy is positive. The final property to be established is the charge carried by a hole, whether it is negative or positive. Referring to Figure 6.1, consider applying an electric field such that the electrons move rightward making $\hbar k$ more positive. As they slide along the dispersion curve they carry the hole with them. This leads to the value of $-\hbar k$ carried by the hole becoming more negative. Consequently a hole carries a charge equal and opposite to that of an electron.

The semiconductors of greatest interest here are silicon with bandgap 1.1 eV and the $\mathrm{Al}_x\mathrm{Ga}_{1-x}\mathrm{As}$ alloys with bandgaps ranging from 1.43 eV to 2.16 eV as x ranges from zero to unity. At 300 K the average thermal excitation is $\sim$0.025 eV so that very few electrons cross these band gaps. The *intrinsic* conductivity is supplemented by *doping* semiconductors with atoms of elements that supply additional electrons or holes. Take the case of silicon, a group IV element with four valence electrons, and consider adding gallium atoms with three valence electrons. Each gallium atom provides an empty level about 0.01 eV above the top edge of the valence band. This is known as *p-doping* using *acceptor* atoms and is shown on the right in Figure 6.2. Those states occupied by electrons are shaded in the two adjacent figures. Electrons from silicon atoms are thermally excited into the acceptor levels. The holes left behind in the valence band are mobile and available to carry current. Conversely *n-doping* with arsenic, a group V element having five valence electrons, provides occupied levels in the band gap about 0.01 eV below the bottom of the conduction band. This case appears on the left in Figure 6.2. Electrons thermally excited from these *donor* levels into the conduction band are available for conduction. In n-doped (p-doped) silicon the electrons (holes) are the *majority* charge carriers. One dopant atom in 10^6 silicon atoms is sufficient to double the intrinsic conductivity. With heavy doping of one in 10^4 the conductivity of silicon approaches that of a metal, and this provides a simple way to make electrical contacts.

6.3 Diode operation

Diodes are the simplest semiconductor circuit elements, having low resistance for one sense of applied voltage (forward bias) and extremely high resistance for a voltage in the other sense (reverse bias). They are thus rectifiers, only allowing current to flow in one direction. Figure 6.3 shows the energy levels of a semiconductor diode at the junction between the p-doped and n-doped regions. In the upper panel no voltage has been applied across the junction. Electrons flow from the n-doped region to the p-doped region until the Fermi levels are equal on either side of the junction. In this way an equilibrium contact potential V_j is established: it then requires an energy eV_j to move an electron from the top of the valence band in the n-doped region to the top of the valence band in the p-doped region. The lower panel shows the effect of applying a positive voltage V_b, of order $1\,\text{V}$ to the p-doped region. This *forward bias* flattens the voltage ramp: the result is that electrons entering the junction from the n-doped region fill holes arriving from the p-doped region. The resulting current is called the *recombination* current. Current due to thermal excitation of electrons across the junction is called *thermal* current. Electron and hole number densities across the junction are shown for forward biasing in Figure 6.4 using a logarithmic scale. Here the notation follows this pattern: n_p is the electron density in the p-doped region where the electrons are the minority carriers; and p_n is the hole density in the n-doped region. The grey shading marks the *depletion layer* in which the applied field has removed most of the free electrons and holes, leaving the dopant ions. In equilibrium the densities are constrained by the semiconductor equations[1]

$$n_p p_p = n_n p_n = n_i^2, \qquad (6.3)$$

where n_i is the carrier density in the undoped *intrinsic* semiconductor. In silicon n_i is $10^{16}\,\text{m}^{-3}$ so that a doping of typically $10^{21}\,\text{m}^{-3}$ heavily suppresses the minority carriers. At large distances from the junction the densities are constant.

The current depends on the bias voltage and temperature. This relationship is deduced by applying Boltzmann's equation[2] to the number densities of holes and electrons at the edges of the depletion layer. We ignore generation and recombination of carriers in the depletion layer. Without any bias

$$n_p = n_n \exp\left[\frac{-eV_j}{k_B T}\right], \qquad (6.4)$$

and with a forward bias V_b

$$n_p + \Delta n_p = n_n \exp\left[\frac{-e(V_j - V_b)}{k_B T}\right]. \qquad (6.5)$$

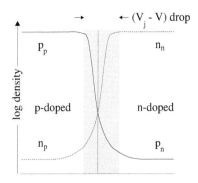

Fig. 6.4 Carrier distributions across a p-n junction under forward bias. The depletion layer is shaded.

[2]The energies are sufficiently large compared to $k_B T$ that the Fermi–Dirac distribution converges to the classical distribution.

---

[1]The proof is supplied in one of the exercises.

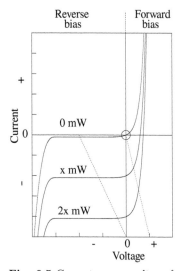

Fig. 6.5 Current versus voltage for a photodiode. The upper curve shows the variation in the absence of illumination, corresponding to a standard diode. The lower curves are for two levels of illumination.

Eliminating n_n from these two equations gives

$$\Delta n_p = n_p \left(\exp \left[\frac{eV_b}{k_\mathrm{B}T} \right] - 1 \right), \qquad (6.6)$$

with a similar expression for Δp_n at the n-doped edge of the depleted region. The excess charge diffuses steadily into the bulk giving a current proportional to the excess charge density

$$I_0 = I_r \left(\exp \left[\frac{eV_b}{k_\mathrm{B}T} \right] - 1 \right). \qquad (6.7)$$

Figure 6.5 shows the current versus voltage plots for a diode. The upper red curve corresponds to the response described by eqn. 6.7. Current increases exponentially with increasing forward bias. Under low reverse bias voltages the reverse current, I_r is tiny in magnitude and fairly constant. However at some critical reverse bias Zener breakdown occurs: the charges are accelerated sufficiently so that they acquire enough energy to initiate ionization of the semiconductor atoms. This leads to an avalanche of ionization and a rapid rise in current. As a result the reverse bias voltage is held at the breakdown value. *Zener diodes* are designed with specific breakdown voltages to limit applied voltages to this predetermined value.

6.3.1   Photon detectors and LEDs

All the absorption coefficients for electromagnetic radiation in silicon, GaAs, and $In_{0.5}Ga_{0.5}As$ rise sharply when the photon energy increases through the bandgap energy. An electron in the valence band can absorb a photon of higher energy and enter the conduction band. This electron and the hole it has left in the valence band are available to carry current. The resulting diode current is proportional to the number of photons incident with energy above the threshold. In the cases of silicon, GaAs and $In_{0.5}Ga_{0.5}As$ the threshold wavelengths lie in the near infrared at 1.1 μm, 0.9 μm and 1.6 μm. This makes diodes constructed from these semiconductors ideal for detecting radiation transmitted over communications optical fibres: serependitiously optical fibres can be made to have low absorption across the wavelength range 0.8 to 1.8 μm. Such diodes are in constant use detecting the radiation that carries terabits of data per second on optical fibre across the Atlantic and Pacific oceans.

Refering back to Figure 6.5 the two blue response curves in Figure 6.5 show the characteristics of a silicon photodetector for two levels of the incident light intensity. Reverse bias is applied so that the current is limited to I_r in the absence of any illumination. Incident light produces electron-hole pairs in the junction region which flow as current in the external circuit. The broken blue line appearing in the lower left quadrant of the figure shows the current versus voltage *load line* when a fixed voltage is applied across a resistance in series with the diode. In this simple circuit the current is linearly proportional to the incident photon flux over a wide range of intensity. A second load line in the bottom right hand quadrant corresponds to a circuit in which a load is connected directly across the diode. In this simpler configuration the diode is operated as a photovoltaic cell. Photovoltaic cells which produce current and voltage when exposed to sunlight are operated in this mode with the load generally chosen to maximize the power (IV) output.

Light emitting diodes, LEDs, utilize the reverse of the detection process. An applied forward bias voltage causes majority carrier holes and electrons to enter the depletion layer, and there they annihilate giving photons. Figure 6.6 compares such processes occuring in silicon and gallium arsenide. The visible and near-visible photons emitted have relatively small momenta and wave numbers ($\sim 10^6\,\mathrm{m}^{-1}$) compared to electrons and holes, ($\sim 10^{10}\,\mathrm{m}^{-1}$). Hence if momentum is to be conserved in the annihilation the electron and hole must have almost equal and opposite wave numbers. In the case of GaAs, shown in the upper panel of Figure 6.6, the transition is of this *direct* type, with the electron dropping vertically in the diagram to reach the hole. On the other hand, the transition in silicon, shown in the lower panel, requires a wave-vector change of $10^{10}\,\mathrm{m}^{-1}$ which means that a *phonon* must accompany the photon in order to conserve overall momentum. The direct transition in GaAs is fast taking $\sim 1\,\mathrm{ns}$ and dominates other processes. By contrast the *indirect* transition in silicon is much slower, taking $\sim 1\,\mathrm{ms}$ so that most of the injected charge annihilates in non-radiative processes. It follows that neither LEDs nor lasers can be made from silicon. This is a significant limitation for modern silicon electronics. Ideally communication between electronics modules would be carried over optical fibre by photons produced as well as detected in those modules themselves. Unfortunately the electron-to-photon conversion requires devices made from non-silicon material, either GaAs or similar alloys.

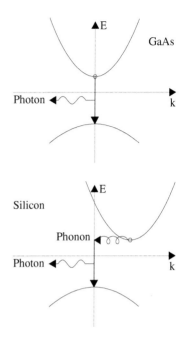

Fig. 6.6 The dispersion relations of GaAs in the upper panel and silicon in the lower panel. Radiative transitions are shown.

6.4   MOSFETs and HEMTs

MOSFET (metal-oxide-semiconductor field effect transistors) and HEMTs (high electron mobility transistors) are devices whose function involves control of a two-dimensional electron gas. Figure 6.7 shows a section through an n-channel MOSFET field effect transistor. This has a p-doped silicon substrate, [S], which isolates heavily n-doped regions (labelled n^+) below the source and drain electrodes. The substrate is biased negative with respect to the source and drain, and the drain is held a few volts positive with respect to the source, which is typically at ground potential. Current flow between the source and drain is controlled by the metal gate electrode, [M], separated from the substrate by an insulating silicon dioxide layer, [O]. If the gate is slightly positive with respect to the source and substrate (nominally at ground) this drives the holes from the adjacent substrate, leaving a depletion layer of negatively charged acceptor ions. As a result there is no conduction between source and

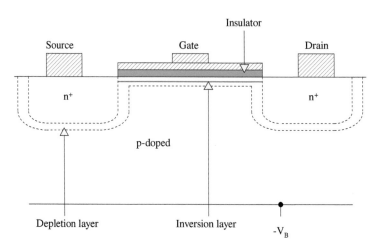

Fig. 6.7 n-channel MOSFET transistor structure.

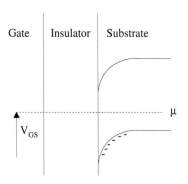

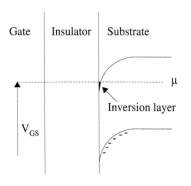

Fig. 6.8 The formation of the inversion layer below the gate in a MOSFET. In the upper panel V_{GS} is small; in the lower panel the V_{GS} is raised to bring the Fermi level into the conduction band.

drain. The corresponding energy diagram at the gate-oxide-substrate is shown in the upper panel of Figure 6.8. If the gate voltage is then made sufficiently positive the Fermi energy in the substrate under the gate will rise and enter the conduction band at the interface with the insulator. The result is that shown in the lower panel of Figure 6.8: an inversion layer populated by electrons develops at the insulator-substrate interface below the gate. This layer is a two-dimensional electron gas, *2DEG*, which is available to carry current between the source and drain. Because the conduction channel is produced by the electric field of the gate the device is called a field effect transistor (FET). The curvature in the potential near the interface in Figure 6.8 is caused by the charge on the ionized acceptor atoms. Both the depletion layer and the inversion layer are indicated in Figure 6.7.

Figure 6.9 shows typical conductance curves for a MOSFET. With low values of the drain-source voltage, V_{DS}, the current is nearly linear with V_{DS}, making this mode of operation suitable for analogue applications. Then the drain current is given by

$$I_D = \mu Q V_{DS}, \tag{6.8}$$

where μ is the electron mobility in $\mathrm{m^2 s^{-1} V^{-1}}$ and Q the charge per unit length in the channel under the gate. This charge is

$$Q = (C/L)(V_{GS} - V_T), \tag{6.9}$$

where C is the capacitance between channel and gate, L is the gate length, V_{GS} the gate-source voltage, and V_T the threshold voltage at which the channel opens. Inserting this expression for Q in the previous equation gives

$$I_D = \mu(C/L)(V_{GS} - V_T)V_{DS}. \tag{6.10}$$

At the other extreme, at large values of V_{DS}, the channel saturates and in this mode the MOSFET is used for digital applications where two stable state settings only are needed.. Because the gate is insulated from the channel its input resistance is large, a significant advantage over other transistor types: very low drive currents become practical and the power drain on the driver is correspondingly small. Complementary MOS (*CMOS*) circuits, using both n-doped and p-doped MOSFETs dominate digital electronics. CMOS devices have been scaled down to have 32 nm long gates, making it possible to build compact processors containing billions of transistors. The tolerances on the voltage levels for switching are less severe than for current controlled logic: voltages in the range 5–15 volts will switch CMOS FETs.

Cell phones and satellite TV receivers handling multi-gigahertz signals require transistors with very high electron mobility. These HEMTs are MOSFETs made from GaAs, which has an electron mobility of 8500 cm^{2}V^{-1}s^{-1} at 300 K. This is several times higher than the mobility in silicon, 1400 cm^{2}V^{-1}s^{-1}. Collisions with donor ions in GaAs would reduce the mobility markedly, so the requirements on the one hand of doping to give sufficient carriers and, on the other hand, of purity to give high mobility are clearly at odds. These requirements are reconciled in HEMTs by a technique known as *modulation doping*. Figure 6.10 shows the valence and conduction bands across a section through the gate, channel and substrate. A thin spacer layer of undoped AlGaAs is grown directly on the substrate: then come in turn the doped n-AlGaAs layer, the insulator shaded grey and the metal gate. This configuration gives a *heterojunction* between GaAs with a bandgap of 1.42 eV and GaAlAs with a bandgap of 2 eV. Under positive gate bias a potential well forms at the interface in GaAs. Electrons originating in the doped n-AlGaAs fall into this potential well just below the surface in the undoped GaAs where they form a 2DEG and benefit from the high mobility in pure GaAs. Modulation doping has the useful additional property that a 2DEG can be maintained even in the absence of a gate potential: devices discussed in the following sections rely on this property.

The combination of semiconductors employed in the HEMTs described here is special. Firstly GaAs and AlAs have very similar lattice constants, 0.565 nm and 0.566 nm, as does any intermediate alloy AlGaAs. Therefore, these alloys can be laid down epitaxially on one another with negligible strain developing at the interfaces. A strained interface would have defects which scatter electrons. Secondly, the bandgap of the alloy AlGaAs can be tuned from 1.42 eV for pure GaAs to 2.16 eV for pure AlAs. Using these properties in designing heterojunctions is known as *bandgap engineering*. The Si/SiO$_2$ interfaces in MOSFETs by contrast have defects due to the mismatch of the lattices which cause scattering and reduced electron mobility.

We can estimate the frequency range over which GaAs HEMTs can

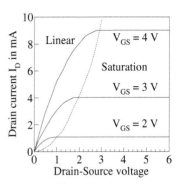

Fig. 6.9 Characteristics of a typical MOSFET for various gate voltages. The broken line separates the regions of linear operation and saturation.

Poisson's equation relates electric potential, ϕ, to charge density, ρ,

$$\mathrm{d}^2\phi/\mathrm{d}x^2 = -\rho/(\varepsilon_0\varepsilon_r),$$

taking the x-coordinate perpendicular to the interface in Figure 6.10 and where ε_r is the dielectric constant (relative permittivity). Thus the positively charged donor ions in AlGaAs and the stored electrons in GaAs cause the opposite curvatures of the bands on the two sides of the heterojunction.

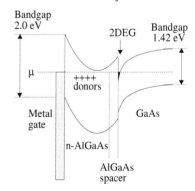

Fig. 6.10 HEMT biased to produce a 2DEG. The gray area is the insulator.

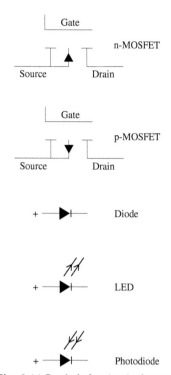

Fig. 6.11 Symbols for circuit elements.

faithfully amplify electrical signals as follows. With a channel length of $10\,\mu m$ and source-drain voltage of $1\,V$ the electric field acting along the channel, E is $10^3\,Vcm^{-1}$. Using an electron mobility, μ, for GaAs of $8500\,cm^2V^{-1}s^{-1}$, the drift velocity along the channel, μE, is about $10^7\,cms^{-1}$. The corresponding kinetic energy is about $0.03\,eV$, equal to thermal excitation at $300\,K$, enough to excite optical phonons. Consequently the drift velocity does not increase appreciably for applied electric fields greater than $10^3\,Vcm^{-1}$. The transit time through the channel is then $10^{-11}\,s$. In practice commercial HEMTs handle binary signals up to frequencies $\sim\!100\,GHz$.

6.5   Mesoscopic scale effects

The ability to fabricate nanometre-sized devices from ultrapure semiconductors and to maintain them at liquid helium temperatures has opened the way to explore quantum properties on a scale larger than that of atoms or molecules. Quantum properties at this intermediate *mesoscopic* scale are described here. Three length scales are crucial to understanding mesoscopic devices. Their importance is discussed here before going on to describe some examples of quantum effects at this scale.

One length scale is the *phase coherence length*, L_ϕ, that is the length over which an electron maintains its phase in a material. Inelastic scattering from other electrons and from phonons both alter the phase in crystals. Elastic scattering from defects and impurities will not usually do so, though if the electron spin is flipped in the scattering the phase is reset. As a result the phase coherence length is mesoscopic only at temperatures low enough to suppress electron-electron and electron-phonon collisions. In pure, defect-free GaAs the phase coherence length is $1\,\mu m$ ($200\,nm$) at $1\,K(5\,K)$. Therefore, at these temperatures interference between electron waves can be expected on the mesoscopic scale in GaAs.

The next significant length is the Fermi wavelength in a 2DEG. Using eqn. B.9 the two-dimensional density of states is $kdk/(2\pi)$ per electron spin state. Thus the count of electron states below the Fermi level is

$$n = 2\int_0^{k_F} kdk/(2\pi) = k_F^2/(2\pi) = 2\pi/\lambda_F^2. \tag{6.11}$$

Within the GaAs channel in an HEMT the electron density is typically $2\,10^{15}\,m^{-2}$ making the Fermi wavelength, λ_F, equal to $56\,nm$. If the width of the potential well at the GaAs/AlGaAs interface in an HEMT is comparable to λ_F the motion of the electrons transverse to the plane of the gas can be quantized at cryogenic temperatures. Feature sizes now produced by photo- and electron beam lithography of semiconductors can be as small as tens of nm making quantization in two or all three dimensions achievable: these devices are known respectively as quantum wires and quantum dots. By contrast the Fermi wavelength in silicon is only 0.5 nm, which makes it a less favourable material for observing this type of quantization.

The third important length scale is the mean free path, L_e, for elastic scattering. Semiconductors, which are *weakly disordered*, that is with few defects and impurities have $L_e \gg \lambda_{\mathrm{F}}$ so that the interference effects just mentioned are not jeopardized. If L_e is much less than the device size, L, the electron motion is *diffusive*. However, if L_e exceeds the device size, L, the only collisions possible are those with the surfaces of the device. The motion of electrons is then called *ballistic*.

6.6   Quantum point contacts

Lithographic methods used to construct MOSFETs and HEMTs have been refined to make devices showing quantum properties on the mesoscopic scale. An example of one such device is shown in Figure 6.12. This HEMT-like structure has gates across the channel, isolated by an insulating layer. A 2DEG covers the whole GaAs/AlGaAs interface when the gates are at the source potential. When a negative bias is applied to these gates the electron gas is excluded from the areas of the channel below them. These areas of exclusion are the white vertical strips midway across the upper panel of the figure. In this way the channel is pinched off to a width of order tens to a few hundred nanometres making the width of the constriction, through which electrons can pass, of order the Fermi wavelength. Consequently it only supports transmission of a few transverse electron modes, making it a *quantum point contact, QPC,* between the pools of two-dimensional electron gas, left and right, at the source and drain potentials respectively. Figure 6.13 shows how the conductance of a cryogenically cooled GaAs/AlGaAs quantum point contact varies with the gate voltage. The QPC conductance rises in steps of equal height with slightly rounded transitions; the step height measured in such experiments being close to $[12.9\,\mathrm{k\Omega}]^{-1}$. As the gate voltage grows less negative the QPC becomes wider, allowing more transverse electron modes to pass through. Each step up the conductance staircase corresponds to the opening of a new transverse mode. The electron current through the QPC for a single transverse mode with longitudinal wave-vector component in the range k to $k+\mathrm{d}k$ is

$$\mathrm{d}I = 2evg(k)\mathrm{d}k, \qquad (6.12)$$

where the factor 2 comes from the two spin states of the electron. The one-dimensional density of states $g(k)$, is given in eqn. B.8, and the electron velocity by $v = [\mathrm{d}E/\mathrm{d}k]/\hbar$. Inserting these values in the above equation gives

$$\mathrm{d}I = [2e/h]\mathrm{d}E. \qquad (6.13)$$

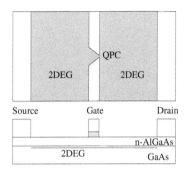

Fig. 6.12 Quantum point contact in an HEMT-like structure.

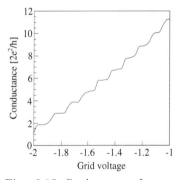

Fig. 6.13 Conductance of a quantum point contact as a function of the gate voltage choking off the constriction. Adapted from B. J. Wees, H. van Houten, C. W. J. Beenakker, L. P. Kouwenhoven, D. van der Merel and C. T. Foxon, *Physical Review Letters* 60, 848 (1988). Courtesy Professor Wees.

Two points need making: first that the energies of the electron in the reservoirs on either side of the QPC range from zero up to the chemical potential; second that the current can flow in both directions. As a result the net current is

$$I = [2e/h] \left(\int_0^{\mu+\Delta\mu} \mathrm{d}E - \int_0^{\mu} \mathrm{d}E \right), \qquad (6.14)$$

where μ and $\mu + \Delta\mu$ are the chemical potentials of source and drain respectively. $\Delta\mu = eV$ where V is the drain-source voltage. Thus the net current through the QPC in one transverse mode is

$$I = [2e/h]\Delta\mu = [2e^2/h]V, \qquad (6.15)$$

giving a conductance *per spin state*

$$G_0 = [e^2/h]. \qquad (6.16)$$

e^2/h is $[25.8\,\mathrm{k\Omega}]^{-1}$, making this the *quantum of conductance*. This simple result requires first of all that the electrons do not scatter within the QPC, which means that the dimensions of the QPC are all smaller than the mean free path for elastic scattering. Secondly the profile of the QPC must vary sufficiently slowly that any transverse mode enters without reflection. Departures from these two conditions give a conductance per transverse mode

$$G = 2G_0T, \qquad (6.17)$$

where the *transmission factor*, T, is less than unity. This is *Landauer's equation* for the conductance of a single mode. Where the constriction is wide enough to allow several transverse modes the conductance is $2G_0 \sum_1^n [T_j]$, which reduces to $2nG_0$ for perfect transmission of n transverse modes through the QPC. Each step upward in Figure 6.13 occurs when an additional quantum channel opens, and along any horizontal flat region the transmission coefficient of all the open channels is unity. At the constriction electron shielding is weaker than in the sea so that the electron-electron scattering increases, and this in turn reduces the phase coherence length. It is sufficiently long however for quantization to be seen in GaAs QPCs of dimension $\sim 100\,\mathrm{nm}$ at cryogenic temperatures.

We shall meet the quantum of conductance again in connection with the quantum Hall effect. Quantum Hall devices provide the most precise, easily reproducible conductance standards. What luck that e^2/h is of a practical magnitude.

Quantization of conductance can also be demonstrated in special circumstances at room temperatures. A metal wire is strained so that a neck forms and then ruptures. If a voltage is applied and the current through the wire is measured continuously throughout this process, steps are seen in the conductance. The last few steps in conductance before the wire fractures are reported to be the most distinct.[3]

[3]See N.Garcia, *Physics Today*, February 1996.

The above analysis can be extended to cover transmission through wires thin enough that only a few transverse modes exist. These are called *one-dimensional* wires. Phase coherence is retained along wires shorter than the phase coherence length, and the electrons go undeflected if the length is shorter than the mean free path for all scatterings. When the wire is this short then the electrons travel without collisions apart from those with the wire surface, that is to say they travel *ballistically*. They also retain phase coherence so that modes are well defined and for each open mode the conductance is $2G_0$ *irrespective* of the wire length.

Surprisingly, quantum ballistic transport can be observed at room temperature in single walled carbon nanotubes, which have radii of a few nm. These can be near perfect crystalline structures in which the mean free path for scattering is of order $1\,\mu$m. There are exactly two transverse electron modes in these nanotubes, independent of the precise radius of the nanotube. The conductance of single walled carbon nanotubes was measured by Li et al.[4] and found to follow the empirical formula

$$G = [4e^2/h]/(1 + L/\lambda), \qquad (6.18)$$

for nanotubes as long as several mm. Here L is the length of the nanotube, λ the mean free path of $1\,\mu$m.

To summarize the requirements for quantum conductance are three-fold: a quantum degenerate Fermi gas; a ballistic transport channel; a constriction of dimensions comparable to λ_{F}.

6.7 Coulomb blockade

Coulomb blockade is observed at metal/insulator/metal junctions where conduction occurs through quantum barrier penetration. The structure could be aluminium contacts separated by surface oxide. If the cross-sectional area is small enough, and the temperature low enough, the energy required to move a single electron across the junction can be the key parameter in determining the conductivity. The charging energy, E_c, required to carry a single electron across the junction is $e^2/(2C)$. Here C is the junction capacitance

$$C = \varepsilon_0\varepsilon_r A/d, \qquad (6.19)$$

with A being the area, d the thickness of the barrier, and ε its dielectric constant (relative permittivity). Taking an area $0.01\,\mu$m^2, ε_r equal to 10.0 and gap of 10 nm the capacitance is $\sim 10^{-16}$ F and E_c is ~ 1.0 meV. Thermal excitation across the junction can be suppressed to much below this energy by lowering the temperature: for example, at $1\,$K $k_{\mathrm{B}}T$ is only $86\,\mu$eV. Then at this temperature and at voltages lower than $E_c/e = e/2C$, that is ~ 1 mV, no current will flow. Figure 6.14 illustrates this *Coulomb blockade*. A further requirement for observing the blockade is that quantum fluctuations in the energy are also much less than E_c. We can infer a lower limit on such energy fluctuation by using the uncertainty principle. First note that the time constant of these

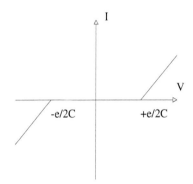

Fig. 6.14 Coulomb blockade across a mesoscopic heterojunction at cryogenic temperatures.

[4]S. Li, Z. Y. C. Rutherglen and P. J. Burke, *Nano Letters* 4, 2003 (2007)

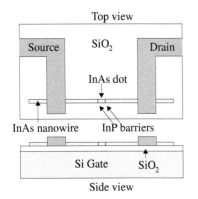

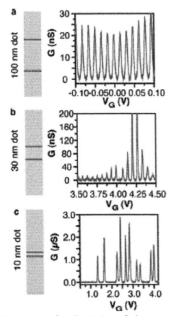

Fig. 6.15 Quantum dot on a nanowire. Adapted from Figure 1: M. T. Bjork, C. Thelander, A. E. Hansen, L.E. Jensen, M. W. Larsson, L. R. Wallenberg and L. Samuelson, *Nano Letters* 4, 1621 (2004). Courtesy Professor Samuelson and The American Chemical Society.

Fig. 6.16 Conductivity of three quantum dots versus the gate voltage. From the top the quantum dot lengths are 100 nm, 30 nm and 10 nm. Adapted from Figure 2: M. T. Bjork, C. Thelander, A. E. Hansen, L.E. Jensen, M. W. Larsson, L. R. Wallenberg and L. Samuelson, *Nano Letters* 4, 1621 (2004). Courtesy Professor Samuelson and The American Chemical Society.

fluctuations is RC where R is the total series resistance of the circuit containing the junction. Hence the energy fluctuation

$$\delta E \geq \hbar/(2RC). \qquad (6.20)$$

Then in order that quantum fluctuations do not wash out the blockade we must have

$$E_c \gg \hbar/(2RC), \qquad (6.21)$$

that is

$$R \gg h/e^2 = 1/G_0 = 25.8\text{k}\Omega. \qquad (6.22)$$

The conditions for Coulomb blockade were first obtained by Delsing and colleagues[5] using a 0.006 μm² Al/AlO/Al junction on a silicon wafer held at 1.3 K.

6.8   Quantum dots

Quantum dots are structures which confine electrons in all three dimensions to regions typically of order 10–1000 nm. They are in essence very large atoms. One technique for forming quantum dots is to use gates to confine electrons within a nanometre sized area of the channel in a modulation doped GaAs device. Here we discuss a different example shown in Figure 6.15. The quantum dot is a section of a 40 nm diameter InAs one-dimensional quantum wire. The dot is bounded left and right by 5 nm wide InP sections. InP has a much wider bandgap (1.27 eV) than InAs (0.35 eV) so that these sections are thin potential barriers through which an electron can easily tunnel. The whole structure is supported on a silicon substrate covered by a 100 nm thick SiO₂ insulating layer. The silicon substrate is also the gate used to control the potential of the dot. Figure 6.16 shows how the conductivity varies with the gate voltage, at low source/drain bias. Conductivities are shown for three quantum dots of lengths 100 nm, 30 nm and 10 nm along the wire. There are peaks in conductivity separated by intervals over which the Coulomb blockade is at work. The quantum dot is also an island capacitor and this aspect needs to be taken into account in interpreting Figure 6.16.

Over the range of gate voltages separating two conductivity peaks there is therefore a fixed integral number of electrons on the dot. We now show that this number changes by exactly one on passing across a conductivity peak. The energy required to add the nth electron to the dot is

$$\begin{aligned} E_n &= \epsilon_n + [ne]^2/(2C) - [(n-1)e]^2/(2C) - eV_g(C_g/C) \\ &= \epsilon_n + [n-1/2]e^2/C - eV_g(C_g/C) \end{aligned} \qquad (6.23)$$

[5]P. Delsing, K. K. Likharev, L. S. Kuzmin and T. Claeson, *Physical Review Letters* 63, 1180 (1989).

where ϵ_n is the binding energy of the nth electron in the potential well defined by the dot. The second term on the right is the charging energy and the final term is the potential energy due to a voltage V_g on the gate. The capacitance between dot and gate is C_g and C is the total capacitance of the dot to the source, the drain and the gate.

Figure 6.17 shows the electron energy levels of the source, dot and drain for three conditions of the gate voltage: just below, at and just above a conductance peak. In the first case, shown in the left hand panel, the source electrons have insufficient energy to enter the empty $(n+1)$th level and there is a blockade. With the higher gate voltage, shown in the centre panel, electrons can move singly onto the dot from the source and exit to the drain, giving a continuous current. With a still higher gate voltage, shown in the right hand panel, the next open electron level on the dot is inaccessible and the current flow is again blocked.

In Figure 6.16 we see that the conductance peaks are of near constant height and equally spaced for the 100 nm wide dot, but irregular for the shorter dots. This change reflects the increasing importance of the binding energy as the dot shrinks: a point worth pursuing. In making

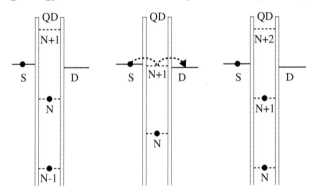

Fig. 6.17 Energy levels in source, drain and quantum dot, with the gate voltage increasing from panel to panel rightward.

an *estimate* for the energy level separation we use the expression for the energy of the nth level above the base of a one-dimensional infinite square potential well of the same width as the dot:

$$\epsilon_n = (\pi^2 \hbar^2 n^2)/(2a^2 m). \tag{6.24}$$

Taking a to be 100 nm, ϵ_n would be $\sim 4\,n^2\,10^{-5}$ eV. This is to be compared to the charging energy for an additional electron, $(n - 1/2)e^2/C$. The capacitance of the InP layers is given by eqn. 6.19, with widths 5 nm, area $\pi\,400$ nm^2, and ε_r of 12.9. The resulting capacitance is ~ 30 aF (attofarad or 10^{-18} F), giving a charging energy $\sim 6n$ meV. Therefore, at low values of n, the binding energy is negligible compared to the charging energy, and the conductance peaks are expected to be equally spaced in gate voltage: this matches the pattern seen in the upper panel of Figure 6.16. The observed spacing of 17 mV can be reconciled with that estimated above from the charging energy (6 mV) if the capacitance quotient C_g/C appearing in eqn. 6.23 is about 1/3. By contrast, when the dot length is reduced to 30 nm the confinement gives a binding energy approaching the charging energy at moderate values of n. As a result the interpretation of conductivity variation seen in the lower panels of Figure 6.16 is less simple.

6.9   Quantum interference

Pioneering experiments that observed electron interference effects in mesoscopic devices were reported by Aronov and Sharvin.[6] In the example to be described here the sample was a lithium film of thickness 0.127 μm evaporated onto a 1 cm long quartz filament of diameter 1.3 μm and held at 1.1 K. In this experiment it is necessary for the material, the lithium film, to be *disordered* so that there are many elastic scatters: this ensures that electrons can follow alternative paths through the film while maintaining phase coherence. The resistance of the film along the filament was measured as a function of the strength of a uniform axial magnetic field B. The results, sketched in Figure 6.18, show periodic oscillations in resistance as the magnetic field strength is varied. We now show that these oscillations are the result of interference between electron waves propagating in opposite senses around the cylinder.

At cryogenic temperatures the phase coherence length in pure lithium is longer than 1 μm. Consequently an electron wave's coherence extends around the circumference of the filament, but not along its length in this experiment. Therefore, alternative electron paths circulating in opposite senses around the magnetized filament acquire a well-defined relative phase. This phase difference is calculated here and is found to depend on the magnetic flux through the filament.

The total momentum appearing in Schrödinger's equation for an electron in the lithium film is given by eqn. 1.78. It is made up of the canonical momentum and an electromagnetic contribution $e\mathbf{A}$ where $\mathbf{A}$ is the vector potential. This is related to the applied axial magnetic field: $\mathbf{B}=\nabla\wedge\mathbf{A}$. Choosing cylindrical polar coordinates (r, ϕ, z) having the polar axis along the axis of the filament, $\mathbf{A}$ only has an azimuthal component. Using eqn. A.8

$$A_\phi = Br/2, \tag{6.25}$$

where B is the magnitude of the magnetic field. The corresponding azimuthal component of the momentum vector, given by eqn. 1.78, in operator form is

$$\hat{p}_\phi = -i(\hbar/r)\partial/\partial\phi + eBr/2, \tag{6.26}$$

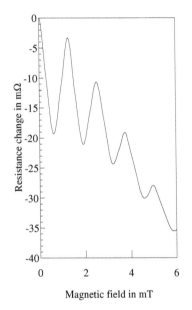

Fig. 6.18 Variation of the resistance of a thin lithium film evaporated on a quartz filament as the axial magnetic field is varied. Adapted from Figure 8 in A. G. Aronov and Yu. V. Sharvin, *Reviews of Modern Physics* **59**, 755 (1987).

---

[6]A. G. Aronov and Yu. V. Sharvin, *Reviews of Modern Physics* **59**, 755 (1987).

where eqn. A.12 has been used. Suppose that the electron wavefunction, $\psi = \exp(it\phi)$, is an eigenstate of p_ϕ, with eigenvalue q_ϕ, then

$$\hat{p}_\phi \psi = q_\phi \psi = (\hbar t/r)\psi + (eBr/2)\psi. \tag{6.27}$$

Rearranging this gives

$$t = rq_\phi/\hbar - eBr^2/(2\hbar), \tag{6.28}$$

so that

$$\psi = \exp[irq_\phi\phi/\hbar]\exp[-ieBr^2\phi/(2\hbar)]. \tag{6.29}$$

In one circuit the electron phase changes by

$$2\pi rq_\phi/\hbar - \pi eBr^2/\hbar. \tag{6.30}$$

The first term is the phase change in the absence of any magnetic field and is consequently some integer m, times 2π: we may note that rq_ϕ is the axial component of the electron's orbital angular momentum $m\hbar$. Therefore the phase change in one circuit reduces to $(\pi eBr^2)/\hbar$ or $\Phi e/\hbar$, where $\Phi = \pi r^2 B$ is the flux threading the filament. Electron paths travelling in opposite senses round the cylinder will have a phase difference twice as large, $2\Phi e/\hbar$. Changes in flux through the filament of $h/2e$ will cause this relative phase to alter by 2π. It follows that the interference between paths in opposite senses around the filament goes through one cycle whenever the enclosed magnetic flux changes by $h/(2e)$. In turn the resistance will cycle with each such change in the enclosed flux. This is the first time we have met a *quantum of magnetic flux*, h/q, where q is the charge of the particles involved. In the present case $h/2e$ is $2.068\,10^{-15}\,\mathrm{Tm^2}$. Superconductivity described in Chapter 15 involves dissipationless currents also carried by pairs of electrons. Measurements with superconductors determine the flux quantum to high precision.

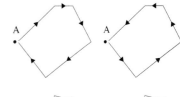

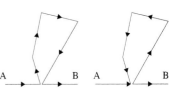

Fig. 6.19 Path pairs responsible for universal conductance fluctuations. The pairs are time-reversed sequences of elastic scatters.

6.10   Universal conductance fluctuations

These are fluctuations that are seen in mesoscopic sized semiconductors at temperatures approaching absolute zero when the magnetic flux threading the sample is varied or when the gate voltage is varied. The fluctuations appear random but are reproducible for any given specimen. Universal conductance fluctuations are caused by interference between electron paths which duplicate one another apart from a portion where they are mirror images of one another. Two examples of such path pairs are shown in figure 6.19. The fluctuations in conductance tend toward a magnitude e^2/h as the temperature approaches absolute zero, *independent of the material, its size, or the degree of disorder of the crystal.*

These universal conductance fluctuations are only seen if the phase coherence length extends over at least a sizable fraction of the sample; consequently liquid helium temperatures are necessary to achieve this.

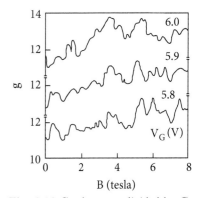

Fig. 6.20 Conductance divided by G_0 versus magnetic field at three gate voltages for an inversion layer segment of a silicon MOSFET. Figure 2: W. J. Skocpol, P. M. Mankiewich, R. E. Howard, L. D. Jackel, D. M. Tennant and A. Douglas Stone, *Physical Review Letters* 56 2865 (1986). Courtesy Professor Stone and The American Physical Society.

Figure 6.20 shows the conductance fluctuations seen in a 300 nm long segment of the inversion layer of a silicon MOSFET held at 4.2 K. Fluctuations are shown as a function of the applied magnetic field and for three gate voltages. Using Landauer's formula for conductance, eqn. 6.17,

$$G = G_0 \sum T_i, \tag{6.31}$$

where T_is are the transmission coefficients for all the N scattering channels, whose number scales like the channel width divided by the Fermi wavelength. Now if the channels were uncorrelated and the T_i just random variables the variance of G would be expected to scale as N and hence depend on the sample dimensions and the mean free path length. Surprisingly, it was shown theoretically that the phase coherence of the electrons across the device, a quantum property, manifests itself through strong correlations among the T_is, so that

$$\text{variance}\,(G) = \text{variance}\,(\sum_i T_i)G_0^2 = \sim G_0^2. \tag{6.32}$$

This result indicates why conductance fluctuations in samples smaller than the phase coherence length are universal and have magnitude $\sim G_0$.

6.11 Further reading

Physics of Semiconductor Devices, 3rd edition. by S. M. Sze and K. K. Ng, published by John Wiley (2006) has been a standard text on the subject.

Introduction to Mesoscopic Physics, 2nd edition by Y. Imry, published by Oxford University Press (2002) provides an advanced account by a noted expert.

Exercises

(6.1) Show that the electron density in the conduction band in an intrinsic semiconductor is

$$n = n_c \exp[(E_{\mathrm{F}} - E_c)/k_{\mathrm{B}}T],$$

where E_c is the energy at the bottom of the conduction band and

$$n_c = 2(2\pi m_e k_{\mathrm{B}}T/h^2)^{3/2}.$$

Assume that the bandgap is large enough so that $(E_{\mathrm{F}} - E_V)$ and $(E_c - E_{\mathrm{F}})$ are much greater than

$k_{\mathrm{B}}T$. Repeat this exercise to obtain the hole density, p, in the valence band. Then show that

$$np = n_v n_c \exp(-E_G/k_{\mathrm{B}}T),$$

where E_G is the bandgap,

$$n_v = 2(2\pi m_h k_{\mathrm{B}}T/h^2)^{3/2}$$

and m_h is the hole effective mass. The integral

$$\int_0^\infty \sqrt{x}\exp(-x/a)\mathrm{d}x = \sqrt{\pi a^3}/2.$$

(6.2) Suppose a 15 nm long ballistic conductor carries two transverse modes. What current is produced when a 1 V potential is applied? The conductor is held near to absolute zero Kelvin and the transmission factor for both modes is unity. What is the current if the conductor is 30 nm long?

(6.3) Estimate the resistance of a defect-free single-walled carbon nanotube of length 1 mm if the mean free path for electrons is 1 μm.

(6.4) Suppose electrons in the valence band have wave-vectors $k\,\mathrm{m}^{-1}$ and energies given by

$$E = -0.3\,10^{-37} k^2\,\mathrm{J}$$

relative to the top of the valence band. Consider removing one electron with wave-vector $10^9\,\mathrm{m}^{-1}$. What is the effective mass, energy, momentum and velocity of the hole? What is the direction of travel of the hole taking the electron to travel in the positive x-direction. Refer to results found in Chapter 5.

(6.5) Let I_l be the photocurrent in a photovoltaic cell due the incident daylight and I_r the saturation current in the absence of any illumination. Then the total current is

$$I = I_r[\exp(eV/k_{\mathrm{B}}T) - 1] - I_l,$$

at a forward voltage V. Show that on open circuit the current is

$$I_{\mathrm{oc}} = I_r \exp[eV_{\mathrm{oc}}/k_{\mathrm{B}}T],$$

and hence that

$$V_{\mathrm{oc}} = (k_{\mathrm{B}}T/e) \ln(I_{\mathrm{oc}}/I_r).$$

Take I_r to be $1\,\mathrm{pA\,cm}^{-2}$, the cell responsivity to be $0.4\,\mathrm{AW}^{-1}$ and the daylight intensity to be $100\,\mathrm{mW\,cm}^{-2}$. What is the short circuit current density and the open circuit voltage. The product of these quantities gives an estimate of the electric power yield. Estimate the efficiency of the cell in converting daylight to energy.

(6.6) When a group IV semiconductor is lightly doped with a group V element the additional electron in the group V atom is weakly bound and its orbit extends into the volume of the group IV atoms. Suppose that the dielectric constant of the this volume is ε_r and that the electron has effective mass m^*. Show that the binding energy is approximately $[m^*/(m\varepsilon_r^2)]E_H$ where m is the mass of a free electron and E_H is the binding in a hydrogen atom. Show that the orbit radius is $m\varepsilon_r/m^*$ larger. If there is p- and n-doping, what happens when the electron and hole wavefunctions overlap?

(6.7) If $A(t)$ is the rate at which photons generate electron-hole pairs at a temperature T in a semiconductor and $npB(T)$ is the rate of recombination for densities n and p of the two species, show that in equilibrium the product np is constant.

(6.8) A junction diode at $300\,\mathrm{K}$ reverse biassed at $0.2\,\mathrm{V}$ carries a current of $4\,\mathrm{\mu A}$. What is the current when it is forward biassed at $0.25\,\mathrm{V}$?

(6.9) In InSb the electrons and holes have respective effective masses 0.014 and 0.18 times the free electron mass. What is the Hall voltage across a specimen $3\,\mathrm{mm}$ wide and $1\,\mathrm{mm}$ deep carrying a current of $150\,\mathrm{mA}$, and across which a field of $0.2\,\mathrm{T}$ is applied. Take the electron and hole densities to be $10^{22}\,\mathrm{m}^{-3}$.

Transitions

7.1   Introduction

Quantum mechanics provides the way to calculate not only such static
properties as the energy levels of electrons in atoms and matter, but
also the rates at which transitions occur. These transitions might be
nuclear decays or reactions between, for example, light and atoms. In
this chapter mainly perturbative processes are discussed, that is to say,
processes in which the energy of the interaction causing the process is
small compared to the total energy. Energy, momentum and angular momentum are conserved in all transitions. Parity and internal quantum
numbers are discussed in the context of elementary particles, in Chapters 18 and 19. Throughout this chapter matter particles are treated
non-relativistically.

   The key tool, of general validity, used in calculating transition rates
is *Fermi's golden rule*. It is derived here using first-order perturbation
theory. The first application here is to calculate the 2p→1s transition
rate in atomic hydrogen. This leads to a discussion of electromagnetic
selection rules for transitions, and of line widths and resonances.

   Next Fermi's golden rule is adapted for reactions, in preparation for
analysing examples of two-body reactions. Each type of reaction can
be quantified by a *cross-section*. This is the effective area of the target
particle as seen by the beam particle for initiating the specified reaction.
An outline of the measurement of cross-sections is included. After this
the perturbative *Born approximation* is introduced, in which the the
incoming and outgoing states are treated as plane waves: Rutherford
scattering is used as an example. A calculation of the rates of nuclear
beta decays is outlined next.

   We show that at sufficiently low energies only s-wave, that is orbital
angular momentum zero, states are scattered. The Ramsauer–Townsend
dip in cross-section in this regime and its quantum mechanical explanation are presented. Finally, the scattering of *cold atoms* at $\sim 10^{-11}$ eV is
analysed in preparation for the description of the properties of gaseous
atomic condensates in Chapter 16.

Quantum 20/20: Fundamentals, Entanglement, Gauge Fields, Condensates and Topology.
Ian R. Kenyon. © 2020. Published in 2020 by Oxford University Press.
DOI: 10.1093/oso/9780198808350.001.0001

7.2   Fermi's golden rule

Fermi's golden rule for calculating transition rates is derived using perturbation theory: this requires the energy of the interaction responsible for the transition, V, to be be small compared to the total energy. We determine the transition rate for process i→f due to V.

Suppose the orthonormal wavefunctions of the initial and final states are $\phi_i\, e^{-iE_i t/\hbar}$ and $\phi_f\, e^{-iE_f t/\hbar}$ respectively. Then to first order in the perturbation the amplitude of the selected final state is given by the overlap integral

$$S_{fi} = \int \phi_f^*\, e^{iE_f t/\hbar}\, V[\phi_i\, e^{-iE_i t/\hbar}]\, \mathrm{d}\mathbf{r}\mathrm{d}t \qquad (7.1)$$

integrated over all space and over the interaction time T. If T is long enough that we can take it as infinite the time integral simplifies

$$\int e^{iE_f t/\hbar}\, e^{-iE_i t/\hbar}\mathrm{d}t = [2\pi/\hbar]\, \delta(E_f - E_i), \qquad (7.2)$$

[1]See Section 1.9.

where conservation of energy is made explicit by the δ-function.[1] Then

$$S_{fi} = [2\pi/\hbar]\, \delta(E_f - E_i)\, V_{fi}, \qquad (7.3)$$

with

$$V_{fi} = \int \phi_f^*\, V\, \phi_i\, \mathrm{d}\mathbf{r}. \qquad (7.4)$$

Thus the transition probability per unit time is

$$
\begin{aligned}
W &= |S_{fi}|^2/T \\
&= [2\pi/\hbar]\, \delta(E_f - E_i)\, |V_{fi}|^2 \int e^{i(E_f - E_i)t/\hbar}\mathrm{d}t/T \\
&= [2\pi/\hbar]\, \delta(E_f - E_i)\, |V_{fi}|^2 \int \mathrm{d}t/T, \qquad (7.5)
\end{aligned}
$$

where the δ-function has been used to reduce $\exp[i(E_f - E_i)]t/\hbar$ to unity. This gives

$$W = [2\pi/\hbar]\, \delta(E_f - E_i)\, |V_{fi}|^2. \qquad (7.6)$$

The density of final states is $\rho_f(E_f)$, that is to say the number of available quantum mechanically distinguishable final states in the energy interval E_f to $E_f + \mathrm{d}E_f$ is $\rho_f(E_f)\, \mathrm{d}E_f$. Then the transition rate into the available final states is

$$
\begin{aligned}
R &= \int W \rho_f(E_f)\mathrm{d}E_f \\
&= [2\pi/\hbar]\, |V_{fi}|^2\, \rho_f(E_i). \qquad (7.7)
\end{aligned}
$$

This prescription has *general validity beyond perturbation theory* when V_{fi} is replaced by the actual amplitude or *matrix element*, M_{fi}, for a transition. Finally,

$$R = (2\pi/\hbar)\, |\, M_{fi}\, |^2\, \rho_f(E_i), \qquad (7.8)$$

which is known as *Fermi's golden rule*. It provides the interface between theory, which predicts the matrix element, and experiment, which measures the reaction rate.[2]

7.3 2p→1s decay in hydrogen atoms

In spectroscopic notation the $n = 2$ states of the hydrogen atom are $2P_{3/2}$, $2P_{1/2}$ and $2S_{1/2}$ while the $n = 1$ state $1S_{1/2}$ is the ground state. Radiation from this decay is the Lyman-α spectral line[3] at 121.6 nm, angular frequency $1.55\,10^{16}\,\mathrm{s}^{-1}$. The decay is through the electric dipole interaction, with energy

$$H' = e\mathbf{E}_0 \cdot \mathbf{r}, \tag{7.9}$$

where $\mathbf{E}_0$ is the electric field and $e\mathbf{r}$ the moment of the electric dipole formed by the electron and proton. Then the matrix element, the amplitude for the transition of the electron from the p-shell to the s-shell, is, like eqn. 7.4, an overlap integral

$$M = \int \Psi_p^* H' \Psi_s\, \mathrm{d}\mathbf{r} = e\mathbf{E}_0 \cdot \int \Psi_p^* \mathbf{r}\Psi_s\, \mathrm{d}\mathbf{r}. \tag{7.10}$$

Ψ_p and Ψ_s are the initial and final state wavefunctions of the electron, and the integral is taken over all space. All the allowed decays are shown in Figure 7.1. The simplest case to evaluate is that of the $2P_{3/2}$ initial state, in which the spin and orbital angular momentum of the electron are aligned parallel. Each of the two wavefunctions involved is then a product of one spatial and one spin term:

$$\Psi_\mathrm{p} = \psi_\mathrm{p}(n = 2; l = 1; m_l = +1)\chi(m_s = +1/2), \tag{7.11}$$

and

$$\Psi_\mathrm{s} = \psi_\mathrm{s}(n = 1; l = 0; m_l = 0)\chi(m_s = +1/2). \tag{7.12}$$

The overlap of the spin wavefunctions is unity. Using Tables 2.1 and 2.2 we have

$$\psi_\mathrm{p} = -r\exp(-r/2a_0)\sin\theta\exp(i\phi)/(8\sqrt{\pi}a_0^{5/2}), \tag{7.13}$$
$$\psi_\mathrm{s} = \exp(-r/a_0)/(\sqrt{\pi}a_0^{3/2}). \tag{7.14}$$

The matrix element squared is

$$|M|^2 = |\,\mathbf{E}_0 \cdot \mathbf{D}\,|^2 \tag{7.15}$$

where the dipole moment

$$\mathbf{D} = e\int \psi_\mathrm{p}^* \mathbf{r}\psi_\mathrm{s}\, \mathrm{d}\mathbf{r}. \tag{7.16}$$

Averaging over all directions of the dipole moment with respect to the electric field gives

$$|M|^2 = E_0^2|\mathrm{D}|^2/3. \tag{7.17}$$

[2]If the measurement does not distinguish between alternative states, the rule is to average over initial states and sum over final states in the calculation.

[3]The copious hydrogen gas in space can be excited by local sources so that this decay provides a useful reference wavelength. The fractional change in wavelength due to cosmological red shift $\Delta\lambda/\lambda = Hd/c$, where d is the distance to the gas, H is the Hubble constant. $H/c \approx 2.5\,10^{-4}$ per megaparsec, a parsec being 3.26 light-years. The furthest objects known are at redshifts exceeding 5.0. Conveniently, the red shift pulls the Lyman-α line across the visible spectrum.

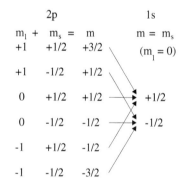

Fig. 7.1 Allowed transitions in hydrogen between the magnetic substates of the 2p and 1s states.

Now we need to understand the electric field involved in order to continue the calculation.

E_0^2 is the mean square fluctuation of the electric field in the vacuum! It was explained in Chapter 1 that the energy of any mode of angular frequency in the vacuum is $\hbar\omega/2$. Summing the energy density in the electric and magnetic fields in the vacuum:

$$\varepsilon_0 E_0^2 V = \hbar\omega/2, \qquad (7.18)$$

where the normalization volume is V. Thus

$$E_0^2/3 = \hbar\omega/(6\varepsilon_0 V). \qquad (7.19)$$

Returning to the evaluation of $\mid D \mid^2$

$$|D|^2 = \left| \int \psi_{\mathrm{p}}^* er\psi_{\mathrm{s}}\, d\mathbf{r} \right|^2 = e^2(|X|^2 + |Y|^2 + |Z|^2), \qquad (7.20)$$

where

$$X = \int \psi_{\mathrm{p}}^* x\psi_{\mathrm{s}}\, d\mathbf{r}, \;\; Y = \int \psi_{\mathrm{p}}^* y\psi_{\mathrm{s}}\, d\mathbf{r}, \;\; Z = \int \psi_{\mathrm{p}}^* z\psi_{\mathrm{s}}\, d\mathbf{r}.$$

Using spherical polar coordinates $d\mathbf{r} = r^2 \sin\theta\, d\theta\, d\phi\, dr$, and the first of these integrals is

$$
\begin{aligned}
X &= [1/(8\pi a_0^4)] \int r^2 \exp\left(-3r/2a_0\right) \sin^2\theta \cos\phi \exp\left(-i\phi\right) d\mathbf{r} \\
&= [1/(8\pi a_0^4)] \left[\int_0^\infty \exp\left(-3r/2a_0\right) r^4 dr \right] \left[\int_0^\pi (\cos^2\theta - 1)d(\cos\theta) \right] \\
&\quad \left[\int_0^{2\pi} \{[1 + \exp\left(-2i\phi\right)]/2\}\, d\phi \right] = 4a_0(2/3)^5.
\end{aligned}
\qquad (7.21)
$$

The Y and Z integrals have the same radial and polar angle components as the X integral so that $Y = X/i$ and $Z = 0$. Collecting terms in eqn. 7.20 gives

$$\mid D \mid^2 = 32(2/3)^{10} e^2 a_0^2. \qquad (7.22)$$

The other ingredient required before applying Fermi's golden rule is the density of photon states. We use eqn. B.11 including both polarization possibilities,

$$\rho(E) = \rho(\omega)d\omega/dE = V\omega^2/(\pi^2\hbar c^3). \qquad (7.23)$$

Applying Fermi's golden rule, eqn. 7.8, gives

$$R = \frac{2\pi}{\hbar} \frac{\hbar\omega}{6\varepsilon_0 V} \frac{\omega^2 V}{\pi^2\hbar c^3} \mid D \mid^2 = 6.26\ 10^8\ \mathrm{s}^{-1}. \qquad (7.24)$$

The decay rates of the six 2p-states of a hydrogen atom, that is the four $2P_{3/2}$ and the two $2P_{1/2}$ states, all have this same value. The $2S_{1/2}$ state does not undergo a dipole transition because the dipole moment between two s-states, $\int e\mathbf{r}\,\mathrm{d}\mathbf{r}$ is identically zero.

Quantum mechanics has provided the means to calculate precisely an atomic decay rate. It was necessary to take a hint from the review chapter and regard the electromagnetic field in the vacuum as responsible for what are known as *spontaneous* decays. This approach is fleshed out in the following chapter where the quantization of the electromagnetic fields is introduced.

7.4   Selection rules

It emerged from the above analysis that the rate of decay is determined by the overlap integral defining $\mid \mathrm{D}\mid^2$. The dipole selection rules outlined in Chapters 1 and 3 simply express the requirement that this integral does not vanish.

We consider the simple case of a hydrogen atom and use the expressions for the wavefunctions given in eqn. 2.35. Taking the electric field direction to define the polar axis, the electron has polar coordinates (r, θ, ϕ) with respect to the nucleus. The *angular* part of the integral is

$$\mathrm{D}_{\mathrm{ang}} = \int Y_{\ell',m'}^*(\theta,\phi)\,\cos\theta\,Y_{\ell,m}(\theta,\phi)\,\sin\theta\,\mathrm{d}\theta\,\mathrm{d}\phi, \qquad (7.25)$$

where (ℓ,m) and (ℓ',m') are the orbital angular momentum quantum numbers of the two states. These integrals vanish identically unless both $\Delta\ell = \ell' - \ell = \pm 1$ and $\Delta m = m' - m = 0, \pm 1$.

A geometric interpretation of these selection rules may be helpful. First note that the vector $\mathbf{r}$ has Cartesian components $x = r\sin\theta\cos\phi$, $y = r\sin\theta\sin\phi$ and $z = r\cos\theta$. Thus $x + iy$, $x - iy$ and z have the same angular dependence as the spherical spherical harmonic functions for unit angular momentum in Table 2.2: $Y_{1,+1}$, $Y_{1,-1}$ and $Y_{1,0}$ respectively. Therefore, we can legitimately treat $\mathbf{r}$ as a spherical harmonic with unit angular momentum in the angular integrations. The integral in eqn. 7.25 is thus the overlap of spherical harmonic functions: one with angular momentum ℓ', one with angular momentum ℓ, and one with unit angular momentum. Angular momenta add vectorially, so that we can infer that the overlap integral vanishes whenever the vector addition is impossible. In this way, the selection rules emerge.

When writing the electric dipole interaction in eqn. 7.9 it was implicitly assumed that the electric field is constant across the atom. The actual spatial variation is

$$\exp(i\mathbf{k}\cdot\mathbf{r}) = 1 + i\mathbf{k}\cdot\mathbf{r} - (\mathbf{k}\cdot\mathbf{r})^2/2 + \cdots$$

in which kr is less than 10^{-3} for light of wavelength $500\,\mathrm{nm}$. The contribution of the $(n+1)$th term is smaller by a factor $(\mathbf{k}\cdot\mathbf{r})^n/n!$ in amplitude, and by a factor $(\mathbf{k} \cdot \mathbf{r})^{2n}/(n!)^2$ in intensity compared to the electric dipole. These successive terms require corresponding increases in the change in the orbital angular momentum of the electron. They are known as *forbidden* transitions. Magnetic dipole and higher order magnetic transitions are weak compared to their electric dipole counterparts. This feature is the topic of an exercise.

7.5   Line widths and decay rates

Suppose the probability per unit time for an atom to decay from an excited state 2 to another state 1 is γ, giving a mean lifetime of $\tau = 1/\gamma$. The wavefunction of one such atom is

$$\psi(t) = \exp[-iE_0 t/\hbar] \exp[-\gamma t/2] \qquad (7.26)$$

where E_0 is its rest energy. Its Fourier transform expressed in terms of energy is

$$\psi(E) \propto \frac{1}{(E - E_0) + i\Gamma/2}, \qquad (7.27)$$

where $\Gamma = \hbar\gamma$. The normalized energy distribution is then

$$I(E) = \frac{\Gamma/\pi}{(E - E_0)^2 + \Gamma^2/4}, \qquad (7.28)$$

which is known as a *Lorentzian* or *Breit–Wigner* distribution. This equation makes explicit the connection between the energy uncertainty of a state and its lifetime, $\tau = 1/\gamma = \hbar/\Gamma$.

Lifetimes are shortened and line widths of atomic transitions increased by the motion of the atom and by its interactions with its neighbours. If the effect is the same for all atoms in the source the broadening is called *homogeneous*. In a perfect crystal the electric fields acting on all atoms would be identical so the broadening would be homogeneous. By contrast in an amorphous solid, like glass, the local fields vary and the broadening is *inhomogeneous*. In this case the transition energy changes from atom to atom, with little or no effect on lifetime. Thermal motion of atoms causes a Doppler shift of the frequency of the radiation emitted, another source of inhomogeneous broadening. Homogeneous broadening leads to a Lorentzian line profile, while inhomogeneous broadening leads to a Gaussian line profile. For reference the corresponding Gaussian profile is

$$I(E) = \frac{1}{\sqrt{2\pi}\sigma} \exp\left[\frac{-(E - E_0)^2}{2\sigma^2}\right], \qquad (7.29)$$

σ being the root mean square deviation of E. The comparison of these two shapes in Figure 7.2 reveals that the Lorentzian has a far longer tail.

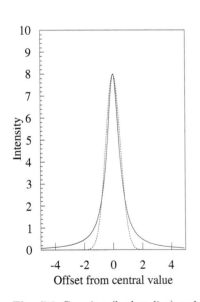

Fig. 7.2 Gaussian (broken line) and Lorentzian (solid line) line shapes.

7.6   Cross-sections

Before we can apply Fermi's golden rule to scattering processes we need to define a cross-section for a reaction and see how it is measured in a scattering experiment. Then Fermi's golden rule will be adapted for the calculation of cross-sections from matrix elements. The examples considered are two-body to two-body reactions, some elastic and some inelastic.

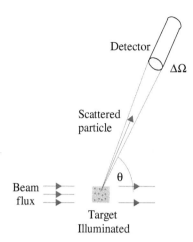

Fig. 7.3 Basic components of a scattering experiment.

Figure 7.3 shows the essential features of a scattering experiment. There is a flux of incoming particles (in blue) $I\,\mathrm{m^{-2}s^{-1}}$ illuminating a volume V with N_T scatterers (in red) per unit volume in a target, which might be a gas or liquid or solid. One or more detectors record the scattered particles. An individual detector records particles scattered within the solid angle $\Delta\Omega$ it subtends at the target, centred on a scattering angle θ. The detectors are placed far enough away from the target that these quantities do not vary appreciably across the target. The rate at which the particles are scattered into a solid angle $\Delta\Omega$ is proportional to the flux of incident particles, and to the number of scatterers illuminated by the incident flux, and also to the solid angle offered by the detector. We have a rate

$$\Delta n \propto IVN_T\Delta\Omega. \tag{7.30}$$

Summing over all directions of emission the total rate is

$$n = \sigma IVN_T, \tag{7.31}$$

where the constant, σ has the dimensions of area. It is the equivalent *area* of one scatterer as seen by an incident beam particle and is called the *cross-section*. Cross-sections depend on the strength of the interaction and this will vary from reaction to reaction. The rate of scattering, n/N_T is given by Fermi's golden rule, eqn. 7.8, so that the cross-section is

$$\sigma = (2\pi/\hbar)\mid M\mid^2\rho/F. \tag{7.32}$$

F is IV and called the *flux factor*. In the non-relativistic regime the flux factor[4] reduces to the difference between the velocities of beam and target particles

$$F = \mid \mathbf{v}_1 - \mathbf{v}_2 \mid. \tag{7.33}$$

In general the scattering intensity can depend on both the azimuthal and polar angles, ϕ and θ respectively. When scattering is measured in a narrow solid angle around one direction the measurement gives a differential cross-section

$$\mathrm{d}\sigma/\mathrm{d}\Omega = (2\pi/\hbar)\mid M(\theta,\phi)\mid^2\rho/F, \tag{7.34}$$

where ρ is now the density of states per unit solid angle in the direction of scattering. If the target is spherically symmetric the scattering is the same for all azimuthal angles, ϕ, around the beam direction and integrating over ϕ simply gives a factor 2π.

[4] ρ and F depend on the normalization of the states used in calculating the matrix element. Here we have for states with momenta $\mathbf{p}$ and $\mathbf{p}'$ $\langle \mathbf{p}|\mathbf{p}'\rangle = \delta(\mathbf{p} - \mathbf{p}')$. A different normalization appropriate to relativistic calculations with Feynman diagrams will be met in Chapter 18.

7.7   The Born approximation

Here the objective is to calculate the number and the angular distribution of particles elastically scattered from a fixed target when the interaction energy is much less than the total energy. The beam energy is taken to be small compared to a beam particle's mass so that the kinematics are non-relativistic. Then the interaction can be represented by a potential $U(\mathbf{r})$ where $\mathbf{r}$ is the particle-target separation. If the potential is spherically symmetric it can be written $U(r)$.

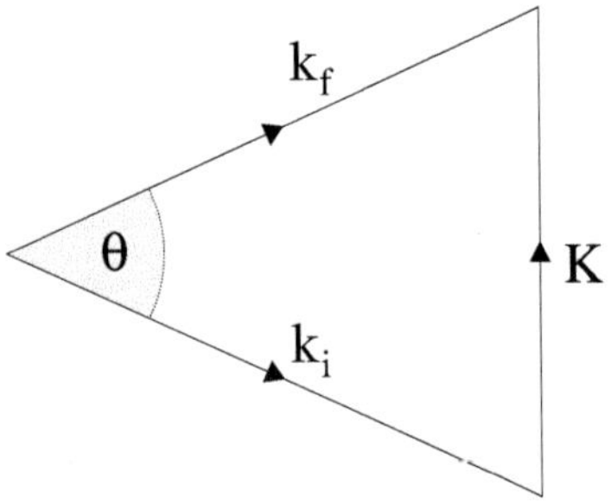

Because the interaction energy is small the ingoing and outgoing particles can be represented by plane waves. This is the first step in making the Born approximation:

$$\psi_i = \exp[i(\mathbf{k}_i \cdot \mathbf{r} - \omega_i t)], \tag{7.35}$$
$$\psi_f = \exp[i(\mathbf{k}_f \cdot \mathbf{r} - \omega_f t)], \tag{7.36}$$

where $\mathbf{k}$ is the wave-vector and ω the angular frequency, and we use unit normalization volume. Conservation of energy requires that $\omega_i = \omega_f$ which we set to ω and that $k_i = k_f$, which we set to k. Then we set

$$\mathbf{k}_f - \mathbf{k}_i = \mathbf{K}. \tag{7.37}$$

For the scattering through an angle θ, shown in Figure 7.4,

$$K = 2k \sin(\theta/2). \tag{7.38}$$

Calculating the matrix element for the interaction to first order in $U(r)$ completes the Born approximation:

$$M = \int \psi_i^* U(r) \psi_f \, d\mathbf{r} = \int U(r) \exp(i\mathbf{K} \cdot \mathbf{r}) \, d\mathbf{r}. \tag{7.39}$$

Its motion is assumed to be non-relativistic so the beam particle's energy $E = \hbar^2 k^2/(2m)$, m being its mass. Note that M has dimensions $\mathrm{kg\, m^5\, s^{-2}}$. The flux factor is simply the incident particle's velocity, $v = \hbar k/m$. Using eqn. B.10 the density of final states appearing in the differential cross-section will be

$$\rho(k, \Omega) \, dk \, d\Omega = \frac{k^2}{2\pi^2} dk \frac{d\Omega}{4\pi}, \tag{7.40}$$

where $d\Omega$ is solid angle presented by the detector at the target. The normalization volume is again unity. The density of final states must be expressed as a function of the total energy in order to be used in Fermi's golden rule. This is

$$\rho(E, \Omega) = \rho(k, \Omega) \frac{dk}{dE} = \frac{mk}{8\pi^3 \hbar^2}. \tag{7.41}$$

Inserting this value and the flux factor into eqn. 7.34 gives

$$d\sigma/d\Omega = \mid M \mid^2 [m^2/4\pi^2\hbar^4]. \tag{7.42}$$

Fig. 7.4 Wave-vectors in an elastic scatter through an angle θ.

The evaluation of the matrix element is straightforward

$$M = \int_{r=0}^{\infty} \int_{\theta=0}^{\pi} U(r) 2\pi r^2 \exp[iKr\cos\theta] \sin\theta \, d\theta \, dr$$

$$= (4\pi i/K) \int U(r) r \sin(Kr) \, dr. \tag{7.43}$$

The eqns. 7.42 and 7.43 form the template for calculating cross-sections in the Born approximation for spherically symmetric potentials. As an example we take the potential

$$U(r) = C \exp(-r/a)/r, \tag{7.44}$$

the shielded Coulomb potential, or Yukawa potential. Then[5]

$$M = \frac{4\pi i C}{K} \int_0^{\infty} \exp(-r/a) \sin(Kr) \, dr = \frac{4\pi C}{K^2 + a^{-2}}. \tag{7.45}$$

Now take the limit $a \to \infty$ so that we have a Coulomb potential and eqn. 7.43 gives

$$M = 4\pi i C/K^2. \tag{7.46}$$

Inserting this matrix element in eq. 7.42 gives

$$d\sigma/d\Omega = 4C^2 m^2/(\hbar^4 K^4). \tag{7.47}$$

Now $K = 2k\sin(\theta/2) = 2(p/\hbar)\sin(\theta/2)$, where p is the incident particle's momentum. If electrons are being scattered by a nucleus of atomic number Z then $C = Ze^2/(4\pi\varepsilon_0)$ and we get

$$d\sigma/d\Omega = \left[\frac{Ze^2 m}{8\pi\varepsilon_0 p^2}\right]^2 \operatorname{cosec}^4(\theta/2), \tag{7.48}$$

which is the expression for Rutherford scattering.

An important point to grasp is that the matrix element given in eqn. 7.39 is the Fourier transform of the potential. Thus the angular distribution of scattered particles can be used to infer the potential shape. Scattering from a $1/r$ potential will always produce the same $\operatorname{cosec}^4(\theta/2)$ angular dependence in the scattered intensity. In the same way diffraction of light at apertures is the Fourier transform of a two-dimensional slit pattern. A related point concerns the scattering of electrons, photons or phonons from a crystal lattice. Whenever the exchanged wave-vector $\mathbf{K}$ is a reciprocal lattice vector, the scattering from all the atoms in the lattice is in phase and the scattered intensity is strong: this is the Bragg scattering condition. Bragg scattering is the mechanism by which Bloch waves are generated, as described in Chapter 5.

---

[5] $\int \exp(-r/a)\sin(Kr)\,dr = K/(K^2 + a^{-2})$: equation 3.893/1 in *Tables of Integrals, Series and Products*, 5th edition, by I. S. Gradshteyn and I. M. Ryzhik, edited by A. Jeffrey, published by Academic Press (1994).

7.8   Nuclear β-decay

Free neutrons decay with a mean lifetime of 885.6 s. The decay is

$$n \rightarrow p + e^- + \overline{\nu}_e, \tag{7.49}$$

where p is a proton and $\overline{\nu}_e$ is an electron antineutrino, a neutral, nearly massless particle. Neutrinos and electrons are leptons, the particle species that feels the weak force and, if charged, the electromagnetic force, but not the strong force. As a result the mean free path of neutrinos with MeV energies is many light-years in material like the Earth. Consequently the electron antineutrino is not usually detected and its energy has to be inferred from the kinematics of the other decay products. The energy release, Q, is calculated from the difference between the rest mass energy of the proton and that of the charged products, giving 0.782 MeV. Nuclei are unstable to β-decay where this is kinematically allowed:

$$^A_Z X \rightarrow ^A_{Z+1} Y + e^- + \overline{\nu}_e. \tag{7.50}$$

In other nuclei a proton may convert to a neutron when energetically favourable.

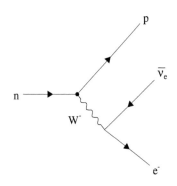

Fig. 7.5 Neutron decay process.

The underlying process is shown in Figure 7.5: a W-boson is emitted, which then decays to a electron and anti-neutrino. The W-boson, a carrier of the weak force, has a mass of 80.379 GeV/c², and lifetime $\hbar/\Gamma_W = \hbar/(0.012\,\text{GeV})$ that is $5.5\,10^{-23}$ s. Consistent with the uncertanty principle the energy-momentum imbalance can be tolerated for this short period. As a consequence weak processes are of short range, comparable to the nuclear size. Fermi made the approximation that the interaction is pointlike. Then the matrix element M_0 for the reaction in eqn. 7.50 collapses to an overlap integral between the parent nucleus and the decay products multiplied by G_F, *Fermi's coupling constant* for weak decays.

We write the kinetic energy and momenta of the daughter nucleus, the electron and the antineutrino as $(E, \mathbf{P})$, $(E_e,\mathbf{p}_e)$ and $(E_\nu,\mathbf{p}_\nu)$ respectively. The energy release in β-decay is a few MeV, hence the momenta of the decay products will be a few MeV/c. Then the daughter nucleus has energy $E = P^2/2M_N$ where M_N is the nucleus mass. E is thus of order 10^{-4} MeV, which is negligible. By contrast the motion of the neutrino with a mass less than $2\,\text{eV}/c^2$ is relativistic: to an excellent approximation $E_\nu = cp_\nu$. The neutrino interacts weakly so its wavefunction can be taken to be that of a free particle. On the other hand the Coulomb potential of the daughter nucleus, carrying charge Z'e, modifies the electron wavefunction, bringing a multiplicative *Fermi factor* $F(Z', p_e)$ into the decay rate. The other component of the matrix element is the overlap integral between the parent and daughter nuclei.

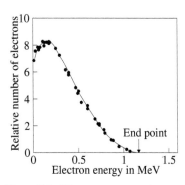

Fig. 7.6 The energy spectrum of electrons from the β-decay of ^{210}Bi; adapted from G. J. Neary, *Proceedings of the Physical Society (London)* A175, 71 (1940).

Turning to the density of final states, this is the product of densities for the leptons given by eqn. B.10,

$$\frac{p_e^2\,\mathrm{d}p_e}{2\pi^2\,\hbar^3}\frac{p_\nu^2\,\mathrm{d}p_\nu}{2\pi^2\,\hbar^3}:\tag{7.51}$$

while the momentum of the daughter nucleus is fixed once the other momenta are specified. Ignoring the tiny energy of the daughter nucleus, conservation of energy requires that

$$cp_\nu = Q - E_e.\tag{7.52}$$

Performing the integral over the neutrino momentum

$$\int\frac{p_\nu^2}{2\pi^2\,\hbar^3}\delta(Q - E_\nu - cp_\nu)\,\mathrm{d}p_\nu = \frac{(Q - E_\nu)^2}{2\pi^2\,\hbar^3\,c^3},\tag{7.53}$$

so that the density of states reduces to

$$\frac{1}{4\pi^4\,\hbar^6\,c^3}\,p_e^2\,(Q - E_e)^2\,\mathrm{d}p_e.\tag{7.54}$$

Inserting this in Fermi's golden rule, eqn. 7.8, the differential decay rate

$$\mathrm{d}R/\mathrm{d}p_e = \frac{G_F^2\,|\,M_o\,|^2}{2\pi^3\,\hbar^7\,c^3}p_e^2\,(Q - E_e)^2\,F\,(Z', p_e)\tag{7.55}$$

in terms of the electron momentum. The corresponding distribution in kinetic energy is shown in Figure 7.6 for the decay of ^{210}Bi. The attraction of the daughter nucleus on the electron, expressed by the Fermi factor, pulls the energy spectrum towards lower values of electron energy. A crude estimate of the overall rate of decay is obtained by putting $E_e = p_e c$, as if the electron mass were negligible, then

$$\int_0^{Q/c} p_e^2(Q - cp_e)^2\,\mathrm{d}p_e = Q^5/(30c^3).\tag{7.56}$$

In this approximation an estimate of the total decay is

$$R = G_F^2\frac{|\,M_o\,|^2\,Q^5 f(Z', Q)}{60\pi^3\,\hbar^7\,c^6},\tag{7.57}$$

where $f(Z', Q)$ is the factor needed to correct for the distortion of the electron wavefunction by the electric field of the daughter nucleus. There is a class of *allowed* decays in which the spins of the products equal vectorially that of the parent, and the spatial overlap of the daughter nucleus with its parent is large. In forbidden decays, the difference between the parent and daughter nuclear wavefunctions reduces the overlap integral by factors of over 10^3. Figure 7.7 shows the dependence of the transition rates on the energy release for some allowed transitions: these rates span nine orders of magnitude. The line corresponds to dependence $Q^{4.5}$ rather than Q^5, reflecting the impact of the Fermi factor in eqn. 7.57.

One method for making an estimate of the neutrino mass involves making very precise measurements of the end point of the electron energy

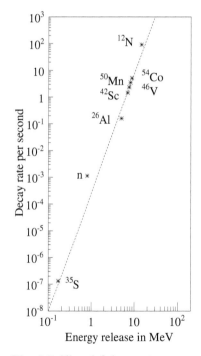

Fig. 7.7 Allowed β-decay rates versus energy release.

spectrum. So far the result from studying the decay $^3\text{H} \rightarrow {}^3\text{He} + \text{e}^- + \overline{\nu}_e$ is only a limit: the electron neutrino mass is less than a $2\,\text{eV}/\text{c}^2$.

The Fermi constant appearing in neutron decay is common to weak decays involving W-bosons. The cleanest determination of G_F is made by measuring the lifetime of the μ-*lepton* or *muon*. This is another lepton with charge $-e$ and spin $1/2$ like the electron. There is a companion neutrino with its antineutrino; all distinct from the electron family. The muon has a mass of $105.66\,\text{MeV}/\text{c}^2$, so it behaves like a heavy electron. Muons decay weakly to electrons via the decay mode

$$\mu^- \rightarrow \text{e}^- + \overline{\nu}_e + \nu_\mu, \tag{7.58}$$

with a lifetime of $2.20\,10^{-6}\,\text{s}$. No nuclei are involved so that the uncertainties connected with the calculation of the overlap integral of nuclei and the Fermi factor are eliminated. The muon lifetime is given by an expression that closely resembles eqn. 7.57,

$$1/\tau_\mu = G_\text{F}^2 \frac{(m_\mu c^2)^5}{192\pi^3 \hbar^7 c^6}, \tag{7.59}$$

where m_μ is the muon mass. This yields a value for G_F of $1.16637\,10^{-5}$ $\text{GeV}^{-2}(\hbar c)^3$ in the customary units of elementary particle physics, or $1.428\,10^{-62}\,\text{Jm}^3$.

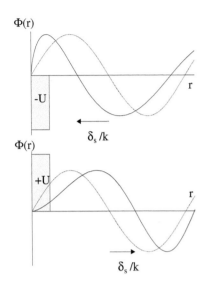

Fig. 7.8 The upper (lower) panel shows the phase advance (lag) for an attractive (repulsive) spherically symmetric potential. Broken lines are undisturbed waves.

7.9   Low-energy scattering

At the lower end of the energy scale, for example when scattering neutrons with sub-MeV energies from protons, a different approach is appropriate. The first step taken here is to show that only waves with zero relative angular momentum about the target are affected. With this simplification the scattering can be quantified by a phase shift or equivalently a scattering length. This formalism proves useful later when characterizing the interactions between cold atoms.

When the energy of the beam particle is small enough that its de Broglie wavelength is much larger than the size of the scatterer, scattering from all parts of the scatterer are in phase and it appears pointlike. In terms of the beam particle wave number k and target size R this condition requires $kR \ll 1$. Hence the orbital angular momentum of the beam particle at the surface of the scatterer is $kR\hbar \ll \hbar$. It follows that only the component of the incoming wave with orbital angular momentum zero, the s-wave component, is scattered.

In the absence of any scatterer the wavefunction of a particle in a beam of momentum $\mathbf{k}\hbar$ and energy $\hbar\omega$ is a plane wave $\exp[i(\mathbf{k}\cdot\mathbf{r} - \omega t)]$. The s-wave component of such a wave is spherically symmetric,

$$\psi(r)\exp(-i\omega t) = (i/2kr)\{\exp(-ikr) - \exp(ikr)\}\exp(-i\omega t). \tag{7.60}$$

Here the first term on the right is ingoing and the second term the outgoing wave. When the scattering potential is present the outgoing wave is affected and now

$$\psi(r) = (i/2kr)\{\exp(-ikr) - \exp[i(kr + 2\delta_\mathrm{s})]\}, \qquad (7.61)$$

where the common time-dependence is omitted from here onward. This equation can be rewritten

$$\psi(r) = \exp(i\delta_\mathrm{s})\sin(kr + \delta_\mathrm{s})/kr, \qquad (7.62)$$

where δ_s is known as the *phase shift* of the outgoing s-wave. The total wave intensity at a distance r from the scatterer is $4\pi r^2|\psi(r)|^2$: consequently the radial waveform $\Phi(r) = r\psi(r)$ is also useful in showing the effect of scattering. The panels of Figure 7.8 show the resultant phase shifts of $\Phi(r)$ for attractive and repulsive potentials. A second parameter used to quantify the low-energy scattering is the *scattering length*, a_s, defined as the radius at which the wavefunction passes through zero. At this location

$$\sin(ka_\mathrm{s} + \delta_\mathrm{s}) = 0. \qquad (7.63)$$

Figure 7.9 shows the waveforms as modified by attractive and repulsive potentials. Note that with a repulsive potential it is always the case that $a_\mathrm{s} > 0$. With a weak attractive potential $\delta_\mathrm{s} < \pi/2$ and $a_\mathrm{s} < 0$. On increasing the potential δ_s eventually reaches $\pi/2$; at which point eqn. 7.61 becomes a standing wave and there can be a bound state. Further increases in potential depth make a_s positive which, unexpectedly, makes the s-wave response a weak repulsion. It may help to recall the behaviour of a driven mechanical oscillator. If the drive frequency is well below the oscillator's resonance frequency the motion produced is in phase with the drive, while if the drive frequency is well above resonance the motion is in antiphase with the drive. When δ_s reaches π eqn. 7.61 reduces to eqn. 7.60 so that there is no scattering and the target appears transparent. This is known as the Ramsauer–Townsend effect. The corresponding sharp drop in cross-section, illustrated in Figure 7.10, was first observed in electron scattering from noble gases (Xe, Kr, Ar) at energies a little below $1\,\mathrm{eV}$. It had to await the arrival of quantum mechanics before a satisfactory explanation could be given.[6]

The scattered wave is the difference between eqns. 7.61 and 7.60

$$\begin{aligned}
\psi_\mathrm{scatt}(r) &= (i/2kr)\exp(ikr)[1 - \exp(2i\delta_\mathrm{s})] \\
&= (1/kr)\exp[i(kr + \delta_\mathrm{s})]\sin\delta_\mathrm{s}. \qquad (7.64)
\end{aligned}$$

Summing the scattered outward flux over all directions

$$I_0 = (4\pi r^2)\,|\,\psi_\mathrm{scatt}(r)\,|^2\,v = (4\pi v/k^2)\sin^2\delta_\mathrm{s}, \qquad (7.65)$$

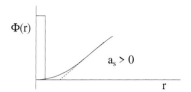
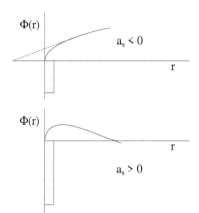
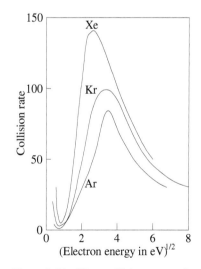

Fig. 7.9 Scattering lengths: $a_\mathrm{s} > 0$ for a repulsive potential; and two possibilities with attractive potentials.

Fig. 7.10 The collision rate for low-energy electrons in noble gases; adapted from R. B. Brode, *Reviews of Modern Physics* 5, 257 (1935).

---

[6] An undergraduate experiment to demonstrate the Ramsauer–Townsend effect using the xenon-filled 2D21 thyratron is described by Kukolich in *The American Journal of Physics*, 36, 701 (1968).

where $v = \hbar k/m$ is the particle velocity. Now the beam flux is simply $I = v$ so the s-wave cross-section

$$\sigma = I_0/I = \frac{4\pi}{k^2}\sin^2\delta_s. \tag{7.66}$$

Then if δ_s is sufficiently small, and using eqn. 7.63

$$\sigma \approx 4\pi a_s^2. \tag{7.67}$$

This formalism can be applied to neutron scattering from protons at sub-MeV energies.

Condensates of gases of the alkali metals, ^{87}Rb, ^{23}Na, or ^{7}Li are described in Chapter 16. The kinematics are unusual: the atoms have kinetic energies $\sim 10^{-11}$ eV with a corresponding free space wavelengths $\sim 1\mu$m. At these energies atom-atom interactions have potentials like those shown in Figure 14.3, $\sim$meV deep, far larger than the kinetic energy of the incident atoms. It follows that kinetic energy inside the potential is orders of magnitude greater than outside. Consequently the wavelength inside the potential well is correspondingly very much shorter, with the waves undergoing many cycles within the potential well. The same mismatch means that most of the incident wave is reflected and the amplitude of the wave within the well is thus relatively small. These features are roughly indicated in the panels of Figure 7.11: the wave amplitude inside the well is tiny compared to that outside the well, and the scattering lengths can be a hundred times the well radius.

For the cases of ^{87}Rb and ^{23}Na, illustrated in the lower panel of Figure 7.11, a_s is positive and there is weak repulsion. This fact is critical to maintaining stability in gaseous Bose–Einstein condensates.

Scattering in a gas condensate is the scattering of like bosons. Exchanging their identities generates a second valid contribution to the s-wave amplitude. The $L=0$ state is symmetric under interchange of like particles, which makes these contributions equal: $f(-\theta, \phi+\pi) = f(\theta, \phi)$. The matrix element is then proportional to $|f(\theta, \phi) + f(\theta, \phi)|/\sqrt{2}$ and the cross-section is doubled:

$$\sigma \approx 8\pi a_s^2, \tag{7.68}$$

Note that like fermion pairs are in an overall antisymmetric states so their spin state would be antisymmetric. In this low-energy regime the Born matrix element from eqn. 7.39 reduces to

$$M = \int_0^\infty U(r)\mathrm{d}\mathbf{r}. \tag{7.69}$$

An approximation useful in the context of Bose–Einstein condensates is to take the potential as point-like, $W_0\delta(\mathbf{r})$, and then M equals W_0. U has the units of energy, $\mathrm{kg\,m^2\,s^{-2}}$, W_0 has the dimensions of energy$\times$volume,

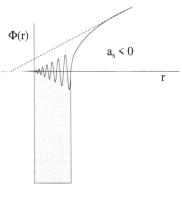

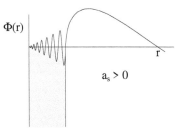

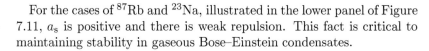

Fig. 7.11 Scattering lengths: for alkali metal collisions at microKelvin temperatures.

i.e. $\mathrm{kg\,m^5\ s^{-2}}$. Inserting this matrix element into eqn. 7.42 gives a total cross-section

$$\sigma = 4\pi \frac{\mathrm{d}\sigma}{\mathrm{d}\Omega} = \frac{W_0^2 m^2}{\pi \hbar^4}. \qquad (7.70)$$

By comparing the above equation with eqn. 7.67 we can get a relation between the point-like potential and the scattering length. We should use the reduced mass, which for identical particles colliding is $m/2$,

$$W_0 = 4\pi \hbar^2 a_\mathrm{s}/m. \qquad (7.71)$$

In a gas condensate the effect of all the atoms must be taken into account. Hence the potential felt by each atom within a condensate with n atoms per unit volume is

$$U_0 = \frac{4\pi \hbar^2 n a_\mathrm{s}}{m}. \qquad (7.72)$$

The results of this section are used in Chapter 16 where Bose–Einstein gas condensates are studied.

7.10   Further reading

More details on nuclear β-decay can be found in *Introductory Nuclear Physics*, by K. S. Krane, published by John Wiley, New York (1988), or *Nuclear Physics, Principles and Applications*, by J. Lilley, published by John Wiley (2001).

The use of Partial Wave Analysis in elementary particle physics is described in *Elementary Particle Physics* by I. R. Kenyon, published by Routledge Kegan-Paul (1987).

A more complete analysis of potential scattering is available in *Lectures on Quantum Mechanics*, by Steven Weinberg, published by Cambridge University Press (2013).

Exercises

(7.1) The potential given in eqn. 7.44 can be used to parametrize the strong force binding nucleons into nuclei. Take a nuclear size of around $1.5\,10^{-15}$ m to define the range of the force. What is the mass of the postulated particle in MeV/c²? Use $\hbar = 6.58\,10^{-22}$ MeVs.

(7.2) Show that the angular distribution produced by scattering from a point target is isotropic.

(7.3) Use this equation for the centre of mass cross section:

$$\mathrm{d}\sigma/\mathrm{d}\Omega = E_\mathrm{CM}^2 \mid M \mid^2 /[64\pi^2 (\hbar c)^4] \qquad (7.73)$$

Show that an estimate of the cross-section for inverse β-decay

$$\bar{\nu}_e + p \to e^+ + n,$$

is

$$\sigma = G_\mathrm{F}^2 s/[16\pi (\hbar c)^4],$$

where s is the cm energy squared. Reines and Cowan in 1956 used this reaction to to detect neutrinos emitted by a nuclear reactor. It is being proposed as a tool to measure neutrino flux from reactors in order to monitor reactor useage. The reaction is isotropic in the centre of mass frame. You may take the matrix element to be G_F, that is $1.166\ 10^{-11}$ MeV^{-2} $(\hbar c)^3$ and the proton mass to be 938 MeV/c^2. Calculate the cross-section in the laboratory frame at a neutrino energy E_ν. Use $\hbar c = 197$ MeV fm.

(7.4) What potential represents an impenetrable (hard) sphere? Calculate the s-wave scattering cross-section for elastic scattering from such a sphere of radius a.

(7.5) At energies below 10 keV the neutron-proton cross-section is elastic and constant. Both particles have spin 1/2 like the electron. The singlet scattering length is -23.5 fm and the triplet scattering length is 5.4 fm. Calculate the cross-section for unpolarized particles. The deuteron is a state of neutron plus proton bound by 2.22 MeV. This is a triplet spin unity state and there is no corresponding bound proton-neutron singlet state. Is this consistent with the scattering lengths?

(7.6) In a scattering process involving identical particles

$$a + a \to a + a,$$

the matrix element for scattering through an angle θ is $f(\theta)$, with labels assigned to follow particles through the Feynman diagrams. What is the full

matrix element squared if they are bosons, and also if they are fermions?

(7.7) The intensity of electric dipole radiation varies with the square of the electric dipole moment; that of magnetic dipole radiation with the magnetic dipole moment squared. Estimate the suppression of atomic magnetic dipole radiation compared to electric dipole radiation. Take the magnetic dipole moment to be iA, where i is the electron current in its orbit and A the orbit area.

(7.8) In the typical weak decay

$$\mu^- \to e^- + \bar\nu_e + \nu_\mu$$

the energy release is 105 MeV and the lifetime is $2.2\ 10^{-6}$ s. In the electromagnetic decay

$$\pi^0 \to \gamma + \gamma$$

the energy release is 125 MeV and the lifetime is $0.8\ 10^{-16}$ s. In the strong process in which a Δ resonance of mass 1232 MeV/c^2 is produced and decays

$$p + \pi^+ \to \Delta^{++} \to p + \pi^+$$

the width of this resonance is 120 MeV/c^2. Compare the strengths of the weak, electromagnetic and strong forces from this data. You can use the value for $\hbar$ of $6.8\ 10^{-22}$ Mev s.

(7.9) Use the Born approximation to calculate the angular distribution of elastic scattering from a fixed spherically symmetric potential $U(r) = A/r^2$. The definite integral $\int_0^\infty x^{-1}\sin x\,dx$ has the value $\pi/2$.

Field quantization

8

8.1 Introduction

The comprehensive theory of processes involving electrons and electromagnetic radiation is *quantum electrodynamics* a theory in which the fields themselves are quantized. This theory is an element of the standard model of particle physics encompassing strong, weak and electromagnetic interactions. The success of this quantum field theory culminated in the recent observation of the Higgs boson. In this chapter elements of the quantum theory of electromagnetism are developed.

Heretofore the electromagnetic interactions have been treated semiclassically with only the atomic states quantized. In the following section of this chapter the complementary step of quantizing the electromagnetic fields is presented. This is known as *second quantization* to distinguish it from the familiar quantization of kinematic quantities. The new quantum operators create or destroy photons.

Einstein deduced that for radiation and matter to attain thermal equilibrium there must exist an additional process beyond the absorption and spontaneous emission, called *stimulated* emission. The close relationship between the three processes is deduced here using the creation and annihilation operator formalism. Next the way matter can be manipulated so that stimulated emissions produce astronomical number of coherent photons is described, a process whose description, *Light Amplification by the Stimulated Emission of Radiation*, led to the acronym *laser*. Widely used semiconductor laser diodes are discussed in detail.

Unexpectedly the correlations between photons in thermal light are stronger than those in laser beams. The pioneering experiments on correlations by Hanbury Brown and Twiss using photons from thermal sources are described. How photon and other boson correlations can give information about the size of the source is reviewed. A discussion of the corresponding properties of fermions in the non-relativistic limit is the final topic of the chapter. Dirac's relativistic equation for the electron is detailed in Appendix G.

Quantum 20/20: Fundamentals, Entanglement, Gauge Fields, Condensates and Topology.
Ian R. Kenyon. © 2020. Published in 2020 by Oxford University Press.
DOI: 10.1093/oso/9780198808350.001.0001

8.2   Second quantization

The electric and magnetic fields (always real) in a plane wave are written in a form convenient to our purpose:

$$
\begin{aligned}
E_x &= 2\zeta_\omega a \cos(kz - \omega t) \\
&= \zeta_\omega \left[a(t) \exp(ikz) + a^*(t) \exp(-ikz) \right], \quad (8.1) \\
B_y &= E_x/v, \quad (8.2)
\end{aligned}
$$

where $\omega = kv$, $v = c/\sqrt{\varepsilon_\mathrm{r}}$ and ε_r is the relative permittivity. $a(t) = a \exp(-i\omega t)$. The constant $\zeta_\omega = \sqrt{\hbar\omega/2\varepsilon_0\varepsilon_\mathrm{r}V}$ is in volts/metre. V is a reference volume.[1] Then the total energy in a volume V of the electromagnetic field:

$$
H = \int_V \varepsilon_0\varepsilon_\mathrm{r}E^2 \, \mathrm{d}V = \hbar\omega \; a^*(t)\,a(t). \quad (8.3)
$$

A first pointer to quantizing the electromagnetic field is that if $a^*(t)a(t)$ were taken to be the number of photons this equation would match Planck's formula. Real quantities, $q(t)$ and $p(t)$, are introduced

$$
a(t) = [\omega q(t) + ip(t)]/\sqrt{2\hbar\omega}, \quad (8.4)
$$

so that the energy equation, eqn 8.3, becomes

$$
H = (1/2)[\omega^2 q(t)^2 + p(t)^2]. \quad (8.5)
$$

Here we have a second pointer: this equation duplicates the energy equation for a simple harmonic oscillator (SHO) where $q(t)$ is its displacement times the square root of the mass and $p(t)$ its momentum divided by the square root of the mass. Consequently we can take over the quantization procedure given in Section 2.3. The equal steps between energy levels of the quantized SHO are mirrored here by equal energy increments as successive photons are added to a mode of the electromagnetic field.

Referring to eqn. 2.29, an immediate inference from the analysis of the harmonic oscillator is that a mode containing n photons has an energy $E_n = (n+1/2)\hbar\omega$. This matches what we anticipated apart from an unexpected energy of $\hbar\omega/2$ in an empty mode: a point we return to shortly.

The electromagnetic field quantities $q(t)$ and $p(t)$ correspond to *position* coordinates and to the *momenta* respectively. Quantum mechanical operators $\hat{q}(t)$ and $\hat{p}(t)$ can therefore be associated with them, which must obey the usual commutator relation connecting position and momentum operators:

$$
[\hat{q}(t), \hat{p}(t)] \equiv \hat{q}(t)\hat{p}(t) - \hat{p}(t)\hat{q}(t) = i\hbar. \quad (8.6)
$$

In turn new operators $\hat{a}$ and $\hat{a}^\dagger$ can be defined using eqn. 8.4 to be the operator equivalents of a and a^*. Then from eqn. 8.6 we deduce the commutation relation

$$
[\hat{a}, \hat{a}^\dagger] = 1. \quad (8.7)
$$

[1]The spatial extent of actual wavepackets is finite, while the plane waves in their Fourier analysis extend to infinity. Hence we need a finite reference volume V to contain the wave packets. V cancels out from all measured quantities.

Operators for different modes always commute.

The operators act on state vectors (see Section 1.15) describing modes of the electromagnetic field. A state vector is written as a ket $|n\rangle$, in which the argument is the number of photons in the mode. The corresponding bra is $\langle n|$. Where necessary, labels can be attached to identify the mode. The state vector of the electromagnetic field is the product of state vectors for all modes. Replacing classical variables by operators in eqn. 8.5 gives the energy operator for a mode:

$$\hat{H} = (1/2)(\omega^2\hat{q}\hat{q} + \hat{p}\hat{p}) = \hbar\omega\,(\hat{a}^\dagger\hat{a} + 1/2). \tag{8.8}$$

Then

$$\hat{H}|n\rangle = \hbar\omega\,(\,\hat{a}^\dagger\hat{a} + 1/2\,)|n\rangle. \tag{8.9}$$

Comparing this result with the mode energy $(n + 1/2)\hbar\omega$ reveals that the operator $\hat{a}^\dagger\hat{a}$ counts the number of photons in the mode.

$$\hat{a}^\dagger\hat{a}|n\rangle = n|n\rangle. \tag{8.10}$$

We now show that $\hat{a}$ and $\hat{a}^\dagger$ themselves are operators respectively annihilating and creating single photons in the mode. Consider the effect of $\hat{a}$ acting on the mode when it contains n photons. The energy of the new state $\hat{a}|n\rangle$ is given by:

$$\hat{H}\,\hat{a}|n\rangle = \hbar\omega\,(\hat{a}^\dagger\hat{a} + 1/2)\hat{a}|n\rangle = \hbar\omega\,\hat{a}\,(\hat{a}^\dagger\hat{a} - 1/2)|n\rangle,$$

where we have used the commutation relation, eqn. 8.7, to move one $\hat{a}$ operator past the $\hat{a}^\dagger$ operator. Continuing,

$$\hat{H}\,\hat{a}|n\rangle = \hbar\omega\,\hat{a}\,(n - 1/2)|n\rangle = (n - 1/2)\hbar\omega\,\hat{a}|n\rangle. \tag{8.11}$$

Evidently the new state, $\hat{a}|n\rangle$, contains one less photon, which justifies the identification of $\hat{a}$ as an annihilation operator. The new state $|n-1\rangle$ including a normalization factor is

$$|n - 1\rangle = \hat{a}|n\rangle/\sqrt{n}. \tag{8.12}$$

Similarly the operator $\hat{a}^\dagger$ creates an additional photon, and we have[2]

$$|n + 1\rangle = \hat{a}^\dagger|n\rangle/\sqrt{n + 1}. \tag{8.13}$$

The corresponding bras are

$$\langle n - 1| = \langle n|\hat{a}^\dagger/\sqrt{n} \text{ and } \langle n + 1| = \langle n|\hat{a}/\sqrt{n + 1}. \tag{8.14}$$

Then for example

$$\langle n - 1|n - 1\rangle = \langle n|\hat{a}^\dagger\hat{a}|n\rangle/n = 1. \tag{8.15}$$

It follows that the state containing exactly n photons, called a *Fock state*, can be built up from the vacuum state $|0\rangle$ as follows

$$|n\rangle = [\hat{a}^\dagger]^n|0\rangle/\sqrt{n!} \tag{8.16}$$

[2]Suppose

$$\hat{a}|n\rangle = k_n|n - 1\rangle,$$
$$\hat{a}^\dagger|n\rangle = l_{n+1}|n + 1\rangle,$$

making no assumptions about k_n and l_n. However, it is the case that

$$[\hat{a}, \hat{a}^\dagger]|n\rangle = [l_{n+1}k_{n+1} - l_nk_n]|n\rangle = |n\rangle,$$

for all n. This is only possible if $k_n = l_n = \sqrt{n}$.

Note that individually neither a nor $a^\dagger$ are observables.

The electric field operator for a given mode is constructed from the classical electric field given in eqn. 8.1,

$$\hat{E} = \hat{E}^+ + \hat{E}^- = \zeta_\omega \left[\hat{a} \exp\left(i\xi_\omega \right) + \hat{a}^\dagger \exp\left(-i\xi_\omega \right) \right], \qquad (8.17)$$

where $\xi_\omega = kz - \omega t$, and as before $\zeta_\omega = \sqrt{\hbar\omega/2\varepsilon_0\varepsilon_\mathrm{r}V}$ and $\omega = kv$. $\hat{E}^+$ only contains *annihilation* operators $\hat{a}$, and is by convention called the *positive frequency* component of the field operator $\hat{E}$. $\hat{E}^-$ contains only *creation* operators $\hat{a}^\dagger$ and is called the *negative frequency* component. The magnetic field is obtained from eqn. 8.2 with operators replacing simple fields. Applying eqn. 8.7 we get the commutation relation between components of the same angular frequency,

$$[\, \hat{E}^+, \hat{E}^- \,] = \zeta_\omega^2. \qquad (8.18)$$

A comment is in order here to explain a change in the way that time-dependence is being handled. The time-dependence has been moved to the operators $\hat{E}^-$ and $\hat{E}^+$, while the modes of the electromagnetic field on which they operate are unchanging. This is called the *Heisenberg* representation, in contrast to the *Schrödinger* representation, in which the time-dependence is all in the wavefunctions, and the operators are time independent.

For reference, in the mechanical harmonic oscillator we can recover the position and momentum operators thus:

$$\hat{q}/\sqrt{m} \;=\; \sqrt{\hbar/(2m\omega)}\,[\hat{a}^\dagger + \hat{a}] \qquad (8.19)$$

$$\sqrt{m}\hat{p} \;=\; i\sqrt{\hbar m\omega/2}\,[\hat{a}^\dagger - \hat{a}] \qquad (8.20)$$

8.2.1   Vacuum energy

Earlier the appearance of an energy $\hbar\omega/2$ in each empty mode was noted for future reference. It follows that the vacuum, the state with zero photons in any mode, has energy:

$$H_0 = \sum_\omega \hbar\omega/2, \qquad (8.21)$$

where the sum runs over all modes of the electromagnetic field. Therefore, *in quantum field theory the vacuum is not a static passive state, but has what is called a zero point energy*. This energy is not accessible to direct measurement because energies that are measurable are always differences in energy, eliminating the zero point energy. In terms of the electric field the total energy in a single mode,

$$\varepsilon_o\varepsilon_\mathrm{r}V\langle \mathbf{E}^2 \rangle = (n + 1/2)\hbar\omega, \qquad (8.22)$$

showing that in the vacuum the mean square values of the fields do not vanish.

From the existence of this *zero point* energy Casimir inferred that two closely-spaced, neutral, conducting surfaces in vacuum should feel a mutual attractive force.[3] Casimir noticed that within the gap the modes must be restricted to the discrete wavelengths for which **E** will vanish at the surfaces; whereas, outside the gap, any wavelength is allowed. This means that the total zero point energy density of the electromagnetic field within the cavity is less than that outside. Hence there is a net inward force on the conductors:

$$F = -\frac{\hbar c \pi}{240 d^4},\tag{8.23}$$

in Nm^{-2} for a separation of d m. Though small, this *Casimir force* has been measured and found to agree with the prediction to within the experimental uncertainty of 1 per cent.

The *vacuum fluctuations* of the electromagnetic field produce small changes in the energy levels of atomic states, and the change is slightly different for each different atomic configuration. Most famous of all is the Lamb shift in the hydrogen atom: the predicted displacement upward of the $2s_{1/2}$ level by $1054\,MHz\,\hbar$ relative to the $2p_{1/2}$ level.[4] The displacement was measured by Lamb and Retherford in 1947. The agreement with the prediction to within $0.2\,MHz\,\hbar$ was crucial early evidence of the precision and reliability of the predictions of quantized field theory.

The upper bound on the frequency and hence the energy of modes of the electromagnetic field is a matter of speculation. This limit must exceed the maximum cosmic ray energy, namely $10^{20}\,eV$. Then by implication the total zero point energy in the vacuum must be enormous. Now it is accepted that all forms of energy contribute to the stress-energy tensor in the fundamental equation of General Relativity:

$$\text{Curvature tensor} \;-\Lambda g = \;[8\pi G/c^4]\;[\text{Stress} - \text{energy tensor}]\tag{8.24}$$

where G is the gravitational constant, Λ is the *cosmological constant* and g the metric tensor from eqn. 1.71. Astrophysical measurements determine the curvature tensor and the cosmological constant. The result implies that the stress-energy content of the universe is 10^n times smaller than that expected when the zero point energy is included in the stress-energy tensor, with n being anywhere from 50 to 100! This factor depends on which plausible assumption is made for the mode cut-off frequency. This is possibly the biggest unsolved problem in physics.[5]

8.3   Absorption and emission processes

The three processes by which matter emits or absorbs electromagnetic radiation are sketched in Figure 8.1. Of these the upper pair are already familiar: the absorption and the *spontaneous* emission of a photon whose energy equals that gained or lost by the atom changing its quantum

[3]See R. J. Jaffe 'Casimir effect and the quantum vacuum', *Physical Review* D72 021301 (2005).

[4]See pages 60 and 159 in *Relativistic Quantum Mechanics* by J. D. Bjorken and S. D. Drell, published by McGraw-Hill Book Company (1964).

[5]When Planck solved one UV catastrophe in physics he unwittingly replaced it with another. See S. Weinberg, 'The Cosmological Constant Problem', *Reviews of Modern Physics* 61, 1 (1989).

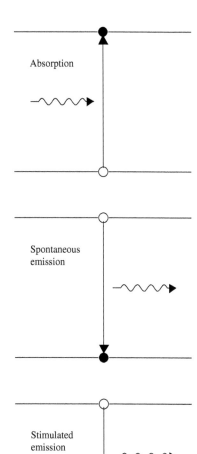

state. In 1916 Einstein realized that *stimulated* emission exhibited on the lower panel of the figure is required to maintain thermal equilibrium. In this process a photon is incident on an atom in an excited state, and with an energy equal to the atom's excitation energy. The incident photon stimulates the atom to drop into the lower energy state, and to emit a second photon with identical energy, direction, polarization and phase. The two photons are thus fully coherent and share the same wavefunction. The relationships between the processes illustrated are deduced here using the quantized radiation field. An electric dipole transition is considered but the conclusions will be seen to be quite general.

The emission rate given by Fermi's golden rule, eqn. 7.8, per unit angular momentum, rather than per unit energy, is

$$\gamma = (2\pi/\hbar^2)|M_{\mathrm{fi}}|^2 \rho(\omega) \tag{8.25}$$

where $|M_{\mathrm{fi}}|$ is the matrix element for the transition considered. $\rho(\omega)$ is the density of photon states in angular frequency given by eqn. B.11, $\omega^2 V/(\pi^2 c^3)$, for a normalization volume V. The matrix element for an electric dipole transition to a lower energy state of the atom accompanied by photon emission is

$$M_{\mathrm{fi}} = \langle m+1| \mathbf{E} |m\rangle \cdot \mathbf{D}_{21}, \tag{8.26}$$

where the field mode involved contains initially m photons and finally $m+1$ photons inside the normalization volume. The electric dipole moment

$$\mathbf{D}_{21} = e \int \psi_2^* \mathbf{r} \psi_1 \, dV,$$

where ψ_1 and ψ_2 are the respective wavefunctions of the atom before and after emission. Now using eqn. 8.17 to re-express the electric field as an operator

$$
\begin{aligned}
\langle m+1| \hat{\mathbf{E}} |m\rangle &= \zeta_\omega \langle m+1|[\hat{a} \exp(+i\xi_\omega) + \hat{a}^\dagger \exp(-i\xi_\omega)]|m\rangle)\boldsymbol{\varepsilon} \\
&= \zeta_\omega \sqrt{(m+1)} \exp(-i\xi_\omega)\,\boldsymbol{\varepsilon},
\end{aligned}
$$

where $\boldsymbol{\varepsilon}$ is a unit vector in the direction of the electric field. Then

$$|M_{\mathrm{fi}}|^2 = \zeta_\omega^2 (m+1)\mu^2, \tag{8.27}$$

where μ^2 is the mean squared value of the electric dipole moment. Thus the rate for emission is

$$\gamma_{\mathrm{em}} = Z(\omega)\,(m+1)\,\rho(\omega), \tag{8.28}$$

where $Z = \pi\mu^2\omega/(\hbar V\varepsilon_0\varepsilon_{\mathrm{r}})$. A similar calculation for absorption starting from a state that contains m photons again gives

$$\langle m-1|\hat{\mathbf{E}}|m\rangle = \zeta_\omega \sqrt{m} \exp(+i\xi_\omega)\,\boldsymbol{\varepsilon}. \tag{8.29}$$

Fig. 8.1 The three photon–matter interactions: in the upper panel absorption; in the centre panel spontaneous emission; and in the lower panel stimulated emission.

Then the *absorption* rate is

$$\gamma_{ab} = Z(\omega)\, m\, \rho(\omega). \tag{8.30}$$

When there are no photons in the initial state eqn. 8.28 collapses to

$$\gamma_{sp} = Z(\omega)\rho(\omega), \tag{8.31}$$

which is therefore identified as the *spontaneous* emission rate. Comparing eqn. 8.31 with eqn. 8.28, we see that the difference is

$$\gamma_{st} = Z(\omega)\, m\, \rho(\omega). \tag{8.32}$$

This is called the rate of *stimulated* emission. Stimulated emission is thus driven by the number of photons already present through the factor m, while spontaneous emission is driven by the vacuum, or to be more explicit, by the electromagnetic field fluctuations in the vacuum. This was the rate calculated semi-classically in Section 7.3 for the $2p \rightarrow 1s$ transition in a hydrogen atom.[6]

In terms of the spectral energy density per unit volume $W(\omega)$, and recalling that there are m photons present,

$$\rho(\omega) = W(\omega)V/(m\hbar\omega). \tag{8.33}$$

Substituting this expression for $\rho(\omega)$ in the three rates calculated above we obtain expressions defining the *Einstein coefficients* A_{21}, B_{21} and B_{12}:

$$\gamma_{st} = g_2 B_{21} W(\omega), \tag{8.34}$$
$$\gamma_{ab} = g_1 B_{12} W(\omega), \tag{8.35}$$
$$\gamma_{sp} = A_{21}. \tag{8.36}$$

The respective degeneracies g_1 and g_2, of the lower and upper atomic states, have been omitted previously to simplify the presentation. B_{21} is the probability for emission per unit spectral energy density, B_{12} the corresponding absorption probability, and A_{21} the spontaneous emission rate. Then in thermal equilibrium (see exercises)

$$g_1 B_{12} = g_2 B_{21} \text{ and } A_{21} = \frac{\hbar\omega^3}{\pi^2 c^3} B_{21}. \tag{8.37}$$

The expressions for γ_{abs}, γ_{st}, and γ_{sp} have in common the factor $Z(\omega)$. When interactions other than the electric dipole are considered this factor alone changes: hence the relations just proved between the rates are valid in general.

8.4 Prerequisites for lasing

Lasing exploits the coherence between the parent and daughter photon in stimulated emission. If this process can be continued through generations with each parent and daughter continuing to stimulate further

[6] If we try to interpret the quantum processes for a classical driven pendulum there is a surprise. The drive does work on the pendulum for one choice of phase and if the phase is shifted by π energy is drawn from the pendulum. These are the classical counterparts of absorption and stimulated emission. Spontaneous emission is the component missing in the classical picture and not, as we might expect, stimulated emission.

Table 8.1 Coherence times and lengths.

Source	τ_c	ℓ_c
Sunlight 0.4–0.7 µm	3 fs	900 nm
Hg lamp 435 nm line	100 fs	30 µm
Low pressure sodium lamp	5 ps	1.5 mm
Multimode He:Ne laser	0.6 ns	18 cm
Single mode He:Ne laser	1 µs	300 m

emissions the outcome is a pulse of photons that are coherent. They share the same wavelength, polarization and phase and have a common wavefunction: the source *lases*. In a material in thermal equilibrium at temperature T K the ratio of stimulated to spontaneous emission is

$$\gamma_{\mathrm{st}}/\gamma_{\mathrm{sp}} = m = 1/[\exp(\hbar\omega/(k_{\mathrm{B}}T)) - 1], \tag{8.38}$$

where m is the mean number of photons in a mode. This is negligible even at 10^3 K for optical transitions with $\hbar\omega$ of order 2 eV. Stimulated emission only becomes important when there is a *population inversion* with more atoms in the excited state than in the ground state: stimulated emission then dominates. The stimulated pulses depend on the presence of a seed photon: hence such pulses are random in time and terminate when they emerge from the material. The final ingredient required to give sustained coherent emission is to enclose the active material in a reflective, *Fabry–Perot*, cavity. Then a standing wave can develop, and if one mirror has a small transmission coefficient the emerging radiation will be a laser beam. If the cavity length is d, and the material has refractive index μ, then the possible cavity resonances have wavelengths, λ, given by

$$n\lambda = 2\mu d. \tag{8.39}$$

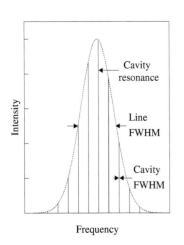

Fig. 8.2 The atomic transition line width and cavity resonance positions. Cavity lines are generally far narrower than drawn here.

The integer n will be very large with a 1 mm long cavity. Several cavity resonances may lie within the natural linewidth, as shown in Figure 8.2, with a pulse amplification per pass proportional to the corresponding line height.

The coherence times and coherence lengths for various sources are shown in Table 8.1; this makes clear the huge improvement in coherence that lasers bring. The simplest *transverse* mode, the TEM_{00} mode, is plane polarized, the *magnitude* of the electric field has azimuthal symmetry and falls off with a Gaussian dependence on the radial distance from the beam axis. This is the mode that lasers are generally designed to produce because of its compactness and symmetry around the beam axis. The lateral spread of many laser beams is due almost entirely to diffraction: that is to say they are *diffraction limited*. Beams from a single longitudinal mode laser can be orders of magnitude brighter than those from sources previously available.

8.4.1 Double heterostructure lasers

The basic semiconductor lasers which operate efficiently in ambient conditions have the generic structure illustrated in Figure 8.3. The material used, here GaAs, has direct bandgaps (see Section 6.3.1) so that electron transitions from conduction to valence band result in a photon emission without any accompanying phonons. This diagram shows a cross-section taken perpendicular to the direction of the laser beam. Lasing occurs in the GaAs region sandwiched between layers of n- and p-doped GaAlAs. There are two *heterojunctions* separating pairs of dissimilar semiconductors, n-GaAlAs/GaAs and GaAs/p-GaAlAs, thus making a *double heterostructure* (DH). The few micron width of the positive electrode in contact with the p-GaAlAs determines the region through which the electric current flows and hence the width of the region in the GaAs layer which lases. This active *stripe* has a rectangular cross-section, ∼100 nm deep. The device length along the beam axis, which lies perpendicular to the diagram, is of order 0.3 mm. The GaAs crystal is cleaved along crystal planes lying parallel to the diagram in order to form a reflective Fabry–Perot cavity. The reflectance, R, at the air interface at normal incidence is $[(\mu - 1)/(\mu + 1)]^2$, where μ is the refractive index. In GaAs μ has the high value 3.6, giving a reflectivity of 0.32. The corresponding cavity line width,[7] for a cavity of length d is $\Delta\lambda = \lambda^2(1 - R)/(2\pi\mu d\sqrt{R})$. This gives $\Delta\lambda = 0.1$ nm.

Figure 8.4 shows the level structure when the DH is strongly forward biased, that is with the p-region made positive with respect to the n-region. The alloy GaAlAs has a larger band-gap than GaAs, therefore the GaAs layer forms a well in potential energy for electrons in the conduction band and for the holes in the valence band. Electrons and holes are injected, under the forward bias, producing in the thin GaAs layer a dense population of electrons in the conduction band and holes in the valence band. This population inversion fuels lasing, with photons being emitted when electrons drop into the holes. Another useful property of the structure is that the refractive index of GaAlAs is less than that of GaAs. Consequently, the GaAs layer forms a waveguide confining the radiation laterally, as illustrated in the lower panel of Figure 8.4.

The alloys on either side of each junction must have crystalline structures that match to high precision. Defects at the junctions would provide locations at which non-radiative electron/hole annihilations would proceed rapidly and so suppress the slower stimulated emission. Fortunately, $Ga_{1-x}Al_xAs$ crystals are all interpenetrating face centred cubic lattices whose lattice constant changes negligibly, from 0.564 to 0.566 nm, as x varies from zero to unity. This makes it possible for

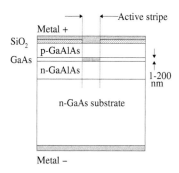

Fig. 8.3 A cross-section through a double heterostructure laser perpendicular to the laser beam's optical axis.

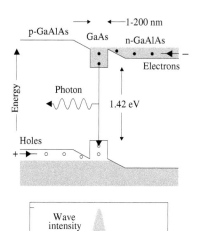

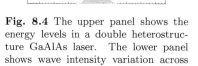

Fig. 8.4 The upper panel shows the energy levels in a double heterostructure GaAlAs laser. The lower panel shows wave intensity variation across the same section.

[7]See eqns 5.55 and 5.56: *The Light Fantastic; A Modern Introduction to Classical and Quantum Optics*, second edition (2011), by I. R. Kenyon, published by Oxford University Press.

Table 8.2 Ranges over which lasers can be constructed from various semiconductor alloys.

Alloy	Range in wavelength
GaAs	0.82–$0.88\,\mu m$
$In_x Ga_{1-x}N$	0.36–$0.60\,\mu m$
$In_x Ga_{1-x}As_y P_{1-y}$	1.0–$1.7\,\mu m$
$Al_x Ga_{1-x}As$	0.68–$0.92\,\mu m$
$Cd_x Pb_{1-x}S$	1.9–$4.2\,\mu m$
$Cd_x Hg_{1-x}Te$	3.2–$17\,\mu m$

crystals to be grown with few defects at the boundary between different GaAlAs alloys. As x changes the band-gap changes and hence the lasing wavelength. This *band-gap engineering* provides GaAlAs lasers with wavelengths anywhere in the interval 750–870 nm.

Silica based optical fibres exhibit extremely low absorption near wavelengths of 1310 nm and 1550 nm. Semiconductor lasers covering these wavelengths are made from *quaternary* compounds $Ga_{1-x}In_xAs_{1-y}P_y$. Table 8.2 gives the wavelengths available with lasers made from a selection of semiconductor alloys. Semiconductor lasers have properties that make them ideal sources for optical fibre transmission in telecom applications:

[8]Etendue is the product of the beam area and angular spread. The components can be altered by lenses but their product is invariant. This makes etendue a useful quantifier.

- The etendue[8]of a semiconductor laser beam is well matched to that required for efficient injection into optical fibres;

- Their compactness and low voltage requirements are compatible with standard modern fast electronics;

- The wavelengths of semiconductor lasers can be chosen to minimize simultaneously both the absorption losses and dispersion along optical fibres.

As a result, $\sim$100 digital signal channels each at $\sim$10Gb/s can be simultaneously transmitted on one telecom fibre supporting a single optical mode. Individual channels are modulated laser-sourced beam modes spaced at $\sim$0.8 nm intervals with negligible cross-talk. These channels can be faithfully recovered after transmission across the Atlantic, refreshed at purely optical repeater stations every few hundred kilometres on the sea bed. This is the basis of modern communications.[9]

[9]see the second edition of *The Light Fantastic: a Modern Introduction to Classical and Quantum Optics* by I. R. Kenyon, published by Oxford University Press (2011).

8.5   First-order coherence

Beams from lasers and thermal sources show fluctuations of amplitude and frequency due to variations in the number and energy of the photons emitted per unit time. If the internal physical state of the source (meaning its chemical content, temperature and pressure) are stable then the character of these fluctuations will not change with time. Such sources are called *stationary* and give beams whose statistical properties observed over a time long compared to the wave period are independent of when the sampling is done. The average value of any measurable quantity taken in this way for a given source is thus equal to the quantum mechanical expectation value of the observable. For such a stationary source, repeating the average after an interval τ,

$$\langle \hat{E}^-(t)\hat{E}^+(t) \rangle = \langle \hat{E}^-(t+\tau)\hat{E}^+(t+\tau) \rangle, \tag{8.40}$$

independent of τ. Usually the sources considered here are stationary.

The rate at which photons are recorded by a photodetector can be obtained using Fermi's golden rule, eqn. 7.8. The matrix element splits into an atomic term involving the emission of an electron and an electromagnetic field term involving the annihilation of a photon. Thus

$$M_{\mathrm{fi}} = \langle\,[\text{electron} + \text{ion}]|\hat{O}|[\text{atom}]\,\rangle \, \langle f|\hat{E}^+|i \rangle \tag{8.41}$$

where $|i\rangle$ and $|f\rangle$ are respectively the initial and final states of the electromagnetic field alone, and $\hat{O}$ is the electron operator. Then the rate at which photons are absorbed by a detector is

$$\gamma = G \sum_f |\langle f|\hat{E}^+|i \rangle|^2, \tag{8.42}$$

where the contribution of the detector is absorbed into a constant G. G contains the product of the efficiency of the detector to detect radiation of the frequency incident and a geometric factor determined by the sensitive area of the detector and its location. The detector will need to be chosen so that the detection efficiency is high, but thereafter G cancels from the quantities of interest used in measuring degrees of coherence. The above equation can be rewritten:

$$\gamma = G \sum_f \langle i|\hat{E}^-|f \rangle \, \langle f|\hat{E}^+|i \rangle. \tag{8.43}$$

The sum over accessible final states can be replaced by the sum over all states without affecting the equality because the additional states all have $\langle i|\hat{E}^-|f \rangle$ equal to zero. Then applying the closure relation eqn. 1.61 gives

$$\gamma = G \langle i|\hat{E}^-\hat{E}^+|i \rangle. \tag{8.44}$$

This is the key, simply interpreted result. It shows that the rate of detecting photons is proportional to the expectation value of the field operator product $\hat{E}^-\hat{E}^+$, and so to the number of photons in the initial state. For compactness, expectation values like $\langle i|\hat{E}^-\hat{E}^+|i \rangle$ will generally be truncated to $\langle \hat{E}^-\hat{E}^+ \rangle$.

The quantum *degree of first-order coherence* between two fields E_1 and E_2 is defined so that the experiment-dependent factor G is cancelled out:

$$g^{(1)}(1,2) = \frac{\langle \hat{E}_1^- \hat{E}_2^+ \rangle}{\sqrt{\langle \hat{E}_1^- \hat{E}_1^+ \rangle \langle \hat{E}_2^- \hat{E}_2^+ \rangle}}. \tag{8.45}$$

The value of the degree of first-order coherence within a stationary beam at times t and $t + \tau$ is

$$g^{(1)}(\tau) = \frac{\langle \hat{E}^-(t)\hat{E}^+(t+\tau)\rangle}{\langle \hat{E}^-(t)\hat{E}^+(t)\rangle}. \tag{8.46}$$

8.6   Second-order coherence

In 1954, Hanbury Brown and Twiss (HBT) made the quite surprising prediction that the arrival times at a detector of photons from a thermal source would show strong correlations. Explicitly the probability of successive photons arriving within a time short compared to the coherence time would be double that expected if the distribution of arrival times were random. Figure 8.5 shows results of a modern measurement of the time intervals between successive photons, in one case from a laser, and in another from a standard thermal source. A rise in probability is seen as the time interval between successive photons tends to zero, but only for the thermal source.

It seems paradoxical that in the case of the laser, whose first-order coherence extends over a much longer period of time, the distribution of the arrival times of the photons is entirely random. Evidently second-order coherence between pairs of photons is very different from first-order coherence in which a photon interferes with itself. These correlations will be explained in quantum mechanical terms. Two preliminary steps toward the explanation are required: the correlations measured have to be expressed quantum mechanically, and the properties of thermal and laser light also need to be expressed in the quantum framework.

The matrix element for detecting photons at locations labelled 1 and 2 is $\langle f|\hat{E}_2^+ \hat{E}_1^+|i\rangle$. Therefore, the rate at which *coincidences* occur between photons arriving at the two separate detectors is given by

$$
\begin{aligned}
\gamma &= G \sum_f |\langle f|\hat{E}_2^+ \hat{E}_1^+|i\rangle|^2 \\
&= G \sum_f \langle i|\hat{E}_1^- \hat{E}_2^-|f\rangle\langle f|\hat{E}_2^+ \hat{E}_1^+|i\rangle
\end{aligned}
\tag{8.47}
$$

where G includes the effect of the efficiency and the geometric area illuminated of each detector. Using the closure relation this becomes[10]

$$\gamma = G\,\langle i|\hat{E}_1^- \hat{E}_2^- \hat{E}_2^+ \hat{E}_1^+|i\rangle. \tag{8.48}$$

---

[10]This ordering of the operators with the annihilation operators preceding (to the right of) the creation operators has emerged naturally and is known as *normal ordering*. In addition, the annihilation operators are in chronological order and the creation operators in the reverse chronological order. This is known as being *time ordered*.

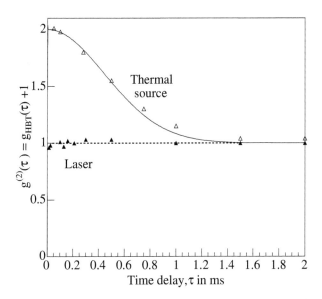

Fig. 8.5 The correlation between the arrival times of successive photons. Adapted from Arecchi, Gatti and Sona, *Physics Letters*, 20, 27 (1966). Courtesy Professor Arecchi.

In order to determine whether coincidences are occurring more or less frequently than the random rate, their rate is divided by the random rate. The random rate is obtained by taking the product of the individual count rates

$$\gamma_0 = \langle i|\hat{E}_1^- \hat{E}_1^+|i\rangle \langle i|\hat{E}_2^- \hat{E}_2^+|i\rangle. \tag{8.49}$$

The *degree of second-order coherence* is then defined (leaving out the initial state label i) as:

$$g^{(2)}(1,2) = \gamma/\gamma_0 = \frac{\langle \hat{E}_1^- \hat{E}_2^- \hat{E}_2^+ \hat{E}_1^+\rangle}{\langle \hat{E}_1^- \hat{E}_1^+\rangle\langle \hat{E}_2^- \hat{E}_2^+\rangle}. \tag{8.50}$$

Experimental factors such as the efficiency and size of each detector cancel out when this ratio is taken. When a single detector is used to detect correlations between photons separated by a fixed time interval, τ, the corresponding result for a stationary source, which is independent of t, is

$$g^{(2)}(\tau) = \frac{\langle \hat{E}^-(t)\hat{E}^-(t+\tau)\hat{E}^+(t+\tau)\hat{E}^+(t)\rangle}{\langle \hat{E}^-(t)\hat{E}^+(t)\rangle^2}. \tag{8.51}$$

8.7   Laser light and thermal light

Laser beams are highly coherent, with photons having the same wavelength, polarization and phase over long distances. By contrast, black body radiation consists of radiation of all wavelengths having short wavepackets with random phases. The radiation from most sources has limited spectral range, and behaves statistically like black body radiation. This radiation is therefore known as *thermal* or *chaotic* radiation. States of the electromagnetic field that contain a specific number of photons in a mode i are known as *Fock states*: mathematically simple but difficult to produce.

8.8   Coherent (laser-like) states

The coherence time of a very well-stabilized laser can last for one second or longer, so that the beam emitted is nearly a pure sinusoidal wave extending over a distance that would reach to the Moon. The exact form of the normalized state vector for a coherent state is a sum of Fock states

$$|\alpha\rangle = \exp\left(-|\alpha|^2/2\right) \sum_{n=0}^{\infty} (\alpha^n/\sqrt{n!})|n\rangle, \qquad (8.52)$$

where $\alpha = |\alpha| \exp(i\phi)$ can be any complex number. Surprisingly, this coherent state vector describes an ensemble of states containing differing numbers of photons. We need to look at its properties more closely.

The effect of the annihilation and creation operators are

$$\hat{a}_\omega |\alpha\rangle = \alpha |\alpha\rangle, \quad \text{and} \quad \langle\alpha| \hat{a}_\omega^\dagger = \alpha^* \langle\alpha|. \qquad (8.53)$$

Using eqn 8.17, the effect of the electric field operator $\hat{E}^+$ on a coherent state is

$$\hat{E}^+|\alpha\rangle = \zeta_\omega \exp(i\xi_\omega)\,\hat{a}_\omega |\alpha\rangle = \alpha\,\zeta_\omega \exp(i\xi_\omega)|\alpha\rangle. \qquad (8.54)$$

Therefore, a coherent state is an eigenstate of the annihilation operator and hence of the positive frequency component of the electric field operator: it is *not* an eigenstate of the electric field operator. In fact the expectation value for the electric field is

$$\langle\alpha|\hat{E}^+ + \hat{E}^-|\alpha\rangle = 2\zeta_\omega |\alpha| \cos(\xi_\omega + \phi). \qquad (8.55)$$

making it a stable sinusoidal wave. The probability that there are exactly n photons in this state is

$$P(n) = |\langle n|\alpha\rangle|^2 = \exp\left(-|\alpha|^2\right)|\alpha|^{2n}/n!, \qquad (8.56)$$

which is a Poissonian distribution with a mean of $|\alpha|^2$. Consequently the variance

$$\langle n^2\rangle - \langle n\rangle^2 = \langle n\rangle. \qquad (8.57)$$

The variance quantifies the *particle noise* at a detector due to the variation in the number of photons arriving in a given time interval. The electric field vector and its uncertainty are shown on an Argand diagram in Figure 8.6. Now

$$
\begin{aligned}
-i\frac{\partial |\alpha\rangle}{\partial \phi} &= \exp[-|\alpha|^2/2] \sum_0^\infty \frac{n\alpha^n}{\sqrt{n!}}|n\rangle \\
&= \exp[-|\alpha|^2/2] \sum_0^\infty \frac{\alpha^n}{\sqrt{n!}}\hat{n}|n\rangle \\
&= \hat{n}|\alpha\rangle,
\end{aligned}
\tag{8.58}
$$

whence n and ϕ are conjugate observables in the same sense as momentum and position. For states of the electromagnetic field in general

$$
\Delta\phi\,\Delta n \geq 1/2,
\tag{8.59}
$$

the corresponding form of Heisenberg uncertainty principle. The definition of this quantum optical phase is full of pitfalls and the reader should refer to advanced texts for details.[11] In the case of Fock states the number of photons is exact but the phase of the field is indeterminate. Coherent beams can be manipulated to make the phase or amplitude uncertainty smaller while keeping the product equal to $1/2$. These are called *squeezed states*. Phase squeezing has the potential for improving the precision of interferometry, in particular in the detection of gravitational waves, a topic developed in Chapter 11.

When a mode contains a large number m, of coherent photons, then the equivalent classical electric field $E_{\mathrm{cl}}\cos{(kx - \omega t)}$ can be deduced using eqn. 8.55,

$$
E_{\mathrm{cl}} = 2\zeta_\omega\sqrt{m} = \sqrt{(2m\hbar\omega/\varepsilon_0\varepsilon_{\mathrm{r}}V)}.
\tag{8.60}
$$

Then $2\zeta_\omega$ is in some sense the *electric field per photon*.

Input for determining the degree of first-order coherence of a coherent state is obtained using eqn. 8.54. Then

$$
\langle \hat{E}^-(t)\hat{E}^+(t+\tau)\rangle = |\alpha|^2\zeta_\omega^2\exp{(-i\omega\tau)}.
\tag{8.61}
$$

Thus the degrees of coherence defined in eqns. 8.46 and 8.50 are

$$
g^{(1)}(\tau) = \exp{(-i\omega\tau)}, \text{ and } g^{(2)} = 1.
\tag{8.62}
$$

As expected the coherence is that of an infinite plane wave.

8.9 Thermal light

Thermal light includes black body radiation as well as radiation from sources with a restricted frequency range in which the photons at any

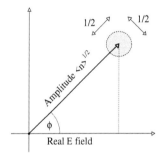

Fig. 8.6 Argand diagram showing the uncertainty in the electric field amplitude for a coherent state.

[11]See *Measuring the Quantum State of Light* by Ulf Leonhardt, Cambridge University Press (1997).

frequency have random phase relative to each other. Using eqn 3.39, the probability of having n photons in a mode in thermal equilibrium at temperature T K:

$$P(n) = \exp(-nx)\left[1 - \exp(-x)\right], \qquad (8.63)$$

where x is $\hbar\omega/(k_B T)$ and $\langle n \rangle$ is $1/[\exp(x) - 1]$. Then[12] the variance

$$\langle [\Delta n]^2 \rangle \equiv \langle n^2 \rangle - \langle n \rangle^2 = \langle n \rangle^2 + \langle n \rangle. \qquad (8.64)$$

[12]We have

$$
\begin{aligned}
\langle n^2 \rangle &= (1 - e^{-x})\sum n^2 e^{-nx} \\
&= (1 - e^{-x})\frac{d^2}{dx^2}\sum e^{-nx} \\
&= (1 - e^{-x})\frac{d^2}{dx^2}\frac{1}{1 - e^{-x}} \\
&= \frac{2e^{2x}}{(e^x - 1)^2} = 2\langle n \rangle^2 + \langle n \rangle.
\end{aligned}
$$

The second term in the variance is the particle noise, which also appeared in the variance of the photon number count of coherent states, while the first term is called the *wave noise*. Figure 8.7 shows the distribution in the number of photons from a laser and a thermal source in a single mode when they both have an average photon count of five in that mode.

A simple relation can now be derived between the degrees of first- and second-order coherence of a stationary thermal source. The numerator in the second order correlation is

$$\text{Num} = \sum_{i,j,k,m} \langle \hat{E}_i^-(t)\hat{E}_j^-(t+\tau)\hat{E}_k^+(t+\tau)\hat{E}_m^+(t)\rangle, \qquad (8.65)$$

where i, j, k, m are mode labels. The only mode pairings that will connect the initial state to itself are $i = k$ and $j = m$, or $i = m$ and $j = k$. There are many modes so that rarely will $i = j$. Then each term factorizes so that

$$
\begin{aligned}
\text{Num} &= \sum_{i,j}[\langle \hat{E}_i^-(t)\hat{E}_i^+(t+\tau)\rangle\langle \hat{E}_j^-(t+\tau)\hat{E}_j^+(t)\rangle] \\
&\quad + \langle \hat{E}_i^-(t)\hat{E}_i^+(t)\rangle\langle \hat{E}_j^-(t+\tau)\hat{E}_j^+(t+\tau)\rangle \\
&= |\langle \hat{E}^-(t)\hat{E}^+(t+\tau)\rangle|^2 + \langle \hat{E}^-(t)\hat{E}^+(t)\rangle|^2. \qquad (8.66)
\end{aligned}
$$

Mean count of 5

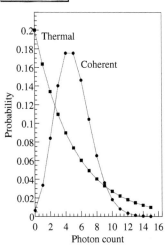

Fig. 8.7 The distribution in counts for a thermal and a coherent source when the average count is five in each case.

Then using eqns. 8.51 and 8.46

$$g^{(2)}(\tau) = \text{Num}/\langle \hat{E}^-(t)\hat{E}^+(t)\rangle^2 = 1 + |g^{(1)}(\tau)|^2. \qquad (8.67)$$

When line broadening is inhomogeneous with a Gaussian distribution of width σ[13]

$$g^{(2)}(\tau) = 1 + \exp\left(-\sigma^2\tau^2\right). \qquad (8.68)$$

[13]The Gaussian and Lorentzian cases are handled as exercises.

For a source with a Lorentzian line shape

$$g^{(2)}(\tau) = 1 + |g^{(1)}(\tau)|^2 = 1 + \exp\left(-\gamma|\tau|\right). \qquad (8.69)$$

At zero time difference the correlation coefficients are exactly 2 in both cases, a common property of thermal light.

8.10   Observations of photon correlations

An experiment of the type originally carried out by Hanbury Brown and Twiss to observe correlations between photons from a thermal source is shown in Figure 8.8. Light from a source is filtered to isolate a spectral line and focused on a pinhole. The emerging light is divided by a beam splitter into two equal beams that illuminate separate photomultipliers. The electronic signals from the detectors are passed through separate bandwidth limited amplifiers, which removed the steady time-average component of the signals. The size of the aperture, the sensitive area of the detectors, and their spacing is such that the detectors cover the area of one transverse mode of the optical field, and their views coincide throughout the measurements. A variable electronic delay is imposed between the outputs from PM_1 and PM_2. These outputs are then multiplied together and time averaged. One signal is delayed by a time τ so that instantaneously the multiplier output is $[I(t) - \langle I \rangle][I(t + \tau) - \langle I \rangle]$ and its time average is $\langle I(t)I(t + \tau) \rangle - \langle I \rangle^2$. This is compared to the mean squared intensity at one detector $\langle I \rangle^2$, giving

$$g_{\mathrm{HBT}}(\tau) = \frac{\langle I(t)I(t + \tau) \rangle}{\langle I \rangle^2} - 1. \tag{8.70}$$

This is seen to be $g^{(2)}(\tau) - 1$. Then for a chaotic source with a Gaussian spectrum we can use the prediction of eqn. 8.68

$$g_{\mathrm{HBT-thermal}}(\tau) = \exp\left(-\sigma^2 \tau^2\right), \tag{8.71}$$

while for a coherent source eqn. 8.62 gives

$$g_{\mathrm{HBT-coherent}}(\tau) = 0. \tag{8.72}$$

These predictions are shown in Figure 8.5 together with experimental data; the agreement is excellent. There are no correlations in the case of the coherent beam so there is no bunching. When the time delay is reduced to zero the degree of second-order coherence, $g^{(2)}(0)$, for a thermal source is thus exactly twice that for a laser. At the other extreme, when a time delay longer than the coherence time is imposed the correlations disappear.

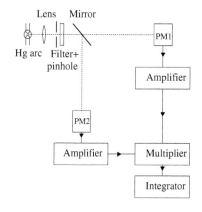

Fig. 8.8 The experimental apparatus used by Hanbury Brown and Twiss to observe the photon correlations from a thermal source.

8.11   Correlation interferometry

Photon correlations were used by Hanbury Brown to measure the size of stars; a method that has been superseded by Michelson interferometry. However, the method is now used to measure the size of the interaction region, typically a few fermis across, in nuclear collisions at high energy. Not only is the scale vastly different but the correlations sought are between pions instead of photons. Such correlations are expected for any boson species, for example, between like-sign pions: pions are $\pi^{\pm}$-mesons with spin zero and rest mass $140\,\mathrm{MeV/c^2}$. Pions are produced copiously in individual high-energy collisions of ions at accelerators. All the

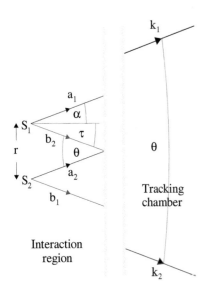

Fig. 8.9 Correlation interferometry using like-sign pions emitted from a nuclear interaction. Pions emerge from source at S_1 and S_2 on the left and their tracks are registered in a detector on the right.

charged particles emitted from one collision are simultaneously observed by tracking detectors within a magnetic field. Their vector momenta at emission can then be reconstructed. Refering to Figure 8.9, S_1 and S_2 are where a pair of like-sign pions emerge from the interaction region, several fermis (femtometres) in size. On the right are the tracks of these pions in the detector and of order cm apart. The alternatives that pion 1 is emitted from S_1 and pion 2 from S_2, or pion 2 from S_1 and pion 1 from S_2 are indistinguishable. Hence the total amplitude for observing a pair of like-sign pions is the sum of the two distinct contributions. With the notation of the figure, where the wave-vectors are written as k and distances from source to detector are written as a or b with appropriate subscripts, the amplitude for observing like-sign pion pairs is

$$
\begin{aligned}
A &= \exp[i(k_1 a_1 + k_2 b_1)] + \exp[i(k_1 a_2 + k_2 b_2)] \\
&= \exp[i(k_1 a_1 + k_2 b_1)] \left\{ 1 + \exp(i[k_1(a_2 - a_1) + k_2(b_2 - b_1)]) \right\} \\
&= \exp[i(k_1 a_1 + k_2 b_1)] \left\{ 1 + \exp(i[k_1 r\alpha + k_2 r\tau]) \right\}.
\end{aligned}
$$

The magnitudes of the wave-vectors are nearly equal so that the enhancement of the simultaneous observation of a like-sign pair over random pairs is

$$
\begin{aligned}
I = |A|^2/2 &= |1 + \exp[ik(r\alpha + r\tau)]|^2/2 \\
&= |1 + \exp[ikr\theta]|^2/2.
\end{aligned}
$$

This reduces to

$$
I = 1 + \cos[kr\theta] = 1 + \cos[qr/\hbar], \tag{8.73}
$$

where q is the relative momentum between the pions. Integration over an extended spheroidal source leads to an enhancement

$$
I = 1 + \lambda[2\hbar J_1(q_i \Delta r_i/\hbar)/(q_i \Delta r_i)]^2, \tag{8.74}
$$

where J_1 is the first-order Bessel function: q_i is the component of the relative momentum along the i-direction and Δr_i the radius of the source perpendicular to this axis. λ would range from unity for an incoherent to zero for a fully coherent source.

Figure 8.10 shows the correlations observed between negative pions from collisions between gold nuclei at a center of mass energy of 200 GeV. The emitted particles were recorded in the STAR detector at the RHIC collider. Correlations are shown projected along the direction of motion of the colliding gold nuclei, at bottom right, and along two orthogonal directions. In each plot the full symbols refer to raw data, while the open symbols are data corrected to remove the effect of the Coulomb repulsion between the negative pions. Equation 8.74 is used to fit the observed correlation between negative pions. This approach provides an estimate of the volume and shape of the emission region. The radius of the region found from this and similar experiments is ~6 fm. This value is comparable to the root mean square radius of the gold nucleus measured by electron scattering, namely 7.3 fm.

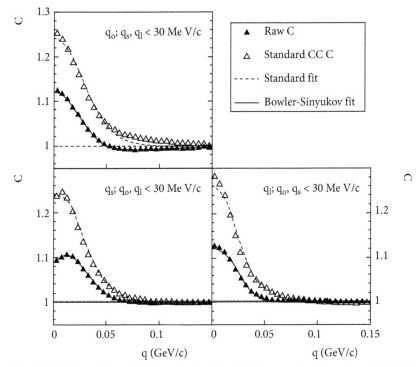

Fig. 8.10 Three dimensional correlation function for negative pions from 200 GeV gold-gold collisions. Figure 4: J. Adams et al (STAR Collaboration), *Physical Review* C71 044906 (2005). Courtesy Professor P. Jones and The American Physical Society.

8.12   Fermions

Second quantization of the electron field required finding a relativistic wave equation for electrons and then solving it. The way that Dirac achieved this is recounted in Appendix G. From this basis Feynman, Tomonaga and Schwinger developed the relativistic theory of electrodynamics, known as *Quantum Electrodynamics (QED)*, which encompasses the properties of electromagnetic fields, electrons and their interactions. Use will be made of Feynman's formulation of QED in terms of diagrams when we come to Chapter 18. Dirac received the Nobel prize in physics in 1933, and the other three theoreticians shared the 1965 Nobel prize in physics.

Here we restrict discussion to the non-relativistic case. Suppose the creation (annihilation) operator for a specific state is $\hat{c}^\dagger$ ($\hat{c}$). Only the states with zero or one occupant exist, explicitly $|0\rangle$ and $|1\rangle$. Hence $\hat{c}^\dagger\hat{c}^\dagger|0\rangle$ is zero, while

$$(\hat{c}^\dagger\hat{c} + \hat{c}\hat{c}^\dagger)|n\rangle = |n\rangle \tag{8.75}$$

for both n values. This operator is the *anti-commutator* written $\{\hat{c}, \hat{c}^\dagger\}$ and we have

$$\{\hat{c}, \hat{c}^\dagger\} = 1, \tag{8.76}$$

whereas for photon (boson) creation and annihilation operators

$$[\hat{a}, \hat{a}^{\dagger}] = 1. \tag{8.77}$$

The whole machinery for creating coherent states is therefore absent for fermions. From Section 3.6

$$\langle n^2 \rangle = \langle n \rangle, \tag{8.78}$$

so that the variance

$$\langle [\Delta n]^2 \rangle = \langle n^2 \rangle - \langle n \rangle^2 = \langle n \rangle - \langle n \rangle^2. \tag{8.79}$$

Collecting the variances:

$$\langle [\Delta n]^2 \rangle = \langle n \rangle \text{ laser light}, \tag{8.80}$$
$$\langle [\Delta n]^2 \rangle = \langle n \rangle + \langle n^2 \rangle \text{ thermal light}, \tag{8.81}$$
$$\langle [\Delta n]^2 \rangle = \langle n \rangle - \langle n^2 \rangle \text{ electrons}. \tag{8.82}$$

Thus the counting statistics for lasers is Poissonian, for thermal light *super-Poissonian* and for electrons *sub-Poissonian*. When average occupation numbers are very much smaller than unity the statistical distributions converge to Poissonian.

8.13   Further reading

The Quantum Theory of Light, third edition by R. Loudon, published by Oxford University Press (2000), a standard text, develops topics introduced here and in the previous chapter more fully and rigorously.

Introductory Quantum Optics, by C. Gerry and P. Knight, published by Cambridge University Press (2005) covers quantum optics at greater depth.

Quantum Optics, by M. O. Scully and M. S. Zubairy, published by Cambridge University Press (1997) gives thorough coverage of modern developments by two international experts. An advanced but comprehensible text.

Exercises

(8.1) When the counting time in the HBT experiment for a thermal source is extended beyond the coherence time several modes, m in number, will be involved. Show that the variance in the photon count is $\langle N \rangle + \langle N \rangle^2 / m$ where N is the total photon count.

(8.2) If the quantum efficiency of the photon detector is η the average count changes from $\langle M \rangle$ to

$\langle m \rangle = \eta \langle M \rangle$. Show that the variance in the photon count for a thermal source becomes $\eta^2 \langle \Delta M^2 \rangle + \eta(1 - \eta)\langle M \rangle$. The first term is the variance expected for perfect efficiency $\times \eta^2$ and the second is the *partition noise*.

(8.3) Prove that $(\hat{a})^n |n\rangle = \sqrt{n!}\,|0\rangle$ and that $(\hat{a})^n |n-1\rangle = 0$.

(8.4) Deduce that the magnetic field operator $\hat{\mathbf{B}}(t) = \sum_\omega \mathbf{k} \wedge \mathbf{e}_\omega (\zeta_\omega/\omega) [\hat{a}_\omega \exp(-i\xi_\omega) + \hat{a}_\omega^\dagger \exp(i\xi_\omega)]$ where $\mathbf{e}_\omega$ is the unit polarization vector of the electric field component with frequency ω and wavevector $\mathbf{k}$.

(8.5) $|\{\alpha\}\rangle$ is a state made up of many coherent states in different modes of the electromagnetic field. In the case that there are two of these modes with frequencies $\omega(1)$ and $\omega(2)$, $|\{\alpha\}\rangle = |\alpha_{\omega(1)}\rangle|\alpha_{\omega(2)}\rangle$. Show that $g^{(2)}(\tau) = 1$ for such states.

(8.6) In Young's double slit experiment the initial state can be expressed as $|in\rangle = (1/\sqrt{2})(\hat{a}_1^\dagger + \hat{a}_2^\dagger)|0\rangle$ where $\hat{a}_i^\dagger$ creates a photon at slit i. The intensity on the screen is proportional to $\langle in|\hat{E}^- \hat{E}^+|in\rangle$. Show that the intensity distribution across the screen is $[1 + \cos(\mathbf{k} \cdot \Delta\mathbf{r})]$ where $\Delta\mathbf{r}$ is the vector difference between the paths from the slits to the point on the screen being considered.

(8.7) By considering the case of a two state system in thermal equilibrium with black body radiation

prove eqn. 8.37.

(8.8) The normalized Lorentzian line shape for emission from isolated atoms is

$$P(\omega) = (\gamma/2\pi)/[(\omega - \omega_0)^2 + \gamma^2/4],$$

where ω_0 is the central angular frequency and γ the full width at half maximum. Deduce the time-dependent correlation and hence the first and second-order correlation in this case. Use the result, p and a being positive,

$$\int_{-\infty}^{\infty} \frac{\exp(-ipx)\mathrm{d}x}{a^2 + x^2} = \frac{\pi \exp(-ap)}{a}. \qquad (8.83)$$

(8.9) The spectrum of sources with inhomogeneous broadening due for example to thermal Doppler shifts has the form

$$P(\omega) = \exp[-(\omega - \omega_0)^2/(2\sigma^2)]/[\sqrt{2\pi}\sigma],$$

where ω_0 is the central angular frequency and σ the root mean square deviation. Deduce the time-dependent correlation, and hence the first- and second-order correlation.

(8.10) Write down the effect of the creation and annihilation operators for electrons, $\hat{b}^\dagger$, $\hat{b}$, on states with various electron content. Hence show that the anticommutator $\hat{b}^\dagger \hat{b} + \hat{b}\hat{b}^\dagger$ is 1.

Entanglement

9.1   Introduction

The quantum state of two photons corresponding to the classical state, with photon i in a mode X and a photon j in a mode Y, X and Y being orthonormal, is the *product state* $|i\rangle_X|j\rangle_Y$. A quantum state that has no classical counterpart is $[|i\rangle_X|j\rangle_Y - |i\rangle_Y|j\rangle_X]/\sqrt{2}$. In such states the two photons are said to be *entangled*: if a measurement is made to learn whether the eigenstate of i is X or Y, the outcome is indeterminate, but the result fixes the eigenstate of j. Thus if i is found to be in eigenstate X then when j is measured the result is always Y. The description of this or any other fully entangled state is as complete as it can be, but the state of the components is indeterminate. If the outcome of measurement made on one component is fully indeterminate but the result of the measurement determines fully the state of the other component then there is complete entanglement. When either the result of the first measurement is not fully indeterminate or the outcome of the second measurement is not fully determined then the system is only partially entangled: for example, $[|i\rangle_X|j\rangle_Y/2 + |i\rangle_Y|j\rangle_X]/\sqrt{3}/2]$. Fully entangled states are the most useful for presenting entanglement in an uncluttered way, but partially entangled states are the general rule.

Entanglement can involve different quantum systems. For example, a photon and an atom can be entangled, or three or more photons can be mutually entangled, as in $|i\rangle_V|j\rangle_V|k\rangle_V + |i\rangle_H|j\rangle_H|k\rangle_H$, where V and H are orthogonal polarizations. Entanglement has been achieved between a systems as complex for example as two spin aligned Cs gas samples each containing 10^{12} atoms, the state lasting for milliseconds.[1] Chapter 18 describes the use of entangled beauty mesons in demonstrating CP violation in particle physics.

Section 9.2 presents the implication from entanglement that quantum mechanics is non-local. Then the use of Bell states in the analysis of two particle entanglement is described. The technique of spontaneous parametric down conversion, for producing a useful flux of entangled photons, is described next. An important application is described in Section 9.5: the Hong–Ou–Mandel experiment, demonstrating the indistinguishability of photons. Teleportation is introduced, and finally, an example of

---

[1] B. Julsgaard, A. Kozhekin and E. S. Polzik, *Nature* 41, 400 (2001).

Quantum 20/20: Fundamentals, Entanglement, Gauge Fields, Condensates and Topology.
Ian R. Kenyon. © 2020. Published in 2020 by Oxford University Press.
DOI: 10.1093/oso/9780198808350.001.0001

the experimental entanglement of remote atoms is outlined.

9.2   Non-locality of quantum mechanics

Suppose measurements are made on two entangled objects i and j so that the measurements have a space-like separation: that is to say their separation in time Δt and space Δr satisfy $|\Delta r| > c\Delta t$. Then no causal connection between the measurements is possible. Despite this, the correlation inherent in quantum entanglement is still observed. For example, in the case of $[|i\rangle_X|j\rangle_Y - |j\rangle_X|i\rangle_Y]/\sqrt{2}$, a measurement finding i in state X ensures that when j is measured it is always in state Y. The observation of such correlations over a space-like separation demonstrates unequivocally that quantum mechanics is *non-local*. A feature of having a space-like separation between i and j that emphasizes the lack of a causal connection is the following: the measurement on i can precede that on j in one inertial frame, while in another inertial frame the measurement on j can precede the measurement on i.

Schrödinger called quantum entanglement 'the characteristic trait of quantum mechanics, one that enforces its entire departure from classical lines of thought'. Quantum entanglement is also the property that Einstein, Podolosky and Rosen used to argue that quantum mechanics, though correct, is incomplete. Chapter 10 describes how the analysis and experiments, that this questioning evoked, has led to a sharper understanding of quantum mechanics. All experimental tests are consistent with the usual *Copenhagen interpretation* of quantum mechanics outlined in Chapter 1.

9.3   Bell states

The use of *Bell states* simplifies the analysis of two particle entanglement. Two photons entangled in polarization provide a suitable example and we take $|i\rangle_V$ and $|i\rangle_H$ to be states of photon i with vertical and horizontal polarization respectively. Then the four fully entangled mutually orthogonal Bell states are

$$\Psi^\pm = [|i\rangle_H|j\rangle_V \pm |i\rangle_V|j\rangle_H]/\sqrt{2}, \tag{9.1}$$
$$\Phi^\pm = [|i\rangle_H|j\rangle_H \pm |i\rangle_V|j\rangle_V]/\sqrt{2}. \tag{9.2}$$

All fully entangled bipartite (two-component) states can be expressed as linear superpositions of the Bell states. If instead the entangled particles are electrons with spins up and down relative to some quantum axis, then

$$\Psi^\pm = \left(\begin{bmatrix} 1 \\ 0 \end{bmatrix}_i \begin{bmatrix} 0 \\ 1 \end{bmatrix}_j \pm \begin{bmatrix} 0 \\ 1 \end{bmatrix}_i \begin{bmatrix} 1 \\ 0 \end{bmatrix}_j \right)/\sqrt{2} \tag{9.3}$$

$$\Phi^\pm = \left(\begin{bmatrix} 1 \\ 0 \end{bmatrix}_i \begin{bmatrix} 1 \\ 0 \end{bmatrix}_j \pm \begin{bmatrix} 0 \\ 1 \end{bmatrix}_i \begin{bmatrix} 0 \\ 1 \end{bmatrix}_j \right)/\sqrt{2} \tag{9.4}$$

The three states Ψ^+, Φ^+ and Φ^- are spin states symmetric under interchange of the photons or electrons, while Ψ^- is antisymmetric. The symmetry of the spatial components of the wavefunction under particle interchange has to be such that the overall (spatial × spin) wavefunction is symmetric (antisymmetric) for photons (electrons). Evidently a pair of electrons sharing a quantum state in an atom are entangled, but this entanglement is not available as a resource in experiments on entanglement.

Pure eigenstates of photon polarization, of electron spin or of a two-state atom can be represented by points on a Bloch sphere, as explained in Appendix D. The photon states $|\rangle_H$ and $|\rangle_V$ used to form the Bell states in eqns. 9.1 and 9.2 are conventionally taken to lie at opposite poles on the Bloch sphere: this is true also for the electron states used in eqns. 9.3 and 9.4. A key point to grasp is that changing choice of quantization axis to one with polar angles θ, ϕ leaves each Bell states unchanged. For example:[2]

$$\sqrt{2}\Psi^-(\theta,\phi) = \begin{bmatrix} \cos(\theta/2) \\ e^{i\phi}\sin(\theta/2) \end{bmatrix}_i \begin{bmatrix} \sin(\theta/2) \\ -e^{i\phi}\cos(\theta/2) \end{bmatrix}_j$$

$$- \begin{bmatrix} \sin(\theta/2) \\ -e^{i\phi}\cos(\theta/2) \end{bmatrix}_i \begin{bmatrix} \cos(\theta/2) \\ e^{i\phi}\sin(\theta/2) \end{bmatrix}_j$$

$$= -e^{i\phi}\left[\cos^2\frac{\theta}{2} + \sin^2\frac{\theta}{2}\right]\left[\begin{bmatrix}1\\0\end{bmatrix}_i\begin{bmatrix}0\\1\end{bmatrix}_j - \begin{bmatrix}0\\1\end{bmatrix}_i\begin{bmatrix}1\\0\end{bmatrix}_j\right]$$

$$= -e^{i\phi}\sqrt{2}\Psi^-(0,0). \tag{9.5}$$

This result shows that the state of polarization of either one of the pair of entangled electrons, photons, etc. is indeterminate. Measure it with respect to any reference axis on the Bloch sphere and either outcome, up or down, is equally probable. However, the polarization of its partner is then determined unambiguously: it lies at the diametrically opposite point on the Bloch sphere. This behaviour typifies fully entangled quantum states: the overall quantum state is well defined but that of the components is indeterminate.

9.4 Spontaneous parametric down conversion

A well-tried means for generating entangled pairs of photons at an adequate rate for experiments is spontaneous parametric down conversion (SPDC). Intense coherent radiation from a laser, the *pump*, impinges on a non-linear crystal whose susceptibility increases with the electric field strength. The process of interest involves the absorption of a *pump* photon (p) and the simultaneous emission of a pair of photons. The latter are called the *signal* (s) and *idler* (i) using the convention $\omega_s \geq \omega_i$.

[2] In order to prove this first resolve each product into its components:
$$\begin{bmatrix} \cos(\theta/2) \\ 0 \end{bmatrix}_i \begin{bmatrix} 0 \\ -e^{i\phi}\cos(\theta/2) \end{bmatrix}_j,$$
that is
$$-e^{i\phi}\cos^2(\theta/2)\begin{bmatrix}1\\0\end{bmatrix}_i\begin{bmatrix}0\\1\end{bmatrix}_j,$$
plus seven other pieces.

Conservation of energy requires that the angular frequencies satisfy the relation

$$\omega_{\mathrm{p}} = \omega_{\mathrm{s}} + \omega_{\mathrm{i}}, \tag{9.6}$$

and conservation of momentum requires in addition that the wave-vectors satisfy the relation

$$\mathbf{k}_{\mathrm{p}} = \mathbf{k}_{\mathrm{s}} + \mathbf{k}_{\mathrm{i}}. \tag{9.7}$$

These two equations give *phase matching* of the pump and outgoing waves: in classical terms, the overall phase of the product of the outgoing waves matches that of the input wave. β-barium borate (BBO), a negative uniaxial crystalline material, often used for SPDC, has many practical advantages. Phase matching with high yields per pump photon is possible over the range 210–3000 nm and it has low absorption (< 5 per cent/cm) over most of this range. BBO crystals withstand high pump photon fluxes and can be made very large (5 cm long) and uniform, and are stable in air when anti-reflection coated. Another comparably useful material is potassium di-hydrogen phosphate (KDP).

The pump beam is usually polarized in a plane containing the optic axis of the crystal. Then alternative choices for the polarization of the signal and idler can be selected. In type-I SPDC the signal and idler are both plane polarized perpendicular to the pump. In type-II SPDC one photon has the same polarization as the pump (extraordinary ray), the other is polarized perpendicularly (ordinary ray). It is possible in addition, with an appropriately cut crystal, to choose the beam directions so that the signal and idler have the same frequency. This is called *degenerate* SPDC.

Figure 9.1 shows the layout for type-II SPDC, for example in a BBO crystal pumped by 351.1 nm UV from an argon-ion laser. It shows rays emerging from the conversion point. Momentum matching requires that the signal photon direction must lie somewhere on a cone whose apex is the conversion point, and the idler follows a complementary path on another such cone. The cone angles depend on the signal and idler wavelengths. In the figure a particularly important configuration is shown in which the cones for degenerate type-II SPDC cross one another. Pinholes in an opaque screen placed at the cross-over points select signal and idler photon pairs. Interference filters placed behind the pinholes then limit the selection to photons of nearly the same frequency (degenerate).

One pinhole will pass an idler photon and the other a signal photon; but which goes through which hole is undetermined. The photons are therefore entangled and because they have different polarization (ordinary and extraordinary) they are said to be *polarization* entangled. Choosing a thin crystal, so that no phase lag develops between the ordinary $|i/s\rangle_{\mathrm{O}}$ and extraordinary rays $|i/s\rangle_{\mathrm{E}}$, their overall state vector is

$$[\,|i\rangle_{\mathrm{E}}|s\rangle_{\mathrm{O}} + \exp(i\phi)|i\rangle_{\mathrm{O}}|s\rangle_{\mathrm{E}}\,]/\sqrt{2}. \tag{9.8}$$

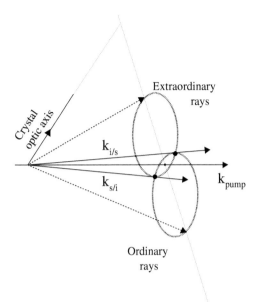

Fig. 9.1 Type-II degenerate SPDC. The circles are where the ordinary and extraordinary rays of a given, in this case equal, wavelength cross the screen. The entangled pair selected by pinholes are labelled with wave-vectors $\mathbf{k}_{i/s}$ and $\mathbf{k}_{s/i}$. The dotted lines indicate the plane containing the optic axis and beam.

Any phase lag ϕ due to the difference between the phase velocities of the ordinary and extraordinary rays can be compensated by passing the signal and idler beams first through half-wave plates so that they exchange polarizations and then through a second BBO crystal identical to that used for SPDC: in this second crystal the phase difference is then correspondingly reversed. The beauty of SPDC is that it provides entangled pairs of photons at a high, controllable rate at wavelengths the experimenter can select in the visible, near-UV or near-IR region.

In order for there to be second-order coherence (see Chapter 8) between the two beams, the coherence time should be longer than the time interval between the detection of the signal and idler photons. A typical filter bandwidth employed is 10 nm, giving a coherence time of $\sim$100 fs.

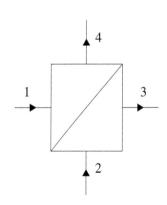

Fig. 9.2 Beam splitter.

9.4.1 Beam splitters

A symmetric, 50:50, non-absorbing beam splitter is a representative *passive* device used in studying the states of entangled photons. Figure 9.2 shows the input and output *ports* with labels. We use the notation of Chapter 8 for field operators. The outgoing electric field operators at ports 3 and 4 are related to the incoming field operators at ports 1 and 2 by

$$\hat{a}_1^\dagger = (\hat{a}_3^\dagger + i\hat{a}_4^\dagger)/\sqrt{2} \text{ and } \hat{a}_2^\dagger = (\hat{a}_4^\dagger + i\hat{a}_3^\dagger)/\sqrt{2}, \qquad (9.9)$$

where the operator $\hat{a}_i^\dagger$ creates a photon at port i, and $\hat{a}_i$ is the corresponding annihilation operator. The factor i comes from the $\pi/2$ phase shift between the reflected and transmitted light. As usual

$$[\hat{a}_i, \hat{a}_i^\dagger] = 1, \qquad (9.10)$$

with field operators at different ports being independent. Correspondingly for the annihilation operators

$$\hat{a}_1 = (\hat{a}_3 - i\hat{a}_4)/\sqrt{2} \text{ and } \hat{a}_2 = (\hat{a}_4 - i\hat{a}_3)/\sqrt{2}. \qquad (9.11)$$

9.5   The HOM interferometer

The interferometer shown in Figure 9.3 was used by Hong, Ou and Mandel to show directly the *indistinguishability* of photons. They injected degenerate signal and idler photons from type-I SPDC into the opposite entry faces of a 50:50 beam splitter BS. Thus they have the same polarization and are indistinguishable. Using eqn. 9.9 we have

$$\begin{aligned}
\hat{a}_1^\dagger \hat{a}_2^\dagger &= (\hat{a}_3^\dagger + i\hat{a}_4^\dagger)(\hat{a}_4^\dagger + i\hat{a}_3^\dagger)/2 \\
&= i(\hat{a}_3^\dagger \hat{a}_3^\dagger + \hat{a}_4^\dagger \hat{a}_4^\dagger)/2. \qquad (9.12)
\end{aligned}$$

Thus the state prepared when the photons exit from the beam splitter will be

$$\hat{a}_1^\dagger \hat{a}_2^\dagger |0\rangle_3 |0\rangle_4 = i[|2\rangle_3 |0\rangle_4 + |0\rangle_3 |2\rangle_4]/2. \qquad (9.13)$$

Thanks to their being indistinguishable the photons should both exit through the same face of the beam splitter, and Hong, Ou and Mandel sought to verify this prediction.

This argument can be made more succinctly. The light reflected and transmitted at a beam splitter differs in phase by $\pi/2$ so that the amplitude for both photons being reflected and the amplitude for both being transmitted differ by π, and thus cancel. Therefore, the photons will exit from the same face of the beam splitter.

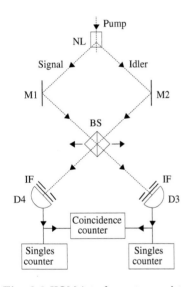

Fig. 9.3 HOM interferometer used to study photon correlations. NL is the non-linear crystal. BS is a 50:50 beam splitter with arrows showing the motion used. The interference filter IF and pinholes select signal and idler.

351.1 nm radiation from an argon-ion laser induces type-I SPDC in a KDP crystal labelled NL. The signal and idler have the same (ordinary) polarization. M1 and M2 are mirrors and BS is a beam splitter that can be displaced laterally to alter the relative path lengths. D3 and D4 are fast photomultipliers. In front of each of these are pinholes and interference filters, labelled IF, which define the degenerate signal and idler beams. The electronic pulses associated with photons arriving at D3 and D4 were counted separately, and also the coincidences between pulses at D3 and D4.

When the beam splitter is exactly centred so that the apparatus is left–right symmetric, both photons are expected to arrive at D4 or both at D3, never one at D3 and one at D4. When the beam splitter is displaced sideways the path lengths differ, so that photon wavepackets do not fully overlap in time, and D3/D4 coincidences can occur. Figure 9.4 shows the D3/D4 coincidence rate as a function of the beam splitter position. This shows the expected drop in the D3/D4 coincidence rate when the paths are equal. Residual coincidences at that point indicate that the beams arriving at the beam splitter do not fully overlap spatially.

When the beam splitter is moved sideways the alternative paths for light reflected there will differ in length by some amount $s = c\tau$. In this case, using eqn. 8.68, the coincidence rate becomes

$$P(\tau) \propto 1 - \exp\left[-(\tau\Delta\omega)^2\right], \qquad (9.14)$$

where $\Delta\omega$ is the bandwidth of the interference filters. In Figure 9.4 the coincidence rate rises as the path difference increases from zero until eventually when the time difference is greater than the coherence time, $1/\Delta\omega$, the rate becomes constant. The fit with eqn. 9.14 shown in Figure 9.4 gives the bandwidth of the signal and idler radiation. At half-height the width of the dip in the figure is $16\,\mu\mathrm{m}$, corresponding to a time delay of 100 fs, which is the coherence time, or equivalently the time duration of the photon wavepackets. An advantage in measuring coherence time using two-photon correlations is that path differences do not need to be held constant to a fraction of a wavelength during the measurement, as would be the case for single-photon interferometry. This advantage carries through to the experiments studying entanglement and teleportation.

In modern versions of the HOM interferometer the light is steered by monomode optical fibres, rather than mirrors, while the beam splitter is replaced by a 50:50 fibre coupler. Signal and idler photons are selected by pinholes and focused by lenses onto the fibre ends.

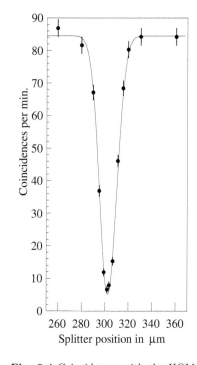

Fig. 9.4 Coincidences with the HOM interferometer as a function of the beam splitter position. The experimental results are compared to the predictions described in the text. Diagram adapted from Figure 2 in C. K. Hong, Z. Y. Ou and L. Mandel, Physical Review Letters 59 2044 (1987).

9.6   Teleportation

Teleportation refers to the transport of a quantum state from one object in one location to another object at another location without the physical transfer of the first object. Quantum states are generally transfered using the polarization of photons carried on optical fibre. Photons carrying quantum information can naturally interact with the carrier medium, so that their state changes, and this degrades the information transmitted. Free space transmission is therefore best, for example, via satellites. At any node in the fibre network, or at a satellite, onward transmission by making a direct copy of the arbitrary quantum state of the input photon is not feasible thanks to the no-cloning theorem discussed in Section 1.13. However, teleportation of this state onto a target photon beyond the node is feasible, but, inevitably, altering the quantum state of the input photon. One application lies in *quantum key distribution QKD*, to be described in Chapter 10. This technique is safe from third-party attempts to access the encryption key being transmitted.

Teleportation makes use of entangled states, which we have seen are non-local. It might be inferred, incorrectly, that information can travel across a space-like separation, faster than the speed of light. However, in order to access the teleported quantum information, some classical information must also be transmitted, travelling at best at the speed of light.

Figure 9.5 shows the components of a successful teleportation scheme introduced by a group at the University of Innsbruck.[3] The information to be transfered is the polarization state (some linear superposition of H and V) of the *input* photon, which is to be transfered to the polarization of the *target* photon. The experimenters produced two pairs of

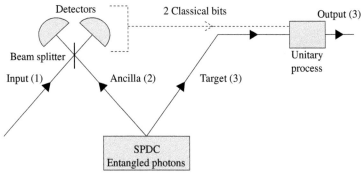

Fig. 9.5 The components of a teleportation scheme using polarization entangled photons.

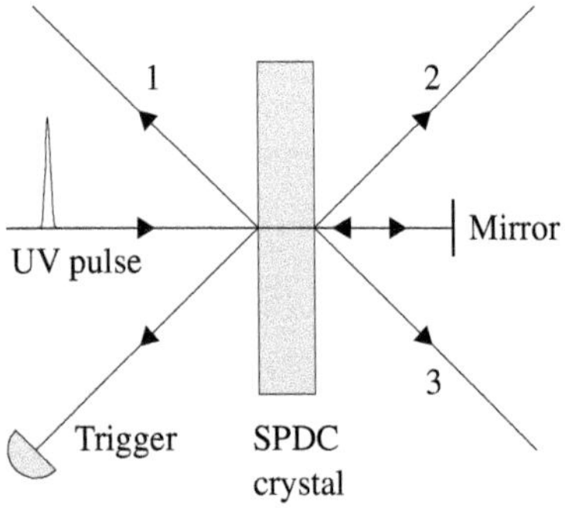

photons by type II SPDC as shown in Figure 9.6. A 200 fs laser pulse passes through the crystal producing one pair of entangled photons, the target, labelled 3, and the *ancilla*, labelled 2. After reflection from the mirror shown the pulse produces a second pair, one of which is the input, labelled 1, while the other is simply used to provide a trigger to initiate data-taking. The polarization state of the ancilla (2) and target (3) resulting from SPDC is the Bell state

$$|\Psi^-\rangle_{23} = [|2\rangle_\mathrm{H}|3\rangle_\mathrm{V} - |2\rangle_\mathrm{V}|3\rangle_\mathrm{H}]/\sqrt{2}. \qquad (9.15)$$

The input photon (1) and the ancilla photon (2) are taken via optical fibre and their quantum state is measured by the beam splitter and the two photodetectors drawn in Figure 9.5. If there is a coincidence between hits in these detectors then either both photons are reflected

Fig. 9.6 Production of two pairs of entangled photons: input (1) and trigger; ancilla (2) and target (3).

[3]D. Boumeester, J.-W. Pan, K. Mattle, M. Eibl, H. Weinfurter and A. Zeilinger, *Nature* 390, 575 (1997).

or both transmitted at the beam splitter. Earlier it was shown that the amplitudes for these two processes differ by a minus sign, making them antisymmetric under interchange of the two photons. As a result the polarization state of input (1) and ancilla (2) must be anti-symmetric in order that the overall spin×spatial state of the two photons is symmetric. The unique possibility for their polarization state is then the Bell state

$$|\Psi^-\rangle_{12} = [|1\rangle_H|2\rangle_V - |1\rangle_V|2\rangle_H]/\sqrt{2}. \tag{9.16}$$

The unknown polarization state of the input photon (1) can be written

$$|\psi\rangle_1 = \alpha|1\rangle_H + \beta|1\rangle_V, \tag{9.17}$$

with $|\alpha|^2 + |\beta|^2 = 1$. Now, the overall polarization state of input (1), ancilla (2) and target (3) initially *before* the Bell state measurement is

$$|\psi\rangle_{123} = |\psi\rangle_1|\Psi^-\rangle_{23}. \tag{9.18}$$

This can be re-expressed in terms of Bell states of the input (1) and ancilla (2) photons thus:

$$\begin{aligned}|\psi\rangle_{123} \quad &= [|\Psi^-\rangle_{12}(-\alpha|3\rangle_H - \beta|3\rangle_V) + |\Psi^+\rangle_{12}(-\alpha|3\rangle_H + \beta|3\rangle_V) \\ &+ \quad |\Phi^-\rangle_{12}(\beta|3\rangle_H + \alpha|3\rangle_V) + |\Phi^+\rangle_{12}(-\beta|3\rangle_H + \alpha|3\rangle_V)]. \end{aligned} \tag{9.19}$$

Whenever coincidences are observed betweeen the detectors, it follows that the input (1) and ancilla (2) are in the Bell state $|\Psi^-\rangle_{12}$. Hence the state of the target (3) can be obtained by taking the scalar product of $\langle\psi|_{123}$ with $|\Psi^-\rangle_{12}$. This gives

$$|\psi\rangle_3 = \langle\Psi^-_{12}|\psi\rangle_{123} = -[\alpha|3\rangle_H + \beta|3\rangle_V], \tag{9.20}$$

showing that the polarization state of the input has been transfered to the target. Note, too, that the quantum state of the input photon state has been changed and cannot be usefully interrogated any longer.

In order to be able teleport the quantum state of the input in every case, not just when the input and ancilla are in $|\Psi^-\rangle_{12}$, it is necessary to be able to detect any of the four Bell states. This requires a more complex experiment.[4] Once the Bell state of the input and ancilla has been determined the state of the target can be infered from eqn. 9.19; this target state can then be converted by linear processes to $[\alpha|3\rangle_H + \beta|3\rangle_V]$. In order to perform the appropriate transformation the identity of the Bell state must be made known. This requires the transfer of two classical bits of information which can only travel as fast as the speed of light. Thus, although the input quantum state can be teleported over a space-like interval, the transfer of *information* never exceeds the speed of light, complying with the special theory of relativity.

---

[4]See Y. H. Kim, S. P. Kulik and Y.Shih, Physical Review Letters 86 1370 (2001).

The Vienna Quantum Optics groups[5] have successfully transmitted pairs of SPDC entangled photons in the Ψ^+ and Ψ^- states over 143 km through air between La Palma and Tenerife despite a 64 dB loss in transmission. Teleportation has been achieved over the uplink to a satellite at 1400 km distance by J.-G. Ren et al.[6]

When teleportation is carried through with all Bell states of the input and ancilla utilized it is known as *deterministic*. Deterministic teleportation has been achieved between atoms in magneto-optical traps,[7] and between ensembles of Cs atoms in glass cells.

9.7   Entanglement between remote atoms

Figure 9.7 shows the experiment carried out at the University of Munich[8] to entangle the quantum states of two rubidium atoms in rooms separated by 20 m. Single rubidium atoms are stored for periods of sec-

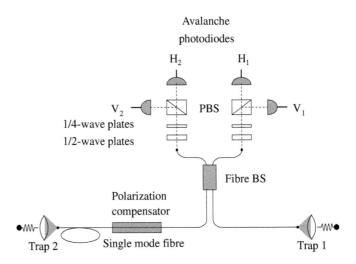

Fig. 9.7 Entanglement of two atoms in traps in separate laboratories. PBS indicates the 50:50 polarizing beam splitters. Detection of a coincidence between the avalanche phodiodes signals a Bell state measurement. Adapted from Figure 1A, J. Hofmann, M. Krug, N. Ortegel, L. Gerard, M. Weber, W. Rosenfeld and H. Weinfurter, *Science* 337, 72 (2012) published by the American Association for the Advancement of Science. Courtesy Professor Weinfurter.

onds in electromagnetic traps at less than 1 mK. The atoms are brought to a specific excited state by short laser pulses, timed so that each atom

[5] A Fedrizzi, R. Ursin, T. Herbst, M. Nespoli, R. Prevedel, T. Scheidl, F. Tiefenbacker, T. Jennewein and A. Zeilinger, *Nature Physics* 5, 389 (2009).
[6] J.-G. Ren et al., *Nature* 549, 70 (2017).
[7] See H. J. Kimble and S. J. van Enk, *Nature* 429, 712 (2004).
[8] J. Hofmann, M. Krug, N. Ortegel, L. Gerard, M. Weber, W. Rosenfeld and H. Weinfurter, *Science* 337, 72 (2012).

emits a single photon. In a proportion of cases both photons emitted are captured by lenses and focused so that they enter and travel along optical fibre to an arrangement of polarizing beam splitters and photon detectors. This arrangement of optical components is used to make a Bell state measurement on pairs of photons, arriving together, one from each atom. Measurement on the two photons, in turn, entangles the spin states of the atoms. When entanglement is heralded by coincidences between the photodetectors this triggers measurements on the atoms to validate their entanglement.

9.8   Interference

Single particle interference could well be described as quantum entanglement, but generally the term entanglement is reserved for correlations between two or more particles. In Young's two slit experiment, if the flux is kept low enough so that a single photon is present in the apparatus at a given moment, the electromagnetic field at the slits is

$$|\psi\rangle = [\,|1\rangle_1\,|0\rangle_2 + |0\rangle_1\,|1\rangle_2\,]/\sqrt{2}, \qquad (9.21)$$

where the subscripts indicate the slit (1 or 2) and the arguments indicate the number of photons at each slit (0 or 1). The entities *entangled* are modes of the electromagnetic fields at the two slits. If measurements are made to locate which slit a photon passes through, then the interference pattern disappears: if the photon is located at slit 1 then the overall state vector collapses to $|1\rangle_1|0\rangle_2$ and the interference pattern becomes that of a single slit. This illustrates Feynman's contention that basic difficulties in understanding quantum phenomena are inherent already in Young's two slit experiment.

9.9   Further reading

Entanglement the Greatest Mystery in Physics, by A. D. Aczel, published by John Wiley and Sons (2002). A more popular account of the background to the subject of entanglement.

Quantum Mechanics, the Theoretical Minimum, by L. Susskind and A. Friedman, published by Allen Lane (2014). This recent book gives a very lucid account of density matrices and entanglement. It is not at all simplistic.

Exercises

(9.1) Photons are separated with equal probability between two paths, 1 and 2, which might be the two ouput ports of a beam splitter. If the total number of photons incident is n, what is the expectation for the product $n_1 n_2$ of the numbers entering each arm?

(9.2) Show that the phase mismatch in the generation of second harmonic waves in the forward direction in a distance dx is $\Delta k dx = (2\omega/c)(n_{2\omega} - n_\omega)dx$ where $n_{2\omega}$ is the refractive index of the crystal for the pump and n_ω that for both idler and signal. In a crystal of length L show that the intensity falls of with the phase mismatch like $\text{sinc}^2(\Delta k L/2)$.

(9.3) Show that for the square matrices A and B $\text{Trace}(AB) = \text{Trace}(BA)$, whether $AB = BA$ or $AB \neq BA$.

(9.4) Show whether the bipartite state $(VV + VH + HV + HH)/2$ is an entangled state or not.

(9.5) What is the state of the light field at the exit ports of a 50:50 beam splitter when only one input port is illuminated by a single photon?

(9.6) What processes, some involving wave-plates, can be used to change one Bell state of photons into any one of the other three Bell states? Note that a half wave plate with its fast axis horizontal will transform $H \rightarrow H$ and $V \rightarrow -V$.

(9.7) In type-II degenerate SPDC a pump laser at 350 nm is being used. If the vertically polarized photon travels at $2.5°$ with respect to the incident laser beam, what are the wavelengths of both the photons produced? In an HOM experiment a half wave plate is used to rotate the the idler photon's plane of polarization by $90°$. What changes are to be expected in the coincidence plot shown in Figure 9.3? Repeat the exercise for a rotation of $45°$.

(9.8) Calculate the polar coordinates on the Bloch sphere for these states:

1)	$[|0\rangle - |1\rangle]/\sqrt{2}$,
2)	$[|0\rangle - i|1\rangle]/\sqrt{2}$,
3)	$|0\rangle\sqrt{3}/2 + |1\rangle/2$,
4)	$|0\rangle/2 + i|1\rangle\sqrt{3}/2$.

(9.9) Alice and Bob share an ensemble of pairs of photons. Half the pairs are in the Bell singlet state Ψ^- and half are in the Bell triplet state Ψ^+. Alice and Bob have no knowledge of which pairs are in the states Ψ^- and which pairs are in Ψ^+. Work out the density matrix that describes Bob and Alice's knowledge of the ensemble. Is this state an entangled state?

EPR and Bell's theorem, and quantum algorithms

10

10.1 Introduction

Using quantum mechanics we can make predictions for the results of measurement of observables, with the restriction that predictions are imprecise to the degree expressed by the uncertainty principle. This restriction is one factor that has led to speculation that quantum mechanics, though proved over a century to be perfectly reliable, is incomplete. The first part of this chapter is used to outline the theoretical and experimental studies that have, thus far, shown no evidence for a deterministic theory underlying relativistic quantum mechanics. The variables of any such putative theory are given the name *local hidden variables* or *elements of reality*. Entanglement has been a key ingredient in this programme, a programme which has both sharpened the understanding of quantum mechanics and brought into play new experimental techniques. After this account, the exploitation of quantum mechanics' non-classical behaviour in practical cryptography is described, and the potential for constructing quantum computing machines with capabilities beyond those of classical computers.

10.2 Local realism and determinism

In 1935 Einstein, Podolsky and Rosen[1] posed questions about aspects of quantum mechanics that, then and now, have made many uneasy. These aspects are the lack of determinism mentioned in the last section and the lack of local realism.

Einstein, Podolsky and Rosen, *EPR*, defined *local realism* in this way: if an observable of a system can be predicted with certainty from some measurement made out of causal contact with the system then this observable should have a real, that is to say definite value in a complete theory. EPR saw that an entangled[2] state was ideal for revealing the absence of local realism in quantum mechanics. At that time modern optical sources were unknown so the example visualized in their *thought experiment* was contrived but nonetheless a valid entangled state. It

[2]The term entangled was first used a little later by Heisenberg.

---

[1]A. Einstein, B. Podolsky and N. Rosen, *Physical Review* 47, 777 (1935)

Quantum 20/20: Fundamentals, Entanglement, Gauge Fields, Condensates and Topology.
Ian R. Kenyon. © 2020. Published in 2020 by Oxford University Press.
DOI: 10.1093/oso/9780198808350.001.0001

involved a pair of particles, A and B, in an eigenstate of two compatible observables: these being their total momentum $P = p_1 + p_2$, and their separation $\Delta x = x_2 - x_1$. EPR argued that if a measurement of the momentum of A is made when A and B are out of causal contact the momentum of B can be predicted $(P - p_1)$ with certainty. Alternatively, by measuring the position of A when A and B are out of causal contact the location of B $(x_1 + \Delta x)$ can be predicted with certainty. Thus both the momentum and coordinates of B should have real (definite) values in any complete theory. However quantum mechanics requires that the product of the precisions in determining the momentum and position of B must exceed $\hbar$. The inference the authors drew was that quantum mechanics is incomplete.

To reiterate: quantum mechanics successfully predicts the correlation of measurements at space-like separation on the components of an entangled system. Einstein had doubts about this 'spooky action at a distance'. It clearly violates the requirement that a physical theory should be *local*. That is to say, measurements made on one component of a system should not affect another component if they are out of causal contact.

Alongside these concerns Einstein was dissatisfied that quantum mechanics only offers probabilities for the generality of measurements; it is not determinate. He famously said that 'God does not play at dice', while in quantum mechanics dice are cast everywhere and all the time.

Suggestions have been made on how to retain the successes of quantum mechanics and at the same time incorporate local realism and determinism. A plausible proposal is that *local hidden variables* exist. These would be properties that fix observables deterministically and might at some future date themselves be shown to exist by experiment. The profound analysis of such ideas by John Bell, and the experiments it led to, are now described. Up to the present all experimental results support quantum mechanics without any need for local realism or determinism.

David Bohm (*Physical Review* 85, 166 (1952)) invented the sole known viable hidden variable theory, albeit *non-relativistic*. Its structure is close enough to *non-relativistic* quantum mechanics that there is no known way to distinguish between them in an experiment. Like quantum mechanics it is a *non-local* theory. In this theory, the wavefunction of all the particles, wherever they are, $\psi = |\psi_0| \exp(i\phi)$, is regarded as a *pilot wave*. Bohm's model generalizes Schrödinger's equation for multiple particles. Then the motion of any particle obeys a first-order differential equation giving the velocity, $\mathbf{v} = (\hbar/m)\nabla\phi$, where ϕ must be evaluated instantaneously for all particles. The fields of particles thus respond to changes in each other's environment, irrespective of separation, which facilitates entanglement.

If the quantum state corresponds exactly to the real physical state, then quantum mechanics is termed *ontic*. If the quantum state only represents accessible knowledge of the real physical state, it would be *epistemic*.

10.3 Bell's inequalities

A simple example reveals the incompatibility of quantum mechanics and local realistic theories with local hidden variables. Suppose that a source emits pairs of photons that are entangled with orthogonal polarizations; one member of each pair travelling right, the other travelling left. Also suppose that for one particular pair the local hidden variable theory fixes the photon polarizations to be at $+45°$ and $-45°$. For clarity let us ignore earlier and later pairs of entangled photons. Let the photon pair fall on *perfect* detectors one to the left, one to the right. In front of these detectors are polarizers oriented at $0°$ and $90°$ respectively. When one photon is detected by one detector local hidden variable theory predicts a probability of one half for the second photon to be detected. However, quantum mechanics requires the polarizations to be orthogonal but otherwise totally indeterminate until one or other photon polarization is measured. Once one photon is detected by one detector, and only then, is its polarization determined; and equally that of the second photon. Hence, according to quantum mechanics the second photon will be detected by the other detector with unit probability. Evidently we cannot know the decision of the hidden variable theory for any given pair, but this example suggests (post hoc) that correlations in detection may hold some hope for discrimination between quantum mechanics and local hidden variable theories.

In the early 1960s it was a widely held view that investigation of the basis of quantum mechanics was unlikely to bear fruit. Then in 1964 John Bell in a seminal paper showed how to discriminate between quantum mechanics and local hidden variable theories using entangled photons. The analysis given is that developed by Clauser, Horne, Shimony and Holt[3] from Bell's work.

Suppose there are a set of local hidden variables that determine the outcomes of all measurements in any correlation experiment involving entangled states. For simplicity the set of local hidden variables can be coalesced into a single one, λ. This would have some distribution $\rho(\lambda)$ such that $\int d\lambda \rho(\lambda)$ is unity. In the case considered here measurements are made of the polarization state of two fully entangled photons with parallel polarization, both V or both H, in the Bell state Φ^+. The arrangement is shown in Figure 10.1. Photons travelling from the source in opposite directions pass through polarizing beam splitters and fall

---

[3]J. F. Clauser, M. A. Horne, A. Shimony and R. A. Holt, *Physical Review Letters* **23**, 880 (1969).

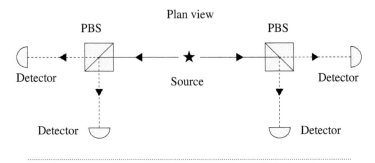

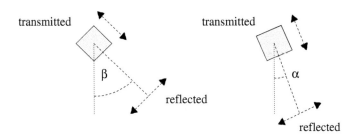

Fig. 10.1 Outline of generic apparatus to test for violations of Bell's inequalities. PBS indicates a polarizing beam splitter.

thereafter on photodetectors that record all the photons without noise. Knowing which detectors record hits determines the polarization of the photons. In each arm the beam splitter and detectors could be rotated freely, independently of the other arm to any selected azimuthal angle around the line of flight from the source. This makes it possible to set the orientations of the polarization required for transmission and reflection differently for the two photons. Let the azimuths be set to α and β for the right-hand and left-hand paths, respectively. Also, we denote detection of a transmitted photon by $+1$ and detection of a reflected photon by -1. Then the outcomes predicted by local hidden variable theories are precise and limited to these values

$$A(\lambda, \alpha) = \pm 1 \ , \ B(\lambda, \beta) = \pm 1, \tag{10.1}$$

for the right- and left-hand paths respectively. A correlation function $E(\alpha, \beta)$ is defined to be $+1$ whenever the outcomes of the measurements on the two arms are the same , either both $+1$ or both -1, and $E(\alpha, \beta)$ is defined to be -1 whenever they differ. Measurements of this correlation must be made for two choices of the azimuthal angle for each photon path in order to bring out the difference between the predictions of quantum mechanics and local hidden variable theories. Then the distinction between these predictions is exposed using the correlator

$$S = E(\alpha, \beta) - E(\alpha, \beta') + E(\alpha', \beta) + E(\alpha', \beta'), \tag{10.2}$$

where α and α' are azimuthal angles selected for the left-hand path; β and β' for the right-hand path. The experimental programme involves making a set of measurements for each of the choices (α, β), (α, β'), (α', β) and (α', β') of the analyser settings. In each case, the correlator is determined by averaging values from a long sequence of photon pairs from the source. Local hidden variable theories predict that

$$
\begin{aligned}
S_{\mathrm{LHV}} &= A(\lambda, \alpha)B(\lambda, \beta) - A(\lambda, \alpha)B(\lambda, \beta') \\
&\quad + A(\lambda, \alpha')B(\lambda, \beta) + A(\lambda, \alpha')B(\lambda, \beta') \\
&= A(\lambda, \alpha)[B(\lambda, \beta) - B(\lambda, \beta')] \\
&\quad + A(\lambda, \alpha')[B(\lambda, \beta) + B(\lambda, \beta')].
\end{aligned} \tag{10.3}
$$

Now B is restricted to the values ± 1, hence, the two terms in square brackets are either 0 and ± 2, or ± 2 and 0, respectively. Similarly, because A is also restricted to the values ± 1, the only outcomes possible are $S_{\mathrm{LHV}} = \pm 2$. It follows that when a statistically significant set of measurements is made the mean value of the correlator must lie in the range

$$
-2 \le \overline{S}_{\mathrm{LHV}} \le +2, \tag{10.4}
$$

which is *Bell's inequality*. Notice that it is not necessary to perform the integration $\int \rho(\lambda) S_{\mathrm{LHV}} \mathrm{d}\lambda$ because each value of S_{LHV} is restricted to lie in the range between -2 and $+2$, and hence, the average will also lie within that range. The prediction using quantum mechanics for identical polarization outcomes left and right with the choice (α, β) is simply given by Malus' law, $[\cos^2(\alpha - \beta)]$, and the prediction for orthogonal polarizations must be $[\sin^2(\alpha - \beta)]$. Thus the correlator predicted by quantum mechanics is

$$
\overline{E}_{\mathrm{QM}} = \cos^2(\alpha - \beta) - \sin^2(\alpha - \beta) = \cos[2(\alpha - \beta)], \tag{10.5}
$$

so that

$$
\overline{S}_{\mathrm{QM}} = \cos[2(\alpha - \beta)] - \cos[2(\alpha - \beta')] + \cos[2(\alpha' - \beta)] + \cos[2(\alpha' - \beta')]. \tag{10.6}
$$

These contrasting predictions are shown in Figure 10.2 for equal angular steps:

$$
\beta - \alpha = \alpha' - \beta = \beta' - \alpha'. \tag{10.7}
$$

The orientations shown in Figure 10.3 with equal steps in angle, of $22.5°$ and $67.5°$, produce the extremal values of S_{QM} of $+2\sqrt{2}$ and $-2\sqrt{2}$ respectively, well outside the range compatible with local hidden variable theories. Bell pointed out that such correlation experiments could rule out local hidden variable theories. The crucial tests culminated in an experiment by Aspect, Grangier and Roger[4]

---

[4]A. Aspect, P. Grangier and G. Roger, *Physical Review Letters* **49**, 91 (1982).

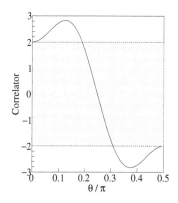

Fig. 10.2 The solid line shows $\overline{S}_{\mathrm{QM}}$ calculated for equal steps of θ in the angular settings in the AGR experiment. Predictions from local hidden variable theories have to lie within the shaded area.

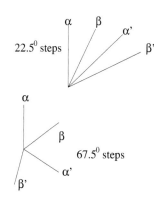

Fig. 10.3 Orientations giving the maximal violation of Bell's inequality.

10.4   The AGR experiment

Entangled photons at wavelengths 551.3 and 422.7 nm were produced in the sequential decay of calcium

$$(4p^2)\,{}^1S_0 \to (4s4p)\,{}^1P_1 \to (4s^2)\,{}^1S_0, \tag{10.8}$$

shown in Figure 10.4. In order to conserve linear momentum the photons travel back to back. Conservation of angular momentum requires that the decay chain is

$$
\begin{aligned}
[0,0]_{Ca} \;\to\; & [1,-1]_{Ca}\,(1,+1)_1 + [1,+1]_{Ca}\,(1,-1)_1 \\
\to\; & [0,0]_{Ca}\{(1,-1)_2(1,+1)_1 + (1,+1)_2(1,-1)_1\},
\end{aligned}
$$

where the round brackets contain single photon states, subscripted 1 and 2; the square brackets contain the atomic state: in both cases the first symbol is the particle spin and the second is its projection along the quantization axis. The quantization axis is taken along the line of flight of the first photon, so that the polarization states, when rewritten as right- and left-handed, are

$$(1,-1)_1 = |R\rangle_1,\;\; (1,+1)_1 = |L\rangle_1,\;\; (1,-1)_2 = |L\rangle_2,\;\; (1,+1)_2 = |R\rangle_2. \tag{10.9}$$

Thus the entangled state of the photons is

$$|R\rangle_1|R\rangle_2 + |L\rangle_1|L\rangle_2 = |V\rangle_1|V\rangle_2 + |H\rangle_1|H\rangle_2, \tag{10.10}$$

which is one of the Bell states, Φ^+, introduced in Section 9.3. The second photon is emitted nearly synchronously with the first because the intermediate atomic state has a lifetime of only 5 ns. As shown in Figures 10.4 and 10.5, a beam of calcium atoms in their ground state is excited by laser beams at 406 nm and 581 nm with waists of 50 nm diameter where they intersect the calcium beam. Absorption of one photon from each laser takes the atom via a virtual state (the grey band in Figure 10.4) to the $(4p^2)\,{}^1S_0$ state. Large aperture aspheric lenses accept photons within a cone of semiangle 32° and focus them onto the beam splitters. Narrow band filters select photons of wavelength 442.7 nm in one arm and 551.3 nm photons in the other arm, thus reducing the extraneous light falling on the detectors. The reflection/transmission coefficients of the polarizing beam splitters for the selected polarizations are 0.94, and only 0.007 for the orthogonal rejected polarizations. Coincidences between photons were registered if the delay between the photon arrival times at the detectors was less than 20 ns. We call the number of coincidences when both photons are transmitted N_{tt}, the number of coincidences when both are reflected N_{rr} and so on. The mean value of the correlation function $E(\alpha, \beta)$ is then

$$\overline{E}_{\exp}(\alpha,\beta) = [N_{tt} + N_{rr} - N_{tr} - N_{rt}]/N, \tag{10.11}$$

where N is the total number of coincidences. The rate for all coincidences was typically $4\,10^7\,s^{-1}$ using a few mW laser power. In addition

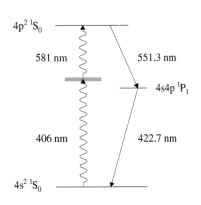

Fig. 10.4 Energy levels in calcium utilized in tests of Bell's inequality by Aspect, Grangier and Roger.

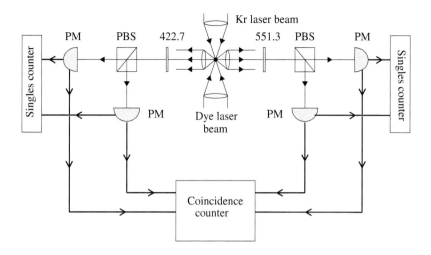

Fig. 10.5 Experiment for observing violations of Bell's inequality. PM indicates a photomultiplier, PBS a polarizing beam splitter. Optical fiters are labelled by their transmission wavelength. The atomic beam travels perpendicular to the diagram where the beams focus. Adapted from Figure 2 of A. Aspect, P. Grangier and G.Roger, *Physical Review Letters* 49, 91 (1982). Courtesy Professor Aspect.

the accidental coincidences occuring at a rate proportional to N^2 are only 100 per second. Each data-taking run lasted only 100 s. With the ordering used in Figure 10.3 and choosing equal steps

$$\beta - \alpha = \alpha' - \beta = \beta' - \alpha' = 22.5° \qquad (10.12)$$

the mean value of the correlator, $\overline{S}$ defined in eqn.10.2, was measured to be 2.697 ± 0.015. This value is far outside the range of −2.0 to +2.0 demanded by any local hidden variable theory. It agrees well with the quantum mechanical prediction evaluated for the experimental conditions, 2.70±0.05.[5]

[5]The raw prediction without taking account of acceptances, accidentals, etc. is $2\sqrt{2}$, or 2.82.

10.4.1 Closing loopholes

Objections were raised about such tests of any local hidden variable theory using Bell's inequality. Three loopholes were identified: first, the efficiency for detecting optical photons is only a few per cent and depends on the wavelength, so it is necessary to assume that this gives a fair sample of the totality of the entangled pairs (*detection loophole*); second, the detectors are close enough that causal contact between components is possible (*locality loophole*); third, the orientation of the beam splitters should be set randomly and independently (*freedom of choice loophole*). All three loopholes have been closed in three recent experiments at Delft University, the National Institute of Standards and Technology in Boulder and at Vienna.[6] Some relevant features of the experiment carried out in Vienna are given next.

In this experiment the photons are polarization entangled pairs at 810 nm produced in SPDC as described in Section 9.4. In each arm the polarization of the photon is randomly and independently rotated ahead of the beam splitter. Only the photons transmitted are detected, using *transition edge sensors* (TES). These detectors contain a thin film of superconducting tungsten kept at 100 mK, below the temperature for transition to normal conduction at 178 mK. When a photon is absorbed the excited electron heats the tungsten enough to cause it to locally go normal. The resistance change is detected by a series array of 100 SQUIDs.[7] In this way a device detection efficiency of 0.95 is achieved. Overall apparatus efficiency for detecting a photon in one arm heralded by one in the other arm is 0.77, indicating low losses in the optical fibre coupling used. Eberhard had earlier showed that, provided this detection efficiency exceeds 0.66, it is possible to avoid the detection loophole. Surprisingly, he also showed that less than full entanglement gives better discrimination between the predictions made by quantum mechanics and local realistic theories. The authors used the entangled state

$$[|V\rangle_1 |H\rangle_2 + r|H\rangle_1 |V\rangle_2]/\sqrt{1 + r^2}, \qquad (10.13)$$

with $r = -2.9$. Any local hidden variable theory should then satisfy the Clauser–Horn–Eberhard inequality

$$P_{++}(\alpha, \beta) \le P_{+0}(\alpha, \beta') + P_{0+}(\alpha', \beta) + P_{++}(\alpha', \beta'), \qquad (10.14)$$

where, for example, P_{+0} signifies detection in the first arm but no photon detected in the second arm. The experimentally observed deviation was 11 standard deviations above this limit, well outside the range allowed by any local hidden variable theory. Figure 10.6 shows the emission, polarization settings and measurements in space-time. The slanting arrowed lines are paths for light in free space. Setting precedes the arrival of the photons. Equally measurements precede the arrival of any possible signal from setting in the other arm. All thanks to the 60 m separation of the detectors.

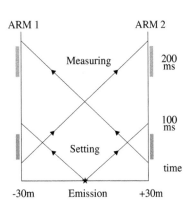

Fig. 10.6 Timing sequence in the measurement of the Clauser–Horn–Eberhard inequality. The grey bars mark the periods when polarization setting and measurement took place on the two arms. The arrowed lines show light lines in free space.

10.5   Quantum Computing

A first remark that needs to be made is that calculations whether performed by classical or by quantum methods have to obey the laws of physics. In the latter case it is wavefunctions that are processed rather than intensities. The logic elements are physical objects whose quantum state is described by a location on the Bloch sphere shown in Appendix

[6]L. K. Shalin et al., *Physical Review Letters* 115, 250402 (2015); B. Hansen *et al.*, Nature 526, 682 (2015); M. Giustina *et al.*, *Physical Review Letters* 115, 250401 (2015).

D: this complex two-dimensional state is the *qubit* which replaces the on/off (0/1) bit in classical computers. Atoms, quantum dots and superconducting Josephson junctions have all been used as elements of devices containing a few qubits. Registers are formed from strings of qubits rather than from strings of bits. Where a 3-bit word can at any given moment contain only a single value (001, for example) a 3-qubit word can contain 2^3 independent superposed orthogonal words/states: for example, $(|000\rangle + |001\rangle +...+ |111\rangle)/\sqrt{8}$. The quantum mechanical development of some physical state, say, according to Schrödinger's time-dependent equation, is unitary. That means that the eigenstates that start orthogonal remain orthogonal throughout a quantum calculation. This fact opens the prospect of making calculations in parallel: n qubits could carry 2^n orthogonal channels undergoing the same processing in parallel.

In order to obey the laws of physics hardware processing in a quantum computer must therefore in general be unitary and reversible.[8] The obvious exception is readout of results which is a measurement; this collapses the wavefunction onto an eigenstate of the observable selected for measurement. The result of a calculation is a generally a distribution rather than a unique number, which means that the quantum algorithm must be run enough times to give this distribution with adequate precision. Cases of special interest are those where the solution is a unique eigenstate among many. The examples of Grover's algorithm to search for a particular entry in an unsorted database, and Shor's algorithm for factorizing products of prime numbers, are of this type. The latter is particularly important because it reveals that a quantum computer, if and when constructed, could solve a problem that is effectively impossible for classical computers to solve in any reasonable length of time. The problem is of great commercial and political interest because solving it would change the most popular current encryption method, used where high security is demanded, from being opaque to being transparent.

[8] Quantum processing is equivalent to a summation over Feynmann diagrams, the complexity possible being determined by the hardware available.

10.6   Basic tools

The gates in quantum computers have to be unitary. The basic set includes gates that load single qubits into registers or read out from registers, and others that produce selective phase rotations of qubits on the Bloch sphere. There must also be control gates involving a target qubit and a control qubit. The CNOT gate, like the classical NAND gate in conventional circuitry, is a gate from which all logical elements could be built. The hardware that would form the registers and the gates is at present available only in a primitive evolving form.

The gate required to load a single qubit is the *Hadamard gate* with the matrix form

$$\text{H}\begin{bmatrix} a \\ b \end{bmatrix} = \frac{1}{\sqrt{2}}\begin{bmatrix} 1 & 1 \\ 1 & -1 \end{bmatrix}\begin{bmatrix} a \\ b \end{bmatrix} = \begin{bmatrix} a+b \\ a-b \end{bmatrix}/\sqrt{2}, \qquad (10.15)$$

with $|a|^2 + |b|^2 = 1$. When n Hadamard gates are applied to a register containing n qubits each initialized to $|0\rangle$, the result is

$$\begin{aligned} \text{H}^{\otimes n}|0\rangle_n &= \text{H}|0\rangle\,\text{H}|0\rangle\,\text{H}|0\rangle...... \\ &= [(|0\rangle + |1\rangle)\sqrt{2}]\,[(|0\rangle + |1\rangle)\sqrt{2}]\,[(|0\rangle + |1\rangle)\sqrt{2}]..... \\ &= \sum_{i=0}^{N-1} |i\rangle_n/\sqrt{N}, \qquad (10.16) \end{aligned}$$

Table 10.1 Input (side) and output (top) qubits with the CNOT gate. The control bit precedes the target bit.

CNOT	00	01	10	11
00	1	0	0	0
01	0	1	0	0
10	0	0	0	1
11	0	0	1	0

where $N = 2^n$. The strings $|i\rangle$ include all possible combinations of 0s and 1s in each qubit along the register. One string has all qubits set to zero; other strings have a single qubit set to 1, the rest being 0; other strings have two qubits set to 1; and so on, ending with the string with all qubits set to 1. *All these qubit combinations are thus simultaneously entangled in one register, and can be processed in parallel.* This latent parallel processing capability distinguishes quantum from classical computing.

The action of a single qubit phase rotation gate is the following

$$|0\rangle = \begin{bmatrix} 1 \\ 0 \end{bmatrix} \rightarrow \begin{bmatrix} 1 \\ 0 \end{bmatrix}; \quad |1\rangle = \begin{bmatrix} 0 \\ 1 \end{bmatrix} \rightarrow \begin{bmatrix} 0 \\ \exp(i\phi) \end{bmatrix} \qquad (10.17)$$

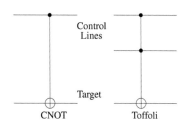

Fig. 10.7 CNOT and Toffoli gates.

A gate rotating the phase through $\pi/4$ is called a T-gate. This gate, together with the Hadamard gate and the CNOT-gate, form a universal set from which all other gates can be built. The controlled NOT gate or CNOT gate has two input qubits, called the control and target qubits as shown in Figure 10.7. The control qubit is unaffected; the target qubit is changed if the control qubit is 1, but not if the control qubit is 0. Thus

$$\text{CNOT}|c\rangle|t\rangle = |c\rangle|t \oplus c\rangle, \qquad (10.18)$$

where $\oplus$ signifies addition modulo 2 of the control qubit value c and the target qubit value t: for example $1 \oplus 1 = 0$. The actions of the CNOT gate are summarized in Table 10.1. Also shown in Figure 10.7, the *Toffoli gate* has two control lines: the target is only changed if both control qubits are $|1\rangle$. Note that the combination of Hadamard gates and a CNOT gate shown in Figure 10.8 interchanges the roles of the control and target lines: see the first exercise. When any quantum gate is applied during a calculation each of the superposed orthogonal states in the input register (equivalently the wavefunction) responds independently of all the others.

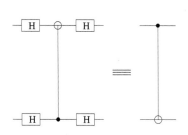

Fig. 10.8 Interchange of the roles of the target and control lines to a CNOT gate when sandwiched between Hadamard gates.

10.7   Shor's algorithm

The classical RSA method of data encryption was invented by Clifford Cocks and re-invented by Rivest, Shamir and Adleman. Details are given in Appendix F. It relies for its security on the astronomical time required by any classical computer to factor out the product of two prime numbers, typically more than 200 decimal digits long. Using classical computing techniques, the only way to penetrate the RSA encryption is to divide the public key n by each prime in turn up to the largest less than $\sqrt{n}$. At some point the factors will be found. The number of primes that are less than $\sqrt{n}$ is around $(n)/\ln(n)$. In the case of the digital public key RSA-130 with 130 decimal digits it required of order 1000 years of MIPS processing (million instructions per second) to uncover the primes: in real time it took several months. None of the keys beyond RSA-232 has been factored mathematically.[9] Shor's algorithm is a quantum algorithm that would, if it could be fully implemented, factor such a product in possibly a matter of hours. Thus far, only a few low products of primes have been factored with quantum techniques.[10]

[9]Side-channel attacks may succeed by making use of information leaking from the packages performing decryption.

Shor's approach will be illustrated using simple products of primes, firstly 15. The sequence $2^x \bmod 15$ with $x = 1, 2,\cdots$ has a repetitive pattern: 1,2,4,8, 1,2,4,8, 1,2,4,8$\cdots$ cycling every four steps, while (4+1) is a prime factor of 15. The next product of primes is 21, and in this case the sequence $2^x \bmod 21$ has a repetitive pattern: 1,2,4,8,16,11, 1,2,4,8,16,11$\cdots$ This sequence cycles every six steps while (6+1) is a prime factor of 21.

These patterns are quite general: Euler showed that if $n = pq$ with p and q being primes, then for $1 < a < n$ the sequence $a^x \bmod n$ for $x = 1$, $2,\cdots$ repeats with a period τ that is a factor of the number $(p-1)(q-1)$.

Shor's algorithm for factoring n seeks to determine the period of the sequence $a^x \bmod n$ for random choices of a. Each result is a factor of $p - 1$ or $q - 1$, so that with enough passes the values of $p - 1$ and $q - 1$ will emerge: hence n itself. Shor recast this as the problem of finding the period of the function

$$f(x) = a^x \bmod n \qquad (10.19)$$

using a quantum computer. Two quantum registers are required, each with a number of qubits, m, greater than the number being factorized, n. Their overall state is initialized as

$$|s\rangle_0 = \frac{1}{\sqrt{m}} \sum_{k=0}^{m-1} |x\rangle|0\rangle, \qquad (10.20)$$

where $|x\rangle$ is the qubit string $|x_0\rangle|x_1\rangle \cdots |x_{m-1}\rangle$. This string is obtained by acting on each of the m initial $|0\rangle$ qubits separately with Hadamard gates: the register content thus constructed is a superposition of strings with all possible permutations of the values 0 and 1 in each qubit. Then

[10]See for example J. J. Vartiainen, A. O. Niskanen, M. Nakahara and M. M. Salomaa, *Physical Review A*70, 012319 (2004).

the second register is loaded with the function $f(x)$ giving the entangled state

$$|s\rangle = \frac{1}{\sqrt{m}} \sum_{k=0}^{m-1} |x\rangle |f(x)\rangle. \tag{10.21}$$

If $f(x)$ is a periodic function with period τ so that $f(x + r\tau) = f(x)$, where r is any integer, the period can be obtained in two steps. The first step is to express $|x\rangle$ in terms of its Fourier transform $|j\rangle$:

$$|s\rangle = \frac{1}{m} \sum_{k=0}^{m-1} \sum_{j=0}^{m-1} \exp[2\pi i x j/m] |j\rangle |f(x)\rangle. \tag{10.22}$$

Step two is to measure both registers. The measurement of $|f(x)\rangle$ projects out the state (remembering that $f(x + r\tau) = f(x)$),

$$\frac{1}{m} \sum_{j=0}^{m-1} \sum_{r=0}^{R} \exp[2\pi i (x + r\tau) j/m] |j\rangle. \tag{10.23}$$

where R is the integer m/τ. The amplitude for any state $|j\rangle$ is

$$
\begin{aligned}
A_j &= \frac{1}{m} \sum_{r=0}^{R} \exp[2\pi i (x + r\tau) j/m] \\
&= \frac{1}{m} \exp[2\pi i x j/m] \sum_{r=0}^{R} \exp[2\pi i r\tau j/m], \tag{10.24}
\end{aligned}
$$

A probability distribution is built up by making repeated passes through the algorithm, giving for the state $|j\rangle$

$$P_j = |A_j|^2 = \frac{1}{m^2} \left| \sum_{r=0}^{R} \exp(2\pi i r\tau j/m) \right|^2. \tag{10.25}$$

For most states $|j\rangle$ there would be cancellations between the various phase factors $\exp(2\pi i r\tau j/m)$ with different values of r. This makes P_j small. However, if $\tau j/m$ is an *integer* then $r\tau j/m$ is also an integer and there is constructive interference between the terms in eqn. 10.25. Their sum is then large: $\exp(i\phi)[R + 1]$ where ϕ is the common phase. In this case the probability is

$$P_j = \left[\frac{R+1}{m} \right]^2. \tag{10.26}$$

The measurement therefore has a high probability of giving the qubit a value $j = Nm/\tau$ where N is some unknown small integer. The algorithm must be used repeatedly with different values of the seed a. By comparing the different values of j that are found m/τ and hence τ can be extracted. Small phase shift errors are to be expected but even so the result is robust.[11]

---

[11] see D. Coppersmith 'An approximate Fourier transform useful in quantum computing' arxiv.org/abs/quant-ph/0201067.

Table 10.2 Dephasing time T_2 in seconds, the time required to change the qubit state T_{op} in seconds, and number of operations possible in the dephasing time N_{op}. The ion is isolated in a magnetic trap. The transmon incorporates a superconducting Josephson junction.

Qubit type	T_2	T_{op}	N_{op}
Qdot	10^{-7}	10^{-12}	10^5
Transmon	10^{-5}	10^{-10}	10^5
Ion	1	10^{-6}	10^6

10.8 Challenges of quantum computing

Atoms, ions and quantum dots have all been tried as qubits with microwave or optical photons exciting transitions. Qubits based on superconducting Josephson junctions appear to have the greatest potential. Details on possible qubits are given in Chapter 15. The conversion between matter and photon qubits presents difficulties because of the low overall rate and randomness inherent in spontaneous emission. Placing a matter qubit in an optical cavity tuned to the transition wavelength of the qubit, and of dimensions a few wavelengths, can be used to enhance the spontaneous emission as explained in Section 12.9. In the case of superconducting qubits the coupling between qubits can be made through short transmission lines carrying microwave photons.

Qubits are more fragile than the states of circuit elements that carry bits in classical computers. The lifetime T_1 of the excited state is one limitation. Qubits possess phase information and this makes them more susceptible to environmental disturbance. The interaction with the environment leads to a loss of coherence, measured by the dephasing time T_2. The crucial measure of the usefulness of a physical qubit is the number of operations possible in the time it takes for decoherence to develop. Table 10.2 shows the dephasing time, the single step processing time, and the number of steps possible before coherence is lost. Superconducting systems have an advantage in the existence of relevant experience in constructing voltage sources consisting of tens of thousands of Josephson junctions, but come with the need to work at cryogenic temperatures.

10.9 Quantum cryptography

Ultimately, encryption methods such as RSA encryption using a public key may be susceptible to quantum computer attacks. However, quantum-based protocols[12] for the private distribution of keys used

[12]For a much fuller account of the techniques of quantum key distribution see 'Quantum Cryptography' by N. Gisin, G. Ribordy, W. Tittel and H. Zbinden, *Reviews of Modern Physics* **74**, 145 (2002).

for cryptography have the potential for being unconditionally secure against any eavesdropper. The features of quantum-based key distribution are illustrated here using the BB84 protocol proposed by Bennett and Brossard in 1984.[13] The technique described depends solely on the impossibility of cloning quantum states as discussed in Section 1.13; variants employ entangled quantum states.

Cryptography *keys* are long strings of bits with which a confidential data stream is scrambled to form a *cryptogram*. After transmission from one party to another, the key is used to unscramble and recover the original data from the cryptogram. Provided the key itself can be transmitted securely between the parties involved, any message it is used to encrypt can be equally secure. Quantum mechanics forbids the cloning of arbitrary quantum states and it is this property that makes it possible to detect any attempt to eavesdrop on the exchange of a cryptography key.

A setup for the BB84 protocol is shown in Figure 10.9. Using the customary names, Alice sends Bob the key in the form of a string of single photons whose polarizations carry the bit information. Eve, the would-be eavesdropper, intercepts the photons. She also attempts to copy the photons and pass these copies on to Bob with the intention of concealing the interception. If the connection between Alice and Bob is made over an optical fibre link then Eve might break the fibre at some point and insert a repeater there.

Alice's photon source provides plane polarized photons. First, Alice rotates the polarization vector of each photon through an angle $\theta_A + \theta_D$ using the Pockels cell PCA. Alice sets θ_D to $0°(90°)$ when the current data bit in the key is $0(1)$. The way that Alice exploits the quantum property of light is by resetting θ_A randomly for each pulse to either $0°$ or $45°$. Alice records the value of θ_A used for each bit transmitted. At Bob's receiver the polarization is rotated through a further angle θ_B by the Pockels cell PCB. The choice of θ_B is also made randomly between the two values $0°$ or $-45°$. This choice is made quite independent of Alice's choice of θ_A. Finally, the photon enters the polarizing beam splitter PBS and travels to one or other photodetector. For each bit received Bob records which photodetector fired and the corresponding setting of θ_B.

When the transmission of the key is complete Alice and Bob openly exchange information on the random settings that they used for each bit, namely, the values of θ_A and θ_B. Whenever they find that the analyser and polarizer settings are complementary, namely that $\theta_B = -\theta_A$, then the bit that Bob receives will be exactly that sent by Alice. If Alice sent

[13]C. H. Bennett and G. Brassard, *Proceeedings of the IEEE International Conference on Computer Systems and Signal Processing*, pages 175-9, published by IEEE (1984).

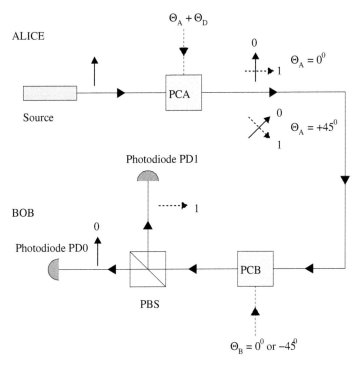

Fig. 10.9 An implementation of the BB84 protocol for secure quantum key exchange using photon polarization to carry the bit information.

a '0' bit the photodetector PD0 would have fired, and if a '1' bit PD1 would have fired. Bob and Alice accept these bits. However Bob and Alice reject bits whenever $\theta_B \neq -\theta_A$. The bits accepted, known as *sieved bits*, are used to form the cryptographic key.

Now, we can examine how quantum mechanics works to foil Eve when she intercepts the photons with a setup identical to Bob's and tries to transmit copies to Bob. We can suppose that Eve is lucky and infers that Bob sets his analyser θ_B randomly to $0°$ or $-45°$, uses those angles herself, but still has to guess which is Bob's choice each time. The four possible combinations of Bob's and Eve's analyser settings are listed in Table 10.3 for one of Alice's settings. These four combinations are equally probable. The essence of the matter is that when Bob and Alice make complementary choices and Eve makes the opposite choice to Bob for her analyser's alignment (for example, $\theta_B = \theta_A = 0°$, while Eve has $\theta_E = -45°$) Eve will randomly get a 0-bit or a 1-bit. In such cases the bit Eve then sends to Bob has a 50 per cent chance of being the same as that sent by Alice. This is the case for the third row in Table 10.3. On the other hand, when Eve uses the same analyser setting as Bob he will receive a correct bit from Eve, as in the first row in the table.

Alice and Bob can detect Eve's presence by comparing a subset of the

Table 10.3 Possible combinations of Bob and Eve's analyser settings for a given polarization set by Alice. Bob's success rate during the interception is shown in the final column.

Alice	Eve	Bob	Success rate
$\oplus$	$\oplus$	$\oplus$	100%
$\oplus$	$\oplus$	$\otimes$	not used
$\oplus$	$\otimes$	$\oplus$	50%
$\oplus$	$\otimes$	$\otimes$	not used

sieved bits, for which they ought to have identical bit values if there is no eavesdropper. The sieved bits come from the analyser combinations in the first and third rows of Table 10.3. Bob and Alice will find that they disagree over the bit value in one quarter of all the subsets they compare, and this tells them that the transmission has been intercepted. With Eve absent there would be 100 per cent agreement. Obviously all sources of error must be eliminated to a level well below 25 per cent in order that Bob and Alice can reliably detect Eve's interception.

The effectiveness of the BB84 protocol vividly illustrates the principle that it is impossible to determine the general quantum state of a single photon. This principle ensures that photons cannot be cloned, which is Eve's undoing. Eve measures the polarization in one basis and in doing so collapses the photon polarization state into one in her basis. This is all the information that Eve can extract and in the face of the random choice of basis made by Alice it cannot fully characterize the incoming photon's polarization.

Table 10.4

x=	0	1
f_0	0	0
f_1	0	1
f_2	1	0
f_3	1	1

10.9.1   Recent applications

The range of quantum key distribution, QKD, over installed optical fibre links is $\sim$50 km at rates above $100\,\mathrm{kb\,s^{-1}}$. Several companies offer systems for QKD, including ID Quantique of Geneva. This company provided secure transmission of election results in the Swiss national elections from the canton of Geneva in 2007. Various institutes have their internal systems. In 2016 the BB84 protocol was used to establish secure communications between Xingling in China and Graz in Austria over 7600 km apart via the Chinese Micius satellite.[14] This approach exploits the negligible absorbtion and induced decoherence when photons pass through empty space. The satellite orbits at 500 km altitude and circles the earth every 94 minutes, and provides $\sim$300 s of downlink time per pass over the ground stations at Graz and Xingling. Secure QKD is obtained as follows. First, two random bit strings, the keys, are transmitted using the BB84 protocol, one to Graz, one to Xingling. Call these MG and MX, respectively. Onboard Micius an exclusive OR of the two strings is taken, MX⊕MG, and the result transmitted to both ground stations using an open radiofrequency channel. Each station recovers the other's key by taking a second exclusive OR: thus at Xingling (MX⊕MG)⊕MX gives back MG, as illustrated in the table 10.5. The secure keys of 100 KB were then successfully used for one-time pad encoding between Graz and Xingling over a terrestrial optical fibre link. In addition, video links within China over 280 km were made over fibre connections, with six relays on route, using the same protocol.

Transferring quantum keys through relays in a network or via a satel-

---

[14]S.-K. Liao *et al.*, *Physical Review Letters* **120**, 030501 (2018).

Table 10.5 Graz–Xingling link

MX	MG	MX⊕MG	(MX⊕MG)⊕MX
0	0	0	0
0	1	1	1
1	0	1	0
1	1	0	1

lite still offers possibilities for malign interference at the nodes. In order to remove this weakness qubit states would need to be teleported across each node. This cannot yet be implemented with current technology.

10.10   Further reading

Speakable and Unspeakable in Quantum Mechanics, by J. S. Bell, published by Cambridge University Press (1987). An outstanding account of Bell's groundbreaking studies presented with depth and humour.

Quantum Computer Science: An Introduction by N. D. Mermin, published by Cambridge University Press (2007).

Quantum Computing Devices: Principles, Designs and Analysis by G. Chen, D. A. Church, B. G. Englert, C. Henkel, B. Rohwedder, M. O. Scully and M. S. Zubairy. Published by Chapman and Hall/CRC, Boca Raton Florida (2007). An authorative text by practitioners about progress toward quantum computing using a wide range of technologies.

Roads towards Fault-tolerant Universal Quantum Computation by E. T. Campbell, B. M. Terhal and C. Vuillon in *Nature* 549, 172 (2017) outlines the challenges ahead in this field.

Exercises

(10.1) Show that $\text{CNOT}|++\rangle = |++\rangle$, $\text{CNOT}|+-\rangle = |--\rangle$, $\text{CNOT}|-+\rangle = |-+\rangle$, $\text{CNOT}|--\rangle = |+-\rangle$. These results reveal that in the Hadamard basis $(+,-)$ the roles of the control and target states are reversed.

(10.2) Prove the equivalence shown in Figure 10.8.

(10.3) Show that the result of two NOT gates (inversion of state) followed two Hadamard gates acting on

$|00\rangle$ is

$$|\psi_D\rangle = [|00\rangle - |10\rangle - |01\rangle + |11\rangle]/2 = |--\rangle.$$

(10.4) A black box (oracle) can produce transformations

$$U|x\rangle|y\rangle = |x\rangle|y + f(x)\rangle,$$

where the function f can be any one of the four transformations shown in Table 10.4. Note that if

$y = 1$ and $f(0) = 1$ then $|x + f(x)\rangle = |0\rangle$. Deutsch's problem is to distinguish between operations that give the same result for either input 0 or 1, and those that give opposite results for 0 and 1: that is, to distinguish between, on the one hand f_0 and f_3, and on the other hand f_1 and f_2. Any classical computer requires two passes at least in order to make the distinction. Defining $\overline{f} = f \oplus 1$, show that if $f(0) \neq f(1)$, then it follows that $\overline{f}(0) = f(1)$ and $\overline{f}(1) = f(0)$.

(10.5) Using the information from the last two exercises, show that when $f(0) = f(1)$

$$|\psi_=\rangle = U|\psi_D\rangle = [|0\rangle - |1\rangle]\,[|f(0) - \overline{f}(0)\rangle]/2,$$

and that when $f(0) \neq f(1)$

$$|\psi_{\neq}\rangle = U|\psi_D\rangle = [|0\rangle + |1\rangle]\,[|f(0) - \overline{f}(0)\rangle]/2,$$

(10.6) From the result of the last exercise show that a single gate acting on the first of the two qubits in $U|\psi_D\rangle$ will distinguish between on the one hand f_0 and f_3, and on the other f_1 and f_2.

(10.7) Use combinations of Pauli X and Z gates (σ_x and σ_z) acting on a single qubit to generate the Bell states Φ^-, Ψ^+ and Ψ^- from the Bell state

$$\Phi^+ = [|0\rangle|0\rangle + |1\rangle|1\rangle]/\sqrt{2}.$$

(10.8) Using equal steps θ in the expression 10.6 calculate and confirm the variation of $\overline{S}_{QM}$ versus θ shown in Figure 10.2.

(10.9) Suppose that A and B share a pair of qubits in the overall state

$$|\psi\rangle = \frac{1}{\sqrt{12}}[3|00\rangle + |01\rangle + |10\rangle - |11\rangle]$$

$$= \frac{1}{\sqrt{3}}[2|00\rangle - H_A H_B |11\rangle].$$

What is the probability that when A and B measure their qubits they find $|11\rangle$? Now, suppose starting from $|\psi\rangle$, A applies a Hadamard gate to the first qubit; what is the probability that they find $|01\rangle$? If instead starting from $|\psi\rangle$, this time B applies a Hadamard gate to the second qubit, what is the probability that they then find $|10\rangle$? Finally, starting from $|\psi\rangle$ both A and B apply Hadamard gates to their qubits, what is the probability that they find $|11\rangle$? Are these results consistent with a deterministic theory? This is a version of Hardy's paradox.[15]

(10.10) *GHZ* states of three qubits such as

$$\psi_{\text{GHZ}} = |000\rangle - |111\rangle$$

were first considered by Greenberger, Horne and Zeilinger.[16] Such states can be used to reveal a direct contradiction between the predictions of local hidden variable theories and quantum mechanics. Their predictions are mutually exclusive, which is a step beyond the probabilistic differences in violations of Bell's inequalities. ψ_{GHZ} is an eigenstate of several products of the Pauli matrices σ_x, σ_y and σ_z. Calculate the eigenvalues for $[\sigma_y]_1[\sigma_y]_2[\sigma_x]_3$, $[\sigma_y]_1[\sigma_x]_2[\sigma_y]_3$, $[\sigma_x]_1[\sigma_y]_2[\sigma_y]_3$ and $[\sigma_x]_1[\sigma_x]_2[\sigma_x]_3$. Hence show that if the values of the first three measurements are set by local hidden variables they cannot be consistent with the final measurement made with the product of three σ_x operators.

[15]L. Hardy, *Physical Review Letters* 68, 2981 (1992).
[16]D. M. Greenberger, M. Horne and A. Zeilinger, pp 73-6 in *Bell's Theorem, Quantum Theory and Conceptions of the Universe* edited by M. Kafatos, published by Kluwer Academic (1989).

Quantum measurement

11

11.1   Introduction

A hallmark of quantum theory is uncertainty in the outcome of measurements additional to any instrumental errors that degrade precision. Heisenberg showed that the product of uncertainties from measuring two conjugate observables has a lower limit, $\hbar/2$. Experimenters often wish to measure one observable with the least error compatible with the uncertainty relation. A technique of growing interest, known as *squeezing*, is to shrink the error on one observable at the expense of making its conjugate observable more uncertain. Squeezing can be applied on the large scale in gravitational wave detectors of kilometre dimensions and in nanoscale experiments.

Other measurements of considerable value are *quantum non-demolition (QND)* measurements. Such measurements preserve the value of the observable measured, which requires that there is no *back action* from the measuring device.

The chapter commences with a section to clarify the distinction between the intrinsic uncertainty inherent in quantum states and the uncertainty due to back action. Next an account is given of von Neumann's analysis of quantum measurement, and quantum non-demolition measurements are explored.

In the remainder of the chapter the example of interferometers employed in gravitational wave detection is used to illustrate the general principles of quantum effects on measurement. As preliminaries the limitation imposed by shot noise in interferometry and of back action in determining position are discussed. Then the advanced LIGO gravitational wave detectors are described: in 2015 these were used to make the first direct detection of gravitational waves. This leads to a presentation of the *standard quantum limit* for interferometer precision. The technique of squeezing and its potential for improving precision is described: for example phase squeezing has been successfully applied in the GEO600 gravitational wave detector. The principles are equally applicable to mesoscopic devices.

Quantum 20/20: Fundamentals, Entanglement, Gauge Fields, Condensates and Topology.
Ian R. Kenyon. © 2020. Published in 2020 by Oxford University Press.
DOI: 10.1093/oso/9780198808350.001.0001

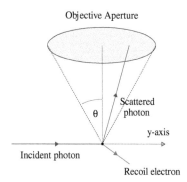

Objective Aperture

θ

Scattered photon

y-axis

Incident photon

Recoil electron

Fig. 11.1 Heisenberg's ultraviolet microscope for measuring an electron's position.

11.2 Uncertainty

Heisenberg took the example of measuring an electron's position with an ultraviolet microscope to show that there is back action in the form of the electron recoil. Figure 11.1 shows the scattering of a photon from the electron. The resolution in position is given by the classical diffraction formula

$$\Delta y = \lambda / \sin\theta, \tag{11.1}$$

where λ is the wavelength and the lens subtends an semi-angle θ at the electron. Now the direction of the scattered photon lies in this angular range so that the uncertainty in the y-component of momentum transfered is

$$\Delta p_y = p \sin\theta, \tag{11.2}$$

where p is the momentum of the incident photon. Thus

$$\Delta y \Delta p_y = \lambda p = h, \tag{11.3}$$

which is the lower limit on the product of the uncertainties. Taking Gaussian distributions of momentum and position this becomes the familiar

$$\Delta y \Delta p_y \geq \hbar/2, \tag{11.4}$$

where $(\Delta y)^2$ and $(\Delta p_y)^2$ are variances. Heisenberg's uncertainty principle is independent of the measurement procedure.

A parallel result involving the *intrinsic* uncertainty inherent in quantum observables was proved by Robertson. Any pair of conjugate variables, F and G, satisfy

$$[\hat{F}, \hat{G}] = i\hbar. \tag{11.5}$$

Writing δF for $\hat{F} - \langle \hat{F} \rangle$ the variance $(\Delta F)^2 = \langle \hat{F}^2 \rangle - \langle \hat{F} \rangle^2 = \langle (\delta F)^2 \rangle$; similar equations hold for G. δF and δG are Hermitian so that

$$\langle \delta F \delta G \rangle^* = \langle (\delta F \delta G)^\dagger \rangle = \langle \delta G^\dagger \delta F^\dagger \rangle = \langle \delta G \delta F \rangle. \tag{11.6}$$

Hence the imaginary part of $\langle \delta F \delta G \rangle$ is

$$\mathrm{Im}\, \langle \delta F \delta G \rangle = \langle \delta F \delta G - \delta G \delta F \rangle / 2 = \langle [\hat{F}, \hat{G}] \rangle / 2 = i\hbar/2, \tag{11.7}$$

so it follows that

$$|\langle \delta F \delta G \rangle|^2 \geq \hbar^2/4. \tag{11.8}$$

Using the definitions given above and applying the Schwarz inequality gives

$$(\Delta F)^2 (\Delta G)^2 = \langle (\delta F)^2 \rangle \langle (\delta G)^2 \rangle \geq |\langle \delta F \delta G \rangle|^2. \tag{11.9}$$

Combining the results of the last two equations we get Robertson's uncertainty relation

$$(\Delta F)^2 (\Delta G)^2 \geq \hbar^2/4. \tag{11.10}$$

This result comes from the uncertainty inherent in wavefunctions and relates the fluctuations whether measured or not. It says nothing about back action in which one measurement disturbs another or about noise in measurements. By contrast Heisenberg's uncertainty relation is the outcome of back action in the measurement process. More comprehensive analyses reformulate Heisenberg's uncertainty relation including all factors. M. Ozawa,[1] and P. Busch, P. Lahti and R. F. Werner[2] have reformulated Heisenberg's uncertainty relation so as to include intrinsic error, noise and back action. If $\sigma(F)$ and $\sigma(G)$ are the intrinsic uncertainties, $\varepsilon(F)$ is the noise in measuring F and $\eta(G)$ is the back action on G, then Ozawa proved that

$$\varepsilon(F)\eta(G) + \varepsilon(F)\sigma(G) + \sigma(F)\eta(G) \geq \hbar/2, \qquad (11.11)$$

The two approaches of Ozawa, and of Busch and colleagues were compared experimentally by Sulyok and Sponar using two spin variables of thermal neutrons.[3] It seems they are in most respects equivalent. The reader will find that in most situations Heisenberg's result is adequate.

11.3   Analysis of quantum measurement

Von Neumann provided a simple analysis of quantum measurements in terms of a measuring device (the probe or meter) and the system whose observable is being measured. *Both* are treated as quantum objects. In a first step the probe interacts with the system so that they become entangled. This requires a suitable choice of the probe and its initial condition. Then their overall state is

$$|\psi\rangle = \sum_j a_j |\psi_{s,j}\rangle |\psi_{p,j}\rangle, \qquad (11.12)$$

where $|\psi_{s,j}\rangle$ and $|\psi_{p,j}\rangle$ are respectively the system and probe eigenstates for the observables $\hat{v}_s$ and $\hat{v}_p$, and $a_j a_j^*$ is the probability of outcome j. The evolution of $|\psi\rangle$ in this phase is unitary. In the interaction representation

$$H_s\psi = i\hbar\partial\psi/\partial t, \qquad (11.13)$$

where H_s is the component of the Hamiltonian (*i.e.* energy) due to the interaction of the probe with the system. In a second, final step the value of v_p is read out, which collapses the overall state to

$$|\psi_{s,k}\rangle |\psi_{p,k}\rangle, \qquad (11.14)$$

where the subscript k labels the eigenstate following the collapse.[4] If the observable v_p has continuous values then this becomes an integral over a range determined by the precision of the probe. Von Neumann's

---

[1] *Physics Letters* A318 21 (2003).
[2] *Reviews of Modern Physics* 86, 12611 (2014).
[3] *Physical Review* A96, 022137 (2017).
[4] The collapse of the wavefunction is discussed in Section 1.11.

analysis confines classical behaviour to the readout of the probe's observable v_p, which determines v_s and the state of the system.

In general a measurement using a probe will affect the system measured and in particular it will cause the observable measured to change with time: then a second measurement of the same observable will give a different result. This effect is known as *back action*. The classic example, discussed above, is that of measuring an electron's position by scattering photons from it. However if after the measurement the system remains in the same eigenstate of the observable measured the result of a second measurement will be the same as the first. Then the measurement is called a *quantum non-demolition, QND, measurement*. This requires that the part of the Hamiltonian (energy) due to the interaction between probe and system commutes with the observable measured:

$$[H_\mathrm{s}, v_\mathrm{s}] = 0. \tag{11.15}$$

The state of the probe can however be affected by the interaction. Important examples of QND measurement will be described here and in Chapter 12. One example involves detecting which of two states an ion occupies, and is the basis for the construction of optical clocks, whose precision exceeds that of atomic clocks. The other example is used to measure the gyromagnetic ratio of the electron to parts in 10^{11}: a measurement that is a critical test of the theory of quantum electrodynamics. For the most part this chapter deals with non-QND measurements of mirror positions in gravitational wave interferometers. Plans are in train for QND measurements in which mirror velocities rather than positions would be measured. One of this chapter's exercises is used to familiarize the reader with this technique.

11.4   Shot noise in interferometry

Figure 11.2 shows a Mach–Zehnder interferometer. The device measures the phase difference ϕ between light arriving at the detector after travelling along the alternative paths. This difference could be due to the difference in path length between the two arms or due to material inserted in one arm. For simplicity we assume that a beam splitter reflects and transmits equally and has no loss. If a single photon enters port $A1$ and none enters port $A2$ then the output from the first beam splitter, using eqn. 9.9, is

$$[|1,0\rangle + i|0,1\rangle]/\sqrt{2}, \tag{11.16}$$

where the first index is the photon count in arm $A3$ and the second is the count in arm $A4$. On entry to the second beam splitter the state has become

$$[|1,0\rangle + i|0,1\rangle \exp i\phi]/\sqrt{2}, \tag{11.17}$$

where the first index refers to arm $B1$ and the second to arm $B2$. Finally, the amplitude of the state emerging from port $B4$ is $i(1 + \exp i\phi)/2$ and

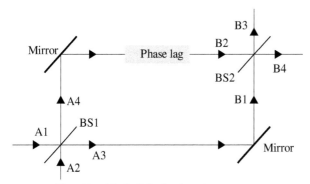

Fig. 11.2 Mach–Zehnder interferometer.

the intensity there is

$$p(\phi) = [1 + \cos \phi]/2 = \cos^2(\phi/2). \qquad (11.18)$$

Using eqn. 8.59 with N photons the precision obtained in measuring ϕ is $1/\sqrt{N}$. This is the *shot noise limit*. In principle this can be improved by injecting N entangled photons at the entry ports. These are the so-called *NOON* states $(|N,0\rangle + |0,N\rangle)/\sqrt{2}$. Such a state arriving at the second beam splitter has evolved to

$$[|N,0\rangle + i \exp(iN\phi)|0,N\rangle]/\sqrt{2}. \qquad (11.19)$$

This has a smaller phase uncertainty in ϕ of $1/N$, and is known as the *Heisenberg limit*. Generating such states is technically difficult: using SPDC, Nagata and colleagues produced $|4,0\rangle + |0,4\rangle$ states with a matching improvement in the phase sensitivity limit.[5]

11.5   Back action on interferometer mirrors

The gravitational wave detectors are giant Michelson interferometers with freely suspended massive mirrors. The passage of the gravitational waves causes the lengths of the two arms to oscillate in antiphase giving a corresponding fringe oscillation that can be measured. We now explore the uncertainties in measuring these displacements as they change with time. Suppose a first measurement of position has an root mean square uncertainty Δx_0 then back action gives a root mean square uncertainty in momentum of Δp_0 greater than $\hbar/(2\Delta x_0)$. Note that $\Delta x_0 \geq 0$ and $\Delta p_0 \geq 0$. The consequent velocity uncertainty is $\Delta p_0/M$ where M is the mirror mass. This in turn leads to an additional uncertainty in position after a time τ of $\Delta p_0 \tau/M$. Mirrors are therefore made as massive as is

[5]T. Nagata, R. Okamoto, J. L. O'Brien, K. Sascki and S. Takeuchi, *Science* 316, 726 (2007).

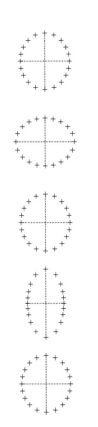

Fig. 11.3 Motion of free masses when a gravitational wave arrives perpendicular to the diagram. One complete cycle of the quadrupole oscillation is shown.

practical to reduce this uncertainty. After time τ the total uncertainty in the mirror's position Δx_τ is given by

$$(\Delta x_\tau)^2 = (\Delta x_0)^2 + (\Delta p_0)^2 \tau^2 / M^2. \tag{11.20}$$

Now

$$(\Delta x_0 - \Delta p_0 \, \tau/M)^2 \geq 0 \tag{11.21}$$

so that

$$(\Delta x_0)^2 + (\Delta p_0)^2 \tau^2 / M^2 \geq 2 \Delta x_0 \, \Delta p_0 \, \tau/M. \tag{11.22}$$

Using this result with eqn. 11.20

$$\Delta x_\tau^2 \geq 2 \Delta p_0 \, \Delta x_0 \, \tau/M \geq \hbar \tau/M, \tag{11.23}$$

where the second inequality makes use of the uncertainty principle. This reveals that any attempt to monitor the position with high accuracy by sacrificing potential knowledge of the momentum is doomed to failure. It won't work because errors from both contribute to Δx_τ.

It is interesting that a measurement of mirror momentum, or equivalently of mirror speed, is a QND measurement: there is no back action (see the first exercise below). This has led to designs of interferometers measuring mirror speeds, which yield the position after integration over time.

In measuring the motion of a mass, for example, a micron-sized cantilever, that has been cooled close to $0\,\mathrm{K}$, there are irreducible zero point fluctuations. A mass M oscillating at angular frequency Ω has position $x_0 \exp(i\Omega t)$ and using eqn. 2.32 the zero point fluctuation in position is

$$x_{\mathrm{zpf}}^2 = \hbar/(2M\Omega). \tag{11.24}$$

11.6   The advanced LIGO interferometer

On the 14th September 2015 the advanced LIGO Michelson interferometers at Hanford in Washington state and at Livingstone in Louisiana, detected a burst of gravitational waves from the inspiralling and merger of two black holes occuring around 1.2 billion years ago.[6] Each rotation of the black holes about their centre of mass generates two cycles of gravitational radiation. As they spiral inward the rotations become ever more rapid and the frequency of the waves rises in a characteristic *chirp* that terminates when the black holes merge. The time difference between observation at the two sites was 7 ms; this is less than the light path time of 10 ms between the detectors, and established the correlation between the two events. The black holes' masses were estimated to be around 30 solar masses, and their merger to have released three solar

---

[6]B. P. Abbott et al. (LIGO Scientific Collaboration and Virgo Collaboration), *Physical Review Letters* 116, 061102 (2016).

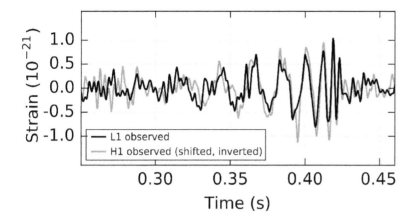

Fig. 11.4 Strain versus time plot from the first gravitational wave signal observed by the advanced LIGO at Hanford (H1) and Livingstone (L1). LIGO Open Science Center at https://losc.ligo.org/events/GW150914.

masses as gravitational wave energy, sufficient to produce a measurable signal on earth. Figure 11.3 illustrates how space-time oscillates over a complete cycle of a gravitational wave. The wave direction of travel is perpendicular to the diagram. Each cross represents a freely suspended mass. Then the arms of a Michelson interferometer indicated in the diagram would lengthen and shorten in antiphase, so that the phase difference between light travelling along the two arms would oscillate too. Observing the resulting interference fringe displacements constitutes the measurement method. Figure 11.4 shows the oscillations of space-time at the two detectors over the crucial 150 ms. The oscillation frequency rises at an increasing rate to approximately 250 Hz. This is the frequency *chirp* expected in the evolution of gravitational waves from such mergers. At this point the merged single black hole rang down with a response

$$\exp(-t/\tau)\cos(2\pi f t + \phi), \tag{11.25}$$

where $f \approx 260/M$ and $\tau = 0.004M$, M being the mass of the final merged black hole in units of 65 times the solar mass. The masses of the merging black holes and the merged black hole were obtained by modelling the event using general relativity and fitting this to the observed oscillations. Figure 11.5 shows the details of the inspiral and coalescence from this numerical modelling. Space-time stretched by only one part in 10^{21} as each wave crest passed. We can now apply the analysis of quantum measurement made above to understand how the design of the advanced LIGO detectors made it possible to achieve this amazing sensitivity. The motion of the mirrors under the action of gravitational waves is, so far as we know, purely classical: it is in the response of the detector that the quantum aspects of measurement emerge.

Figure 11.6 shows the basic components of the advanced LIGO Michelson interferometers. The 40 kg mirrors are test masses used to transfer

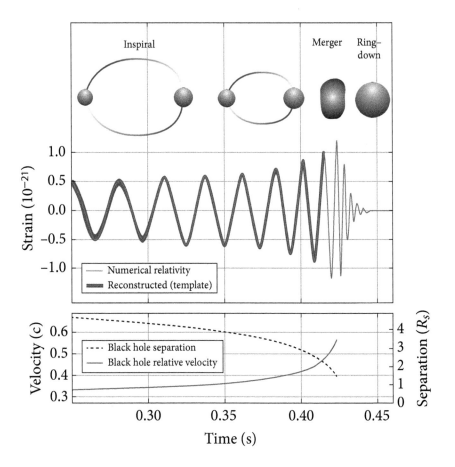

Fig. 11.5 Estimated strain amplitude from GW150914 with numerical relativity models of the black hole horizons as the holes coalesce. In the lower panel the separation of the black holes is given in units of the Schwarzschild radius $2GM/c^2$ and the relative velocity is divided by c. LIGO Open Science Center at https://losc.ligo.org/events/GW150914. The work is reported in B. P. Abbott et al. (LIGO Scientific Collaboration and Virgo Collaboration), *Physical Review Letters* 116, 061102 (2016). Courtesy LIGO Collaboration.

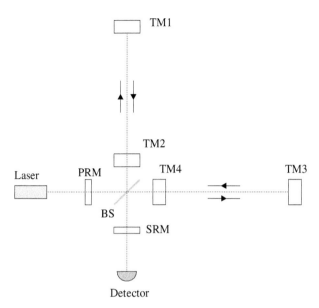

Fig. 11.6 Sketch of components of the advanced LIGO detectors. The mirrors are TM1, TM2, TM3 and TM4. PRM is the power recycling mirror, SRM the signal recycling mirror. BS is the beam splitter.

the classical motion of space to the electromagnetic field stored in the arms of the interferometer. Surprisingly, it is the quantum properties of the electromagnetic field that determine the precision achievable. For the moment we may take this for granted and re-examine the role of the mirrors later. A stabilized laser beam at 1064 nm illuminates the 50/50 near lossless beam splitter. Each of the two 4 km long arms has in addition to the far mirror a second similar massive mirror placed close to the beam splitter. This second mirror converts each arm into a Fabry–Perot cavity and the laser is tuned to a cavity resonance. Radiation in resonance with the cavity passes to and fro about 300 times before its intensity falls significantly. As a result the phase resolution of the interferometer is improved by a similar factor. It is arranged that in the absence of gravitational waves the reflected beams arrive out of phase at the photodetector, in other words it views a dark fringe. An important consequence is that almost all the radiation entering through the input face of the beam splitter ends up exiting through this face toward the laser. In general the electric field at the detector is the real part of

$$E = E_{\text{in}}[\exp(i\phi_x) - \exp(i\phi_y)]/2, \qquad (11.26)$$

where $\phi_{x,y}$ are the phase changes in the two arms. When the interferometer is set on a dark fringe a gravitational wave travelling perpendicular to the plane of the interferometer produces an electric field at the detector

$$E_{\mathrm{GW}} = E_{\mathrm{in}}[\exp(i\phi_{\mathrm{GW}}) - \exp(-i\phi_{\mathrm{GW}})]/2 \approx i\phi_{\mathrm{GW}}E_{\mathrm{in}}, \qquad (11.27)$$

where ϕ_{GW} is the oscillating phase difference induced by the gravitational wave. In order to measure this tiny phase better the output in the absence of any disturbance is offset from the dark fringe by a small angle so that a constant and larger DC field is added into the signal:

$$E_T = E_{\mathrm{GW}} + E_{\mathrm{DC}}. \qquad (11.28)$$

The power is then

$$\begin{aligned} P &= E_{\mathrm{DC}}^2 + 2E_{\mathrm{DC}}E_{\mathrm{GW}} \\ &= E_{\mathrm{DC}}[E_{\mathrm{DC}} + 2\phi_{\mathrm{GW}}E_{\mathrm{in}}] \end{aligned}$$

where the negligible term E_{GW}^2 has been ignored. This power is converted to a current in the photodetector and from the current the oscillating phase ϕ_{GW} is extracted.

11.7   The standard quantum limit

The role of the mirrors is to transfer the fluctuations of space-time to the electromagnetic field in the interferometer.[7] As a result it is the quantum fluctuations of the radiation in the interferometer that impose the ultimate limit on the precision of the phase measurement. There are two ways in which they affect the precision: noise from statistical variation of the number of photons arriving at the detector, *shot noise*; and noise from the fluctuations of radiation pressure on the mirrors.

First consider the shot noise. If the optical power is P at angular frequency ω and wave number k then during a measurement taking a time τ the integrated photon count at the detector, N is $P\tau/\hbar\omega$. The corresponding minimum phase uncertainty $\delta\phi_{\mathrm{s}}$ is $1/\sqrt{N}$, and this leads to an uncertainty in the path length of

$$\delta x_{\mathrm{s}} = \delta\phi_{\mathrm{s}}/k = \sqrt{\frac{\lambda\hbar c}{2\pi P\tau}}, \qquad (11.29)$$

where the radiation wavelength $\lambda = 2\pi/k$. It is the intensity ($\sim$ power) spectrum of the noise that quantifies its impact. The shot noise has a *spectral density*, that is an intensity per unit frequency, of

$$S_{\mathrm{s}} = (\delta x_{\mathrm{s}})^2\tau = \frac{\lambda\hbar c}{2\pi P}, \qquad (11.30)$$

---

[7]This matter is pursued more fully by V. B. Braginsky, M. L. Gorodetsky, F. Ya. Khalili, A.B. Matsko, K. S. Thorne and S. P. Vyatchanin, *Physical Review D*67, 092001 (2003).

with units $m^2 Hz^{-1}$ This imposes a limit on the detectable strain measured over a distance L of

$$h_s = \frac{\sqrt{S_s}}{L} = \frac{1}{L}\sqrt{\frac{\hbar c \lambda}{2\pi P}}. \tag{11.31}$$

Next consider the effect of fluctuations in the radiation pressure on the mirrors. This is simply the *back action* considered earlier in the general case. The force due to each photon striking a mirror of mass M during the measurement time τ is $2\hbar k/\tau$. With an average of N photons arriving at the mirror in time τ the fluctuation on this force is

$$\delta F = 2\sqrt{N}\hbar k/\tau. \tag{11.32}$$

When the freely suspended mirror responds to a gravitational wave of angular frequency Ω_{GW} its equation of motion is

$$F\cos(\Omega_{GW}t) = Md^2x/dt^2 = -M\Omega_{GW}^2\, x\, \cos(\Omega_{GW}t). \tag{11.33}$$

Then the arm length fluctuation due to the radiation pressure fluctuation is

$$\delta x_r = \delta F/M\Omega_{GW}^2 = \frac{2\sqrt{N}\hbar k}{\tau M\Omega_{GW}^2}. \tag{11.34}$$

The corresponding spectral noise intensity is

$$\begin{aligned}
S_r = (\delta x_r)^2\tau &= N\tau\left[\frac{2\hbar k}{\tau M\Omega_{GW}^2}\right]^2 \\
&= \left(\frac{P\tau^2}{\hbar\omega}\right)\left[\frac{2\hbar k}{\tau M\Omega_{GW}^2}\right]^2 \\
&= \frac{8\pi\hbar P}{\lambda c}/(M^2\Omega_{GW}^4).
\end{aligned} \tag{11.35}$$

Expressed as a strain this gives

$$h_r = \frac{\sqrt{S_r}}{L} = \frac{1}{LM\Omega_{GW}^2}\sqrt{\frac{8\pi\hbar P}{\lambda c}}. \tag{11.36}$$

The limitimg strains due to the shot noise, eqn. 11.31, and the radiation pressure noise, eqn. 11.36, have inverted dependences on the laser power P. Hence the minimum detectable strain is obtained when these contributions are equal:

$$\frac{\lambda\hbar c}{2\pi P} = \frac{8\pi\hbar P}{\lambda c M^2\Omega_{GW}^4}, \tag{11.37}$$

whence we have an optimum optical power

$$P = \frac{\lambda c M\Omega_{GW}^2}{4\pi}. \tag{11.38}$$

Substituting this value the minimum total spectral noise power density is

$$S = S_s + S_r \geq \frac{4\hbar}{M\Omega_{GW}^2}, \tag{11.39}$$

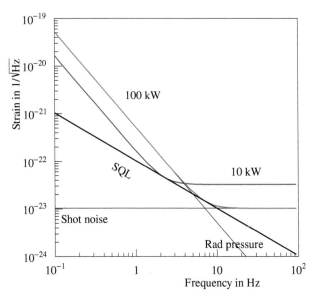

Fig. 11.7 Limit on detectable strain with a simple Michelson interferometer having 10 km arms, 10 kg mirrors and laser wavelength 1064 nm. Examples are shown for 10 kW and 100 kW circulating power.

with a corresponding strain

$$h = \frac{1}{L}\sqrt{\frac{4\hbar}{M\Omega_{\mathrm{GW}}^2}}. \tag{11.40}$$

We have here the *standard quantum limit*, SQL. Putting in the parameters of a simple Michelson interferometer with the dimensions similar but not identical to LIGO, the limiting strain detectable is plotted in Figure 11.7. The masses of the mirrors are 10 kg, the laser wavelength is 1064 nm and the arms are 10 km long. Two cases are shown, for a circulating optical power in the arms of 10 kW and 100 kW. Raising the power lowers the shot noise and improves the sensitivity for detecting high frequency gravitational waves. However, the radiation pressure fluctuations increase with increasing power, which reduces the sensitivity to low frequency waves. The frequency at which the SQL is attained rises as the circulating optical power is increased.

Then our ideal detector with 100 kW power would have adequate sensitivity to detect the event at ∼100 Hz observed by advanced LIGO. Evidently very high optical powers are required, much greater than the wattage available from the well-stabilized 1064 nm NdYAg lasers as used in LIGO.

11.7.1   Cavity enhancement

Mirrors are inserted into both arms of the LIGO interferometers near the beam splitter making each arm into a 4 km long Fabry–Perot cavity. Fabry–Perot cavities are characterized by their *finesse*, $\pi\sqrt{R}/(1-R)$, where R is the reflectance of the mirrors. Finesse is the ratio of fringe separation to the fringe width, which with the high reflectance mirrors used in advanced LIGO is ~ 300. This has the effect that the light at a cavity resonance traverses the cavity around 300 times before its intensity falls off appreciably. The phase resolution of the interferometer is improved by a similar factor and the energy in the radiation stored in the interferometer is similarly boosted, cutting down the shot noise. However a drawback is that gravitational waves with a period shorter than the lengthened storage time will be undetectable. In operation the laser is tuned close to a resonance of the cavity so as to benefit from maximum phase resolution.

Even with the Fabry–Perot cavities the power circulating inside the interferometer arms is still far less than the 100 kW required to detect the waves from astronomically distant black hole mergers. What saves the day is that there is the opportunity to recycle wasted power. The detection mode requires a dark fringe on the detector in the absence of any gravitational waves, that is negligible power entering the detector. Consequently almost all the radiation exits through the entry port toward the laser. Placing a *power recycling mirror, PRM* between the laser and the beam splitter a new cavity is formed between this mirror and an effective mirror formed by the interferometer. The returning power is recycled in this cavity. In this way the power at the beam splitter is enhanced to over 100 kW. Standard mirrors would overheat in such beam intensities. Those in LIGO are highly homogeneous silica with reflective coatings that are non-absorbing and flat to better than λ/F. The far mirrors have reflection coefficients of 0.999995 and the near mirrors a more modest 0.985 to facilitate entry and exit of beams. The bandwidth of the composite interferometer-PRM cavity is only 1 Hz which, with mode cleaning, renders the radiation entering the interferometer uniquely noise free. Another recycling mirror is placed between detector and beam splitter to enhance the signal.

A further refinement is a mode cleaning cavity that is inserted before the power recycling mirror. This cavity is tuned to transmit only the principal TEM_{00} laser mode which has a clean Gaussian profile of 5 cm diameter. At the same time it disperses all other modes in the laser beam, thus eliminating most of the noise in the beam.

Figure 11.8 shows the noise floor achieved by the advanced LIGO detectors. At low frequencies seismic noise is suppressed by the mirror damping and the suspension. Individual spikes correspond to mechanical resonances of the components of the mirror assemblies. The phase

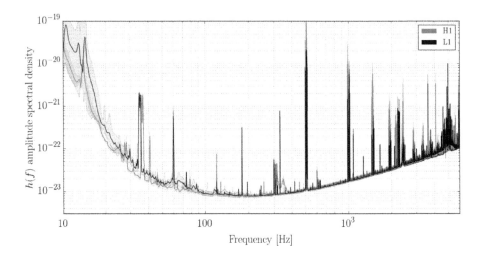

Fig. 11.8 Strain noise level in advanced LIGO. Figure 3 from LIGO Open Science Center at https://losc.ligo.org/events/GW150914. Courtesy of the LIGO Collaboration.

[8]For more details see C. Bond, D. Brown, A. Friese and K. A. Strain, *Living Reviews in Relativity*, December 2016, 19:3.

amplification coming from the Fabry–Perot cavities is flat out to around 300 Hz and then falls off steadily with increasing frequency.[8] The frequency at which this knee in the response occurs is roughly the inverse of the storage time in the Fabry–Perot etalons that are formed by the interferometer mirrors. There is an accompanying rise in the minimum detectable strain at higher frequencies in Figure 11.8. This effect was not included in the analysis used to produce Figure 11.7.

11.8   Squeezing

Heisenberg's uncertainty principle gives a lower limit to the product of the precisions with which a pair of conjugate observables can be determined. If only one observable is of importance, such as the phase in an interferometer, then in principle it could be measured to high precision while sacrificing potential information about its conjugate, the radiant energy at the interferometer detector. The uncertainty principle requires that the uncertainties in the phase ϕ and the photon count, N, satisfy eqn. 8.59. This yields a SQL

$$\Delta\phi = \Delta N = 1/\sqrt{2}. \tag{11.41}$$

It is however possible to have $(\Delta\phi)^2 < 1/2$ while $\Delta n \Delta \phi$ remains greater than $1/2$. This is an example of *squeezing* and gives the prospect of beating the SQL. Phase determination is at the heart of interferometry so that using light with a squeezed phase uncertainty can give improved resolution in metrology, in optical component testing, and in spectroscopy. Tests with the GEO600 gravitational wave detector have demonstrated a

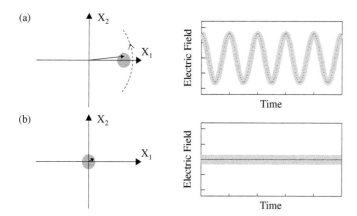

Fig. 11.9 Instantaneous electric field and electric field evolution for unsqueezed light showing range of uncertainty: for (a) a coherent laser-like source and (b) the vacuum.

noise reduction of 3 dB in strain measurements. Details of squeezing and its application to gravitational wave inteferometry will now be described.

Simplifying eqn. 8.17, the electric field in a single mode is

$$\hat{E} = \hat{a}\exp(-i\omega t) + \hat{a}^{\dagger}\exp(i\omega t)/\sqrt{2}$$
$$= [\hat{X}_1 \cos(\omega t) + \hat{X}_2 \sin(\omega t)], \tag{11.42}$$

where $\hat{a}$ annihilates and $\hat{a}^{\dagger}$ creates a photon in the mode of angular frequency ω. The new operators are

$$\hat{X}_1 = (\hat{a}^{\dagger} + \hat{a})/\sqrt{2} \text{ and } \hat{X}_2 = i(\hat{a}^{\dagger} - \hat{a})/\sqrt{2}. \tag{11.43}$$

$\hat{X}_1 \cos(\omega t)$ and $\hat{X}_2 \sin(\omega t)$ are known as *quadratures*. From eqn. 8.7, $[\hat{a}, \hat{a}^{\dagger}] = 1$, so it follows, using the results of Section 11.2 that

$$[\hat{X}_1, \hat{X}_2] = i \text{ and hence } (\Delta X_1)^2(\Delta X_2)^2 \geq 1/4. \tag{11.44}$$

A fully coherent beam attains the minimum uncertainty limit with $(\Delta X_1)^2 = (\Delta X_2)^2 = 1/2$. This is the example illustrated on the left of the upper Figure 11.9(a): a typical state vector is drawn, and the range of uncertainty in its endpoint is shaded. The time variation of the electric field amplitude, produced by rotating this state vector is shown in the right hand panel. The corresponding plots for the vacuum field, itself by definition a coherent state, appear in Figure 11.9(b). In the case of both coherent states, laser or vacuum, the areas of the uncertainties in the quadratures have equal magnitude, namely 1/4. Figure 11.10 shows the wave-vectors and the time variation of the electric fields for two squeezed states. The upper panels show the effect of squeezing phase for a coherent laser source, while the lower ones show the effect of squeezing the vacuum. The product of uncertainties $(\Delta X_1)^2(\Delta X_2)^2$ remains 1/4.

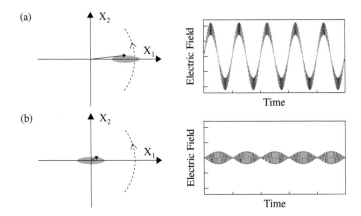

Fig. 11.10 Instantaneous electric field and electric field evolution for phase squeezed light showing range of uncertainty: for (a) a coherent laser-like source and (b) the vacuum.

This enhancement needs to be put in words. In the upper panel of Figure 11.10 the effect is to give a sharper definition of the phase, while the amplitude variation worsens. Equally in the lower panel the vacuum variation is suppressed at the part of the cycle defining the phase.

The mechanism for squeezing a mode of the electromagnetic field is introduced here. Squeezing the vacuum lightly produces a change of state of the mode of the electromagneic field of interest:

$$|0\rangle \rightarrow |\psi\rangle \equiv |0\rangle + s|2\rangle, \tag{11.45}$$

with $s \ll 1$ being the amplitude corresponding to two coherent photons being injected into the mode. Then

$$
\begin{aligned}
\langle\psi|\hat{X}_1^2|\psi\rangle &= \langle\psi|(\hat{a}^\dagger + \hat{a})^2|\psi\rangle/2 \\
&= [\langle 0| + s\langle 2|]\,[\hat{a}^\dagger\hat{a}^\dagger + \hat{a}\hat{a} + \hat{a}^\dagger\hat{a} + \hat{a}\hat{a}^\dagger]\,[|0\rangle + s|2\rangle]/2 \\
&= (1 + 2\sqrt{2}s)/2 \tag{11.46}
\end{aligned}
$$

to order s. Also

$$\langle\psi|\hat{X}_1|\psi\rangle = [\langle 0| + s\langle 2|][\hat{a}^\dagger + \hat{a}][|0\rangle + s|2\rangle]/\sqrt{2} = 0. \tag{11.47}$$

Thus the variance

$$(\Delta X_1)^2 = \langle\psi|\hat{X}_1^2|\psi\rangle - |\langle\psi|\hat{X}_1|\psi\rangle|^2 = 1/2 + \sqrt{2}s \tag{11.48}$$

Similary it follows that

$$(\Delta X_2)^2 = 1/2 - \sqrt{2}s, \tag{11.49}$$

which demonstrates squeezing[9] in X_2. It requires a nonlinear process to generate the required pairs of entangled photons appearing in eqn.

---

[9]The product of ΔX_1^2 and ΔX_2^2 appears to violate the uncertainty principle. If the expansion is continued to higher orders in s this violation disappears.

11.45. One device that has been used to squeeze phase uncertainty in the gravitational wave detectors is an *optical parametric oscillator OPO*, represented schematically in Figure 11.11. The laser pump has double the frequency of the mode resonant in the Fabry–Perot cavity of the OPO. Its action on the non-linear crystal is to produce a pair of entangled photons in so-called *signal* and *idler* modes. This process closely resembles SPDC described in Section 9.4. The requirements of energy and momentum conservation lead to a strong response when these modes are collinear and have frequency equal to that of the cavity resonance. These pairs of entangled photons are in the same mode and at the correct frequency to apply squeezing.

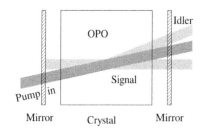

Fig. 11.11 Optical parametric oscillator.

11.9 Squeezed light for LIGO

The noise in gravitational wave interferometers comes not only from seismic sources, from vibration of the mirror assembly components, and from thermal effects: noise also comes from the vacuum that *illuminates* the apparently *unused* entry face of the beam splitter, the same surface through which the signal exits. We have seen that the quantum noise in the vacuum is as substantial as that in any other coherent source, and must contribute to the noise in the electromagnetic field in the interferometer. Tests have been performed with squeezed vacuum from an OPO injected into the *unused* port. In order to squeeze the vacuum the OPO pump intensity is held just below threshold for parametric oscillation. Effectively the input to the OPO is then the vacuum and its ouput is the squeezed vacuum. The squeezing in Figure 11.10 is set to squeeze the phase defined by the interferometer signal. Tests with the GEO600 interferometer achieved a 3 dB reduction in noise over the audio portion of the interferometer spectrum. This translates into doubling the distance from earth at which a given type of gravitational wave source becomes detectable. For black hole mergers this would extend advanced LIGO's reach to around 2.5 billion light-years.

11.10 Further reading

Quantum Measurement by V. B. Braginsky and F. Ya. Khalili, edited by K. S. Thorne, published by Cambridge University Press (1990) gives a compact account of the basics of quantum measurement, written by pioneers in the understanding of quantum measurement. An essential read for further studies.

Quantum Optics by M. O. Scully and M. S. Zubairy, published by Cambridge University Press (1997) contains a insightful discussion of optical squeezing.

Quantum Noise in Mesoscopic Physics, edited by Y. V. Nazarov, published by Kluwer, Dordrecht (2003) contains several useful articles in

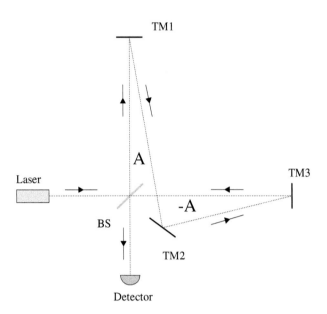

Fig. 11.12 Sketch of components of a possible speed interferometer. The mirrors are TM1, TM2 and TM3. BS is the beam splitter. The path of light reflected at the beam splitter is indicated. This diagram is used in an exercise below.

which noise in mesoscopic devices is reviewed.

Exercises

(11.1) The planned LISA interferometer would be space-based with arms 10^6 km long designed to detect gravitational waves at mHz frequencies from massive black hole mergers. Taking the laser to emit 1 W power at wavelength as 1064 nm, the received power along a single pass in one arm would be 40 pW. What would be the minimum strain detectable? Shot noise dominates in this frequency range at this power.

(11.2) Figure 11.12 shows an interferometer with arrows showing the path followed by light reflected at the beam splitter. Light transmitted traces the reverse of this path inside the interferometer. Let λ be the wavelength of the laser light. Calling the path lengths in the two arms x_1 and x_2, the distance

along the arrowed path is $(x_1(t) + x_2(t+\delta t))$ where δt is the time spent in one arm. Show that the phase difference between light transmitted and reflected at the beam splitter is

$$\delta\phi = (2\pi/\lambda)\delta t[\mathrm{d}x_1/\mathrm{d}t - \mathrm{d}x_2/\mathrm{d}t].$$

How can the phase difference between the arms when a gravitational wave passes be measured continuously? Why is this a QND measurement?

(11.3) Calculate the quadratures for an n-photon state (a Fock state). Hence show that, unlike coherent (laser-like) states, these n-photon states are not minimum uncertainty states.

(11.4) Show that the quadrature $(\Delta X_1)^2$ is 1/2 in a coherent state.

(11.5) Damped harmonic motion is described by the equation

$$m(\mathrm{d}^2x/\mathrm{d}t^2) + \gamma(\mathrm{d}x/\mathrm{d}t) + kx = F_0\exp(i\omega t).$$

where m is the mass, k the spring constant, γ the damping constant and F_0 the applied force at angular frequency ω. Show that at the resonance frequency the zero point fluctuation is $\sqrt{\hbar/(2\gamma)}$.

(11.6) Check eqn. 11.43.

(11.7) A nanomechanical oscillator is in the form of a wire spanning a microwave cavity. The mass is 11 pg and its essentially undamped natural oscillations are at 1.04 MHz. It is held at 130 mK. Calculate its zero point fluctuations. How many phonons does the oscillator carry in thermal equilibrium at 130 mK?

(11.8) A system has an initial wavefunction

$$|\psi_0\rangle = \sum_n c_n |E_n\rangle,$$

where $|E_n\rangle$ are orthonormal energy eigenstates. At a later time t its state is

$$|\psi_t\rangle = \sum_n c_n \exp(-iE_n t/\hbar)|E_n\rangle.$$

Show that $S(t) = \langle\psi_0|\psi_t\rangle$ has a real part:

$$\mathrm{Re}[S(t)] \geq 1 - \frac{2Et}{\pi\hbar} + \frac{2}{\pi}\mathrm{Im}[S(t)].$$

Hence deduce that the earliest a transition to an orthogonal state can occur is $h/4\overline{E}$ where $\overline{E}$ is the mean energy of the states. Use the inequality

$$\cos x \geq 1 - \frac{2}{\pi}(x + \sin x).$$

(11.9) In the case of a mass m deduce that the autocorrelation of the position at different times is $[x(0), x(t)] = i\hbar t/m$ and show that $\Delta x(0)\Delta x(t) \geq \hbar t/m$.

Cavity quantum physics

12

This chapter describes theoretical and experimental studies of the interaction between an atom and the electromagnetic modes within a reflective cavity enclosing the atom; and of experiments where instead of an atom, it is an ion, or an electron that is similarly confined. Ideally the cavity provides isolation from external influence, making the system rather simple. By cryogenically cooling the cavity, the number of photons in a cavity mode can be reduced to of order one, so that simple quantum effects are revealed. When the energy of a photon in a cavity mode matches that of an atomic transition, energy may flow to and fro between the mode and the atom, in what is called *Rabi flopping*. Having a cavity small compared to the mode wavelength concentrates both the mode energy and electric field strength thus enhancing the interaction strength with the atom. *Cavity quantum electrodynamics, CQED*, is the term used to denote the application of quantum electrodynamics to such systems. A simplification that is often useful is to consider an atom with just two energy levels enclosed in a cavity with a single mode. A quantum model due to Jaynes and Cummings will be introduced which applies to a two-state atom interacting with a single electromagnetic mode that might be a cavity mode or the mode of a laser.

There are three processes of importance, illustrated here in Figure 12.1. The interaction between an atom and a cavity mode with strength g; the escape of photons from the cavity at a rate $\kappa = 1/\tau_{\mathrm{c}}$; and the decay of the excited state of the atom at a rate $\gamma = 1/\tau$. τ_{c} and τ are lifetimes of the cavity mode and excited state respectively. When the interaction is strong enough that $g \gg \kappa$ and $g \gg \gamma$ there is Rabi flopping, with energy transfering continually, to and fro, between the atom and the cavity mode: the system becomes a hybrid quasi-particle state known as *dressed* atomic state. On the other hand, if the interaction is relatively weak a perturbative description using Fermi's Golden Rule to calculate transition rates is appropriate. Curiously the result of enclosing the atom in a cavity may, depending on the circumstances, reduce or increase its lifetime by what is called the *Purcell effect*.

The initial topic is an experiment by Haroche[1] and colleagues at College de France and Ecole Normale Superieure, using *circular Rydberg atoms* traversing a cryogenically cooled evacuated microwave cavity. Rydberg atoms have large electric dipole moments, which enhances the interaction with the microwave cavity mode. The experimenters observed

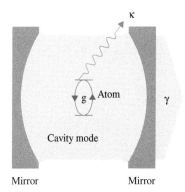

Fig. 12.1 A two-level atom in a resonant cavity. The competing processes are indicated. g is the cavity mode to atom coupling, γ is the rate of loss of photons from the cavity and κ is the rate of decay of the excited state.

[1]Nobel Laureate awarded for this and related research in 2012.

Quantum 20/20: Fundamentals, Entanglement, Gauge Fields, Condensates and Topology.
Ian R. Kenyon. © 2020. Published in 2020 by Oxford University Press.
DOI: 10.1093/oso/9780198808350.001.0001

Rabi flopping, its decay, and the purely quantum recovery of these oscillations. They were also able to determine the statistical distribution of the number of microwave photons in the cavity.

An equally challenging set of experiments described here were performed by Wineland[2] and his colleagues at the National Institute for Standards and Technology. They held single ions in a three-dimensional harmonic well formed by an evacuated micron sized electromagnetic *Paul trap*. The ion can be cooled by laser beams to the point that there are zero *phonons* of mechanical motion. Such experiments expose directly the collapse of a quantum state. This ability was exploited in developing optical clocks of a precision not attainable with atomic clocks.

One of the most sensitive tests of quantum electrodynamics, the theory touched on in Chapter 8, is to compare the prediction for the magnetic moment of the electron with its measured value: these now agree to parts in 10^{12}. The current measurement was made by Gabrielse and colleagues on single electrons held in an electromagnetic *Penning trap*. This built on methods and measurements made by Dehmelt and Van Dyck.[3] *Quantum non-demolition* measurements were necessary to achieve the required level of experimental precision. Outlines of the experiment and of the theoretical calculation are given. This leads to a discussion of the *Purcell effect*.

12.1   The Jaynes–Cummings model

This model describes how a two-level atom interacts with a single mode of the electromagnetic field; which might be a laser mode or a cavity mode. The energy is the sum of the energies of the atom and mode, isolated from one another, *plus* the interaction energy. The atom's energy levels are taken to be spaced in energy by $\hbar\omega_a$, and the mode energy per photon is taken to be $\hbar\omega_c$. In general these differ so we have a *detuning*

$$\Delta = \omega_a - \omega_c. \qquad (12.1)$$

Assuming an electric dipole interaction, the interaction energy is the scalar product of the atom's electric dipole moment, $\mathbf{d}$, and $\mathbf{E}$ the mode's electric field. The interaction energy with a single photon in the mode is customarily re-expressed in terms of what is called the *Vacuum Rabi angular frequency*, Ω_0:[4]

$$\mathbf{d} \cdot \mathbf{E_0} = \hbar\Omega_0/2, \qquad (12.2)$$

where the magnitude of the vacuum electric field is given in 8.17

$$E_0 = \sqrt{\hbar\omega/(2\varepsilon_0 V)}. \qquad (12.3)$$

Phases are chosen to make Ω_0 real. Collecting the contributions gives a total energy operator

$$\hat{H} = \hbar\omega_a \sigma_z/2 + \hbar\omega_c(\hat{a}^\dagger \hat{a} + 1/2) + \frac{\hbar\Omega_0}{2}[\sigma_+ \hat{a} + \sigma_- \hat{a}^\dagger], \qquad (12.4)$$

[4]In semi-classical calculations of Rabi flopping the Rabi angular frequency is given by $\hbar\Omega_R = \mathbf{d} \cdot \mathbf{E}^{class}$, where $E^{class} = E_0^{class}\cos(\omega t)$. The results converge if E_0^{class} is replaced by $2\sqrt{n}E_0$ where there are n photons in the mode.

where we take advantage of the formal equivalence of the two atomic states to the two electron spin states: this allows us to make use of the equations in Appendix D.

In the first term on the right-hand side of this energy equation the excited state energy is taken to be $\hbar\omega_a/2$ and the ground state energy to be $-\hbar\omega_a/2$. The second term is the energy of the electromagnetic field. For radiation $\hat{a}$ ($\hat{a}^\dagger$) is the photon annihilation (creation) operator introduced in Chapter 8; $\hat{a}^\dagger\hat{a}$ gives the photon count in the mode. The third term accounts for the interaction energy. σ_+ (σ_-) raises (lowers) the atom's energy state, hence $\sigma_+\hat{a}$ simultaneously annihilates a photon and raises the atom from ground to excited state; $\sigma_-\hat{a}^\dagger$ reverses the process, creating a photon and lowering the atom to its ground state.

Corrections to account for decays of the atom by alternative paths and for the escape of photons from the cavity contribute further terms to the above energy equation. For the sake of clarity these are neglected in what follows. Photon loss can be safely ignored for a laser mode.

Total energy states of a non-interacting atom and mode are displayed in the left-hand column of Figure 12.2. The quantum states $|g, n\rangle$ and $|e, n\rangle$ are the ground and excited atomic states respectively, each with n photons. Acting with the energy operator on $|e, n\rangle$ and $|g, n+1\rangle$, the states connected by the interaction, gives:

$$\hat{H}|e, n\rangle = [(n+1/2)\hbar\omega_c + \hbar\omega_a/2]|e, n\rangle$$
$$+ [\sqrt{n+1}\hbar\Omega_0/2]|g, n+1\rangle; \tag{12.5}$$
$$\hat{H}|g, n+1\rangle = [(n+3/2)\hbar\omega_c - \hbar\omega_a/2]|g, n+1\rangle$$
$$+ [\sqrt{n+1}\hbar\Omega_0/2]|e, n\rangle. \tag{12.6}$$

We now re-use the Bloch sphere, identifying the hybrid states $|e, n\rangle$ and $|g, n+1\rangle$ with the up and down states on the sphere:

$$|e, n\rangle = \begin{bmatrix} 1 \\ 0 \end{bmatrix} \quad \text{and} \quad |g, n+1\rangle = \begin{bmatrix} 0 \\ 1 \end{bmatrix}. \tag{12.7}$$

Then the energy operator in eqn. 12.4 can be written more compactly in terms of these hybrid states

$$\hat{H} = \hbar\omega_c(n+1)\begin{bmatrix} 1 & 0 \\ 0 & 1 \end{bmatrix} + \frac{\hbar}{2}\begin{bmatrix} \Delta & \Omega_n \\ \Omega_n & -\Delta \end{bmatrix}, \tag{12.8}$$

where $\Omega_n = \Omega_0\sqrt{n+1}$. Diagonalizing this energy equation gives eigenstates

$$|+\rangle = \begin{bmatrix} \cos(\theta/2) \\ \sin(\theta/2) \end{bmatrix}$$
$$|-\rangle = \begin{bmatrix} \sin(\theta/2) \\ -\cos(\theta/2) \end{bmatrix}, \tag{12.9}$$

Often the zero point energy $\hbar\omega_c/2$ is dropped because it remains constant and inactive throughout the analysis. When this term is left out the energies $(n+1)\hbar\omega_c$ appearing later *in this and the following section* would become $(n+1/2)\hbar\omega_c$.

with energies

$$E^{\pm} = (n+1)\hbar\omega_c \pm \frac{\hbar}{2}\sqrt{\Delta^2 + \Omega_n^2}, \qquad (12.10)$$

and where $\tan\theta = \Omega_n/\Delta$. $|+\rangle$ and $|-\rangle$ are entangled states of one atom and multiple photons. The spectrum of levels is illustrated on the right-

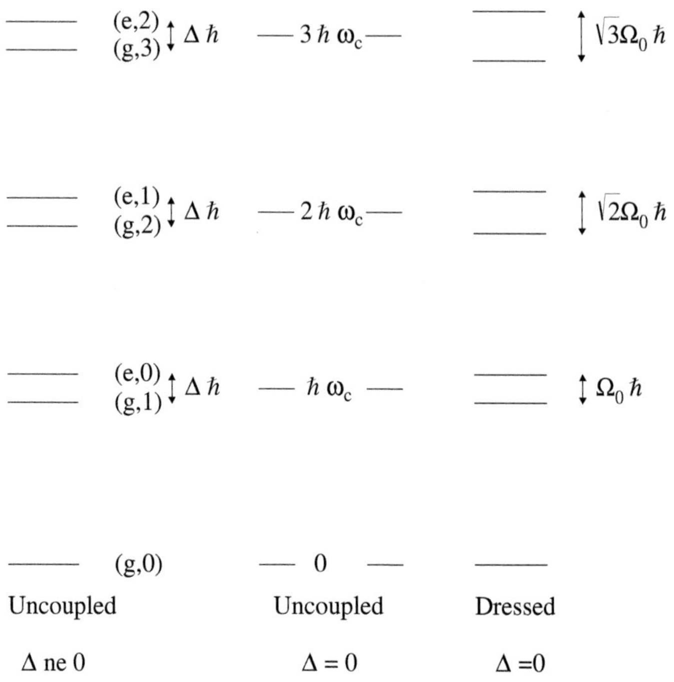

———	(e,2) $\updownarrow \Delta\hbar$	— $3\hbar\omega_c$ —	——— $\downarrow \sqrt{3}\Omega_0\hbar$
———	(g,3)		———
———	(e,1) $\updownarrow \Delta\hbar$	— $2\hbar\omega_c$ —	——— $\updownarrow \sqrt{2}\Omega_0\hbar$
———	(g,2)		———
———	(e,0) $\updownarrow \Delta\hbar$	— $\hbar\omega_c$ —	——— $\updownarrow \Omega_0\hbar$
———	(g,1)		———
———	(g,0)	— 0 —	———
Uncoupled		Uncoupled	Dressed
Δ ne 0		$\Delta = 0$	$\Delta = 0$

Fig. 12.2 Uncoupled and dressed states of photon and two-state atom.

hand panel of Figure 12.2. We show below that if the driving field is near resonance, the system oscillates between the states $|e, n\rangle$ and $|g, n+1\rangle$.

How the energy eigenstates drift apart when the transition and cavity frequencies separate is described by eqn. 12.10. This equation gives the solid lines in Figure 12.3. In the absence of any interaction the two states would follow the broken lines and cross one another, while with the interaction present the states avoid one another. This behaviour is common to coupled systems, the simplest being that of coupled pendulums.[5] With a driving field far off resonance, $\Omega_n^2 \ll \Delta^2$, the state $|e, n\rangle$ suffers an energy shift of magnitude

$$(\hbar/2)\sqrt{(\Delta^2 + \Omega_n^2)} - \hbar\Delta/2 = \hbar\Omega_n^2/[4\Delta], \qquad (12.11)$$

which is known as a *light shift* or the *AC Stark shift*. The light shift of $|g, n+1\rangle$ is equal and opposite. With a more complex level structure the AC Stark shift is state-dependent but still proportional to E_0^2.

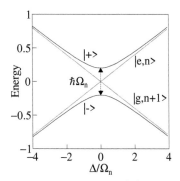

Fig. 12.3 Energies of the $|\pm\rangle$ states relative to $(n+1)\hbar\omega_c$ as a function of the offset, $\hbar\Delta$, between the transition and the cavity mode energies.

[5]See also Figure 4.21 in which a phonon is coupled to a photon.

12.2   Rabi flopping

For simplicity the resonant case is considered, thus $\omega_c = \omega_a$ and hence $\Delta = 0$, and $\theta = \pi/2$. Rearranging eqn. 12.9 gives

$$|+\rangle = \begin{bmatrix} 1/\sqrt{2} \\ 1/\sqrt{2} \end{bmatrix} = [|e, n\rangle + |g, n+1\rangle]/\sqrt{2};$$

$$|-\rangle = \begin{bmatrix} 1/\sqrt{2} \\ -1/\sqrt{2} \end{bmatrix} = [|e, n\rangle - |g, n+1\rangle]/\sqrt{2}. \qquad (12.12)$$

Conversely, the pure states are

$$|e, n\rangle = [|+\rangle + |-\rangle]/\sqrt{2},$$
$$|g, n+1\rangle = [|+\rangle - |-\rangle]/\sqrt{2}. \qquad (12.13)$$

Again for simplicity we move the origin of energy to $(n+1)\hbar\omega_c$, making $E^\pm = \pm\hbar\Omega_n/2$. Then the states $|\pm\rangle$ evolve with time to become $|\pm\rangle \exp(\mp i\Omega_n t/2)$. Consequently a pure state $|e, n\rangle$ prepared at time zero evolves in this way:

$$\begin{aligned} |e, n\rangle &\rightarrow [|+\rangle \exp(-i\Omega_n t/2) + |-\rangle \exp(+i\Omega_n t/2)]/\sqrt{2} \\ &= |e, n\rangle \cos(\Omega_n t/2) - i|g, n+1\rangle \sin(\Omega_n t/2), \qquad (12.14) \end{aligned}$$

which is a superposition of $|e, n\rangle$ and $|g, n+1\rangle$. The probability of a state initially $|e, n\rangle$ remaining in this state varies like $\cos^2(\Omega_n t/2) = [1 + \cos(\Omega_n t)]/2$. This dependence on time is plotted as the solid line in Figure 12.4: the broken line is the complement, the probability of finding it in the state $|g, n+1\rangle$. This behaviour is known as Rabi flopping and has a period of oscillation $2\pi/\Omega_n$. Its discovery in nuclear magnetic resonance led to Isidor Rabi winning the Nobel Prize in Physics in 1944.

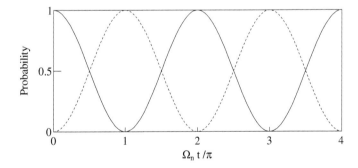

Fig. 12.4 Rabi flopping: the solid curve shows the evolution of the initially pure state $|e, n\rangle$ and the broken line curve that of the state $|g, n+1\rangle$.

Energy is thus continuously transfered between the cavity mode and the atom. Similarly an initially pure state $|g, n+1\rangle$ evolves like this

$$|g, n+1\rangle \rightarrow |g, n+1\rangle \cos(\Omega_n t/2) - i|e, n\rangle \sin(\Omega_n t/2) \qquad (12.15)$$

The entangled energy eigenstates $|\pm\rangle$ are known as *dressed states* of the atom and are shown in the right-hand column on Figure 12.2 for the case that atom and mode are at resonance, that is $\Delta = 0$. Then eqn. 12.10 reduces to

$$E^{\pm} = (n+1)\hbar\omega_{\mathrm{c}} \pm \frac{\hbar}{2}\sqrt{n+1}\Omega_0. \qquad (12.16)$$

When $n = 0$, the oscillation $|e,0\rangle \leftrightarrow |g,1\rangle$ is at a frequency, Ω_0. As noted earlier, this is known as the *Vacuum Rabi angular frequency*.

Pulses of continuous radiation provide a widely used technique to manipulate quantum states. A π pulse is a pulse just long enough to take a system from the $|g\rangle$ state into the $|e\rangle$ state; so that it goes from pointing down on the Bloch sphere to pointing up. A $\pi/2$ pulse is correspondingly a pulse that takes a system initially in the $|g\rangle$ state to $[|g\rangle + |e\rangle]]/\sqrt{2}$. In other words, a $\pi/2$ pulse moves the state from pointing down on the Bloch sphere to a point in the equatorial plane.

Figure 12.5 shows the fluorescence observed when an intense laser beam illuminates an atomic gas at resonance with a transition from the ground state to an excited state. This is the subject of an exercise at the end of the chapter.

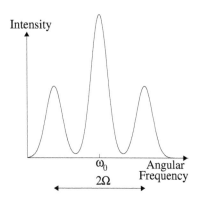

Fig. 12.5 Mollow fluorescence.

12.2.1   Decay and revival

Generally the number of photons, n, has a distribution about a mean value and the Rabi flopping frequency is similarly variable. The corresponding amplitudes for $|e,n\rangle$ slip out of phase as time passes and the Rabi oscillations die away. In classical calculations made with electromagnetic waves Rabi oscillations are also predicted and then decay due to noise. The predictions for subsequent behaviour differ: classically Rabi flopping cannot revive but in the quantum picture it can. Consider the contributions from different photon numbers n and $n+1$: these come back into phase after a time t_r such that

$$\Omega_0[\sqrt{n+1} - \sqrt{n}]t_r = 2\pi. \qquad (12.17)$$

For large n, $\sqrt{n+1}$ tends to $\sqrt{n}[1 + \frac{1}{2n}]$, so that $t_r = 4\pi\sqrt{n}/\Omega_0$. Then provided that the fluctuation in the number of photons is much smaller than the mean number, $\bar{n}$, there will be revival of Rabi flopping after a time $4\pi\sqrt{\bar{n}}/\Omega_0$. Also quite unintelligible from the classical perspective are the Rabi oscillations observed when few photons are present in a mode. These quantum predictions have been tested extensively. Here we present experiments carried out by Haroche and colleagues using Rydberg atoms travelling through a cavity populated with microwaves.

12.3   Microwaves and Rydberg atoms

In a rubidium atom (and in other alkali metal atoms) the single valence electron can be excited to states with large principal quantum number n, say 50, and with the maximum orbital angular momentum quantum number $l = |m_l| = n - 1$. Such an electron has an orbit resembling a circular Bohr orbit and the atom is called *circular Rydberg atom*. The orbit's radius s is 50^2 times the Bohr radius of the hydrogen ground state, that is 120 nm. Hence its electric dipole moment es is ~ 2500 times that involved in optical transitions between states with small principal quantum numbers. With an applied electric field E, equal to $2\,10^{-4}\,\mathrm{Vm}^{-1}$, the vacuum Rabi frequency, $2esE/\hbar$, is around 50 kHz. Circular Rydberg states decay via microwave transitions with unit change in n and these states have a long lifetime. For the transition in Rubidium from $n = 51$ to $n = 50$ the radiation is at 51 GHz or 6 mm wavelength and the lifetime is 36 ms. Haroche and colleagues sent circular Rydberg rubidium atoms through a microwave cavity with the aim of investigating the Rabi flopping between atoms and cavity modes when the number of photons is small. The cavity had a highly polished Niobium surface giving a Q-value of 10^7. Consequently the mean lifetime for a 51 GHz photon against escape from the cavity is $Q/51\,\mathrm{GHz}$ or $\sim 200\,\mu\mathrm{s}$, much longer than the transit time of the atom through the cavity. Strong coupling between the cavity mode and the atomic transition was ensured by the huge dipole moment of the circular Rydberg atom, its long lifetimes, and the long-lived cavity mode. With the notation of the introduction $g \gg \kappa, \gamma$.

Figure 12.6 shows the basic features of the apparatus used by Haroche and colleagues. The rubidium atoms emerging from the oven are in the $F = 2$ or 3 $|5S_{\frac{1}{2}}\rangle$ states. They first cross one laser beam, which pumps all

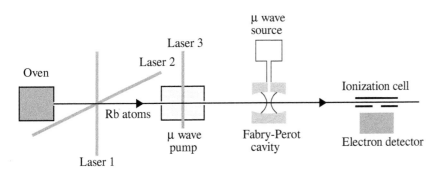

Fig. 12.6 Schematic of CQED experiments with circular Rydberg atoms. Courtesy Professor Haroche.

the atoms into the lower $F = 2$ level. Then the atoms cross, at an acute angle, a second laser beam red-shifted relative to the $F = 2 \rightarrow F = 3$ transition. Those atoms within a narrow range of velocity see the laser beam blue-shifted to the transition frequency and are promoted to the $F = 3$ level. The mean velocity of those promoted could be selected in the range 140 to $600\,\mathrm{ms}^{-1}$ by varying the wavelength offset between the two lasers. The subsequent excitation of these selected atoms into the desired circular Rydberg orbits requires a sequence of laser and microwave pumping.[6] The Rydberg atoms produced then pass through the Fabry–Perot cavity in a region maintained at $0.8\,\mathrm{K}$, thus suppressing thermal photons. The cavity axis is vertical and hence perpendicular to the atoms' path. There is a hole in the upper mirror through which an external microwave source injects $51\,\mathrm{GHz}$ photons. The cavity mode width of $15\,\mathrm{mm}$ at beam height defines the length of the interaction region. This is long enough for the interaction to produce a full Rabi oscillation of the fastest atoms selected.

After exiting the cavity the atoms enter a detector used to distinguish between Rydberg atoms in the upper energy state ($n = 51$) and those in the lower energy state ($n = 50$). The technique exploits the difference between the threshold electric fields required to ionize the atom in these states, respectively 13.4 and $14.5\,\mathrm{kVm}^{-1}$. Any electrons ejected from the beam path by the applied electric field are counted with an electron multiplier. The distribution of count rate versus field strength yields the proportion of atoms in the $n = 51$ and $n = 50$ states.

In order to investigate the strong coupling in the quantum regime Haroche and colleagues worked with small numbers of photons in the cavity. The mean number of photons in the cavity mode was determined by the microwave power injected through the hole in the upper mirror. Measurements were made of the fraction of atoms emerging from the cavity in the $n = 50$ and $n = 51$ states as a function of the duration of the interaction for each selected microwave power. In Figure 12.7 each row exhibits results corresponding to different power settings: reading down from the top, the average numbers of photons in the cavity were 0.06, 0.40, 0.85 and 1.77. Data in the first row was recorded with the microwave source off, 0.06 being the mean number of black body photons expected in the cavity mode at $0.8\,\mathrm{K}$. The fraction of atoms in the $n = 50$ state exiting the cavity is shown in the plots in the first column as a function of the time spent in the interaction region. Rabi flopping is evident. Plots in the second column are the Fourier transforms of the data in the first column. The dotted lines drawn there indicate the frequencies of the expected Rabi oscillations with 0, 1, 2 and 3 microwave ($51\,\mathrm{GHz}$) photons in the cavity. Reading from the left, these are $\Omega/2\pi$ ($47\,\mathrm{kHz}$), $\sqrt{2}\Omega/2\pi$, $\sqrt{3}\Omega/2\pi$ and $\sqrt{4}\Omega/2\pi$. The area of the peaks matching these dotted lines are the contributions from instances when there were zero, 1, 2 and 3 photons respectively within the cavity at the time the Rydberg atom crosses it. Plots in the right-hand column

---

[6]See pages 260 onward in *Exploring the Quantum: Atoms, Cavities and Photons*, by S. Haroche and J.-M. Raymond, published by Oxford University Press (2006).

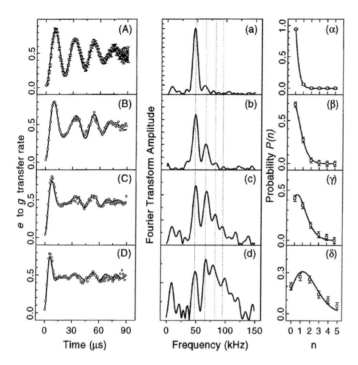

Fig. 12.7 Rabi flopping with few photons in the cavity mode. Details are discussed in the text. Figure 2 from M. Brune, F. Schmidt-Kaler, A. Maali, J. Dreyer, E. Hagley, J . M. Raimond and S. Haroche, *Physical Review Letters* 76, 1800 (1996). Courtesy Professor Haroche and The American Physical Society.

show the probability distribution of the photon count obtained by this Fourier analysis: the superimposed solid curves indicate the Poissonian distributions expected for the inferred mean photon count. There is overall consistency of the quantum interpretation of the data. In the lower left-hand plot another significant quantum feature is just visible: namely, the collapse within 15 μs and later revival of Rabi oscillations[7].

[7]More impressive examples of collapse and revival are to be seen in the article by T.Meunier, S. Glayzes, P. Maioli, A. Auffeves, M. Nogues, M. Brune, J. M. Raimond and S. Haroche, Physical Review Letters 94, 010401 (2005).

12.4   Ions in electromagnetic traps

Another approach to studying cavity quantum interactions has been to use electromagnetic fields to trap single ions in vacuum and then to cool them using laser beams. Such trapped cooled ions can be manipulated in isolation for long periods, even days, rather than the milliseconds available when Rydberg atoms traverse a microwave cavity. Wineland and colleagues trapped and cooled single ^{199}Hg$^+$ ions, and constructed what promises to be an example of the most precise of all clocks, the optical clock. Novel techniques, such as shelving and optical combs, were introduced to implement this device.

[8]This is one way of presenting Earnshaw's theorem.

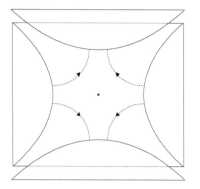

Fig. 12.8 Paul trap showing the direction of the electric force on an ion in one half-cycle of the applied radio frequency field. The star marks where the micromotion vanishes.

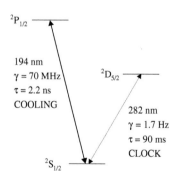

Fig. 12.9 Energy levels of the mercury ion. The transition at 194 nm is an allowed dipole transition, the transition at 282 nm is forbidden as a dipole, but an allowed quadrupole transition.

Ions cannot be trapped electrostatically because it is impossible to form a local minimum in a purely electrostatic potential.[8] However, electric fields oscillating at radio frequencies can produce a stable potential well. Figure 12.8 shows a vertical section containing the axis of a radio frequency trap, a version of the *Paul trap*. Its metal electrodes are a pair of hyperboloid endcaps and a central ring whose cross-section is hyperbolic. The trap would be contained in an evacuated container typically cooled to 4 K by liquid helium. A radio frequency voltage is applied between the endcaps and the ring: in one half-cycle the electric force on an ion in the trap points toward the ring while in the next half-cycle this force is reversed. The net effect is an electric potential well with a minimum at the trap centre. Trap volumes are typically of millimetre dimensions with an electric field of order $100\,\mathrm{Vm^{-1}}$ oscillating at around $10\,\mathrm{MHz}$. With the illustrated hyperboloid electrodes the potential would be exactly quadratic. The electrodes used by experiments with $^{199}\mathrm{Hg^+}$ ions were simplified to a washer-like ring and short cylindrical endcaps as shown in Figure 12.12. The potential is sufficently close to quadratic in both the radial and axial directions that the ion undergoes harmonic motion like that analysed in Section 2.3. We take the vertical axis of symmetry to be the z-axis. Once an ion is cooled sufficiently its remaining *secular motion* is quantized with total kinetic energy

$$E = \sum_{i=x,y,z} \hbar\omega_i(n_i + 1/2), \qquad (12.18)$$

where ω_x, ω_y and ω_z are the angular frequencies for the motion along orthogonal axes, while n_x, n_y and n_z are the numbers of *phonons* of excitation. The azimuthal symmetry of the trap makes ω_y equal to ω_x.

In addition to the secular motion in the well there is residual micromotion at the radio frequency, but this motion vanishes at the well centre. Atoms are injected into the trap from a furnace, and ionized by a low energy electron beam. These hot ions are cooled to thermal velocities by collisions with helium at very low pressure, and then this helium gas is pumped off; after which, one or more stages of laser cooling bring the ions close to rest near the trap centre. Only a single ion is required to provide a reference frequency, and this will reside at the trap centre and will have no micromotion.

Laser cooling of the trapped ions was first proposed by Wineland and Dehmelt in 1975. The relevant energy levels of the $^{199}\mathrm{Hg^+}$ ion are shown in Figure 12.9. The fast transition from $^2\mathrm{P}_{1/2}$ to $^2\mathrm{S}_{1/2}$ at 194 nm wavelength is the *cooling* transition. In a first, *Doppler* step, a laser is used to illuminate the ions tuned to a frequency just below this transition frequency. Ions travelling towards the laser beam see it blue-shifted, so that for some ions the shifted laser frequency equals the transition frequency. These particular ions each absorb a laser photon in a head-on collision and their momenta are thereby reduced. These excited ions then decay with each emitting a photon in some random direction; thus the overall momentum acquired by the cloud of ions due to de-excitation is zero. With counter-propagating laser beams the ions travelling in both senses can be cooled. Motion along all three axes can be cooled simultaneously by inclining the laser beam to make roughly equal angles with all three trap axes. The minimum temperature, T, achievable in this way is limited by the line width, γ, of the transition being used for cooling:

$$k_\mathrm{B} T = \hbar\gamma/2. \qquad (12.19)$$

In the case of $^{199}\mathrm{Hg}^+$ ions cooled using the $^2\mathrm{P}_{1/2} \to {}^2\mathrm{S}_{1/2}$ transition the line width is 70 MHz so that the minimum attainable temperature is around 1 mK. The quantum of vibrational energy of such ion in the trap is hf_s where the frequency of vibration of the ion, f_s is roughly 3 MHz. This means that the ion's energy after Doppler cooling amounts to $70/3 = 23$ quanta (phonons) of vibrational motion.[9]

A lower final temperature is achieved by using a transition that has a narrow line width. In the case of a $^{199}\mathrm{Hg}^+$ ion there is a convenient electric quadrupole transition $^2\mathrm{D}_{5/2} \to {}^2\mathrm{S}_{1/2}$ at 282 nm with a natural line width of only 1.7 Hz and a lifetime of 90 ms. This too is shown in Figure 12.9.

Figure 12.10 shows the absorption spectrum when the frequency of a laser beam illuminating a Doppler cooled mercury ion cloud is scanned through frequencies covering the $^2\mathrm{S}_{1/2} \to {}^2\mathrm{D}_{5/2}$ transition. Note that the expected Doppler broadened absorption peak of width $\sim 2\omega_s$ ($4\pi f_s$) due to secular motion has been replaced by narrow discrete peaks at angular frequencies $\omega_0 - \omega_s$, ω_0 and $\omega_0 + \omega_s$. This comes about because the amplitude of secular motion, 10 nm, is much smaller than the wavelength of the laser beam, λ. As a result the variation of laser phase over the ion's path is negligible, which defines what is called the *Lamb–Dicke regime*. The waveform of the radiation at the ion can be written

$$E = \cos\left(\omega_0 t - 2\pi z/\lambda\right) = \cos\left(\omega_0 t - kz\right), \qquad (12.20)$$

where the position of the ion z, taking account of its secular motion, is

$$z = z_0 \cos\left(\omega_s t\right). \qquad (12.21)$$

In the Lamb–Dicke regime we can take $\sin kz = kz$ and $\cos kz = 1$. Then expanding the right-hand side of eqn. 12.20 at the ion gives[10]

$$
\begin{aligned}
E &= \cos\left(\omega_0 t\right) + kz \sin\left(\omega_0 t\right) \\
&= \cos(\omega_0 t) + kz_0 \cos(\omega_s t) \sin(\omega_0 t) \\
&= \cos\left(\omega_0 t\right) + (kz_0/2)\{\sin\left[(\omega_0 + \omega_s)t\right] + \sin\left[(\omega_0 - \omega_s)t\right]\}.
\end{aligned}
$$

Thus the atom absorbs dominantly at the unshifted frequency, and more weakly by a factor $(\pi z_0/\lambda)^2$ at the sidebands. Effectively the ion behaves as a point-like oscillator vibrating at an angular frequency ω_s. Thanks to the narrow line width of the transition the sidebands are well resolved

[9] The axial frequency f_z is 2.9 MHz and the radial frequency $f_x = f_y$ is 1.46 MHz. For simplicity here the radial motion has been ignored.

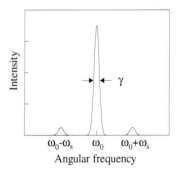

Intensity

γ

$\omega_0 - \omega_s$ ω_0 $\omega_0 + \omega_s$
Angular frequency

Fig. 12.10 Absorption spectrum around the 282 nm transition for Doppler cooled mercury ions. Only the stronger, first sidebands are shown.

[10] This is typical of phase modulation at angular frequency w_s producing sidebands at $\omega_0 \pm \omega_s$.

from the central peak.

In the second stage of cooling the laser frequency is tuned to coincide with that of the lower sideband of the 282 nm transition. Each absorption of a photon of energy $\hbar(\omega_0 - \omega_s)$ is nearly always followed by the spontaneous emission of a photon of energy $\hbar\omega_0$, so that the ion loses overall one quantum of vibrational energy. Most ions end up in the lowest energy state, with *zero point* energy $\hbar\omega_s/2$. The secular frequency is 3 MHz so that the final ion temperature is only $\hbar\omega_s/(2k_B)$ or 100 μK.

After cooling the line width of the central peak in Figure 12.10 is free of first-order Doppler broadening due to secular motion. What remains is second-order Doppler broadening due to time dilation, whose origin is the zero point motion with r.m.s. velocity v_s.[11] This velocity is given by

$$\hbar\omega_s/2 = m_{Hg} v_s^2/2. \tag{12.22}$$

Substituting the values of the secular frequency and the mercury ion mass in this equation gives a velocity of around 0.1 m s^{-1}. This implies a second order Doppler shift of order one part in 10^{18}. As a result the 282 nm transition of a Doppler-cooled, trapped ^{199}Hg$^+$ ion offers a potential frequency and time standard of outstanding precision.

[11] The second-order Doppler shift due to time dilation changes the wavelength from λ to

$$
\begin{aligned}
\lambda_{obs} &= \lambda/\sqrt{1 - v_s^2/c^2} \\
&\approx \lambda(1 + v_s^2/2c^2),
\end{aligned}
$$

making the fractional change in wavelength or frequency to be $v_s^2/2c^2$.

12.5   Quantum jumps, shelving and quantum amplification

The $^2D_{5/2}$ to $^2S_{1/2}$ transition of ^{199}Hg$^+$ ions is an example of an electric-dipole forbidden (electric quadrupole allowed) transition possessing an exceptionally narrow natural line width, 1.7 Hz in 1064 THz. We have seen that with appropriate cooling this is potentially an ultra-precise frequency reference, providing the pendulum of an *optical clock*. A basic advantage of an optical frequency clock over a microwave frequency atomic clock is that if the reference transitions have comparable width, say 1 Hz, then simply counting cycles per unit time offers $\sim 10^5$ higher resolution at optical frequencies. However, the use of such narrow optical transitions presents practical difficulties: how they were resolved is the next topic addressed.

In principle the central frequency and the width of the $^2S_{1/2} \to {}^2D_{5/2}$ transition could be determined by illuminating the mercury ion with a stabilized laser and making a scan in frequency in small steps across the line at 282 nm (1064 THz) and observing the resonance fluorescence at each step. Wherever the fluorescence was maximum would be the line centre. However, because the lifetime of the $^5D_{1/2}$ level is 90 ms, the ion emits few photons per second; of these rare photons a single detector intercepts only a few per cent; and of this subset the detector registers a fraction determined by its quantum efficiency. As a result the fluorescence photon count is swamped by the background count from noise in the detector.

The way around this difficulty was proposed by Dehmelt in 1982. A single cold ^{199}Hg$^+$ ion is simultaneously illuminated by a *pump* laser tuned near 194 nm and a 282 nm *probe* laser: see Figure 12.9. If the 194 nm radiation is sufficiently intense to saturate the $^2S_{1/2} \to {}^2P_{1/2}$ then the ion emits $1/(\tau = 2.2\,\mathrm{ns})$ or $\sim 10^9$ 194 nm fluorescent photons per second. This stream of pump fluorescence photons at 194 nm comes to an abrupt halt on each occasion that a photon from the probe laser at 282 nm is absorbed by the ion. The fluorescence vanishes whenever the ion makes a quantum jump to the $^2D_{5/2}$ state. It remains, *shelved* in that state, for a period determined by its lifetime of 90 ms and then decays to the $^2S_{1/2}$ ground state. Upon which, the fluorescence at 194 nm resumes as abruptly as it ceased. This is illustrated by a plot of the fluorescence intensity at 194 nm versus time in Figure 12.11 for a single trapped mercury ion illuminated in this way.

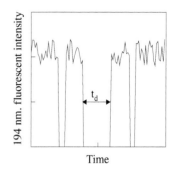

Fig. 12.11 Intensity of the fluorescence at 194 nm from a single cooled trapped Hg ion, showing typical shelving periods. The distribution of the duration of shelving times, t_d, depends on the lifetime of the $^2D_{5/2}$ level.

Observing whether the 194 nm fluorescence is absent or present determines whether the ion is in the $^2D_{5/2}$ state or not. Monitoring the 194 nm fluorescence then provides a *quantum non-demolition measurement*: that is a measurement that does not affect the quantum state of the system (ion) being measured. The probe and pump laser have to be applied *alternately* in order to avoid AC Stark shifting of the clock transition: see eqn. 12.11. Detection is now practicable. A detector intercepting 1 per cent of the 194 nm fluorescence photons with 10 per cent detection efficiency would still record of order 10^6 photons per second.

The central frequency of the envisaged $^2D_{5/2} \to {}^2S_{1/2}$ *clock transition* of the ^{199}Hg$^+$ ion is probed by a well stabilized laser whose frequency is scanned in small steps to cover the line width of the clock transition. After a short clock probe, the absence (presence) of 194 nm fluorescence from the ion exposed to the readmitted pump beam indicates with near unit probability that a clock transition did (did not) take place. This sequence is repeated many times and the total of successful shelvings counted. Plotting such total counts against the frequency of the probe laser reveals the line shape and hence the central frequency of the clock transition. Thanks to this approach the electric dipole forbidden transition at 282 nm can be probed with the statistics of the allowed transition at 194 nm: a statistical leverage dubbed *quantum amplification.*.

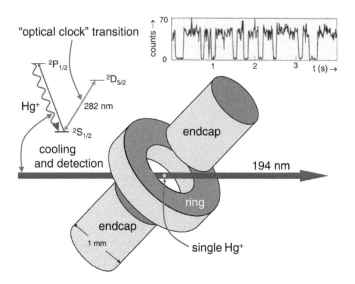

Fig. 12.12 Physical features of an optical clock based on a single cooled, trapped ^{199}Hg ion. From Figure 1 in *Reviews of Modern Physics* 85, 1103 (2013). Courtesy Professor Wineland and The American Physical Society.

12.6   The $^{199}\mathrm{Hg}^+$ optical clock

Figure 12.12 shows the core components of the optical clock design implemented by James Bergquist and co-workers[12] using the ^{199}Hg$^+$ $^2\mathrm{D}_{5/2} \to {}^2\mathrm{S}_{1/2}$ clock transition described above. Following the spectroscopic identification of the line and its width, the technique used is to lock the frequency of a probe laser to the line centre of this transition. This is done by stepping the probe frequency between half intensity points on either side of the resonance and at each setting applying the pump. A lock is effected by steering the frequency so that the shelvings on either side are equal. This sequence is repeated and the probe laser steered ever closer to the centre of the clock transition of the ^{199}Hg$^+$ion.

Finally, the cycles per second of the probe laser must be counted to complete a practical clock, something beyond the capacity of any direct counter. However, a *frequency comb* can be used to convert the optical count to counts at microwave frequencies, for which electronics does exist. The frequency comb in question is a laser-based source emitting frequencies, f_m, equally spaced across a wide spectral range in the visible and near infrared, the interval between the successive lines, f_{rep}, being

---

[12]For details see S.A.Diddens, Th. Udam, J. C. Bergquist, E. A. Curtis, R. E. Drullinger, L. Hollberg, W. M. Itano, W. D. Lee, C. W. Oates, K. R. Vogel and D. J. Wineland, "An optical clock based on a single ^{199}Hg$^+$ ion", *Science* 293 825 (August 2001).

around 1 GHz:

$$f_m = m \times f_{\text{rep}} + f_{\text{ceo}}. \qquad (12.23)$$

The essential features of how the optical comb is used are illustrated by Figure 12.13. In essence beats at three microwave frequencies are measured to determine the probe laser frequency f_{laser}.[13] First, radiation at

[13] f_{ceo} is an offset frequency that depends on the optical comb design, and needs to be determined.

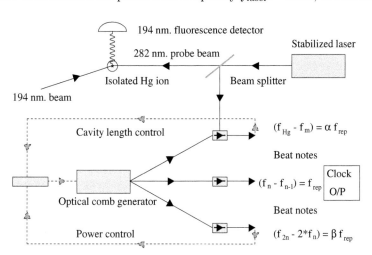

Fig. 12.13 Schematic diagram of an optical clock based on a single cooled, trapped ^{199}Hg ion. The electronic control paths are indicated by broken lines. Courtesy Professor Wineland.

a selected tooth frequency f_m is frequency doubled to $2 \times f_m$ and beat together with radiation at frequency f_{2m}, to give their difference f_{ceo}; second, radiation from many pairs of adjacent teeth f_n and f_{n-1} are beat together to give their difference f_{rep}. These measurements fix the frequency of any individual tooth. The last step in measurement is to beat the radiation of the stabilized laser at frequency f_{Hg} against the nearest tooth of the comb f_n and measure the difference: this determines f_{Hg}. In practice the difference between laser and comb tooth frequencies can be measured to a precision of $f_{\text{rep}}/10^5$ or 10 mHz. Currently precisions for f_{Hg} of one part in 10^{15} are achieved routinely.

12.7   The electron's anomalous magnetic moment

The comparison of the measured value of the electron's magnetic moment with that predicted by quantum electrodynamics, QED, is one of the most stringent tests of the quantum theory of electromagnetic interactions. Both values are known to extraordinary precision and they agree to within that precision.

Chapter 19 describes electron-proton scattering experiments, at several hundred GeV/c momentum transfer, which have not detected any

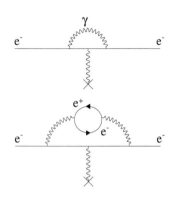

Fig. 12.14 Examples of low order Feynman diagrams contributing to the electron interaction. Straight lines indicate an electron, wavy lines a photon.

[14]Leptons are discussed in Chapter 18.

internal structure of the electron. Dirac predicted that a structureless electron would have a magnetic moment associated to its intrinsic angular momentum (spin) of

$$\mu = -g\mu_B m_s, \qquad (12.24)$$

where $\mu_B = e\hbar/2m$ is the Bohr magneton, with e and m being the electron charge and mass respectively, m_s is the component of electron spin along the quantization axis, while g is exactly 2. However, in QED g is predicted to depart from 2 due to Feynman diagrams which involve more than single photon exchange. Two simpler examples of these Feynman diagrams are exhibited in Figure 12.14. The total matrix element (amplitude) for the process is the sum of contributions from all the permissible diagrams. In the contributions of all diagrams the interaction strength is present through a multiplicative factor $\sqrt{\alpha}$ for each interaction vertex: $\alpha = e^2/(4\pi\varepsilon_0\hbar c)$ the fine structure constant being roughly equal to $1/137$. The more complex a diagram is the smaller its contribution will be. Calculations have been made including diagrams of up to tenth order, that is with ten additional vertices, in all over 10^4 diagrams. Corrections for loops with μ-leptons and τ-leptons,[14] and for strong and weak interactions have also been included. The prediction obtained is[15]

$$-\mu/\mu_B = g/2 = 1.001\,159\,652\,181\,78\,(77), \qquad (12.25)$$

where the bracketed number is the estimated error in the last two decimal places. The measurement of $g/2$ described immediately below by Gabrielse and colleagues is equally precise, and consistent with the prediction at that level[16]

$$-\mu/\mu_B = g/2 = 1.00\,115\,965\,180\,73\,(28). \qquad (12.26)$$

This is like having a measurement and prediction of the distance to the moon that agree to a hairsbreadth. This measurement is therefore a most stringent test of the standard model that provides the comprehensive explanation of the strong, weak and electromagnetic forces.

12.8   Measurement of g–2

Single electrons are stored in an evacuated Penning trap, with quadrupole electrostatic potential $V \sim (z^2 + \rho^2/2)$, where $z(\rho)$ is axial (radial) dimension and a uniform axial magnetic field B of 5 T. The cylindrically symmetric electrodes shown in section in Figure 12.15 are cooled to 4 K. The endcaps and the central ring are held at positive and negative potentials respectively. In this device the trapped electron undergoes simple harmonic motion axially with excursions of about 10 μm at a frequency ν_z, around 300 MHz. The electron motion projected in the plane

[15]T. Aoyama, M. Hayakawa, T. Kinoshita and M. Nio, *Physical Review Letters* 109 111807 (2012).
[16]D. Hanneke, S.F. Hoogerheide and G. Gabrielse, *Physical Review* A83 052122 (2011).

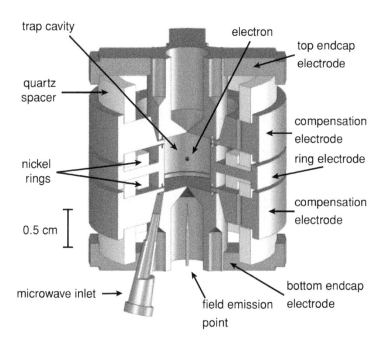

Fig. 12.15 Sectional view of the Penning trap used in the measurement of the electron g–2. Microwaves with 2mm wavelength enter through the tapered inlet. Courtesy Professor Gabrielse.

transverse to the axial 5 T magnetic field has two components. First it undergoes cyclotron motion at frequency ν_c around 150 GHz with an orbit of radius of ∼0.01 μm. Secondly there is magnetron motion, that is to say a slow drift along a circular path under the action of the crossed radial electric and axial magnetic fields. The magnetron motion at frequency ν_m around 100 kHz has an orbit of radius ∼1 μm. Taken together the electron follows a cycloid in the transverse plane, centred on the trap axis. The cyclotron motion is quantized in *Landau levels* with energies that are integral multiples of $h\nu_c$ with $\nu_c = eB/(2\pi m)$. Correspondingly the axial components of the electron's total angular momentum are $(2n_\ell + 1)\hbar$ from the cyclotron motion, n_ℓ being a positive integer, and $m_s\hbar$ from its spin.[17] During measurement the cavity is held at around 100 mK by a He³/He⁴ dilution refrigerator.[18] The equivalent thermal energy $k_B T$, is in appropriate units 2 GHz×h, which renders any thermal excitation of the cyclotron levels negligible.

The electron's orbital angular momentum and spin states couple to the electromagnetic modes of the cylindrical cavity defined by the highly polished internal surfaces of the electrodes. This cavity is leak-tight at radio frequencies and has a Q-value of around 10^5 for the relevant modes.

[17] The Bohr–Sommerfeld condition on the orbital angular momentum of an electron in a cyclotron orbit is $(n_\ell + 1/2)h = \oint(\mathbf{p} \cdot \mathbf{dr})$. Taking into account that its linear momentum is not simply $m\mathbf{v}$ but $m\mathbf{v} + e\mathbf{A}$, with $\mathbf{A}$ being the vector field, we have

$$(n_\ell + 1/2)h = \oint(m\mathbf{v} + e\mathbf{A}) \cdot \mathbf{dr}$$
$$= 2\pi mvr - eB\pi r^2 = \pi mvr,$$

so that mvr is $(2n_\ell + 1)\hbar$.

[18] See section 10 in A. Waele, *Journal of Low Temperature Physics* **164**, 179 (2011)

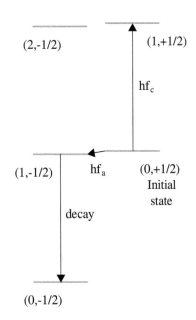

(2,-1/2)

(1,+1/2)

hf$_c$

(1,-1/2)   hf$_a$ (0,+1/2)
Initial
state

decay

(0,-1/2)

Fig. 12.16 Electron energy levels labelled by (n_l, m_s), the axial orbital and intrinsic angular momentum components. The measured transition energies hf_c and hf_a are indicated.

Next consider the spin states of the electron in the axial magnetic field. Using eqn. 12.24 their energy separation is $g\mu_B B$ so that radiation is absorbed and emitted at a frequency ν_s where

$$h\nu_s = g\mu_B B = (g/2)e\hbar B/m = \frac{g}{2}h\nu_c. \qquad (12.27)$$

This is called the *spin precession frequency* at which, in the classical view, the spin vector precesses around the magnetic field. As reviewed above the electron interaction with the vacuum causes the spin precession frequency, ν_s to differ from the cyclotron frequency, ν_c by about one part in a thousand. Measurements of the difference have been used to determine g with increasing precision. In the experiment by Gabrielse and co-workers the quantity measured is

$$g/2 - 1 = (\nu_s - \nu_c)/\nu_c. \qquad (12.28)$$

Figure 12.16 shows the lowest energy levels of the trapped electron labelled by (n_ℓ, m_s). The pure orbital transition indicated on the right has frequency

$$f_c = \bar{\nu}_c - \Delta, \qquad (12.29)$$

where $\bar{\nu}_c$ is the cyclotron frequency in the trap, rather than in free space; Δ is a small, calculable relativistic shift. The frequency of the $(0,+1/2)$ $\rightarrow (1,-1/2)$ *anomaly transition* indicated in Figure 12.16 is the difference between the spin and orbital excitation:

$$f_a = \frac{g}{2}\nu_c - \bar{\nu}_c, \qquad (12.30)$$

After this transition the electron undergoes spontaneous decay to the $(0,-1/2)$ state.

This difference between the cyclotron frequencies of a free and a trapped electron complicates the measurement and analysis. If it did not exist the experimenters could use f_c and f_a directly.

Rearranging the last equation:

$$g/2 = (\bar{\nu}_c + f_a)/\nu_c. \qquad (12.31)$$

At this point in the analysis the Brown–Gabrielse invariance theorem must be invoked: this links the components of the electron motion,

$$\nu_c^2 = (\bar{\nu}_c)^2 + (\bar{\nu}_z)^2 + (\bar{\nu}_m)^2, \qquad (12.32)$$

where $\bar{\nu}_z$, $\bar{\nu}_m$ and $\bar{\nu}_c$ are the cavity modified frequencies. These key frequencies have, by design, a distinct hierarchy which is exploited to make a measurement of $g - 2$ of extreme precision:

$$f_c \text{ and } \nu_s \sim 150\,\text{GHz},$$
$$f_a \sim 175\,\text{MHz and } \nu_z \sim 200\,\text{MHz},$$
$$\nu_m \sim 100\,\text{kHz},$$
$$\Delta \sim 10\,\text{Hz}.$$

Hence, using eqn. 12.29, the invariance theorem may be written to an excellent approximation

$$\nu_c = \overline{\nu}_c + \overline{\nu}_z^2/(2f_c).$$ (12.33)

Starting with eqn. 12.31 we first replace ν_c using eqn. 12.33, and afterwards using eqn. 12.29 replace $\overline{\nu}_c$. The result is that

$$g/2 = 1 + \frac{f_a - (\overline{\nu}_z^2/2f_c)}{f_c + \Delta + (\overline{\nu}_z^2/2f_c)}.$$ (12.34)

It emerges from this analysis that in order to measure $g/2$ to parts in 10^{13} it is adequate to measure f_a and f_c to parts in 10^{10}, and then $\overline{\nu}_z$ conventionally with a thousand times poorer precision.

$$
\begin{aligned}
g/2 &= (\overline{\nu}_c + f_a)/\nu_c \\
&= \frac{\overline{\nu}_c + f_a}{\overline{\nu}_c + \overline{\nu}_z^2/(2f_c)} \\
&= \frac{f_c + \Delta + f_a}{f_c + \Delta + \overline{\nu}_z^2/(2f_c)} \\
&= 1 + \frac{f_a - (\overline{\nu}_z^2/2f_c)}{f_c + \Delta + (\overline{\nu}_z^2/2f_c)}.
\end{aligned}
$$

12.8.1 Another quantum non-demolition measurement

The required exquisite precision in measuring f_a and f_c can be achieved by performing the measurements in such a way that measurements do not affect the quantum state of the electron: this is another type of *quantum non-demolition measurement*. A very weak coupling is induced between the transverse and axial motion of the trapped electron. For this purpose nickel inserts are installed in the cavity walls, as shown in Figure 12.15: the magnetic field is then strong enough to saturate in nickel, changing the uniform axial field to

$$B_z = B_0 + B_2 z^2.$$ (12.35)

Here the second term, which leads to the desired weak coupling, is much smaller than B_0. B_2 was around $1540 \, \text{Tm}^{-2}$. The interaction energy between the transverse and axial motion is

$$\boldsymbol{\mu} \cdot \mathbf{B}_2 = 2\mu_B B_2 (n_\ell + m_s) z^2,$$ (12.36)

$\boldsymbol{\mu}$ being the electron's total magnetic moment, n_ℓ the Landau level and $m_s \hbar$ the electron's axial spin component.[19] Its effect is to shift the axial frequency by about $4 \, \text{Hz}$ when the electron's *Landau level* changes. This frequency change was detected by an external circuit and provided a signal that there had been a transition between cyclotron Landau levels.

[19] Here taking g to be 2 gives adequate precision.

Because the interaction in eqn. 12.36 commutes with the cyclotron energy, there is *no back action from measurement of the axial motion on the cyclotron motion*, making this a non-demolition measurement. This critical feature allows high precision to be obtained.

In the experiment a set sequence is followed. The electron is first pumped into the $(0,+1/2)$ state. Then an external detection circuit is activated to determine the axial frequency, and afterwards turned off. Next $150 \, \text{GHz}$ (f_c) microwave or $175 \, \text{MHz}$ (f_a) radiofrequency pulses are fired into the cavity so as to produce the f_c or f_a transition respectively, as shown in Figure 12.16. After each pulse the external detection circuit is activated and couples to the axial vibration: this process required around 0.25 s to stabilize. In the case of the microwave transition to the $(1,+1/2)$ state at frequency f_c the interaction causes the axial frequency to rise by 4 Hz. In the case of the radiofrequency transition at frequency f_a to the state $(1,+1/2)$ the electron subsequently spontaneously decays to the $(0,-1/2)$ state and the interaction causes the axial frequency to fall by 4 Hz. The external circuit detects the change in the axial frequency and identifies that a change of state has occured. In a measurement cycle the pulse frequency was stepped across the linewidth of the transition and the number of successful transitions detected recorded at each step. Then the success rate peak determines the position of the central frequency, either f_c or f_a. Finally, eqn. 12.34 is used to obtain the quoted value of $g/2$.

Cyclotron levels in free space at 150 GHz have a lifetime of only 75 ms, so that it appears very surprising that such high precision can be attained, given the need for a settling time of 0.25 s for the detection circuit. What makes the experiment viable is the fact that the density of final states in a cavity differs from that in free space. The resultant modification of the decay rate through the *Purcell effect* boosts the lifetime of the cyclotron levels to as long as 15 s, adequate for a measurement cycle.

12.9 The Purcell effect

The suppression or enhancement of spontaneous emission from an atom (or electron) enclosed in a cavity that confines the radiation mode involved is known as the *Purcell effect*. The case considered here is that of weak coupling where the the cavity mode lifetime is much shorter than that of the atomic (electron) state ($\kappa \ll \gamma$). Then the decay is into a cavity empty of radiation at the emission frequency and the emission becomes purely spontaneous. The essential point is that the rate of decay can be enhanced (suppressed) if the phase space accessible at the emission frequency is larger (smaller) than in free space.

If the transition is an electric dipole transition at angular frequency ω, a perturbative calculation of the spontaneous emission rate gives eqn. 8.31

$$\gamma = \frac{\pi \mu^2 \omega}{\hbar \varepsilon_0 V} \rho(\omega). \tag{12.37}$$

where μ is the electric dipole moment. The density of states at angular frequency ω in a cavity mode of angular frequency ω_c and width γ_c is

$$\rho_c(\omega) = \frac{\gamma_c/2\pi}{(\omega - \omega_c)^2 + \gamma_c^2/4}. \tag{12.38}$$

The normalization of $\rho_c(\omega)$ is chosen so that integrating this distribution over ω gives unity, as it should do for a single mode. This density of states is to be compared to the density of states in vacuum given by eqn. B.11,

$$\rho_f(\omega) = \frac{\omega^2 V}{2\pi^2 c^3}. \tag{12.39}$$

Then the ratio of the cavity rate to the rate in free space is

$$R = \frac{\mu_c^2 \rho_c}{\mu_f^2 \rho_f} = \frac{3Q(\lambda/2)^3}{2\pi^2 V} \frac{\gamma_c^2}{(\omega - \omega_c)^2 + \gamma_c^2/4}. \tag{12.40}$$

Note that in free space the atomic dipole is taken to be randomly oriented with respect to the electric field, while it will be aligned with respect to the cavity mode; this makes $\mu_f^2 = (2/3)\mu_c^2$. $Q = \omega_c/\gamma_c$. Exactly on resonance this ratio, R, is known as the *Purcell factor*,

$$F_P = \frac{6Q}{\pi^2} \frac{(\lambda/2)^3}{V}. \tag{12.41}$$

We can see that by having a high Q cavity which is at the same time small compared to the wavelength the decay rate can be enhanced above that observed in free space.

On the other hand if the atomic transition frequency is well away from any cavity resonance the decay can be suppressed through lack of accessible final photonic states. Figure 12.17 illustrates this behaviour. Picture sliding the atomic transition across the cavity resonance: the emission will rise to peak at exact resonance and then decline. When the transition frequency lies in the tail of a cavity resonance eqn. 12.40 reduces to

$$R = \frac{3Q(\lambda/2)^3}{2\pi^2 V} \frac{\gamma_c^2}{(\omega - \omega_c)^2} \sim \frac{3}{2\pi^2 Q} \frac{(\lambda/2)^3}{V}, \tag{12.42}$$

where we have approximated $(\omega - \omega_c)/\gamma_c$ by Q. Evidently a large Q-value can overwhelm the enhancement coming from having a small cavity. This leads to the suppression of spontaneous emission, which was exploited in measuring g–2 of the electron. In a medium of refractive index n_r λ is replaced everywhere in this section by λ/n_r.

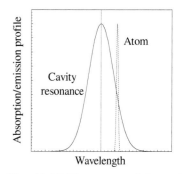

Fig. 12.17 The spectra of the cavity mode and of the atomic (quantum dot) transition.

12.10   Further reading

Exploring the Quantum: Atoms, Cavities and Photons by S. Haroche and J.-M. Raimond, published by Oxford University Press (2006). This gives a thorough survey of experiments with atoms and ions in cavities.

Laser Cooling and Trapping by H. J. Metcalf and P. van der Straten, published by Springer (1989) covers cooling and trapping of neutral atoms rather than ions, and broader physical applications.

'Controlling Photons in a Box and Exploring the Quantum to Classical Boundary', Nobel lecture by S. Haroche, *Review of Modern Physics* 85, 1083 (2013).

'Superposition, Entanglement and Raising Schrödinger's Cat', Nobel lecture by D. J. Wineland, *Review of Modern Physics* 85, 1103 (2013). These two Nobel lectures provide insights into the techniques and experiments with ions and atoms in cavities.

Exercises

(12.1) Prove eqn. 12.15.

(12.2) An atomic gas is exposed to a laser beam whose angular frequency, ω_0 matches that of an atomic transition from the ground state. The spectrum of the fluorescence produced, known as Mollow fluorescence, is shown in Figure 12.5. Explain the features of this spectrum, including the reason why the central peak is twice the area of either sideband. The dressed atomic states contain very large numbers of photons n due to the intensity of the photon flux in a laser beam. To a good approximation the fluctuations in n are small compared to the mean $\bar{n}$.

(12.3) How is it possible for a single laser beam to produce overall cooling in the motion in all three orthogonal directions of a group of trapped ions, and of a single ion in a perfect quadratic trap? If an ion is cooled using a transition with width 5 GHz what temperature is attainable with the first stage of Doppler cooling?

(12.4) Show that the angular frequency of the transition from one circular Rydberg rubidium atom with principal quantum number $n+1$ to one with n is approximately $2\frac{R}{\hbar}\frac{1}{n^3}$, where R is the Rydberg constant for rubidium and hence that for $n=50$ the frequency is tens of GHz.

(12.5) Explain qualitatively why the circular Rydberg atoms are inhibited from decaying to low lying states with emission at optical frequencies.

(12.6) Show that the cyclotron frequency of an electron in a uniform 5 T magnetic field is 140 GHz. Hence obtain the energy quantum in eV and estimate the radius of the orbit with one quantum excitation.

Take the magnetron frequency in the g–2 experiment to be 10^5 Hz and the orbital velocity of the magnetron motion, v_m, to be E/B where E and B are the crossed electric and magnetic fields respectively. Estimate E if the magnetron orbit has radius r_m of 1 μm.

(12.7) An InAs quantum dot inside a GaAs cavity emits at 1μm wavelength and would have a lifetime in free space against spontaneous decay of 1 ns. The cavity volume is 1 μm by 1 μm by 0.3 μm and has a Q-value of 5000. Calculate the lifetime of the exciton in the cavity using 3.5 for the refractive index of GaAs, when the cavity mode coincides with the transition in frequency.

(12.8) The lifetime of a photon in an empty cavity is dependent on how close the reflection coefficient, R, of the walls is to unity. Consider a Fabry–Perot cavity that has parallel mirrors of area much larger than their separation so that we can restrict discussion to modes travelling perpendicular to the mirrors (actual devices have curved mirrors designed to allow only such modes to propagate without loss). The cavity lifetime is

$$\tau = \frac{t}{1-R}, \qquad (12.43)$$

t being the time for radiation to cross the cavity. Calculate the lifetime in a GaAs cavity whose mirrors are 0.1 μm apart and have a reflection coefficient 0.99. Take the refractive index of GaAs to be 3.5. What is the Q-value of the cavity at 1.1 μm wavelength?

(12.9) What are the zero point energies and equivalent temperatures for the spin precession, axial vibration and magnetron motion in the g–2 experiment?

(12.10) A laser whose beam power density is $P\,\mathrm{Wm^{-2}}$ illuminates a gas of atoms. It is tuned to have the same frequency as an atomic transition. The electric dipole moment of the atom is D for the transition. Show that the Rabi frequency is $(2D/\hbar)\sqrt{2P/(\varepsilon_0 c)}$. If D is $2ea_0$ where a_0 is the Bohr radius deduce the power density required to produce Rabi oscillations at $50\,\mathrm{MHz}$. How powerful a laser is required if the target area of the gas sample is $10^{-8}\,\mathrm{m^{-2}}$?

(12.11) Show that the change in the axial frequency used in the measurement of $g - 2$ to detect a change in Landua level is around $4\,\mathrm{Hz}$. The solution is worked through in detail.

Symmetry and topology

13

13.1   Introduction

Symmetries in space-time and the conservation laws for energy, momentum and angular momentum are directly connected, a connection first spelled out formally for classical physics in 1916 by Emmy Nöther. Chapter 2 showed that these conserved quantities are quantized, their values depending on the boundary conditions (potential) inserted in Schrödinger's equation.

The properties and interactions of elementary paricles can be characterized by assigning them internal quantum numbers of which the simplest is electric charge. Correspondingly there are internal spaces and symmetries under transformations in these spaces. The standard model of particle physics has been built by applying quantum theory to take account of these internal spaces and their symmetries.[1] The convenient and simple example used here to introduce features of these *quantum gauge theories* is electromagnetism. All hinges on the fact that electric charge is conserved locally: this will be shown to demand the existence of *gauge fields* extending over all space-time with precisely the properties of the four-vector electromagnetic potentials (ϕ/c, $\mathbf{A}$).

These fields are not uniquely defined: they can undergo *gauge transformations* which leave the observed electric and magnetic fields unaffected. So it might be thought that they are only of computational interest. Experiments suggested by Aharonov and Bohm in 1959 have shown the physical importance of these potential fields: electrons respond to the gauge fields in regions where the electric and magnetic fields vanish!

The Aharonov–Bohm effect provides an example of quantization distinct from that due to symmetries: this is topological quantization. When the wavefunction of an electron is followed round a closed path in space it must return to its initial value modulo 2π in phase because a wavefunction has a unique value at a point. If there is no impediment to shrinking this path until the area it encloses vanishes then the phase change must be zero. However, if there is an area within the initial path where the wavefunction does not exist the loop can't be shrunk until it vanishes, and phase change can be $2n\pi$ with n being a non-zero integer. The Bose–Einstein condensates superfluid ^{4}He and superconductors ex-

[1] This encompasses quantized theories of the strong and electroweak interactions. General relativity, the successful theory of gravitation, defies heroic, decade long efforts to quantize it, and remains a classical theory.

Quantum 20/20: Fundamentals, Entanglement, Gauge Fields, Condensates and Topology.
Ian R. Kenyon. © 2020. Published in 2020 by Oxford University Press.
DOI: 10.1093/oso/9780198808350.001.0001

hibit similar quantization. If magnetic flux threads such a hole then this flux is correspondingly quantized in units h/q, where q is the charge involved. This makes h/q the relevant quantum of magnetic flux.[2]

The quantum effects originating in symmetries are treated in the opening part of this chapter. First the relationship between conservation laws and symmetries will be formulated quantum mechanically. Space-time symmetries and the properties of the group of transformations consistent with special relativity (Poincaré group) are discussed next. Then the internal charge symmetry is introduced, and using this the gauge principles underlying electromagnetism are brought to light. After which the emphasis shifts to topological effects.

Aharonov and Bohm's prediction of the existence of a geometric quantum phase is related first. Then its experimental verification is described. The quantization of magnetic flux is shown to emerge from the topological argument made by Ahoronov and Bohm. The related theme of Berry's phase and its measurement is covered in Sections 13.5. A simple example of the topology of a twisted space is presented in the final section of the chapter.

13.2   Space-time symmetries

A symmetry operation under which observables are unchanged must be described by a *unitary* operator, U such that:

$$U^\dagger = U^{-1}. \tag{13.1}$$

Here $U^\dagger$ is the complex conjugate transpose of U, and U^{-1} is the inverse of U such that $U^{-1}U = I$, I being the identity matrix. For example the expectation value for the presence of a particle is preserved

$$\langle U\psi|U\psi\rangle = \langle\psi|U^\dagger U\psi\rangle = \langle\psi|\psi\rangle. \tag{13.2}$$

Symmetry operations of interest in space-time are displacements and rotations. Each has an inverse: for example, a rotation of α about some axis has as an inverse a rotation through $-\alpha$ about the same axis. Symmetry properties involve pairs of conjugate observables: as for example position and momentum. For this we work in the Heisenberg representation where the operators change and the wavefunctions are constant. Then the rate of change of momentum, conjugate to position, is

$$i\hbar(\mathrm{d}\mathbf{p}/\mathrm{d}t) = [\mathbf{p}, H] \tag{13.3}$$

where H is the Hamiltonian (energy). It is well established that identical experiments carried out at different locations give identical results, for example in measuring physical constants. This means that the Hamiltonian must be unaffected by displacements in space, that is replacing the location of an isolated system $\mathbf{r}$ by $\mathbf{r}+\mathbf{r}_0$. Then the system is said to possess symmetry under translations in space. It follows that the Hamiltonian in the above equation cannot contain any function of absolute position **r**. Consequently the system's momentum **p** and H commute and it follows that the vector momentum is conserved. It is equally true that identical experiments at diffent times give identical results. A fully relativistic analysis puts the symmetry under spatial and time displacements on an equal footing. Therefore, we can infer that symmetry in time implies conservation of energy E. Finally, because experimental results do not depend on the orientation of experimental apparatus there is symmetry under rotations and correspondingly angular momentum is conserved. To summarize this relativistic reworking of Nöther's theorem: each space-time symmetry is mirrored by a conservation law.

The transformations, just described, are members of the Poincaré group of space-time transformations consistent with the special theory of relativity. In addition, the Poincaré group includes Lorentz transformations between two inertial frames given in eqn 1.69. These Lorentz transformations are called *boosts*, and have the mathematical form of rotations involving the time dimension and a space dimension. Formally a set of operations form a group provided that:

- if a and b are any two members of the group $a \star b = c$, is another member of the group;
- a^{-1} the inverse of a is a member of the group;
- the identity I is a member of the group equal to $a^{-1} \star a$, making the transformations *unitary*;
- the operations are both associative and distributive: $a \star (b \star d) = (a \star b) \star d$ and $a \star [b + d] = a \star b + a \star d$, where a, b and d are any three members of the group.

The *operation* $\star$ might be multiplication as for the Poincaré group, or a more complex operation, depending on the group.

The Poincaré group is an example of the very important class of *Lie* groups. These are groups of continuous transformations which can be made up from transformations infinitesimally close to the identity. We return to the Schrödinger representation, in which the wavefunctions change while the operators are constant. As an example, a wavefunction transforms in the following way under an infinitesimal displacement

$$\psi(x_0) \rightarrow \psi(x_0 + \mathrm{d}x) = \psi(x_0) + \frac{\partial \psi}{\partial x}\mathrm{d}x = \psi(x_0) + ip_x\psi(x_0)\mathrm{d}x/\hbar. \quad (13.4)$$

In the case of finite displacements this becomes

$$\psi(x_0) \rightarrow \psi(x_0 + x) = \exp(ip_x x/\hbar)\psi(x_0). \quad (13.5)$$

The quantity $p_x/\hbar$ is called the *generator* of the transformations. Note that the coordinate being transformed, x, and the generator, p_x are conjugate observables with $[x, p_x] = i\hbar$. An important connection is seen here: the group transformations are unitary and their generators are Hermitian, in common with other Lie groups we meet. This simplifies calculations with Lie groups. The generators summarize the group's mathematical structure and correspondingly the relations between the generators form what is called the group *algebra*. The Poincaré group generators include: displacement generators; the rotation generators, which are the angular momentum operators; and the boost generators. The generator of rotations about the x-axis is $J_x = yp_z - zp_y = -iy\hbar\frac{\partial}{\partial z} + iz\hbar\frac{\partial}{\partial y}$. The algebra of the Poincaré group includes these relationships

$$[p_m, p_n] = 0, \quad [J_m, J_n] = i\epsilon_{mnq}\hbar J_q, \quad [J_m, p_n] = i\epsilon_{mnq}\hbar p_q, \quad (13.6)$$

where ϵ_{mnq} is totally antisymmetric in its indices. When mnq are all different and in cyclic order ϵ_{mnq} is $+1$, if all different and in anticyclic order -1, and otherwise zero.

The Poincaré generators of a quantum system form a set of compatible observables which, in the case of an elementary particle defines its space-time properties. These are the intrinsic angular momentum (spin), its component along a quantization axis, and the four-momentum. There are additional invariant combinations of generators for a Lie group called *Casimir invariants*; two in the case of the Poincaré group. One of these is the centre-of-mass energy which is evaluated in any frame by the expression $\sqrt{E^2 - p^2c^2}$. This equates to the (rest mass)$\times c^2$ for a particle. The other Casimir invariant is effectively the product of mass and total angular momentum, so that both a particle's intrinsic angular momentum and its mass are invariants.[3] The basic particle properties, mass and spin, are now seen to be the Casimir invariants required of representations of the Poincaré group, that is to say invariants imposed by special relativity.

A representation of a group is a set of objects which is closed under the group of transformations.[4] The simplest, the *fundamental representation*, has the same dimensions as the group itself. The fundamental particles from which all else is made are the *leptons* and *quarks* introduced in Chapter 18. Not surprisingly, they fit neatly into the fundamental representations of the Lie symmetry groups that underlie the strong and electroweak forces.

Discussion of the discrete symmetries parity, charge conjugation and time reversal is postponed to Chapter 18 where the interest is in the violation of the discrete symmetries observed in particle physics. In addition to the space-time symmetries there are internal symmetries that are the basis of the forces of the standard model of particle physics. These symmetries are described by Lie groups and can be pictured as operating in spaces not accessible to direct observation but existing everywhere. The simplest of these symmetries is charge symmetry which operates in a one-dimensional complex space. The Lie group of operations in this space is given the name U(1). Its importance lies in the fact that this

[3]An extended treatment of this topic can be found from page 187 onward in the third edition of *Quantum Mechanics* by L. I. Schiff, published by McGraw-Hill International (1968).

[4]This means first, that any member of the set can be transformed into any other member of the set by some group transformation. Second, all group transformations take any member of the set into some superposition of members of the set

is the symmetry underlying electromagnetism. The analysis given here for U(1) provides the pattern to be followed when analysing the more complex symmetries underlying the weak and strong forces.

13.3   Charge symmetry

The best understood internal property of a particle, as distinct from the dynamic properties, is its electric charge. This is conserved so we may expect by analogy with Nöther's argument that there exists a corresponding symmetry. We explore the consequence of such a symmetry and find that quantum mechanics then requires the existence of vector potentials having properties that exactly match those possessed by electromagnetic fields. This result brings out the fundamental relationship between charge symmetry and electromagnetism in quantum theory.

We start from the observation that net charge is conserved in the universe. Then each particle's wavefunction, ψ_0 may be replaced as follows

$$\psi_0 \rightarrow \psi = \psi_0 \exp[iq\alpha] \qquad (13.7)$$

where α is a global constant and q is the particle's charge. This is called a *global gauge transformation*. The position, energy and momentum of the particle are easily seen to be unaffected:

$$
\begin{aligned}
\psi^*\psi &= \psi_0^*\psi_0 \\
-i\hbar\,\psi^*\frac{\partial\psi}{\partial x_\mu} &= -i\hbar\,\psi_0^*\frac{\partial\psi_0}{\partial x_\mu}.
\end{aligned}
$$

The subscript introduced here μ, runs from 0 to 3, with $(x_0, x_1, x_2, x_3) = (ct, x, y, z)$. Another shorthand to be used is to write ∂_μ for $\partial/\partial x_\mu$ so that the above momentum equation becomes

$$-i\hbar\,\psi^*\partial_\mu\psi = -i\hbar\,\psi_0^*\partial_\mu\psi_0. \qquad (13.8)$$

The global transformation is seen to be innocuous. Going further we know that net charge is conserved locally. Photons can split into electron positron pairs and these can annihilate to give photons; however, both processes, and similar processes for other particle species, conserve charge locally. In addition, a charged particle jumping a space-like separation would violate the special theory of relativity. This suggests replacing the global constant α by a variable $\alpha(ct, \mathbf{r})$ having a smoothly varying dependence on position and time. Explicitly

$$\psi_0 \rightarrow \psi = \psi_0 \exp[iq\alpha(ct, \mathbf{r})], \qquad (13.9)$$

which leaves the probability of finding the particle at any given point unchanged

$$\psi^*\psi = \psi_0^*\psi_0. \qquad (13.10)$$

This is called a *local gauge transformation* (of the first kind). At this stage there appears to be a snag with energy and momentum because

$$\psi^*(-i\hbar\partial_\mu)\psi = -i\hbar\psi_0^*\partial_\mu\psi_0 + \hbar q\psi_0^*\psi_0\partial_\mu\alpha. \quad (13.11)$$

However, charges do not exist in isolation but interact through electromagnetic fields. We must implement the modification that the energy-momentum of a charge particle undergoes due to the presence of an electromagnetic field. In this *minimal electromagnetic coupling* presented in eqn. 1.78 the momentum and energy become:

$$m\mathbf{v} = \mathbf{p} - q\mathbf{A}, \quad \text{and } E + q\phi, \quad (13.12)$$

where $\mathbf{v}$ is the particle's velocity, $\mathbf{p}$ its canonical momentum, E its kinetic energy, q its charge and $(\phi/c, \mathbf{A})$ is the four-vector potential of the electromagnetic field. It is useful to write the four-vectors $(E, c\mathbf{p})$ and $(\phi/c, \mathbf{A})$ as p_μ and A_μ where the time components are $p_0 = E$ and $A_0 = \phi/c$. The momentum can be made invariant under local gauge transformations by introducing a *covariant derivative*

$$\mathrm{D}_\mu = \partial_\mu - iqA_\mu/\hbar. \quad (13.13)$$

Then taking account of eqn. 13.12 the four-momentum operator replacing $-i\hbar\partial_\mu$ is

$$-i\hbar\mathrm{D}_\mu = -i\hbar\partial_\mu - qA_\mu, \quad (13.14)$$

Correspondingly the expectation value of the four-momentum is

$$\psi^*(-i\hbar\,\partial_\mu - qA_\mu)\psi = -i\hbar\,\psi_0^*\partial_\mu\psi_0 + \hbar q\psi_0^*[\partial_\mu\alpha]\psi_0 - \psi_0^*qA_\mu\psi_0. \quad (13.15)$$

Something else is needed in order to give this expression the same functional form as that before the transformation. That something is to make a *gauge transformation* familiar from classical electromagnetism. We are at liberty to make such transformations because they leave the observable electric and magnetic fields unchanged. The necessary gauge transformation uses $\alpha(ct, \mathbf{r})$, introduced earlier, as the gauge parameter:

$$A_\mu \rightarrow A_\mu + \hbar\partial_\mu\alpha(ct, \mathbf{r}). \quad (13.16)$$

This is called a *local gauge transformation* (of the second kind) and A_μ is called the *gauge field*. With this transformation the energy-momentum expression in eqn. 13.15 becomes

$$\psi^*\mathrm{D}_\mu\psi = \psi_0^*\mathrm{D}_\mu\psi_0, \quad (13.17)$$

exhibiting the invariance under the complementary local gauge transformations of the first and second kinds given in eqns. 13.9 and 13.16 respectively. Maxwell's equations take a compact form when expressed in terms of the gauge fields. Defining $F_{\mu\nu} = \partial_\mu A_\nu - \partial_\nu A_\mu$, Maxwell's equations take the form $\partial_\mu F_{\mu\nu} = j_\nu$, with j_ν being the four-vector current. We can also note that gauge invariance rules out a non-zero photon mass. A mass term in the Hamiltonian would have the form $m^2\mathbf{A}^2$ and this does not transform into $m^2\mathbf{A}^2$ under a gauge transformation.

This analysis has revealed the quantum origin of electromagnetism in the local conservation of charge and the connected symmetry under gauge transformations. These are complementary gauge transformations of the particle wavefunctions and the gauge fields, acting in unison so that observables are unaffected. The gauge transformations of eqn. 13.9 form a Lie group in one complex dimension, and being unitary this group is called U(1). These gauge transformations commute, $\exp(i\alpha)\exp(i\beta)$ is the same as $\exp(i\beta)\exp(i\alpha)$, so U(1) is called an Abelian group. We shall find that the theories underlying the strong and the weak forces involve new symmetry spaces and corresponding new gauge fields. Transformations in these new spaces do not commute and their groups are therefore labelled non-Abelian groups.

13.4   The Aharonov–Bohm effect

Measurements on the electromagnetic field utilize the force on charges to determine the electric and magnetic field strengths, and not the vector potentials $(\phi/c, \mathbf{A})$. These vector potentials can be altered by gauge transformations without changing the values of the electric and magnetic fields. This made it appear that the vector potentials were simply mathematical tools. Then in 1959 Aharonov and Bohm noticed that quantum mechanics requires measurable effects due to the vector potentials in regions of space where the magnetic and electric fields vanish. Underlying the analysis is the quantum view that particles have wavefunctions and can experience interference. Using the expression for the momentum of an electron in eqn. 13.12, and taking the electrostic potential to vanish, the phase change of an electron in travelling a distance $d\mathbf{r}$ in empty space in time dt is

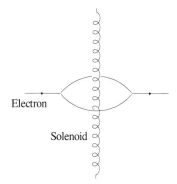

Fig. 13.1 Experiment proposed by Aharonov and Bohm.

$$(\mathbf{p} + e\mathbf{A}) \cdot d\mathbf{r}/\hbar - Edt/\hbar. \qquad (13.18)$$

The terms in energy E and momentum $\mathbf{p}$ make up what is called the *dynamic phase* change. Aharonov and Bohm's[5] insight was that the other term containing the vector potential could be finite in a region where the magnetic field itself vanishes. This adds a *geometric phase change* to the dynamical phase change over the path. Such behaviour is in contradiction to the classical view that an electron is unaffected in a region where the magnetic and electric fields are both zero.

Figure 13.1 shows in outline an experiment proposed by Aharonov and Bohm to test their prediction. Electrons travel on either side of a very long solenoid which *totally* confines its magnetic flux; and the interference between the electron waves beyond the solenoid is observed.

[5]Y. Aharonov and D. Bohm, *Physical Review* 115, 485 (1959).

It simplifies the analysis to take note that the phase difference between the two paths is identical to the change in phase of an electron which follows a closed loop, travelling out on the far-side path in the figure and returning on the near-side path. The geometric phase change in an electron's wavefunction produced by travelling around this closed loop, due to the field $\mathbf{A}$, is

$$\phi_{AB} = (e/\hbar) \oint \mathbf{A} \cdot d\mathbf{r}. \tag{13.19}$$

Using Stokes' theorem, eqn. A.21, this can be converted to a surface integral over the area inside the loop, including of course the interior of the solenoid, where $\mathbf{B}$ is non-zero,

$$\phi_{AB} = [e/\hbar] \int \nabla \wedge \mathbf{A} \cdot d\mathbf{S} = [e/\hbar] \int \mathbf{B} \cdot d\mathbf{S}. \tag{13.20}$$

This reduces to

$$\phi_{AB} = (e/\hbar)\Phi \tag{13.21}$$

where Φ is the *magnetic flux through the solenoid*. In the classical view altering the solenoid current would not affect the electrons because there is no magnetic field along their paths. However, in the quantum interpretation, because ϕ_{AB} changes the interference pattern changes. Many experiments have confirmed the existence of this effect: the first by Chambers used a magnetic whisker in place of the solenoid.

From eqn. 13.21 we can infer that if the closed path is such that the electron wavefunction extends, and is *coherent* over the whole loop then the phase change round the loop must be exactly an integer multiple of 2π and the flux enclosed must be precisely an integral multiple, n_q, of $[h/e]$. This condition was attained in the experiment reported in Section 6.9, where the path was $\sim \mu$m at ~ 1 K. It showed, on the mesoscopic scale, that the magnetic flux through a closed circuit is quantized in units of h/q where the current carrying particles have charge q. The ideal case on the macroscopic scale will be met when the path is superconducting. This type of quantization is not the outcome of rotational symmetry but is topological: states with different values of n_q are distinct because they have a different number of twists in the electron phase around the path.

Permalloy has a typical composition (83 per cent Ni:17 per cent Fe) and is used in magnetic recording heads. It has a low magnetostriction coefficient so that buckling and rupture are avoided in the cooling-heating cycle between room temperature and liquid helium temperatures.

An experiment confirming the Aharonov–Bohm effect was carried out by Tonomura and colleagues in Tokyo.[6] Their apparatus, sketched in Figure 13.2, was installed as the target of an electron microscope. The coherent electron beam travels from the top to the bottom of the page. Aharonov and Bohm's long solenoid is replaced by a continuous closed loop of Permalloy, a permanent magnetic material. This closed magnetic circuit eliminates the field spillage that exists with an open-ended solenoid. The 20 nm thick Permalloy layer lies between SiO layers, the

---

[6]N. Osakabe, T. Matsuda, T. Kawasaki, J. Endo, A. Tonomura, S. Yano and H. Yamada, *Physical Review* A34 815 (1986).

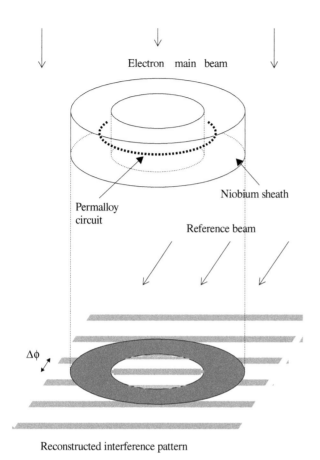

Reconstructed interference pattern

Fig. 13.2 Experimental layout for observing the Aharonov–Bohm effect. Adapted from Figure 1 from N. Osakabe, T. Matsuda, T. Kawasaki, J. Endo, A. Tonomura, S. Yano and H. Yamada, *Physical Review A*34 815 (1986). Courtesy Dr Osakabe.

whole encapsulated in an hermetic shell of niobium at least 300 nm thick. Below a temperature of 9.3 K niobium is superconducting. A final copper coating connected to ground zeroes the local electric field.

Electrons can travel along two alternative paths: either through the 2 μm diameter aperture or around the 6 μm diameter casing. In order to gain phase information a *reference* component of the electron beam, and hence coherent with these *signal* beams, is steered to fall with them on a recording film. An example of the interference pattern is superposed in Figure 13.2 at the imaging plane. The interference pattern displayed there shows an offset between the fringes inside and outside the shadow of the magnet housing. This comes from the difference in phase between the electrons travelling through and past the solenoid, partly due to dynamic and partly from the geometric phase difference.

When niobium enters the superconducting state the current migrates to its surface to form the superconducting loop indicated in Figure 13.3. This traps flux within the closed superconducting loop formed by the

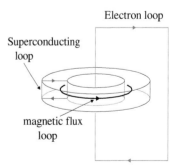

Fig. 13.3 Electron paths, magnetic flux path and current in niobium.

niobium shell (NOT through the aperture threaded by part of the electron beam).

The offset between the fringes in Figure 13.2 inside and outside the solenoid shadow changed abruptly at the superconducting transition. by an integral multiple, n of one half of the fringe repeat distance. This change cannot arise from the dynamic phase, because the electron paths are unaltered, but from the Aharonov–Bohm phase. Hence the Aharonov–Bohm phase change, $\Delta\phi_{\mathrm{AB}}$ was $n\pi$. In turn this implies a change of the flux trapped within the superconducting loop of

$$\Delta\Phi = (\hbar/e)\Delta\phi_{\mathrm{AB}} = [h/2e]n. \qquad (13.22)$$

Surprisingly, the flux quantum observed is $h/2e$ rather than h/e. This is simply because superconducting currents are carried by pairs of electrons and the flux quantum changes accordingly. Superconductivity is discusseed in Chapter 15.

13.4.1 The Aharonov–Casher effect

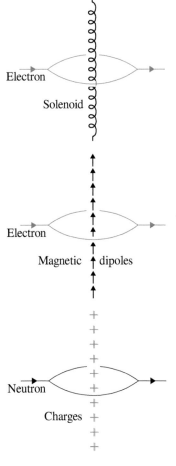

Figure 13.4 displays a sequence which we now explore, starting from the original experimental layout proposed by Aharonov and Bohm, and ending with another layout proposed by Aharonov and Casher. The first step takes us from the Aharonov–Bohm layout in the uppermost panel to the middle panel in which the solenoid is replaced by a line of magnetic dipoles. This mimics the experiment performed by Tonomura and colleagues. The inference is then made that if the phase of a charged particle (electron) is influenced by the vector potential $\mathbf{A}$ in regions where $\mathbf{B}$ is zero, it follows that the phase of a neutral magnetic dipole (neutron) should be influenced by the electric potential ϕ. A notional layout for testing this inference as envisioned by Aharonov and Casher is sketched in the lowermost panel of Figure 13.4. However, it turns out that this inference leads to a prediction shared by classical theory. Although an electric field does not exert a direct influence on a magnetic dipole, nonetheless, classical electromagnetism requires a contribution to the dipole's momentum from the electric field:

$$\mathbf{p} = m\mathbf{v} + \boldsymbol{\mu} \wedge \mathbf{E}/c^2, \qquad (13.23)$$

Fig. 13.4 Sequence of basic diagrams illustrating the inferred steps from the Aharonov-Casher effect to the Aharonov-Casher effect. Magnetic materials are in black, charged in red.

where m is the dipole mass, $\boldsymbol{\mu}$ its magnetic moment and $\mathbf{v}$ its velocity. The second term on the right-hand side is the so-called *hidden momentum*, first spotted in 1967 by Shockley and James.[7] Taking a magnetic dipole in a closed loop around a line of charge is predicted to produce a phase change due to the hidden momentum

$$\phi_{\mathrm{AC}} = \oint \boldsymbol{\mu} \wedge \mathbf{E} \cdot d\mathbf{r}/(\hbar c^2). \qquad (13.24)$$

[7]W. Shockley and R. P. James, *Physical Review Letters* 18, 876 (1967).

This *Aharonov–Casher* effect has been observed experimentally using neutrons.[8] In this case the prediction of quantum theory survives into classical electromagnetism.

13.5 Berry's phase

In 1983 Berry[9] arrived at a general analysis of quantum-specific phase changes in cyclic adiabatic processes. The starting point is the *adiabatic theorem* proved by Born and Fok, which states that if the environment of a quantum system in an eigenstate changes sufficiently slowly the system remains in this eigenstate, which itself changes to accomodate the environmental change. Suppose there exists a second eigenstate with a higher energy ΔE_{ex} having a spontaneous lifetime τ. Then for the process to be adiabatic the rate of change in energy produced in the system by the environment should be very small compared to $\Delta E_{ex}/\tau$: that is

$$\tau \frac{\mathrm{d}E_{en}}{\mathrm{d}t} \ll \Delta E_{ex}. \tag{13.25}$$

If not the environment can stimulate a transition to the second eigenstate. However, the energy change integrated over time can be large, so that in general perturbation theory is not a reliable guide to behaviour.

A simple adiabatic process considered here is that of a neutron in a changing magnetic field. Initially the neutron spin is aligned with the magnetic field. The field direction is made to rotate slowly and the neutron spin orientation follows the field orientation. Berry's insight was to infer that if the adiabatic process was cyclic, returning the system and its environment to their initial condition, then this would result in a geometric phase change of the system in addition to any change in the dynamic phase. The neutron's wavefunction would alter over the course of the cyclic change as follows

$$|\psi\rangle \rightarrow \exp(i\phi)|\psi\rangle = \exp[i(\eta + \phi_B)]|\psi\rangle. \tag{13.26}$$

Here η is the dynamical phase and ϕ_B is *Berry's phase*. In detail the dynamical phase change is

$$\eta = \oint (-E\mathrm{d}t/\hbar), \tag{13.27}$$

where E is the neutron kinetic energy. The environmental parameters affecting the quantum system are summarized by the symbol $\mathbf{R}(t)$. Then according to Berry the process can be re-expressed as

$$|\psi(\mathbf{R}(0))\rangle \rightarrow \exp[i\phi(t)]|\psi(\mathbf{R}(t))\rangle. \tag{13.28}$$

The adiabatic theorem underpins the Born–Oppenheimer analysis of molecular quantum states and renders the analysis tractable. Because the nuclei are so much heavier than the electrons their motion can be treated as adiabatic. In a first calculation the nuclei are taken stationary in their mean locations; later the nuclear motion is analysed taking the electrons' average spatial distribution.

---

[8]A. Cimmino, G. I. Opat, A. G. Klein, H. Kaiser, S. A. Werner, M. Arif and R. Clothier, *Physical Review Letters* 63, 380 (1989).
[9]M. V. Berry, *Proceedings of the Royal Society London* A392, 45 (1984).

Inserting the right-hand side into Schrödinger's time-dependent equation, and dropping the time arguments for conciseness, gives

$$i\hbar \frac{\mathrm{d}}{\mathrm{d}t}[\exp(i\phi)|\psi(\mathbf{R})\rangle] = E(\mathbf{R})\exp(i\phi)|\psi(\mathbf{R})\rangle, \qquad (13.29)$$

that is

$$-\hbar\exp(i\phi)\frac{\mathrm{d}\phi}{\mathrm{d}t}|\psi(\mathbf{R})\rangle + i\hbar\exp(i\phi)\frac{\mathrm{d}}{\mathrm{d}t}|\psi(\mathbf{R})\rangle = E(\mathbf{R})\exp(i\phi)|\psi(\mathbf{R})\rangle. \qquad (13.30)$$

Multiplying this equation by $\langle\psi(\mathbf{R})|\exp(-i\phi)$ gives

$$-\hbar\frac{\mathrm{d}\phi}{\mathrm{d}t} + i\hbar\langle\psi(\mathbf{R})|\frac{\mathrm{d}}{\mathrm{d}t}|\psi(\mathbf{R})\rangle = E(\mathbf{R}). \qquad (13.31)$$

Thus

$$\frac{\mathrm{d}\phi}{\mathrm{d}t} = -E(\mathbf{R})/\hbar + i\langle\psi(\mathbf{R})|\nabla_{\mathbf{R}}|\psi(\mathbf{R})\rangle \cdot \frac{\mathrm{d}\mathbf{R}}{\mathrm{d}t}. \qquad (13.32)$$

The first term on the right-hand side is the dynamic phase. In the other term we define

$$\mathbf{A}(\mathbf{R}) = i\langle\psi(\mathbf{R})|\nabla_{\mathbf{R}}|\psi(\mathbf{R})\rangle \qquad (13.33)$$

a vector taking the rate of change of $\psi(\mathbf{R})\rangle$ with $\mathbf{R}$ and projecting it onto $\psi(\mathbf{R})\rangle$. This vector is then a measure of the curvature of $|\psi(\mathbf{R})\rangle$ in $\mathbf{R}$-space. As such it is called a *connection* and in this instance Berry's connection. The corresponding Berry's phase resulting from a change in the environment from $\mathbf{R}(0)$ to $\mathbf{R}(t)$ in a time t is

$$\phi_{\mathrm{B}} = \int_{\mathbf{R}(0)}^{\mathbf{R}(t)} \mathbf{A}(\mathbf{R}) \cdot \mathrm{d}\mathbf{R}. \qquad (13.34)$$

The choice of symbols here brings out the parallel between Berry's phase and the Aharonov–Bohm phase in eqn. 13.19. However, there is a distinction between these two cases: Berry's analysis depends on the adiabatic condition, whereas the Aharonov–Bohm analysis does not. The comparison reveals that the vector magnetic potential is the connection for an electromagnetic field. Over a closed path

$$\phi_{\mathrm{B}} = \oint \mathbf{A}(\mathbf{R}) \cdot \mathrm{d}\mathbf{R}, \qquad (13.35)$$

which depends on the path followed in $\mathbf{R}$-space from $\mathbf{R}(0)$ out and back to $\mathbf{R}(0)$ again. We now show that Berry's phase for a closed path is a gauge invariant, which renders it physically interesting. The gauge transformation made is to multiply $\psi(\mathbf{R})$ by a phase factor $\exp[i\Lambda(\mathbf{R})]$. Berry's phase then acquires an added term involving

$$\oint \nabla_{\mathbf{R}}\Lambda(\mathbf{R}) \cdot \mathrm{d}\mathbf{R} = \Lambda(\mathbf{R}(0)) - \Lambda(\mathbf{R}(0)) \qquad (13.36)$$

which vanishes. Thus Berry's phase is gauge invariant.

Berry's phase does not depend on the time taken to cover the path, nor the kimematic details, but rather the path in $\mathbf{R}$-space. For this reason Berry called it a *geometric phase* relating to the geometry of the eigenstate space.

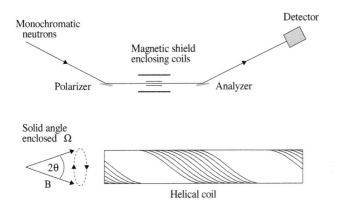

Fig. 13.5 The experiment of Bitter and Dubbers to measure Berry's phase is sketched in the upper panel with the coil shown in more detail below.

13.5.1 Experimental verification

Bitter and Dubbers[10] confirmed these predictions. They measured the Berry's phase accumulated by neutrons travelling slowly through a magnetic field that rotated through 360°. Their experiment is sketched in Figure 13.5. The neutrons originate from the Institut Laue–Langevin high flux reactor at Grenoble. A beam of slow neutrons with velocity $\sim 500\,\mathrm{m\,s^{-1}}$ is spin-polarized by a supermirror polarizer. The polarized neutrons are guided so that they enter and travel along the axis of an open ended Mumetal cylinder 80 cm long and 8 cm in diameter. Mumetal has a very high relative permeability, $\sim 100\,000$, so that the region enclosed is isolated from any external magnetic fields. Inside this magnetic shield the neutrons travel axially through two concentric 40 cm long coils providing a static magnetic field for the experiment. One coil is a solenoid producing an axial magnetic field B_z. The other coil is wound as shown in Figure 13.5, and produces a magnetic field B_1 that lies in a plane perpendicular to the axis and rotates through 360° along the coil length. The variation of the total magnetic field $\mathbf{B}$ is indicated on the left of the figure. The neutrons take 0.8 ms to pass through these coils, travelling slowly enough to make the magnetic field change adiabatic. On emerging the neutrons' polarization is analysed by another supermirror and those retaining their initial polarization are deflected to travel toward a neutron detector. The polarization state of the emerging neutrons provides a means to determine the phase change they have acquired.

The dynamic phase change due to travel for a time t through the

Supermirrors exploit the difference in reflection coefficients between neutrons having their spins aligned parallel and antiparallel to the magnetization direction in ferromagnetic materials. A supermirror consists of alternate layers of for example copper and nickel (ferromagnetic). Layer thickness is varied in order to polarize neutrons over a range of energies.

[10]T. Bitter and D. Dubbers, *Physical Review Letters* **59**, 251 (1987).

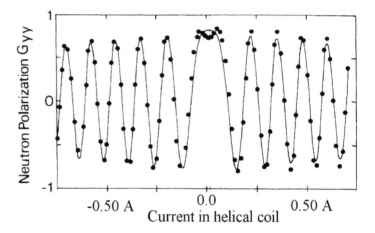

Fig. 13.6 Neutron total phase shift as a function of the current in the helical coil. Adapted from Figure 2(a) in T. Bitter and D. Dubbers, *Physical Review Letters* 59, 251 (1987). Courtesy Professor Dubbers and The American Physical Society.

combined magnetic field of the two coils, $\mathbf{B} = \mathbf{B}_z + \mathbf{B}_1$, is simply

$$\eta = E_{\mathrm{m}}t/\hbar = \gamma\mathbf{B}\cdot\boldsymbol{\sigma}t, \qquad (13.37)$$

where E_{m} is the magnetic field energy, $\gamma = g\mu_{\mathrm{N}}/\hbar$ is the neutron gyromagnetic ratio, μ_{N} being the nuclear magneton, $\boldsymbol{\sigma}$ is the unit length spin vector and $g = 1.913$ corrects for the neutron structure not being point-like. Berry's geometric phase develops as the neutron spin direction follows the rotation of the magnetic field. If ω is the rate of this rotation, then the neutron state with its spin aligned along the magnetic field is

$$|s_+\rangle = \begin{bmatrix} \cos(\theta/2) \\ \exp(i\omega t)\sin(\theta/2) \end{bmatrix}, \qquad (13.38)$$

where we have used eqn. D.2 and put $\phi = \omega t$. Using eqns. 13.33 and 13.35 Berry's geometric phase is

$$\phi_{\mathrm{B}} = \int_0^{2\pi/\omega} \langle s_+|i\frac{\mathrm{d}}{\mathrm{d}t}|s_+\rangle\,\mathrm{d}t, \qquad (13.39)$$

where $2\pi/\omega$ is the period of one rotation. Replacing $|s_+\rangle$ in this equation gives

$$\begin{aligned}\phi_{\mathrm{B}} &= \int_0^{2\pi/\omega} -\omega[\cos(\theta/2)\ \exp(-i\omega t)\sin(\theta/2)]\begin{bmatrix} 0 \\ \exp(i\omega t)\sin(\theta/2)\end{bmatrix}\mathrm{d}t \\ &= -2\pi\sin^2(\theta/2) = -\pi(1-\cos\theta). \qquad (13.40)\end{aligned}$$

This is numerically half the solid angle subtended by the cone traced out in Figure 13.5 by the magnetic field vector as it rotates once. Bitter and Dubbers measured the polarization components of the outgoing neutrons P_α. These are related to the incident polarization components through

$$P_\beta(t) = G_{\alpha\beta}P_\alpha(0) \qquad (13.41)$$

where $G_{\alpha\beta}$ are analytic functions of the phase shift. Results are shown in Figure 13.6 for the transverse spin component parameter G_{yy} as a

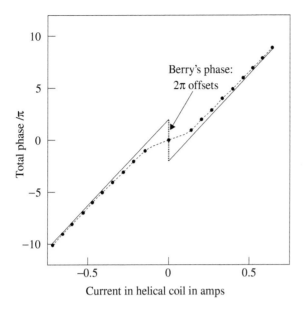

Fig. 13.7 Variation of the total phase shift as a function of the current in the helical coil. The asymptotes to the data as the helical field becomes very large are drawn in and intersect the zero current line at phases $\pm 2\pi$. Adapted from Figure 2(b) from T. Bitter and D. Dubbers, *Physical Review Letters* **59**, 251 (1987). Courtesy Professor Dubbers.

function of the current in the helical coil. The phase extracted is double the sum of the dynamic and geometric phases,

$$\phi = 2\eta - 2\pi(1 - \cos\theta). \tag{13.42}$$

Figure 13.7 shows how this phase varied as the magnitude of the rotating field B_1 was varied. When B_1 is made very large and positive (negative), ϕ tends to $2\eta \pm 2\pi$. The asymptotes to the data for large B_1 are drawn as solid lines and demonstrate their expected offsets of $\pm 2\pi$ at zero current, thus confirming Berry's prediction. Aharonov and Anandan[11] have shown that geometric phases can be evaluated for non-adiabatic processes, and for open paths in the parameter space of the environment. Their results collapse to those of Berry for closed adiabatic loops. Experiments to measure such phase changes in these relaxed conditions are difficult.

[11] *Physical Review Letters* **58**, 1593 (1987).

13.6   Phase and topology

The Hilbert space of quantum states has been presented in earlier chapters as a multidimensional Cartesian flat space. The examples met in this chapter show that Hilbert space can be more complex. It remains true that Hilbert space can always be mapped locally by a flat space. Berry's analysis deals with a Hilbert space in which a parameter affecting the wavefunction phase changes as the system travels. An attempt is made to picture such a Hilbert space in Figure 13.8. At the bottom is a two-dimensional representation of the flat dimensions of Hilbert space

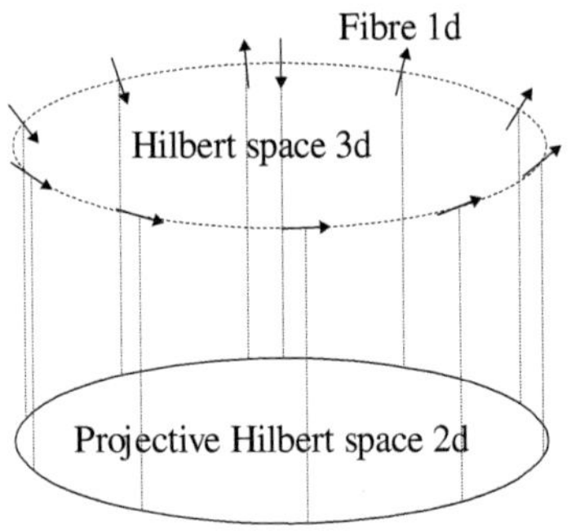

Fig. 13.8 Topology of locally flat Hilbert space with a Moebius twist.

and above it the total Hilbert space with one additional twisted dimension. This dimension is chosen to mark out a Moebius band above a closed circuit in the rest of Hilbert space. The twist in the Hilbert space means that it is not possible to shrink the path shown smoothly to a point. Thus the space has a discontinuity: if this is a hole the space is said to be multiply connected. In the cases of the Aharonov–Bohm effect or a non-zero Berry's phase the path encircles one or more holes. The value of the hole count is a topological property: *it is not associated with any particular element of the path but rather the whole circuit*. In the figure, following the closed path in the flat dimensions of Hilbert space leads to a rotation of π in orientation in the twisted dimension. This departure from the initial value is called a holonomy in mathematical texts and an anholonomy by Berry and other physicists. The axes in the twisted component of space, here only a single axis indicated by an arrow, are called fibres and collectively they are called a fibre bundle.

The topology met in the integral quantum Hall effect in Chapter 17 is that of a wavefunction which is defined over a torus. The count of the discontinuities in the wavefunction is quantized and is the integer of the integral quantum Hall effect.

13.7 Further reading

An Elementary Primer for Gauge Theory by K. Moriyasu, published by World Scientific Publishing (1983) is indeed a useful elementary primer addressing gauge theory from the particle physics viewpoint.

Geometric Phases in Physics edited by A. Shapere and F. Wilczek, published by World Scientific Publishing (1989) contains a selection of key scientific papers and helpful reviews in each section. It gives thorough coverage at the date of publishing.

Quantum Field Theory for the Gifted Amateur by T. Lancaster and S. J. Blundell, published by Oxford University Press (2014) has good sections on symmetries, broken symmetries and simple topology in sections 3, 6 and 7 respectively, at the advertised level.

Topology and Geometry for Physicists by C. Nash and S. Sen published by Dover (2011) provides a comprehensive introduction to topology.

Elements of Group Theory for Physicists, by A. W. Joshi, published by Wiley Eastern Ltd. (1971). A text giving a clear and systematic introduction with many applications.

Modern Differential Geometry for Physicists, by C. J. Isham, published by World Scientific (1989). An excellent, thorough, tough book.

Exercises

Appendix A has many useful formulae needed here.

(13.1) An electron is taken over a closed path round a 200 turn per metre solenoid carrying a current $2\,\mu A$ and of area of cross-section $0.3\,\mathrm{cm}^2$. The path is a circle centred on the solenoid and perpendicular to its axis. What is the geometric phase shift for one revolution?

Suppose the electron path is extended to include a second loop of equal area and in the same plane but forming a figure 8 with the first loop. What is the new geometric phase shift?

(13.2) Is it possible to have a gauge theory based on conservation of mass or of energy-momentum?

(13.3) Suppose that in a two path electron interferometer the paths travel through identical open conducting cylinders. Pulses of electrons are injected into the interferometer. For an interval t, during which the electron wavefunction is entirely within the cylinders, a voltage V is applied between the cylinders. How will this affect the relative phase between the two paths?

Taking a $1\,\mu V$ pulse lasting $1\,\mathrm{ns}$ what fringe displacement can be expected?

(13.4) In a region where there is a constant uniform magnetic field $\mathbf{B}$, show that $\mathbf{A} = \frac{1}{2}\mathbf{B} \wedge \mathbf{r}$.

(13.5) What gauge transformation would convert the vector potential $\mathbf{A}$ of the last exercise to $\frac{1}{2}\mathbf{B} \wedge (\mathbf{r}+\mathbf{s})$, where $\mathbf{s}$ is a fixed displacement?

(13.6) Suppose there exists somewhere a single magnetic monopole so that the magnetic field at a distance $\mathbf{r}$ is

$$\mathbf{B} = m\mathbf{r}/r^3,$$

where m is the magnetic charge. Then there would exist suitable vector potentials pointing azimuthally in the polar coordinate system (r, θ, ϕ) relative to the monopole as origin. Show that there are two solutions

$$\mathbf{A}_{1/2} = m\frac{\mp 1 + \cos\theta}{r\sin\theta},$$

where you should use equation A.4. Discuss what cover these two solutions provide over the whole 4π solid angle. Show that a gauge transformation connects these solutions making them equivalent where they overlap,

$$\mathbf{A}_2 = \mathbf{A}_1 + \nabla\xi,$$

and find ξ. What is the corresponding change in wavefunction, produced by this transformation, for a particle carrying charge q?

Argue from this result that electric charge must be quantized.

(13.7) A vector on the surface of a sphere is carried parallel to itself along a closed path made up of sections of great circles: from the north pole due south to the equator, then along the equator, finally due north to the pole again. What is the angle that the vector will have turned through compared to its initial orientation?

What is the solid angle which the path subtends at the centre of the sphere?

(13.8) Show that if physical properties are unaffected by any rotation around the z-axis then the angular momentum around the z-axis is conserved.

Superfluid ^{4}He

14

14.1 Introduction

The element ^{4}He, whose atoms have spin 0 and are thus bosons, liquifies at $4.2\,\mathrm{K}$. Further cooling through $2.17\,\mathrm{K}$ is accompanied by the sharp λ-like spike in the specific heat shown in Figure 14.1. This rapid shedding of entropy as the temperature drops through $2.17\,\mathrm{K}$ indicates that a new well-ordered liquid phase has been entered. The two liquid phases of ^{4}He are known as He-I above, and He-II below the transition temperature, itself known as T_λ. The phase diagram in Figure 14.2 shows that ^{4}He remains liquid at $0\,\mathrm{K}$ at atmospheric pressure: it requires 25 atmospheres pressure to convert He-II to the solid phase. He-II is also special in possessing *superfluidity*, that is the ability to flow without viscous drag. This happens because a substantial fraction of the atoms in He-II form a Bose–Einstein condensate described in Sections 3.7 and 3.8. When the temperature falls to T_λ the atomic de Broglie wavelength has become as large as the atomic separation so that the condensing atoms enter the lowest energy eigenstate with a common macroscopic wavefunction. In the case of ^{4}He the inferred temperature for condensation given by eqn. 3.50 is $3.3\,\mathrm{K}$. The calculation neglects any effect of interactions between the closely packed atoms in liquid helium which accounts for the deviation from $2.17\,\mathrm{K}$. Superfluid ^{4}He and superconductors were the first examples discovered of macroscopic quantum states. Currently the Large Hadron Collider at CERN uses 130 metric tonnes of He-II at $1.9\,\mathrm{K}$ to cool 1232, $14.3\,\mathrm{m}$ long superconducting dipole magnets which steer the $6.5\,\mathrm{TeV}$ proton beams. The volumes of condensate involved are certainly macroscopic!

Figure 14.3 shows the Lennard–Jones approximation for the potential wells between like pairs of noble gas atoms, all of which are spin-0 bosons,

$$V = \epsilon \left(\left[\frac{r_\mathrm{m}}{r} \right]^{12} - 2 \left[\frac{r_\mathrm{m}}{r} \right]^6 \right), \tag{14.1}$$

where r is the separation, r_m the location of the potential minimum, and ϵ the potential energy there. The first term is the Coulomb repulsion of the electron clouds; the second attractive term is due to the atoms polarizing each other, with the resultant dipoles then attracting one another (the van der Waals force). The depth of the potential well increases markedly from helium to neon to argon. Consequently they liquify in the sequence argon ($87\,\mathrm{K}$), neon ($27\,\mathrm{K}$) and helium ($4.2\,\mathrm{K}$). From the

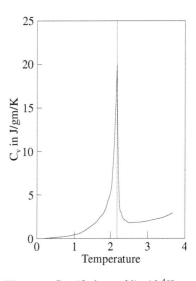

Fig. 14.1 Specific heat of liquid ^{4}He

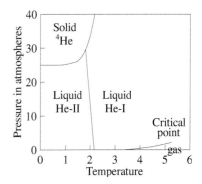

Fig. 14.2 Phase diagram of ^{4}He

Quantum 20/20: Fundamentals, Entanglement, Gauge Fields, Condensates and Topology.
Ian R. Kenyon. © 2020. Published in 2020 by Oxford University Press.
DOI: 10.1093/oso/9780198808350.001.0001

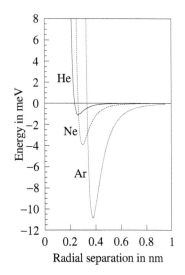

Fig. 14.3 Interatomic potentials for the three lightest noble gases.

[1]Mystifyingly Landau made no use of the concept of a condensate: his analysis is not impaired by this lacuna.

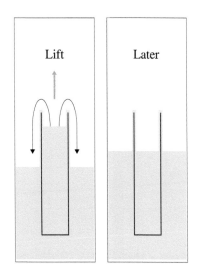

Fig. 14.4 Having lifted the beaker the superfluid flows in the surface film until the He-II levels in the beaker and container equalize.

figure we see that in the trough of the potential well the atom-to-atom separation for helium is about 0.26 nm, comparable to the de Broglie wavelength 0.34 nm at 4 K, so that the condition for Bose–Einstein condensation is attained. In the case of neon the atom-to-atom separation on freezing is 0.3 nm while its de Broglie wavelength is only 0.07 nm, so that Bose–Einstein condensation cannot occur. Helium doesn't freeze at one atmosphere because the atoms are light enough that their zero-point motion overcomes the relatively weak binding potential.

The fraction of atoms in the condensate in He-II only reaches ∼0.1 close to 0 K, held down by interactions between the atoms. At the same time these interactions carry the superfluidity of the condensate to a larger proportion of the liquid. He-II is 100 per cent superfluid at 0 K, while at T_λ, the condensate and superfluid vanish.

In the first section below superfluid properties are described. Fritz London and Lazlo Tizsa were the first to perceive that superfluidity implied that a Bose–Einstein condensate must be present in liquid ^{4}He. Landau developed the modern two fluid theory which is introduced next, with He-II being composed of interpenetrating and inseparable normal fluid and superfluid components.[1] The superfluid carries no heat and hence no entropy. One consequence is the existence of two types of sound waves: the usual *first sound* density waves and *second sound* waves that carry entropy.

A section of this chapter is used to explore the theory of condensate and superfluid behaviour.

The normal component of He-II consists of the atoms involved in thermal excitations from the superfluid. These excitations are studied in low energy neutron scattering experiments. Neutron scattering at higher energies has been used to probe individual atoms, making it possible to detect the fraction of atoms that are at rest and hence can be assigned to the condensate. A section is devoted to describing the results of these experiments.

A basic equality connects the superfluid velocity to the gradient of the phase of the wavefunction of the condensate: $\mathbf{v}_s = (h/m)\nabla\phi$. One topological consequence is the existence of vortices in He-II around which the circulation is quantized. Within the core of a vortex there is no condensate or circulation, which avoids a singularity in flow at the axis of the vortex. Excitations can scatter from vortices and exert a frictional force between the superfluid and normal fluid. Disturbance of He-II generally produces a jumble of vortices so that slow, smooth changes are needed to exclude them. Vortices are discussed in the final section, starting with Vinen's discovery of the quantization of circulation.

14.2 Superfluid effects

The first observations of superfluidity were made by Allen and Misener and by Kapitza, in 1938, three decades after helium was first liquified. Allen and Misener connected two reservoirs of He-II filled to different depths by a single capillary around 1 m in length, either of radius 250 μm or 10 μm. In normal fluids the flow velocity depends strongly on the pressure gradient g_p and viscosity η:

$$v \propto g_p/\eta. \tag{14.2}$$

However Allen and Misener found that fluid velocity changed little on changing capillaries, and by only a factor three under a 55-fold change in pressure gradient. Treating the flow as normal the viscosity came out smaller than that of He-I by a factor of 1500. In fact the normal component is pinned to the wall of a narrow capillary by viscous drag while the superfluid component passes unimpeded through either pipe.

A demonstration of the effect of the superfluidity of He-II is shown in Figure 14.4. When the beaker is raised helium flows out of the beaker through the surface layer shown until the levels inside and outside the beaker are equal. The attraction between liquid and beaker surface establishes this continuous film, which is about 30 nm (100 atoms) thick, and this layer acts like a siphon. If the beaker is lifted clear of the surface of the reservoir the liquid flow continues under the beaker and drips into the reservoir, in time emptying the beaker. Figure 14.5 exhibits the *fountain effect* which shows that the superfluid carries no heat and thus no entropy. A narrow open tube is dipped into a bath of He-II, its lower end closed by a superleak.[2] The superfluid component can pass freely through the fine holes while the normal component is clamped by viscous drag. When the metal powder is heated by incident radiation a stream of liquid helium shoots out of the upper end of the tube. In this unexpected, but nonetheless equilibrium state, the free energy G does not change across the superleak. This means that

$$\Delta G = -V\Delta P + S\Delta T = 0, \tag{14.3}$$

where ΔP and ΔT are the respective pressure difference and temperature difference across the superleak, and S is the entropy carried solely by the normal component in a volume V of He-II. It is the pressure differential

$$\Delta P = (S/V)\Delta T, \tag{14.4}$$

which drives the fountain.

When the ambient pressure above a bath of He-I is reduced it naturally boils. If the temperature is then lowered through T_λ the boiling ceases abruptly and the fluid is quiescent.[3] This is because below T_λ the superfluid component of He-II flows instantaneously to equalize the

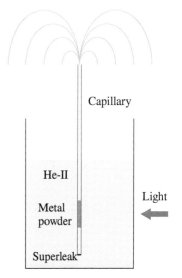

Fig. 14.5 The fountain effect.

[2]A superleak could be a fine powder or a Nuclepore membrane. This is a flexible polycarbonate membrane that has been exposed to charged particles from radioactive decays and then etched to produce an array of micron sized holes where the structure has been weakened by the passage of these particles.

[3]Hence the evaporation is entirely at the surface.

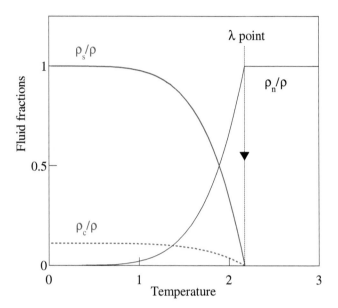

Fig. 14.7 Fractions of normal fluid, superfluid and condensate in He-II.

entropy density and temperature across the fluid. Such behaviour illustrates that the two phases, superfluid and normal, are interpenetrating and not physically separable.

14.2.1   Andronikashvili's experiment

Andronikashvili, in designing an experiment to measure the proportions of the two components of He-II, took advantage of the fact that the normal component can be pinned in narrow spaces where the superfluid component flows freely. His apparatus is sketched in Figure 14.6. A stack of closely spaced identical disks, suspended from one wire is immersed in He-II. Andronikashvili measured the angular frequency, ω of torsional oscillation as a function of temperature. ω equals $\sqrt{\Gamma/I}$, where I is the moment of inertia of the suspended system about the wire axis, and Γ is the torque per unit angular displacement due to the wire. The mass rotating includes that mass of disks plus the mass of normal He-II dragged along between the disks. Andronikashvili's measurement determined the the mass of normal fluid: dividing this by the total mass in the volume between the disks gives the normal fraction. Figure 14.7 shows the measured proportions of normal and superfluid as a function of temperature based on modern measurements. Near 0 K the normal fluid fraction is proportional to T^4. For reference the fraction of the condensate is shown as a broken line. Above T_λ the condensate and superfluid fractions vanish. At the other limit, 0 K, the superfluid

He-II

Rigidly rotating
vane structure

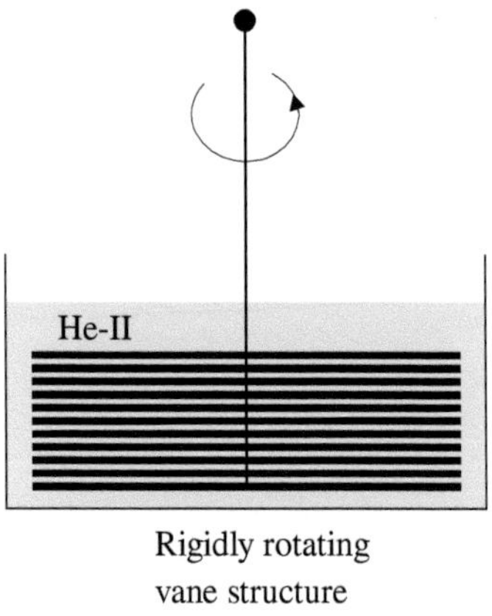

Fig. 14.6 Andronikashvili's experiment to determine the superfluid fraction of He-II.

fraction has risen to 100 per cent, while the condensate fraction is only around 10 per cent. We shall see later that interactions not only cause this difference but also account for why the superfluid fraction inherits properties of the condensate.

14.2.2   Sound and second sound in He-II

The sound waves excited by a piston in He-II are pressure waves in which the normal and superfluid components travel in phase as shown in the upper panel of Figure 14.8. Peshkov discovered that it is also possible to excite entropy waves in which the two components move exactly out of phase, while the density remains constant. These *second sound* waves are illustrated in the lower panel of Figure 14.8. Both sound velocities in He-II were determined by Heiserman *et al.*[4] They measured the resonant frequencies of plane wave modes in closed cylinders having small enough radii that modes with complex radial structure were suppressed. At one end a capacitively coupled diaphragm acts as a piston to excite pressure (sound) waves and at the other end a similar capacitively coupled diaphragm senses the pressure wave. Frequencies at which standing waves develop were determined. If the cylinder is of length L and the frequency is f then the velocity of first sound is

$$c_1 = 2Lf/m, \tag{14.5}$$

where m is an integer. In the measurement of second sound, that is heat/entropy waves, the transducers' diaphragms were Nuclepore membranes. These are impervious to normal fluid but transparent to superfluid. Consequently the waves produced are heat waves in which the normal and superfluid components move in antiphase. The hydrodynamic formulae for the velocities are

$$
\begin{aligned}
c_1^2 &= \left[\frac{\partial P}{\partial \rho} \right]_S, \\
c_2^2 &= \frac{\rho_{\rm s} T S^2}{\rho_{\rm n} C},
\end{aligned}
$$

where ρ, $\rho_{\rm s}$ and $\rho_{\rm n}$ are the respective densities, C is the heat capacity per unit mass, T the temperature, P the pressure and S the entropy per unit mass. The measured velocity of normal sound is close to $240\,{\rm m\,s^{-1}}$ from $0\,{\rm K}$ to T_λ. The velocity of second sound plateaus at $20\,{\rm m\,s^{-1}}$ between 1 and 2 K, then rises to $\sim 100\,{\rm m\,s^{-1}}$ at lower temperatures.

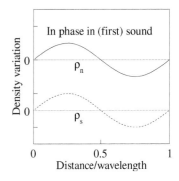

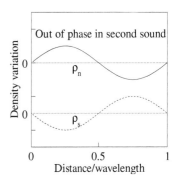

Fig. 14.8 Wave motion of the normal and superfluid components in first and second sound.

[4]J. Heiserman, J. P. Hulin, J. Maynard and I. Rudnick, *Physical Review* B14, 3862 (1976).

14.3   Condensate-linked properties

The properties of the condensate in He-II are explored here analytically, paying attention to the implications for the properties of the much larger superfluid component of He-II. First a statistical argument is made to show that if the equivalent short range potential between atoms is positive this assists condensation into a single state.

The interaction between the ^{4}He atoms is modelled by a point-like potential as in eqn. 7.72

$$U(\mathbf{r}) = U_0\delta(\mathbf{r}) = \frac{4\pi\hbar^2 n a_{\mathrm{s}}}{m}\delta(\mathbf{r}), \qquad (14.6)$$

where a_{s} is the s-wave scattering length, m the atomic mass and n the number density of atoms. When two bosons, a and b, lie in the same quantum state χ, the interaction energy is

$$U_0\int |\chi(\mathbf{r}_a)\chi(\mathbf{r}_b)|^2\delta(\mathbf{r}_a - \mathbf{r}_b)\mathrm{d}\mathbf{r}_a\mathrm{d}\mathbf{r}_b = U_0\int |\chi(\mathbf{r})|^4\mathrm{d}\mathbf{r}, \qquad (14.7)$$

whereas if they are in different states χ_1 and χ_2 the energy is

$$U_0\int \delta(\mathbf{r}_a - \mathbf{r}_b)\left|\left[\{\chi_1(\mathbf{r}_a)\chi_2(\mathbf{r}_b) + \chi_2(\mathbf{r}_a)\chi_1(\mathbf{r}_b)\}/\sqrt{2}\right]\right|^2\mathrm{d}\mathbf{r}_a\mathrm{d}\mathbf{r}_b$$

$$= 2U_0\int |\chi_1(\mathbf{r})|^2|\chi_2(\mathbf{r})|^2\mathrm{d}\mathbf{r}, \qquad (14.8)$$

where a crucial factor two emerges because the state is symmetrized. Hence if the potential, U_0 is positive, which is the case for the short range repulsion between helium atoms, the lowest energy state is that with both atoms in the same quantum state. This favours condensation. Equally an attractive short range potential would inhibit condensation.

For an n-boson pure state with wavefunction $\psi(\mathbf{r}, \mathbf{r}_2, \cdots \mathbf{r}_n; t)$ the single boson density matrix is defined to be

$$\rho_1(\mathbf{r}, \mathbf{r}'; t) = \int \psi^*(\mathbf{r}, \mathbf{r}_2, \cdots \mathbf{r}_n; t)\psi(\mathbf{r}', \mathbf{r}_2, \cdots \mathbf{r}_n; t)\mathrm{d}\mathbf{r}_2 \cdots \mathrm{d}\mathbf{r}_n$$

$$= \langle \hat{\psi}^\dagger(\mathbf{r}; t)\hat{\psi}(\mathbf{r}'; t)\rangle.$$

This specifies the correlation between a boson's wavefunction (complex amplitude) at locations $\mathbf{r}$ and $\mathbf{r}'$. $\psi(\mathbf{r}; t)$ is a single particle wavefunction and $\hat{\psi}(\mathbf{r}; t)$ the corresponding annihilation operator. The limit at points remote from one another

$$\lim_{|\mathbf{r} - \mathbf{r}'| \to \infty} \rho_1(\mathbf{r}, \mathbf{r}'; t) \qquad (14.9)$$

measures the degree of what is called *off-diagonal long range order* in the fluid described by the wavefunction.[5] When non-interacting bosons form a condensate $\rho_1(\mathbf{r}, \mathbf{r}'; t)$ will be exactly unity for all separations Δr.

[5]Diagonal indicates that $\mathbf{r}' = \mathbf{r}$, off-diagonal means they differ.

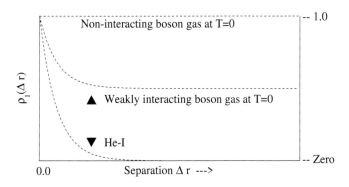

Fig. 14.9 The spatial dependence of the one-particle density matrix for three cases: an ideal non-interacting boson gas, a weakly interacting boson gas and He-I, both at 0 K.

However in an interacting boson system such as He-II the limit is smaller but still finite, while in He-I $\rho_1(\mathbf{r}, \mathbf{r}'; t)$ will quickly fall off to zero over a few atomic spacings. Such behaviours are presented for comparison in Figure 14.9. Off-diagonal long-range order is the hallmark of a condensate, while the weakly interacting boson example also shown in the figure indicates the behaviour of the superfluid component of He-II.

In He-II the interactions between atoms eject atoms from the condensate and this keeps the fraction of atoms in the condensate down to a maximum of ~ 0.1 at 0 K. However these interactions provide the mechanism by which superfluid flow can extend to a larger proportion of He-II atoms, reaching 100 per cent at 0 K. The loss of condensate at 0 K due to interactions is known as quantum depletion: the further loss that comes with thermal excitation as the temperature rises is called thermal depletion.

Bogoliubov analysed weakly interacting boson fluids below the condensation temperature making two principal approximations. First, the wavefunction of the condensate with number density n_0 can be expressed as a classical wave of amplitude $\sqrt{n_0}$; second, the only interactions are those in which pairs of bosons with equal and opposite momentum leave or enter the condensate[6]. His analysis is only indicative for He-II because the interactions between the ^{4}He atoms are not that weak. Predictions about low energy phonon-like excitations are quantitatively correct, but the higher energy excitations (rotons) are not predicted.

Expressing the condensate as a classical wave

$$\Psi(\mathbf{r}, t) = \sqrt{n_0}\psi(\mathbf{r}, t) = |\Psi(\mathbf{r}, t)| \exp[i\phi(\mathbf{r}, t)], \qquad (14.10)$$

with $\phi(\mathbf{r}, t)$ being the phase of the condensate. It follows that the condensate number density

$$\rho_c = |\Psi(\mathbf{r}, t)|^2 \qquad (14.11)$$

[6] This would guarantee that correlations existing in the condensate and the associated superflow both propagate to the rest of the He-II fluid.

and current (boson flux per unit time), using eqn. 1.27 is

$$\mathbf{j}_c = |\Psi|^2 \frac{\hbar}{m} \nabla \phi(\mathbf{r}, t), \tag{14.12}$$

giving a condensate and *superfluid* velocity

$$\mathbf{v}_s = \mathbf{j}_c / \rho_c = \frac{\hbar}{m} \nabla \phi(\mathbf{r}, t). \tag{14.13}$$

This is a crucial result: it links the condensate phase to a macroscopic observable.

14.4   Topological implications

Following a closed path in the condensate returns the wavefunction to its initial value and this requires that the phase ϕ returns to its initial value *or* changes by an integral multiple, n of 2π. This is called the *Onsager–Feynman* condition. It is equivalent to Bohr's original quantization condition for electron orbits in hydrogen, and has the same logical basis: wavefunctions are single valued at any point. Any other choice gives destructive interference and hence no physical state. Thus

$$\oint \nabla \phi \cdot d\mathbf{r} = 2n\pi. \tag{14.14}$$

Then the *circulation* defined as the integral of the velocity round a closed path is, using eqn. 14.13

$$\oint \mathbf{v}_s \cdot d\mathbf{r} = n\frac{h}{m} = n\kappa_0. \tag{14.15}$$

Circulation is therefore quantized in units of h/m (κ_0) in the condensate and so by extension in the superfluid.

There is a fundamental distinction between quantization of the orbital angular momentum quantum numbers in for example the hydrogen atom and quantization of the *winding number n* in the last equation. When there is rotational symmetry the angular momentum associated with wavefunctions is quantized, but not otherwise. On the other hand the quantization condition expressed in eqn. 14.15 is topological: it does not rely on rotational symmetry, and the winding number is unaffected by changes in the path shape that do not leave the condensate.

From eqn. 14.13 it follows that superfluid flow is *irrotational*, that is to say

$$\nabla \wedge \mathbf{v}_s = (\hbar/m)\nabla \wedge \nabla \phi = 0. \tag{14.16}$$

This appears to pose a problem because using Stokes' law, eqn. A.21, and writing $d\mathbf{A}$ for an element of area, we have

$$\oint \mathbf{v}_s \cdot d\mathbf{r} = \int (\nabla \wedge \mathbf{v}_s) \cdot d\mathbf{A} \tag{14.17}$$

which would make the loop integral vanish! However, the integral can be non-zero if within the area enclosed by the loop there is a region where there is no condensate. This can happen in one of two ways. Either the He-II can flow round inside a closed, for example, toroidal container with an obvious hole. Alternatively, there can be vortices in the He-II within whose cores order is lost. In cases where the surface enclosed by the loop has one or more holes mathematicians say the space is multiply connected, and if there are no holes they call it simply connected. The loop integral only vanishes if its area is free of holes. More generally the circulation is $n\frac{h}{m}$ with n being the number of holes the loop encloses. Notice that when it is possible to smoothly close up the loop to a point without crossing a hole then the value of the loop integral is zero and n is zero. Examples of both types of hole are investigated below.

14.5 Spontaneous symmetry breaking

Phase coherence across a condensate parallels the coherence of magnetization across a ferromagnet. The phase ϕ of the condensate is the equivalent of the orientation angle of the magnetization. In the case of a ferromagnet the Hamiltonian, the energy operator, is entirely independent of the orientation (θ, ϕ) of the material, whereas the ground state has all the atomic magnets aligned parallel in some preferred direction. The symmetry is still present in this sense: equivalent ground states differ only in the orientation of magnetization and have the same energy. This is known as *spontaneous symmetry breaking*.

In finite ferromagnets the states with alternative orientations belong to the same Hilbert space and can be superposed. However, if the space is infinite the states with different orientations of magnetization are the ground states of separate Hilbert spaces. We can understand the difference between the finite and infinite systems by considering the practical process by which the orientations would be changed, for example taking one atom at a time and rotating it. Then evidently the energy barrier between ground states is only penetrable in finite time for finite systems. As a consequence, only in a finite system are states with differing orientations of magnetization in the same Hilbert space.

In the case of He-II the phase symmetry possessed by the Hamiltonian is broken in the condensate: the value of the phase at one location fixes that at any other location. Where it does vary, its divergence is the local superfluid velocity. A global shift in phase across the whole superfluid leaves every observable unchanged, and is therefore of no physical significance. This recalls the analysis of the phase of charged particles appearing in Section 13.3. In the present case however, there is nothing corresponding to the minimal electromagnetic coupling which would make it possible to build a local gauge theory.

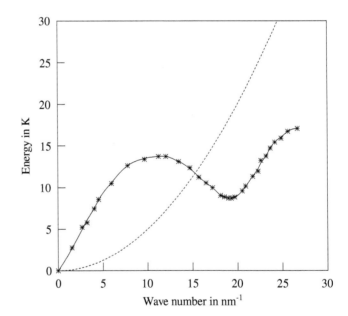

Scattered
neutron

Incident
neutron

excitation

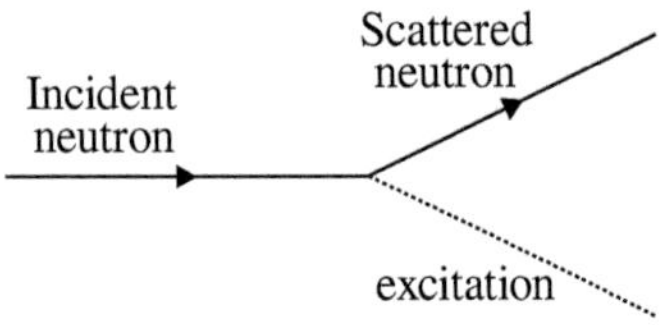

Fig. 14.10 Excitation produced by an incident low energy neutron.

[7]In Chapter 4 the conseqences of a kinematic mismatch were discussed for measurements on phonons.

14.6   Excitations

The preferred process used to probe the excitations in He-I and He-II is low energy neutron scattering. Figure 14.10 shows the dominant process in which an excitation is produced. The energy and momentum of the excitation are simply the energy and momentum lost by the neutron: therefore these observables are measured for the incoming and outgoing neutron. One reason that neutrons are well suited as probes is that they have a long mean free path in matter compared to X-rays so that the bulk properties are probed. Another reason is that because the neutron mass is comparable to that of a ^{4}He atom it follows that the fractional uncertainty in the measurement of the neutron energy/momentum) leads to a comparable fractional uncertainty in the excitation energy/momentum.[7] Reactors produce copious low energy neutrons: these are first collimated and then momentum selected by Bragg scattering from a crystal. The

Fig. 14.11 Excitation spectrum measured at 1.12 K using incident neutrons with wavelength 0.404 nm. The broken line is the free ^{4}He atom dispersion curve. Adapted from Figure 4 in D. G. Henshaw and A. D. B. Woods, *Physical Review* 121, 1266 (1961).

outgoing neutrons after scattering from He-II have a distribution of directions and energies. Collimation and Bragg scattering are used to select and count the neutrons within a narrow range of vector momenta, and so build up a spectrum. In a systematic study Henshaw and Wood[8]

[8]D. G. Henshaw and A. D. B. Woods, *Physical Review* 121, 1266 (1961).

measured the spectrum of excitations at $1.1\,\text{K}$, shown here in Figure 14.11. This displays the distribution of the mean energy of the excitations versus wavevector, $k = p/\hbar$, p being the excitation momentum. There is a narrow spread around the mean energy, which argues, using the uncertainty relation, that the excitations are long-lived and travel $\sim 1\,\text{cm}$ in the fluid. The dispersion relation for the excitations is thus surprisingly simple for a fluid. Beyond $20\,\text{nm}^{-1}$ the energy spread broadens into a diffuse band indicating that excitations are short lived. The broken line curve in Figure 14.11 is the dispersion relation for free helium atoms

$$E = p^2/2M_{\text{He}}. \tag{14.18}$$

The features of the excitation spectrum are broadly consistent with Landau's interpretation. At low wave numbers the excitations are phonons having a dispersion relation

$$E = cp, \tag{14.19}$$

where c is the quasi-particle velocity, E its kinetic energy and p its momentum. The slope of this region is nearly constant making the phase velocity and group velocity very similar with values around $240\,\text{m s}^{-1}$. This agrees with the measured velocity of sound. Not shown in Figure 14.11 is the intensity of the scattering which reflects the relative number of quasi-particles in He-II. The intensity is largest at the pronounced dip and falls off rapidly away from this point along the curve. Landau called the copious excitations around the dip *rotons*. A crude picture to distinguish between phonons and rotons is this: the phonon is a quasi-particle made up of the atom (struck by the neutron) and a cloud of atoms travelling with it; in the roton the atom is travelling with a cloud of atoms that flow around it, from front to back. Fitting Landau's inferred dispersion relation for rotons,

$$E = \Delta + \frac{(p - p_0)^2}{2\mu}, \tag{14.20}$$

to the data gives: $\Delta = 8.65\,\text{K}$ or $0.745\,\text{meV}$, $p_0/\hbar = 19.1\,\text{nm}^{-1}$ and $\mu = 0.16 M_{\text{He}}$. Rotons have, in some sense, the kinematics of quasi-particles of mass μ and momentum $\pi = p - p_0$: those with positive (negative) π being called R^+ (R^-) rotons. Rotons have been studied by generating heat pulses and detecting the atoms they eject on meeting the He-II surface. Viewed as refraction at a surface with the roton being the incident ray and the atom the outgoing ray, some atoms emerge at negative angles of refraction. In these cases the interpretation offered is that the rotons are R^- rotons.[9] It is noteworthy that when ^{4}He crystallizes under pressure the lattice spacing is $0.3\,\text{nm}$, giving a first Bragg peak at $20\,\text{nm}^{-1}$. The similarity to $p_0/\hbar$ suggests that the correlations of the solid state are appearing already in the liquid state as rotons.

[9]See M. A. H. Tucker and A. F. G. Wyatt, *Science* 283, 1150 (1999).

As the temperature is raised toward T_λ the intensity of roton production weakens, finally disappearing in He-I. On the other hand the intensity for producing phonons continues unaffected across the λ-point. Formalism like that deployed in Chapter 4 to calculate the heat capacity reproduces the measured specific heat of He-II using just the roton and phonon contributions at 1 K, showing that these are the significant excitations. It is now reasonable to draw the important conclusion that the normal component of He-II is made up of excitations, or more precisely atoms undergoing these excitations. As noted, when the temperature rises the excitations contribute to thermal depletion of the condensate, on top of the quantum depletion due to the mutual interactions of ^{4}He atoms at 0 K.

14.7   Measurement of the condensate fraction

In a stationary volume of He-I cooled carefully below T_λ the atoms in the condensate are nearly at rest, their mutual weak interaction increases their motion from the zero point motion of an ideal condensate. Inelastic neutron scattering is used to elicit information on the existence and properties of the condensate. The kinematic regime is different from that involved in detecting excitations. The duration of the interaction between neutron and atom should be sufficiently short so that the atom recoils freely. This requires a momentum exchange much larger than the roton momentum. Such an *impulsive* process is shown in Figure 14.12. A representative experiment is that of Woods and Sears.[10] The authors found that the momentum spectrum of helium atoms changes negligibly between 4.2 and 2.2 K, so they inferred that this should replicate the spectrum of non-condensate ^{4}He atoms in He-II. Figure 14.13 shows the outcome of subtracting this presumed spectrum of the normal component from the measured spectrum at 1.1 K. The spike at zero wave-vector expected for stationary condensate atoms is broadened due to the experimental resolution and residual interactions. Data of this and similar experiments has been used to produce the plot of the condensate fraction seen in Figure 14.7.

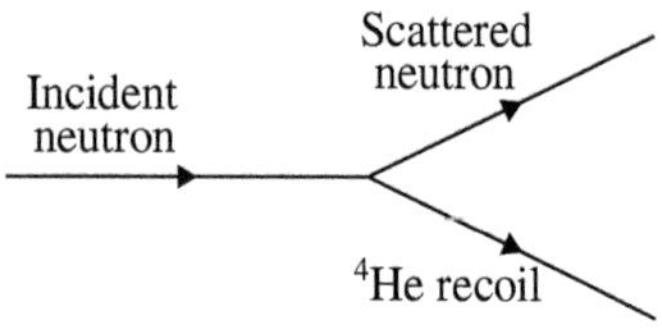

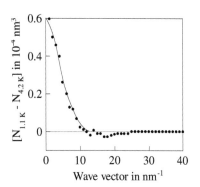

Fig. 14.12 Neutron scattering from a ^{4}He atom in He-II.

Fig. 14.13 Change in wave-vector distribution between 4.2 K and 1.1 K in He-II. The curve indicates the condensate peak. The quoted resolution is $3\,\text{nm}^{-1}$. Adapted from Figure 2: A. D. B. Woods and V. F. Sears, *Physical Review Letters* 39, 415 (1977).

14.8   Critical velocity

If the velocity of He-II relative to its container is large enough then collisions with the container can produce quasi-particles, which simply convert superfluid to normal fluid, and the superflow ceases. The inverse situation of an object of mass M much greater than that of a helium atom moving through stationary He-II is entirely equivalent and easier to analyse. Suppose the moving mass has velocity before (after) exciting a quasi-particle $\mathbf{v}$ ($\mathbf{v}_f$) and that the excitation has momentum $\mathbf{p}$ and

[10]A. D. B. Woods and V. F. Sears, *Physical Review Letters* 39, 415 (1977).

energy E. Then conservation of energy and momentum require that

$$Mv^2/2 = Mv_f^2/2 + E,$$
$$M\mathbf{v} = M\mathbf{v_f} + \mathbf{p}.$$

Eliminating $\mathbf{v_f}$

$$E - \mathbf{p} \cdot \mathbf{v} + p^2/(2M) = 0, \qquad (14.21)$$

where the third term is tiny and can be ignored: thus $E = \mathbf{p} \cdot \mathbf{v}$. For it to be possible to produce any excitation at all we must therefore have

$$v > E/p \qquad (14.22)$$

otherwise $\mathbf{p} \cdot \mathbf{v}$ is always less than E. It follows that superfluid flow is only possible if the velocity is smaller than the lowest value of E/p for any excitation. This critical velocity is

$$v_c = (E/p)_{min}. \qquad (14.23)$$

From Figure 14.11 we can see that it is the excitation of rotons that sets the critical velocity, which is $58\,\mathrm{m\,s^{-1}}$. Measurements by Allum and colleagues[11] showed that negative ions travelling in He-II did not suffer any drag until their velocity reached $48\,\mathrm{ms^{-1}}$. For flow in general, the critical velocity is much lower, typically cm/sec: this is because vortices develop in the superfluid at surface irregularities, and excitations exert a frictional force on them.

The dispersion relation for free ^{4}He atoms in a non-interacting condensate would follow the broken line in Figure 14.11, $E = p^2/2M$. In that case the critical velocity is zero and superfluid motion could not happen.

14.9 Quantized circulation

The quantization of circulation in He-II was first demonstrated in an elegant experiment by Vinen,[12] with the apparatus sketched in Figure 14.14. A 5 cm long, 25 µm diameter wire was stretched along the axis of a vertical cylinder filled with liquid ^{4}He. Circulation was established by rotating the whole apparatus around the vertical axis for 20 min at a temperature above T_λ. Then it was slowly cooled to 1.3 K taking 30 min. After this the cylinder was halted with the result that the normal component of He-II came to rest while the superfluid continued rotating about the wire. A 0.3 T magnetic field was applied transverse to the rotation axis. A brief pulse of current sent through the wire set the wire vibrating in its lowest mode around 500 Hz. The motion of the wire in

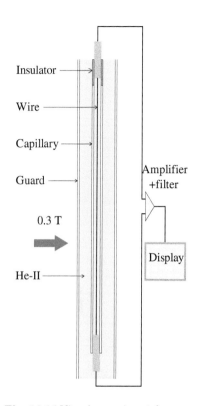

Fig. 14.14 Vinen's experiment demonstrating the quantization of circulation in He-II. Adapted from Figure 1, W. F. Vinen, *Proceedings of the Royal Society* A260 218 (1961).

[11]D. R. Allum, P. V. E. McClintock, A. Phillips and R. M. Bowley, *Philosophical Transactions of the Royal Society London* 248, 179 (1977).
[12]W. F. Vinen, *Proceedings of the Royal Society* A260 218 (1961).

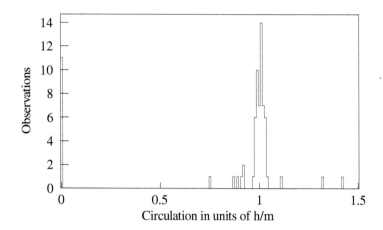

Fig. 14.16 Circulation in He-II in units of h/m. Adapted from Figure 6: W. F. Vinen, Proceedings of the Royal Society A260, 218 (1961).

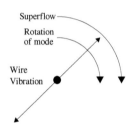

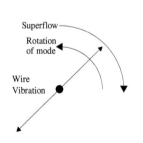

Fig. 14.15 View along the axis of rotation.

the magnetic field induced a voltage across the wire, which was amplified, frequency filtered to select the lowest mode of vibration, and then displayed. The crucial feature of the experiment is that the circulating superfluid exerts a force on the wire, which is in opposite senses for the two circularly polarized components of the wire motion. The two cases are shown in Figure 14.15. This *Magnus force* breaks the degeneracy between the modes, separating them in frequency by

$$\Delta f = \rho_{\mathrm{s}} \kappa / (2\pi m), \tag{14.24}$$

where ρ_{s} is the density of the fluid circulating, in this case the superfliud, m is the mass per unit length of the wire plus half the mass of the fluid it displaces. The circulation appearing here, κ, would be single valued if the whole wire had the same velocity. Instead the velocity is proportional to $\sin(\pi x/L)$ at a point a distance x along the total length L. Thus the equivalent circulation to be inserted in the place of κ in eqn. 14.24 is

$$\frac{2\kappa}{L} \int \sin^2(\pi x/L)\mathrm{d}x \tag{14.25}$$

integrated along the wire. The beat frequency, Δf of the voltage across the wire was measured, and from this the circulation κ was extracted using eqn. 14.24. Early measurements yielded a random distribution of values of the circulation. Vinen deduced that the cause was that the vortex in the superfluid was only attached to the wire over part of the wire length and then trailed off to terminate on the cylinder wall.He was able to detach these partially attached vortices by pulsing a voltage applied between the ends of the wire. Valid measurements with attachment of the vortex along the full length of the wire were therefore those for which repeated pulsing did not alter the value of the circulation.

These values are shown in Figure 14.16. They cluster around the value

$$\kappa = \kappa_0 \equiv h/m, \tag{14.26}$$

demonstrating that the rotating superfluid carries one quantum of circulation. Similar studies have shown excitation of multiple quanta of circulation.

The picture of superflow presented here implies that in an equilibrium state superfluid flow persists indefinitely. This conclusion was tested by Reppy and Depatie[13] in a similar setup to Vinen's. Liquid ^{4}He was contained in a closed vertical cylinder which was suspended on a magnetic bearing; the cylinder being divided internally by closely spaced mica disks to inhibit the flow of normal He-II fluid. After cooling below T_λ the cylinder was set spinning rapidly on its axis, well above the critical velocity, so that all the fluid rotated with the cylinder. Next the cylinder was gently brought to rest and locked in place magnetically. After a short time the normal component would come to rest and, it is presumed, the superfluid component would continue to rotate.

The authors' test of this hypothesis was to wait for a further interval, then release the brake and use a brief intense pulse of light to heat the cylinder and its contents above T_λ. The heat pulse converts the rotating superfluid instantaneously to normal fluid. Applying the principle of the conservation of angular momentum, it follows that angular momentum previously carried by the rotating superfluid must transfer to the cylinder and its contents. Reppy and Depatie observed the recoil of the cylinder under this impulse. Crucially they found no detectable reduction in the magnitude of the recoil despite waiting for intervals as long as 12 hours between halting the cylinder and applying the heat pulse. This convincingly demonstrated the long term persistence of superfluid flow. The angular momentum recoil and hence superfluid fraction was found to depend on the initial temperature of the He-II as shown in Figure 14.17. This is consistent with other measurements of how the fraction of superfluid in He-II varies with temperature.

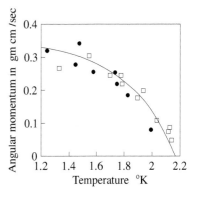

Fig. 14.17 Persistent current angular momentum versus temperature. The squares (circles) indicate that after the formation of the current the temperature was held constant (lowered) before measurement. The curve is the function $0.338\rho_s/\rho$ versus temperature. Adapted from Figure 2 in J. D. Reppy and D. Depatie, Physical Review Letters 12, 187 (1964).

14.10   Vortices

We have seen that the circulation of superfluid helium around a closed loop is quantized so that

$$\oint \mathbf{v_s} \cdot \mathrm{d}\mathbf{r} = nh/m \equiv n\kappa_0, \tag{14.27}$$

where m is the mass of the ^{4}He atom and n is an integer: this integer is zero if the whole area enclosed by the loop contains stationary He-II. A simple vortex involves uniform circular motion around the axis of

---

[13]J. D. Reppy and D. Depatie, *Physical Review Letters* 12, 187 (1962).

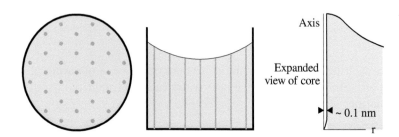

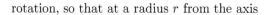

Fig. 14.18 Rotating superfluid in plan view and in vertical section with an expanded view of a single vortex.

rotation, so that at a radius r from the axis

$$v_{\rm s} = n\frac{\hbar}{mr}. \tag{14.28}$$

The superfluid velocity increases as the axis of rotation is approached: once the critical velocity $v_{\rm c}$ is exceeded quasi-particles are excited so that the fluid reverts to normal. Thus a vortex has a core within which order is lost. An estimate of the core radius is $\hbar/(mv_{\rm c})$ or $0.28\,$nm. When He-I is set rotating in a cylindrical container and then cooled through the λ-point the surface retains the concave shape of a normal fluid. What happens is that the angular momentum of the superfluid component is concentrated in vortices. Their mutual repulsion leads to a uniform distribution, and the pattern they form rotates with the normal fluid. Suppose that the container, radius R, rotates with angular velocity ω, then the circulation round the outer rim is $(2\pi R)(\omega R)$ or $2\pi\omega R^2$. Therefore, the corresponding number of vortices needed to carry the circulation is

$$N_{\rm v} = 2\pi\omega R^2/\kappa_0 = 2\pi\omega R^2 m/h, \tag{14.29}$$

and the count per unit area is

$$n_{\rm v} = 2\omega m/h. \tag{14.30}$$

The mutual interaction between vortices in the superfluid and excitations making up the normal fluid brings the normal fluid and vortices into an equilibrium pattern which is idealized in Figure 14.18. Vortices carrying a single quantum of circulation are the rule because a set of such vortices has a lower total energy than one vortex carrying all the quanta.

As noted earlier vortices formed at surface irregularities feel a frictional force through their mutual interaction with the quasi-particle excitations. This mechanism imposes a limit to superfluid flow at velocities far below the critical velocity calculated from the roton dip. At some threshold angular velocity a first vortex forms at the surface and quickly moves to the rotation axis. Angular acceleration beyond this threshold increases the number of vortices to fill the volume, and eventually flow becomes turbulent.

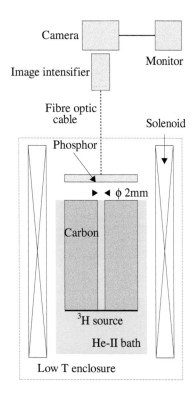

Fig. 14.19 Sketch of apparatus for recording images of vortices in He-II. Adapted from Figure 1 E. J. Yarmchuk and R. E. Packard, *Journal of Low Temperature Physics* **46**, 479 (1981).

The vortices can be pinpointed by *decorating* them with electrons, using a technique pioneered by Packard. Figure 14.19 displays the components used by Yarmchuk and Packard.[14] In order make it easy to record data with a rotating fluid all components within the broken line are rotated about the vertical axis as one rigid structure. This guarantees that the images recorded are observed in the rest frame of the rotating fluid. The tritium source projects electrons into the volume of He-II contained in a tube drilled through a stack of carbon resistors. A potential of $500\,\mathrm{V}$ between the source and the phosphor detector guides the electrons through the tube and onto the detector. The electrons are attracted to the cores of the vortices and then travel along these cores. A solenoid provides a $0.5\,\mathrm{T}$ magnetic field that suppresses electron defocusing along the tube. Electrons leaving the He-II surface strike the phosphor and imprint an image of the pattern of vortices. This image is transfered by a fibre optic array to a region at room temperature where it can be intensified and displayed. Figure 14.20 shows typical results taken at $10\,\mathrm{s}$ intervals over $10\,\mathrm{min}$, $10\,\mathrm{s}$ being the time required to charge a vortex. Turbulence in He-II is made up of tangled vortices. When $^4\mathrm{He}$ is cooled rapidly through T_λ or set in motion the most likely result is a network of tangled vortices. Achieving an array of untangled vortices relies on making smooth, slow changes. Another complication is the appearance of vortices stretching from one wall of the container to another wall; vortices may also form closed loops.

14.11   Further reading

Superconductivity, Superfluidity and Condensates by J.F. Annett, published by Oxford University Press (2004) contains a readable discursive account of these three related topics.

Superfluidity and Superconductivity, second edition, by D. R. Tilley and J. Tilley, published by Adam Hilger (1986) is an older, more formal account with copious experimental detail of important experiments.

Quantum Liquids: Boson Condensation and Cooper Pairing in Condensed Matter Systems by A. J. Leggett, published by Oxford University Press (2006) is a more demanding text by a Nobel Prize winner in the field. It repays time spent on it.

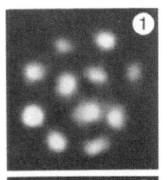
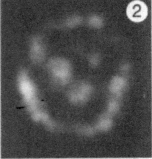
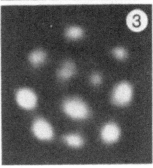

Fig. 14.20 Decoration of vortices in He-II. Figure 3 from E. J. Yarmchuk and R. E. Packard, *Journal of Low Temperature Physics* 46, 479 (1981). Courtesy Professor Packard and Springer.

[14] E. J. Yarmchuk and R. E. Packard, *Journal of Low Temperature Physics* 46, 479 (1981).

Exercises

(14.1) The neutrons used to study excitations in He-II have thermal energies ($\sim 0.025\,\text{eV}$). Does the neutron wavefunction extend over one or more ^{4}He nuclei in He-II?

(14.2) Show that the superfluid flow round a vortex is incompressible, that is that $\nabla \cdot \mathbf{v_s} = 0$.

(14.3) Assume that the mutual interaction between ^{4}He atoms in He-II can be approximated by a point contact potential $U(\mathbf{r}) = W_0\delta(\mathbf{r})$. Make a first approximation for the energy density. Hence show that the pressure is

$$P = W_0 n^2/2,$$

where n is the number of atoms per unit volume. Hence determine the velocity of sound using the expression

$$c_1 = \sqrt{\partial P/\partial \rho},$$

ρ being the density. Now use the measured velocity to infer W_0. Take the value of U_0 to infer the equivalent scattering length for the interatomic interactions. Use eqn. 7.71.

(14.4) By differentiating eqn. 14.16 with respect to time show that

$$\partial \mathbf{v_s}/\partial t = \nabla \Omega(\mathbf{r}, t),$$

where Ω is some scalar function of $\mathbf{r}$ and t. For He-II that is in uniform thermal equilibrium show that $\mathbf{v_s}$ must be constant in time.

(14.5) A stationary vortex is located in a cylinder of He-II of radius R. Show that the kinetic energy per unit length of such a vortex carrying one quantum of circulation, is

$$0.5\rho\frac{\kappa_0^2}{2\pi}\ln(R/a),$$

where ρ is the fluid density and a is the core radius.
Can you make a qualitative argument to explain why parallel vortices repel one another, while antiparallel vortices attract?

(14.6) Use eqn. 14.4 to deduce what column height of He-II is required at $1.4\,\text{K}$ to balance the effect of a $0.01\,\text{K}$ temperature difference in a fountain effect experiment. Take the specific entropy of He-II at $1.4\,\text{K}$ to be $131\,\text{J}\,\text{kg}^{-1}\text{K}^{-1}$.

(14.7) Show that the phonon number density at temperature T K is

$$N = 9.6\pi[k_\text{B}T/hc_1]^3$$

where c_1 is the speed of sound in He-II. The integral

$$\int_0^\infty \frac{x^2}{\exp(x) - 1}\,\mathrm{d}x = 2.404$$

and reference to the parallel topic in Chapter 4 will both be useful.

(14.8) Show that rotons with momentum less than p_0 in eqn. 14.20 have negative group velocity. Hence they travel opposite to the momentum they carry!

(14.9) A cylinder containing He-I rotates at $0.1\,\text{Hz}$ and is gently cooled through the λ-point and further to near $0\,\text{K}$. Calculate the number of vortices per unit area of the upper surface. What happens when the cylinder, still rotating, is warmed slowly to $1.9\,\text{K}$?

(14.10) Consider the following sequence applied to a closed toroid containing He-I. The toroid has a minimum radius of curvature very small compared to the maximum radius, R. Then to an excellent approximation the liquid forms a ring of radius R rotating about the symmetry axis perpendicular to its plane with angular velocity ω. The mass of helium is m. The toroid is now cooled to below the λ-point. What are the allowed velocites of the superfluid? What is the superfluid energy in the rotating frame? The conversion from laboratory energy E to that, E_r in a frame rotating with angular velocity $\boldsymbol{\omega}$

$$E_\text{r} = E - \boldsymbol{\omega} \cdot \mathbf{L} = E - \omega L,$$

where $\mathbf{L}$ is the angular momentum in the laboratory frame. The second equality holds because the vectors are parallel. For convenience set

$$\omega_\text{c} = \kappa_0/(2\pi) = \hbar/m.$$

If $\omega < \omega_\text{c}/2$ does the superfluid rotate? It helps to note that the equilibrium state will the one with the lowest value of the energy E_r.

Superconductivity

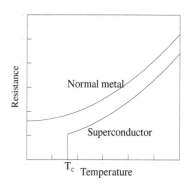

15.1 Introduction

In 1911 Kammerlingh-Onnes cooled mercury with the newly available liquid helium, and discovered that electric current flowed in mercury without resistance. Many other elements also become *superconducting* below their own *critical temperature*, T_c: warmed above T_c the behaviour returns to normal. Figure 15.1 contrasts the resistance at low temperatures of a normal metal like copper with that of a superconducting metal. Measurements of the magnetic field produced by a current circulating in a superconducting ring have shown that the lifetime of the current exceeds the current age of the universe. This supercurrent is analogous to the superflow of He-II met in the previous chapter and heralds that a condensate is present in a superconductor. However, electrons, the carriers of the supercurrent, are excluded from condensing in a lowest energy state by the Pauli principle! The paradox was resolved when it was realized that there can be a boson condensate, with the bosons taking the form of weakly bound *Cooper pairs* of electrons at the top of the Fermi sea. An energy gap, Δ, opens up between the condensate and excited electron (quasi-particle) states. The proportion of electrons in the condensate is of order one in 10^4 of those in the Fermi sea at $0\,\mathrm{K}$ and falls to zero above T_c.

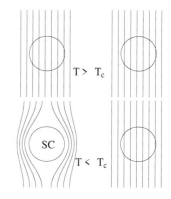

Another surprise is that in sufficiently small magnetic fields (which can be Teslas for some superconductors) the magnetic flux is expelled from a material when it becomes superconducting: a property detected much later in 1933 by Meissner and Ochsenfeld. They measured the magnetic field between two superconducting cylinders and found that below T_c the tubes became diamagnetic. All the superconducting current is confined to a ∼50 nm thick layer at the surface of the superconductor: in the interior the magnetic induction $\mathbf{B}$ is zero. Now recall that $\mathbf{B} = \mu_0(\mathbf{H}+\mathbf{M})$, where $\mathbf{H}$ is the magnetic field and $\mathbf{M}$ is the magnetization due to the surface current. It follows that $\mathbf{M} = -\mathbf{H}$, making a superconductor a perfect diamagnetic material. By contrast if a material becomes a perfect conductor as distinct from a superconductor, Faraday's law requires that $\mathrm{d}\mathbf{B}/\mathrm{d}t = 0$. Thus any existing magnetic flux is frozen in, rather than expelled. This distinction between a perfect conductor and a superconductor is brought out by the sequences shown in Figure 15.2.

A sufficiently large magnetic field, H_c, or a large enough current, dis-

Fig. 15.1 Resistance of a normal metal and a superconductor at low temperature.

Fig. 15.2 Superconductor (left) and perfect conductor (right) responses to an applied magnetic field. The upper panels show the materials initially above the critical temperature and in the lower panels after being cooled below the critical temperature.

Quantum 20/20: Fundamentals, Entanglement, Gauge Fields, Condensates and Topology.
Ian R. Kenyon. © 2020. Published in 2020 by Oxford University Press.
DOI: 10.1093/oso/9780198808350.001.0001

rupts the weakly bound Cooper pairs and terminates superconductivity. In type-I superconductors there is a single transition, illustrated in the upper panel of Figure 15.3. The energy difference between the normal and superconducting states is best expressed by the Gibbs free energy, G, which is the work available in a reversible process:

$$dG = -SdT - \mu_0 \mathbf{M} \cdot d\mathbf{H}, \qquad (15.1)$$

where S is the entropy. For a superconductor the change in Gibbs free energy between zero field and H_c, at a temperature T, is

$$G_s(T, H_c) - G_s(T, 0) = -\int_0^{H_c} \mu_0 \mathbf{M} \cdot d\mathbf{H}. \qquad (15.2)$$

In the superconducting phase $\mathbf{M} = -\mathbf{H}$, so that we have

$$G_s(T, H_c) - G_s(T, 0) = \mu_0 H_c^2/2. \qquad (15.3)$$

When the field value is H_c the normal and superconducting phases have the same Gibbs free energy:

$$G_n(T, H_c) = G_s(T, H_c). \qquad (15.4)$$

In the normal state the magnetic energy is negligible, so we also have

$$G_n(T, 0) = G_n(T, H_c). \qquad (15.5)$$

Using these last three equations together gives the difference in the Gibbs free energy between the two phases in the absence of any applied magnetic field

$$G_n(T, 0) - G_s(T, 0) = \mu_0 H_c^2/2, \qquad (15.6)$$

$\mu_0 H_c^2/2$ is therefore called the *condensation energy*. For lead $B_c = \mu_0 H_c$ is 0.08 T, making a condensation energy density of 2550 J m^{-3}. The lattice spacing in lead is 0.5 nm so it follows that the condensation energy per free electron is only 2 μeV. As pointed out, only one in 10,000 of those electrons in the Fermi sea enters the condensate. This boosts the binding to meV rather than μeV. However, electrons at the Fermi surface have eV energies and would be expected to shatter the weakly bound Cooper pairs. Evidently some subtle process is at work to protect a condensate with such weak binding.

There are also type-II superconductors in which the transition to superconductivity is not sharp: there exists an intermediate range of magnetic fields over which the field partially penetrates the superconductor. This behaviour is illustrated in the lower panel of Figure 15.3. In magnetic fields below H_{c1} there is no penetration and in fields above H_{c2} the material is normal. The ratio of H_{c2}/H_{c1} can be as high as 10^4.

In 1986 Bednorz and Müller discovered a new class of superconductors whose critical temperatures now range up to 138 K, that is above

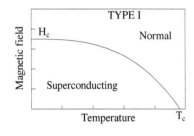

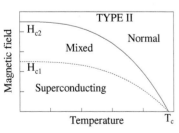

Fig. 15.3 Variation of the critical magnetic field with temperature for type-I and type-II superconductors.

the temperature at which nitrogen liquifies; the technical advantage of avoiding liquid helium is obvious. One group of these *high temperature superconductors* (HTSCs) has layered structures, in which the superconducting electrons appear to travel within layers containing CuO groups. Cooper pairs of electrons are the elemental bosons in both *classical* low temperature superconductors and HTSCs. HTSCs are less well understood than the classical superconductors whose transition temperatures lie below 30 K.[1] From here on, only the classical low temperature superconductors will be discussed.

Section 15.2 puts the London brothers phenomenological framework for describing superconductivity in modern dress. Bardeen, Cooper and Schrieffer discovered the underlying microscopic *BCS* theory of superconductivity in 1956-7[2] and this will be developed in Section 15.3. In common with superfluid He-II, the superconducting condensate has a macroscopic wavefunction that establishes long range order over the material. Again, as with He-II, the definite choice of phase breaks the symmetry possessed by the Hamiltonian (energy). This will lead in Section 15.4 to Ginzburg and Landau's analysis of symmetry breaking.[3] The superflow can form vortices, analogous to those in He-II, whose quantized circulation now carries electrical current. Section 15.5 explains how the vortex cores are threaded by a magnetic flux quantized[4] in units of $h/2e$. This *quantum of magnetic flux* is given the symbol Φ_0 and has a value $2.068\,10^{-13}\,\mathrm{T\,m^2}$. Section 15.6 discusses the penetration of magnetic fields into type-II superconductors.

Shortly after the BCS theory appeared Josephson deduced the existence of two surprising effects, described in Section 15.7. These are observed when a pair of superconductors are in contact through a weak link, for example a few nm thick insulating film. In the first, a DC current flows spontaneously across such a junction without any applied voltage. In the second, when a DC potential is applied across such a weak link an AC current flows.[5] The application of the Josephson effect to provide voltage standards reproducible between laboratories worldwide to better than parts per billion is described in Section 15.8.

Superconducting quantum interference devices, *SQUID*s, that is superconducting rings with one or more weak links, are treated in Section 15.9. These devices are now widely used to measure the weak magnetic fields generated by the human heart and brain. Recent progress towards building quantum computers based on the use of SQUIDs is sketched in Section 15.10.

15.2 London equations

The London brothers realized that there must be normal and superconducting electron phases in a superconductor. From this they deduced

[1] High temperature superconductors have not displaced classical superconductors in most applications because they are ceramics which have required new and costly fabrication processes to make commercially useful wires.

[2] For which they shared the 1972 Nobel prize in Physics.

[3] This analysis was later justified directly from BCS by Gorkov.

[4] Examples of magnetic flux quantization have been met in Sections 6.9 and 13.4: another will appear in the integral quantum Hall effect.

[5] At this time Josephson was a PhD student at Cambridge and subsequently received the 1973 Nobel Physics Prize.

that external magnetic fields can only penetrate short distances into superconductors, thus linking superconduction and the Meissner effect. Consider the case of electrons travelling at velocity v in an alternating field with no damping due to electrical resistance:

$$E = E_0 \exp(-i\omega t) \quad \text{and} \quad v = v_0 \exp(-i\omega t). \tag{15.7}$$

The equation of motion $-eE = m\mathrm{d}v/\mathrm{d}t$ gives $-im\omega v_0 = -eE_0$, and the superconducting current amplitude is

$$j_{s0} = -nev_0 = ine^2 E_0/(m\omega), \tag{15.8}$$

where n is the electron density. Putting this equation in vector form, and taking the curl, gives the first London equation:

$$\nabla \wedge \mathbf{j_s} = \frac{ine^2}{m\omega} \nabla \wedge \mathbf{E} = -\frac{ine^2}{m\omega} \frac{\partial \mathbf{B}}{\partial t} = -\frac{ne^2}{m} \mathbf{B}, \tag{15.9}$$

where Faraday's law has been used. A second London equation results from replacing $\mathbf{B}$ by $\nabla \wedge \mathbf{A}$. Then

$$\mathbf{j_s} = -\frac{ne^2}{m} \mathbf{A}, \tag{15.10}$$

which is simply eqn. 1.83 for a uniform charge distribution. In the static limit one of Maxwell's equations becomes $\nabla \wedge \mathbf{B} = \mu_0 \mathbf{j_s}$ which allows us to replace $\mathbf{j}_s$ in the first London equation, giving

$$\nabla \wedge \nabla \wedge \mathbf{B} = -\frac{\mu_0 ne^2}{m} \mathbf{B}, \tag{15.11}$$

whence using eqn. A.15 and $\nabla \cdot \mathbf{B} = 0$ we have

$$\nabla^2 \mathbf{B} = \frac{\mu_0 ne^2}{m} \mathbf{B}. \tag{15.12}$$

This has the simple solution

$$B = B_0 \exp(-z/\lambda), \tag{15.13}$$

where z is the distance from the superconductor's surface and

$$\lambda = \sqrt{m/\mu_0 ne^2} \tag{15.14}$$

is called the *London penetration length*. Thus, an external magnetic field falls off exponentially with the distance inside a superconductor. For an electron density $10^{29}\,\mathrm{m}^{-3}$, λ is expected to be $\sim 17\,\mathrm{nm}$. It follows that the supercurrent of eqn. 15.10 flows at the surface of the superconductor and its magnetic field exactly cancels the applied magnetic field within the body of the superconductor. As a result penetration depths are generally much larger, $\sim 100\,\mathrm{nm}$ in lead.

By analogy with He-II there can be vortices in a superconductor made up of supercurrents circulating round the vortex axis. The core is threaded by magnetic field lines parallel to the axis. The London analysis yields a magnetic field dependence at a small radial distance ($r \ll \lambda$) from the vortex core

$$B \sim \ln(\lambda/r). \qquad (15.15)$$

The corresponding current flowing round the core is

$$j = \nabla \wedge \mathbf{B}/\mu_0 \sim 1/r. \qquad (15.16)$$

This flow ceases at a radius ξ_0 within which the magnetic field restores conductivity to normal. This parameter ξ_0 is associated with the *coherence length* of a superconductor. At the other extreme of radii large compared to λ

$$B \sim \exp(-r/\lambda)/\sqrt{r}, \qquad (15.17)$$

consistent with the current penetrating a distance of order λ into the surrounding superconductor. The Londons provided a descriptive framework for the properties of superconductivity: the fundamental microscopic theory was provided by Bardeen, Cooper and Schrieffer.

15.3 BCS theory

The mechanism responsible for superconductivity was teased out in 1956–7 by BCS,[6] developing ideas by Fröhlich.[7] Crucially the critical temperature of a set of isotopes depends on the isotope mass, $T_c \propto 1/\sqrt{M}$: Frohlich inferred that superconductivity must involve lattice interactions, meaning that phonons are important. Cooper was able to show that the exchange of phonons (via the lattice) between a pair of electrons just above the Fermi sea could produce a weak attraction between them. This requires the two electrons have opposite spin alignment and equal and opposite momentum. He demonstrated that these *Cooper pairs* are bound in a state separated by an energy gap Δ (per electron) from the Fermi sea of excited electrons (quasi-particles). What is critical is that this binding occurs *however weak* the attraction between them. The thermal energy at the critical temperature is evidently enough to break up a Cooper pair, for example 0.62 meV in the case of lead. This implies that Δ is similarly of order meV, which contrasts to the eV energies of electrons at the top of the Fermi sea in a metal.

Chapter 5 explained that in a metal each free electron is transformed into a quasi-particle and is in this way effectively screened from the electrostatic repulsion of the other free electrons. The process for producing a small net attraction between a pair of electrons via the lattice can be visualized like this. One electron attracts an ion, which responds in a time τ determined by the lattice oscillation period, that is by the phonon frequency. Using Debye's theory the time interval is $1/\omega_D$ namely $\sim 10^{-13}$ s.

[6]L. N. Cooper, *Physical Review* 104, 1189 (1956); J. Bardeen, L. N. Cooper and J. R. Schrieffer, *Physical Review* 108, 1175 (1957).
[7]H. Fröhlich, *Physical Review* 79, 845 (1950).

In this interval the electron itself travels a distance $v_F\tau$, which amounts to about 100 nm. A second electron arriving at the ion after time τ has elapsed feels a force of attraction toward the ion, while the direct repulsion from the now distant first electron is negligible. The effect on the second electron is an attraction to the first conveyed through the ion (lattice) excitation, or in quantum terms, through phonon exchange.

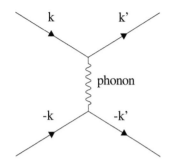

Fig. 15.4 Cooper pair interaction.

Cooper's demonstration of the existence of the energy gap considers what happens to two electrons just above the Fermi sea at $T = 0$. The total energy is $H = H_0 + H_i$ where H_0 is the energy in the absence of phonon exchange and H_i is the energy due to phonon exchange. In operator form

$$\hat{H}_i = -V \sum \hat{c}^\dagger_{\mathbf{k}'} \hat{c}^\dagger_{-\mathbf{k}'} \hat{c}_{\mathbf{k}} \hat{c}_{-\mathbf{k}} \tag{15.18}$$

corresponding to Figure 15.4 with electrons of momenta $\mathbf{k}$ and $-\mathbf{k}$ scattering to momenta $\mathbf{k}'$ and $-\mathbf{k}'$. V specifies the strength of the interaction. $(c_{\mathbf{k}})\hat{c}^\dagger_{\mathbf{k}}$ (annihilates) creates an electron with momentum $\hbar\mathbf{k}$.

The number of states is so large that the chance that the target state $(\mathbf{k}' \uparrow, -\mathbf{k}' \downarrow)$ above the Fermi surface is filled is negligible. In the absence of phonon exchange the energy of each electron is taken to be $\xi_{\mathbf{k}}$. The interaction energy of a phonon is at most around $\hbar\omega_D$, so that the interaction can only bind electrons with energies in the range specified by

$$|\xi_{\mathbf{k}} - \xi_F| \leq \hbar\omega_D. \tag{15.19}$$

The wavefunction of the pair when correctly antisymmetrized is

$$\psi_{\mathbf{k}} = \begin{vmatrix} \psi_{\mathbf{k}\uparrow}(\mathbf{r}_1) & \psi_{\mathbf{k}\uparrow}(\mathbf{r}_2) \\ \psi_{-\mathbf{k}\downarrow}(\mathbf{r}_1) & \psi_{-\mathbf{k}\downarrow}(\mathbf{r}_2) \end{vmatrix}, \tag{15.20}$$

and the state vector for such pairs is

$$|\psi\rangle = \sum_{\mathbf{k}} \phi_{\mathbf{k}} |\psi_{\mathbf{k}}\rangle. \tag{15.21}$$

By definition

$$\hat{H}_0 |\psi_{\mathbf{k}}\rangle = 2\xi_{\mathbf{k}} |\psi_{\mathbf{k}}\rangle, \tag{15.22}$$

and if the Cooper pair energy is E

$$\langle\psi|\hat{H}|\psi_{\mathbf{k}}\rangle = E\langle\psi|\psi_{\mathbf{k}}\rangle, \tag{15.23}$$

whence

$$E\phi_{\mathbf{k}} = 2\xi_{\mathbf{k}}\phi_{\mathbf{k}} - V \sum_{\mathbf{k}'} \phi_{\mathbf{k}'}. \tag{15.24}$$

Then putting $\sum_{\mathbf{k}'} \phi_{\mathbf{k}'} = C$,

$$\phi_{\mathbf{k}} = -CV/[E - 2\xi_{\mathbf{k}}]. \tag{15.25}$$

Summing over $\mathbf{k}$ and cancelling C gives

$$1 = -V \sum_{\mathbf{k}} \left[\frac{1}{E - 2\xi_{\mathbf{k}}} \right]. \tag{15.26}$$

The sum is in fact an integral over the Fermi sea from E_F to $E_F + \hbar\omega_D$ so that

$$1/V = N(0) \int_{E_F}^{E_F + \hbar\omega_D} d\xi_{\mathbf{k}}/[2\xi_{\mathbf{k}} - E], \qquad (15.27)$$

with $N(0)$ being the density of electron states at E_F. Thus

$$\frac{1}{V} = [N(0)/2] \ln \left[\frac{2E_{\mathbf{F}} - E + 2\hbar\omega_D}{2E_{\mathbf{F}} - E} \right], \qquad (15.28)$$

The interaction is weak making $\Lambda = N(0)V/2 \ll 1$ and $\exp(-1/\Lambda) \ll 1$, so that in this limit the equation collapses to

$$E = 2E_F - 2\hbar\omega_D \exp(-1/\Lambda). \qquad (15.29)$$

It follows that $E < 2E_F$, so that the two electrons are bound *irrespective of how weak the interaction may be*. These bound pairs of electrons are known as *Cooper pairs* and are necessarily bosons. The energy gap (per electron) between the Cooper pairs and excited electron states at $T = 0$ is

$$\Delta(T = 0) = \hbar\omega_D \exp(-1/\Lambda), \qquad (15.30)$$

At the critical temperature this falls to zero: $\Delta(T_c) = 0$. The exchange potential V is short range so that we can approximate it by a delta function, $V\delta(\mathbf{r} - \mathbf{r}')$. For this to have any effect the electrons forming the Cooper pair must be in an s-state of relative motion, because only in this state does the wavefunction[8] penetrate to $(\mathbf{r} - \mathbf{r}') = \mathbf{0}$. Finally, the overall Cooper pair (spatial × spin) wavefunction must be antisymmetric under interchange of the electrons. With a spatially symmetric s-state wavefunction the spin state has to be the antisymmetric $(\uparrow\downarrow - \downarrow\uparrow)$ spin zero state.

[8]See for example the wavefunctions of the hydrogen atom in Chapter 2.

Cooper pairs are bosons and form the condensate that carries the superconducting current in both classical and high temperature superconductors. In the rest frame of the condensate the electrons making up a Cooper pair have equal and opposite momentum and opposite spin alignment: $\mathbf{k}_\uparrow$ and $-\mathbf{k}_\downarrow$. The spatial wavefunction of a Cooper pair is generically

$$\psi \left(\frac{\mathbf{r}_i + \mathbf{r}_j}{2} \right) \chi (\mathbf{r}_i - \mathbf{r}_j). \qquad (15.31)$$

ψ describes the motion of the centre of mass of any pair, and is identical for all pairs. If the condensate is in a uniform region then ψ would be equal all over this space. χ is the internal wavefunction of any pair, which we can take to be a Gaussian. The size of this internal wavefunction, that is to say the size of a Cooper pair, can be inferred from the uncertainty principle. These electrons have energies within Δ of E_F so that their momenta have a spread around $\hbar k_F$ of Δ/v_F. Hence the spatial extent of a Cooper pair is

$$\xi \sim \hbar/[\Delta/v_F] = \hbar v_F/\Delta. \qquad (15.32)$$

The physical distance over which the condensate wavefunction can change significantly is thus ξ and this defines what is called the *coherence length* of the Cooper pair condensate. Taking Δ to be $k_B T_c$ at $T = 0$ the corresponding coherence length is given by

$$\xi_0 \sim \hbar v_F / k_B T_c, \tag{15.33}$$

and is typically around 1 μm.

Electrons that make up the condensate have energies within $k_B T_c$ of the Fermi energy. They therefore make up only a small fraction, $k_B T_c / k_B T_F$, of all conduction electrons: namely around 10^{-4}. Hence their density is $\sim 10^{24}\,\mathrm{m}^{-3}$, giving a mean separation between Cooper pairs of around $\sim 10^{-8}\,\mathrm{m}$. This separation can be compared to the size of a Cooper pair evaluated in the last paragraph. It follows that any one pair overlaps on average of order $(10^{-6}/10^{-8})^3$, that is 10^6 other pairs.

In the analysis by Bardeen, Cooper and Shrieffer the state vector of the condensate of Cooper pairs is taken to have the form, now specifying the spin alignments explicitly,

$$|\psi_{\mathrm{BCS}}\rangle = \prod_{\mathbf{k}} [u_{\mathbf{k}} + v_{\mathbf{k}} \hat{c}^\dagger_{\mathbf{k}\uparrow} \hat{c}^\dagger_{-\mathbf{k}\downarrow}] |0\rangle, \tag{15.34}$$

where $\hat{c}^\dagger_{\mathbf{k}\uparrow}$ creates an electron with wave-vector $\mathbf{k}$ and spin-up, and $\hat{c}^\dagger_{-\mathbf{k}\downarrow}$ creates an electron with wave-vector $-\mathbf{k}$ and spin-down. In order to have unit probability for all outcomes $|u_{\mathbf{k}}|^2 + |v_{\mathbf{k}}|^2 = 1$. Initially it surprised Cooper that $|\psi_{\mathrm{BCS}}\rangle$ is a superposition of states with different numbers of Cooper pairs. This unexpected feature can be understood in the following way. Interactions are continually creating Cooper pairs from the Fermi sea and also returning them to this much larger reservoir and the state vector describes the ensemble of possible states with different numbers of Cooper pairs. This indeterminacy in the occupation number is shared by laser state vectors which contain superpositions of states with different numbers of photons (see Section 8.8). If the number of Cooper pairs could be counted instantaneously then a definite number would be observed with a probability equal to the square of the wavefunction amplitude for that number of pairs. Bardeen, Cooper and Schrieffer determined the precise connection between the critical temperature and the energy gap at $T = 0$ to be

$$\Delta(T = 0) = 1.76 k_B T_c. \tag{15.35}$$

which confirms the idea that at the critical temperature thermal excitations are energetic enough to break up any Cooper pair.

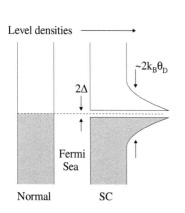

Fig. 15.5 Energy level densities at the top of the Fermi sea in normal conductors and superconductors.

The distribution, $N_s(E)$, of energy levels of the other electrons, the quasi-particles, must suffer some corresponding change around the Fermi energy. BCS argued that the quasi-particle levels are in one-to-one correspondence with the electron levels in the normal phase. Measuring energies relative to the Fermi energy

$$N_s(E)\mathrm{d}E = N_n(\xi)\mathrm{d}\xi, \tag{15.36}$$

where the excitation energy

$$E = \sqrt{\Delta^2 + \xi^2}, \tag{15.37}$$

with ξ being the energy of the corresponding normal state electron. Then

$$\frac{\mathrm{d}E}{\mathrm{d}\xi} = \frac{\xi}{\sqrt{\Delta^2 + \xi^2}} = \frac{\sqrt{E^2 - \Delta^2}}{E}. \tag{15.38}$$

Now $N_\mathrm{n}(\xi)$ is nearly constant over the narrow range of energies involved, hence we have

$$\begin{aligned} N_\mathrm{s}(E) &= N_\mathrm{n}(\xi)E/\sqrt{E^2 - \Delta^2} \text{ for } |E| > \Delta, \\ N_\mathrm{s}(E) &= 0 \text{ for } |E| < \Delta, \end{aligned} \tag{15.39}$$

without quasi-particle states in the gap, $E < \Delta$. Figure 15.5 contrasts the energy level distribution at the top of the Fermi sea in the normal and superconducting phases. The width of the peaks is determined by the available phonon energy, that is $k_\mathrm{B}\Theta_\mathrm{D}$, where Θ_D is the Debye temperature.

Giaever[9] studied the tunnelling current between a metal and a superconductor separated by a nanometre thick insulator. Figure 15.6 compares this current (the full line) with the current through such an insulating layer between two normal conductors (the broken line), all at $T \approx 0$. Adding a single Cooper pair to the superconductor requires an energy 2Δ, and hence, there is no current for a voltage below Δ/e. Mea-

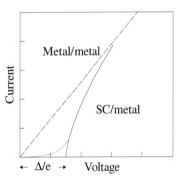

Fig. 15.6 Tunnelling currents: that through a superconductor/insulator-film/metal junction with no applied magnetic field is shown as the solid line, the dotted line indicates the effect of an applied field; that through a metal/insulator-film/metal junction is shown by a broken line.

[9]Nobel Laureate in physics in 1973

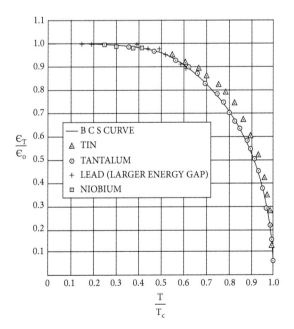

Fig. 15.7 Comparison of the temperature variation of the reduced values of the measured energy gaps with the BCS prediction. Figure 4 in P. Townsend and J. Sutton, *Physical Review* 128, 591 (1962). Courtesy of the American Physical Society.

surements of the gap in this way by Townsend and Sutton as a function of temperature are shown in Figure 15.7. Their results track the BCS prediction, indicated by the solid line.

15.3.1 Dirty superconductors

Superconductors are characterized by three lengths: the coherence length, ξ, the distance over which the condensate wavefunction can undergo significant change; λ, the magnetic field penetration depth; ℓ, the mean free path of an electron in the normal state. In a pure uniform material ℓ is much longer than the other two lengths making a *clean superconductor*. *Dirty superconductors* is the term used to describe superconductors loaded with crystal defects and/or chemical impurities. Elastic scattering from these defects changes the electron wavefunctions markedly from Bloch waves. However, after a small drop in T_c when the first defects are added, the superconductive state remains robust. Anderson pointed out how *time reversal invariance* comes into play in superconductivity. This is the symmetry between processes going forward and backward in time. He noted that the wavefunction $\psi^*(-\mathbf{k}\downarrow)$ is the time reverse of $\psi(\mathbf{k}\uparrow)$, making it the case that both partner states of a potential Cooper pair are simultaneously present in a degenerate Fermi gas. A magnetic field breaks the symmetry, and if this field exceeds H_c it suppresses superconductivity. At superconducting temperatures the only important scattering processes are the said elastic scatters from defects and these preserve time reversal invariance.[10] Dirty superconductors therefore retain time reversal symmetry and remain superconductors; the exception being when the impurities are themselves magnetic. The major practical applications of superconductivity require dirty superconductors.

[10] See Section 17.8.1

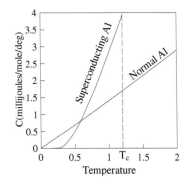

Fig. 15.8 Heat capacity of aluminium below 2 K. The data for the normal metal below T_c was obtained by applying a magnetic field of 0.3 T. All the other data was recorded for zero applied magnetic field. Adapted from Figure 4, N. E. Phillips, *Physical Review* **114**, 676 (1954).

15.3.2 Heat capacity

Figure 15.8 shows the variation of the heat capacity of aluminium measured by Phillips above and below the critical temperature of 1.175 K. The data for the normal state below the critical temperature was obtained by imposing a magnetic field greater that H_c. At low temperatures the dominant contribution to the heat capacity of a normal metal is that of the electrons, and it was shown in Exercise 3.6 to have the form

$$C_n = \pi^2 N k_B^2 T/[2E_F]. \tag{15.40}$$

This linear dependence on temperature is evident in Phillips' data. In the superconducting phase the condensate makes no contribution to the entropy and hence to the heat capacity: only quasi-particle excitation out of the superconducting state can contribute. Now the entropy of the occupied and empty excited states at temperature T is given by the Boltzmann expression

$$S = -2k_B \sum [(1 - f_k) \ln(1 - f_k) + f_k \ln(f_k)], \tag{15.41}$$

where f_k is mean occupation number given by the Fermi–Dirac function $1/[1 + \exp(E_k/k_BT)]$. At very low temperatures $f_k \rightarrow \exp(-E_k/k_BT)$. Hence both the entropy and heat capacity are predicted to vary exponentially with the temperature in the superconducting phase, again consistent with Phillips' data. The large rise in the heat capacity towards threshold marks the loss of ordering in the normal metal.

15.4   Ginzburg and Landau approach

Prior to the microscopic BCS theory Ginzburg and Landau developed a phenomenological approach to superconductivity which provides useful insights on the macroscopic scale. Gorkov showed later that their analysis could be derived directly from BCS theory. The *order parameter* introduced by Ginzburg and Landau, $\psi(\mathbf{r}) = |\psi(\mathbf{r})| \exp[i\phi(\mathbf{r})]$, is equivalent to the wavefunction of the Cooper pairs in eqn. 15.31: here $\mathbf{r}$ is is the centre of mass coordinate of any of the Cooper pairs.[11] The corresponding expression for the supercurrent is obtained by applying eqn. 1.83 for charge carriers which are Cooper pairs with mass $m^* = 2m$ and charge $-e^* = -2e$,

[11]This order parameter is also equivalent to the wavefunction of the condensate in He-II.

$$
\begin{aligned}
\mathbf{j}_s &= -\frac{ie^*\hbar}{2m^*}[\psi\nabla\psi^* - \psi^*\nabla\psi] - \frac{e^{*2}}{m^*}\mathbf{A}|\psi|^2 \\
&= -\left[\frac{e^*\hbar}{m^*}\nabla\phi + \frac{e^{*2}}{m^*}\mathbf{A}\right]|\psi|^2. \quad (15.42)
\end{aligned}
$$

The density of Cooper pairs is $n_s = |\psi(\mathbf{r})|^2$, which is half the electron density. The second term in this equation is identical to Londons' inferred supercurrent[12] in eqn. 15.10; the first term extends the validity of the equation to cover spatial variation of the condensate phase. This expression for the current is invariant under local gauge transformations; any change in ϕ being compensated by a change in $\mathbf{A}$. Ginzburg and Landau expanded the free energy of the condensate in powers of $|\psi|^2$ and its first derivative arguing that, at least near T_c, $|\psi|^2$ should be a small quantity. Their Gibbs free energy density, in the absence of any magnetic field is to second order

$$
G(E) = (\hbar^2/2m^*)|\nabla\psi|^2 + a|\psi|^2 + (b/2)|\psi|^4. \quad (15.43)
$$

The first term is kinetic energy, and the potential terms are taken to be smoothly varying functions of T, with b weakly dependent on T. Evidently b should be positive, otherwise in the lowest energy state ψ would be infinite. Choosing a either positive or negative gives the contrasting behaviours seen in Figure 15.9 using the $|\psi|$ and phase coordinates drawn underneath. With positive a the free energy is at a minimum when ψ is zero and this represents a state at a temperature above the critical temperature. However, if a is negative the potential energy is least, at

[12]The derivation of the London equations was made semi-classically without quantizing the current. It started from eqn. 1.78 rather than eqn. 1.83.

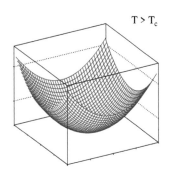

T > T_c

Ginzberg-Landau Potentials

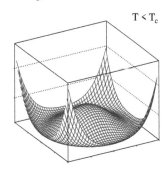

T < T_c

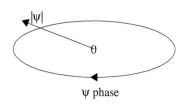

ψ phase

Fig. 15.9 Free energy plots with $|\psi|$ as the radial coordinate and the phase of ψ as the azimuthal coordinate. Upper panel for positive a; lower panel for negative a. At $|\psi| = 0$ the two surfaces touch.

See 'On massive photons inside a superconductor' by R. de Bruyn Ouboter and A. N. Omelyanchouk, *Low Temperature Physics* 43, 889 (2017).

$-a^2/2b$ when $|\psi|^2 = -a/b$, which produces the superconducting state.

Note that in the superconducting state the value of ψ could lie at any of the points with equally low energy in the circular trough. Possible values for the phase ϕ range anywhere from 0 to 2π. In the physical superconductor the fact that there is a specific phase breaks this gauge invariance spontaneously. The parallel example of superfluid He-II is discussed in Section 14.5. We can make links to observables as follows: $n_s \sim -a/b$; while the condensation energy $\mu_0 H_c^2/2 \sim a^2/2b$.

When a magnetic field is present the kinetic term in eqn. 15.43 becomes

$$\frac{1}{2m^*}[(-i\hbar\nabla + e^*\mathbf{A})\psi]^*[(-i\hbar\nabla + e^*\mathbf{A})\psi]. \quad (15.44)$$

This expression contains the term

$$\frac{e^{*2}}{2m^*}|\mathbf{A}|^2|\psi|^2, \quad (15.45)$$

by which the electromagnetic field $\mathbf{A}$ acquires a mass thanks to the condensate. Note that the overall expression for the free energy density is still gauge invariant.

We can also show how the limited penetration of magnetic fields into superconductors implies that within a superconductor the photons acquire mass. The electromagnetic wave equation, eqn. 1.86, when applied to a superconductor gives

$$\nabla^2\mathbf{A} - \partial^2\mathbf{A}/(c\partial t)^2 = -\mu_0\mathbf{j}_s. \quad (15.46)$$

Using London's second equation to replace the supercurrent gives

$$\nabla^2\mathbf{A} - \partial^2\mathbf{A}/(c\partial t)^2 = \frac{\mu_0 n e^2}{m}\mathbf{A} = \mathbf{A}/\lambda^2. \quad (15.47)$$

For comparison we now write the relativistic wave equation, eqn. 1.88,

$$m^2 c^4 \phi = -\hbar^2\partial^2\phi/\partial t^2 + \hbar^2 c^2\nabla^2\phi, \quad (15.48)$$

where ϕ is the wave function describing a massive scalar particle. Rearranging gives

$$\nabla^2\phi - \partial^2\phi/(c\,\partial t)^2 = m^2 c^2\phi/\hbar^2. \quad (15.49)$$

Comparing eqns 15.47 and 15.49 it is apparent that the electromagnetic field within the superconductor behaves as if the photons had a finite mass $\hbar/(c\lambda)$. Putting in a penetration depth of 100 nm the mass is around $2\,\mathrm{eV}/c^2$. The above analyses prefigure the BEH mechanism, described in Chapter 19, which gives mass to the weak vector bosons of the standard model of particle physics.

15.5   Flux quantization

The condensate wavefunction at a given point in a superconductor is unique, so it follows that the change in phase round a closed path entirely within a superconductor is $2n\pi$, where n is an integer. Equation 15.42 supplies an expression for the phase:

$$\frac{e^*\hbar}{m^*}\nabla\phi = -\frac{\mathbf{j}_s}{n_s} - \frac{e^{*2}}{m^*}\mathbf{A}, \tag{15.50}$$

where n_s is the number density of Cooper pairs. In the interior of a superconductor where no current flows this collapses to

$$\hbar\nabla\phi = -2e\mathbf{A}, \tag{15.51}$$

where e^* has been set to $2e$. This equation is directly comparable to eqn. 14.13 for superfluid flow. Integrating round a closed path in the superconductor gives

$$\frac{nh}{2e} = \oint \mathbf{A}\cdot d\mathbf{r}, \tag{15.52}$$

with n an integer. Applying Stokes' eqn. A.21 gives

$$\frac{nh}{2e} = \int \mathbf{B}\cdot d\mathbf{S}, \tag{15.53}$$

where the integral in the magnetic field $\mathbf{B}$ is over the enclosed area. The right-hand side is evidently quantized in units

$$\Phi_0 = \frac{h}{2e} = 2.068\,10^{-15}\,\mathrm{T\,m^2}, \tag{15.54}$$

the quantum of magnetic flux. The same equation emerged in connection with the Aharonov–Bohm effect in Section 13.4.

Deaver and Fairbank[13] confirmed the quantization of flux using the apparatus sketched in Figure 15.10. A superconducting cylindrical shell of tin was formed by depositing tin on a cm long, 13 μm diameter copper wire. This structure was vibrated at 100 Hz with mm amplitude along its axis in an axial magnetic field. The two ten thousand turn coils shown surrounding the ends of the cylinder are wound to give in-phase induced voltages, the sum being measured with an AC voltmeter. The cylinder was exposed to a measured axial magnetic field and then cooled through tin's transition temperature at 3.72 K in order to trap flux Φ inside the cylinder. Next the external field was removed and the AC voltage from the pickup coils measured. This induced voltage is proportional to the combined magnetic moment of the superconducting tin cylinder, M_s, and the flux trapped within it, Φ. A calibration was required to eliminate M_s. In this step the cylinder was cooled below 3.72 K in zero applied field, then a known field applied and the induced voltage measured. Results are presented in Figure 15.11 including the calibration. The calibration magnetic moment gave the lowest solid line, with

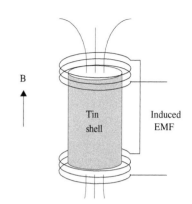

Fig. 15.10 Deaver and Fairbank's experiment to exhibit the quantization of magnetic flux.

[13]B. S. Deaver Jnr. and W. Fairbank, *Physical Review Letters* 7, 43 (1961).

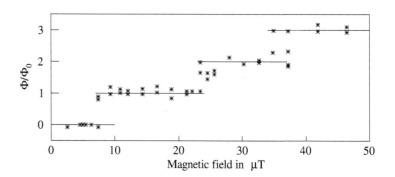

Fig. 15.11 Variation of the magnetic flux trapped in a superconducting solenoid as a function of the applied field. Adapted from Figure 1 in B. S. Deaver Jnr. and W. Fairbank, *Physical Review Letters* 7, 43 (1961).

the other solid lines displaced in turn by unit steps upward. The measurements of the trapped flux are the points: these show the expected quantization, lying on plateaux at integral values of the flux quantum Φ_0. At the steps where the integer changes the phase change around the condensate slips by 2π. The quantum of flux is so small that it was necessary to shield from external field with three orthogonal Helmholtz coils, bringing the ambient field down to 0.1μT. Current measurements yield the value $h/2e = 2.067833831(13)\,10^{-15}\,\mathrm{Tm}^2$.

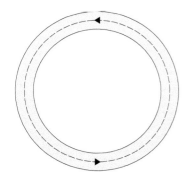

Fig. 15.12 A superconductor that is not simply connected. The broken line indicates the current path on the superconductor surface.

Returning to eqn. 15.53 for a path within the superconductor away from the surface current: if this path can be shrunk smoothly to a point without crossing any material boundary then n is zero. Otherwise, if the region occupied by the superconductor is not simply connected, which happens for the loop in Figure 15.12 n is non-zero. The quantization of flux presented here in eqn. 15.53 involves a topological quantum number like that met in the quantization of circulation in He-II. The quantum number n does not depend on the symmetry of the path but only on its topology. This number counts the number of vortices or physical holes enclosed by the path: it is termed a *winding number* because the count is increased by m if the path loops a hole m times.

We can now refer back to the London predictions for the magnetic field in a vortex given in eqns. 15.15 and 15.17. The normalizing factor left out from these equation contains a term $(\Psi_0/2\pi\lambda^2)$ which indicates that the core contains a flux Ψ_0 spread over an area of radius $\sqrt{2}\lambda$. This fits in well with the quantum picture given here.

15.6 Type-II superconductors

It is the relative magnitudes of the coherence length, ξ and the penetration length, λ that determines whether a superconductor is type-I or type-II. A useful parameter, named for Ginzburg and Landau, is thus

$$\kappa = \lambda/\xi. \tag{15.55}$$

Figure 15.13 illustrates the magnetic field and wavefunction amplitude at the interface between superconducting and normal regions of a metal with an applied field H_c. The upper panel shows the distributions in a type-I superconductor, for which $\kappa \ll 1$, that is when the coherence length is much greater than the penetration length; while the lower panel shows the corresponding distributions in a type-II superconductor for which $\kappa \gg 1$. We analyse the situation for type-I. The energy densities in the bulk on either side of the interface are identical in equilibrium. In the superconducting region the condensate energy density is lowered by $\mu_0 H_c^2/2$ with respect to the normal region. However this only develops over the coherence length ξ. There is a compensating increase in the condensate magnetic energy density with respect to the normal region of $\mu_0 H_0^2/2$ due to the surface current which cancels the applied magnetic flux. This only develops over the penetration depth λ. In the surface layer where these changes develop there is a *net increase* in energy density of $\mu_0 H_c^2[\xi - \lambda]/4$ per unit area of the interface. Therefore in a type-I superconductors the tendency is for the overall surface area between superconducting and normal phases to become as small as possible. In type-II superconductors the reverse will be true, the lowest energy state is that in which the interface area is as large as possible.

We can extend this analysis to explain the existence of the two transitions shown in Figure 15.3 for type-II superconductors. When the applied magnetic field reaches H_{c1} it begins to penetrate the material because it becomes energetically favourable for flux vortices to form. A cross-section of a vortex is sketched in Figure 15.14. At this field value, H_{c1}, there is a reduction in energy due to the response no longer being diamagnetic over an area $\pi\lambda^2$. This balances the energy increase due splitting the Cooper pairs to enter the normal state over an area $\pi\xi^2$. That is

$$\pi\lambda^2(\mu_0 H_{c1}^2) = \pi\xi^2(\mu_0 H_c^2), \tag{15.56}$$

whence $H_{c1} = (\xi/\lambda)H_c$. It can also be shown that the field at which the vortices fill the metal so that normal conductivity is restored is given by $H_{c2} = (\lambda/\xi)H_c$, and $H_{c2} = (\lambda/\xi)^2 H_c^2$.

Vortices can be made visible by spraying the surface of the superconductor with grains of paramegnetic material; these are attracted to the vortex cores where the magnetic field emerges. Another method is to scatter neutrons from the surface: the magnetic field at a vortex produces detectable rotation of the neutron spin. The vortices in a uniform specimen of type-II superconductor form a pattern with triangular symmetry, called an *Abrikosov lattice*.

H_c is of order $0.1\,\mathrm{T}$ for type-I superconductors while for the type-II superconductor Nb_3Sn H_{c2} is boosted by a large value of κ to $24.5\,\mathrm{T}$. This makes compounds such as Nb_3Sn of value in applications such as magnetic resonance imaging. Perfect crystals are to be avoided because they develop a finite resistance in the following way. A vortex in a region

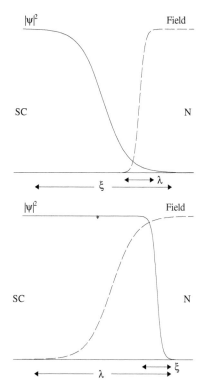

Fig. 15.13 Interfaces between superconducting and normal regions. In the upper (lower) panel for type-I superconductors (type-II superconductors). λ is the penetration depth and ξ is the coherence length.

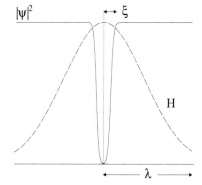

Fig. 15.14 Cross-section through a vortex in a superconductor. The solid line shows how the condensate falls off and the broken line shows the field penetration.

Table 15.1 Common superconductor properties.

	T_c (K)	ξ (nm)	λ (nm)	$H_{c/c2}$ (T)
Al	1.18	1600	50	0.01
Pb	7.19	83	39	0.08
Nb	9.25	40	44	0.20
Nb$_3$Sn	18.2	3.6	124	24.5

with supercurrent $\mathbf{J}$ feels a Lorentz force

$$\mathbf{F} = \mathbf{J} \wedge (\Phi_0 \hat{z}), \tag{15.57}$$

with $\hat{z}$ being a unit vector along the vortex axis. In a perfect lattice the vortices are free to move and travel perpendicular to the current and exit the superconductor. The rate of change of flux generates a corresponding EMF opposing the current flow, and hence an undesirable electrical resistance. Fortunately vortices get trapped on crystal defects and boundaries, or on impurities. In order to take advantage of this *flux pinning* the superconductor is made dirty either by adding impurities, or by work hardening to generate defects. Superconducting magnets are therefore wound from wires of type-II work-hardened superconductor interwoven with copper.[14] This design avoids *quenches* in which local heating above T_c produces a return to normal conductivity and consequent further heating and the subsequent continuous expansion of the volume of normal state material. This sequence can instantly boil off the liquid helium coolant with potentially catastrophic results. The interleaved copper provides an alternative low-resistance path to current and more importantly conducts heat away.

[14]The wires used for winding LHC dipoles are 1 mm diameter and 10 km long. Each contains 6000 Nb-Ti filaments of 6 μm diameter, protected by a Nb barrier, and well separated in a copper matrix. These dipoles develop 8.3 T. Nb-Ti can be reliably processed with uniform quality.

15.7 Josephson effects

These effects were the first unequivocal evidence that macroscopic quantum states do exist. A *Josephson junction* at which these effects are manifest is made of superconductors connected by a *weak link* so that their wavefunctions overlap. One example consists of millimetre wide strips of superconducting aluminium forming a cross and separated by an oxide skin 2 nm thick. The contact area is generally made sufficiently small that current flows over the whole contact area. The time dependent Schrödinger equation[15] for the wavefunction ψ_1 of the first superconductor at the interface is

$$i\hbar \partial\psi_1/\partial t = E_1 \psi_1 + K\psi_2, \tag{15.58}$$

where E_1 is the energy in the bulk and $K\psi_2$ is the energy due to the interaction with the second superconductor having a wavefunction ψ_2.

[15]Following R. P. Feynman, *Lectures on Physics* Vol.3 Chapter 21, published by Addison-Wesley, (1965).

Similarly the wavefunction of the second superconductor obeys an equation

$$i\hbar\partial\psi_2/\partial t = E_2\psi_2 + K\psi_1. \qquad (15.59)$$

Making the substitution $\psi_i = a_i \exp(i\phi_i)$ in each equation, where $n_i = a_i^2$ is the Cooper pair density, gives

$$
\begin{aligned}
i\hbar\dot{a}_1 - a_1\hbar\dot{\phi}_1 &= E_1 a_1 + K a_2 \exp i(\phi_2 - \phi_1), & (15.60)\\
i\hbar\dot{a}_2 - a_2\hbar\dot{\phi}_2 &= E_2 a_2 + K a_1 \exp i(\phi_1 - \phi_2), & (15.61)
\end{aligned}
$$

where a dot over the quantity indicates the time derivative. Continuity of current requires that $\dot{a} = \dot{a}_1 = -\dot{a}_2$. Assuming the two superconductors are physically the same, and taking changes in the total Cooper pair densities to be small, we can put $a_1 \approx a_2 \approx a$, $n = a^2$ and the equations then reduce to

$$
\begin{aligned}
i\hbar\dot{a} - a\hbar\dot{\phi}_1 &= E_1 a + K a \exp i(\phi_2 - \phi_1), & (15.62)\\
-i\hbar\dot{a} - a\hbar\dot{\phi}_2 &= E_2 a + K a \exp i(\phi_1 - \phi_2). & (15.63)
\end{aligned}
$$

Then taking the imaginary part of the difference between these two equations:

$$\hbar\dot{a} = K a \sin\phi, \qquad (15.64)$$

where ϕ is the phase difference $\phi_2 - \phi_1$ between the superconductors. Now because $\dot{a}$ is non-zero then so too is the current across the weak link:

$$I = 2edn/dt = 4ea\dot{a}. \qquad (15.65)$$

Substituting for $\dot{a}$ from the previous equation,

$$I = I_0 \sin\phi \qquad (15.66)$$

where I_0 is $4eKn/\hbar$. This is the first Josephson equation predicting that current flows across a weak junction between superconductors in the absence of an applied voltage. These currents therefore do not dissipate any energy! Taking the real part of the difference between eqns. 15.62 and 15.63 gives

$$\hbar[d\phi/dt] = E_1 - E_2. \qquad (15.67)$$

Josephson also considered what would happen when a voltage V is applied across a weak junction. Then $E_1 - E_2$ becomes $2eV$ and the equation reduces to

$$\frac{\hbar d\phi}{dt} = 2eV, \;\; \text{or} \;\; V = \frac{\Phi_0}{2\pi}\frac{d\phi}{dt}, \qquad (15.68)$$

which is the *second Josephson equation*. This predicts that an applied voltage causes the phase difference across the weak link to oscillate with angular frequency, and frequency,

$$\omega_{\rm J} = \frac{2eV}{\hbar}, \;\; \text{and} \;\; f_{\rm J} = \frac{2eV}{h} \qquad (15.69)$$

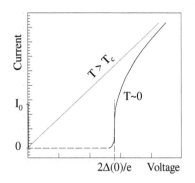

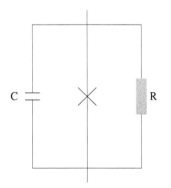

Fig. 15.15 Current-voltage characteristics of a Josephson junction.

Fig. 15.16 Equivalent circuit of a Josephson junction. The multiplication cross is the usual symbol for a Josephson junction.

respectively. Figure 15.15 illustrates the response of a Josephson junction driven by a direct *current source*. As the current is increased from zero to I_0 the measured voltage remains at zero. Any larger current cannot be sustained by the condensate: additional current must be provided by electrons excited across the energy gap. In turn the voltage V across the weak link needed to excite these electrons must be such that $eV \geq 2\Delta(0)$. The transition to this condition is suggested by the broken line; in practice the characteristic follows the load line determined by the external circuit. Thereafter the response follows the solid curve to the right. If the current is increased above I_0, conduction by electrons from the Fermi sea will eventually dominate so that the response asymptotically approaches normal conduction, indicated by the dotted line.

To summarize: the macroscopic wavefunctions of the condensates on either side of a weak junction *superpose* and we observe the consequences of the resulting *interference*. In this way Josephson effects provide unambiguous evidence that condensates exist with this special property: they possess a macroscopic degree of freedom, and an associated wavefunction describing simultaneously the location and motion of the centres of mass of all the Cooper pairs. If the overall wavefunction were simply a product of wavefunctions describing the motion of individual Cooper pairs the interference effects would be uncorrelated and no Josephson effects would be produced. This crucial distinction has been emphasized by Leggett.[16]

In the following sections attention is shifted to the scientific and technological applications of Josephson junctions. According to Josephson's second equation a DC voltage applied to a Josephson junction causes the phase difference between the superconducting reservoirs to oscillate at an angular frequency $2eV/\hbar$. This relationship makes it possible to construct a reproducible international voltage reference. The key is that a voltage can now be determined independently to high precision by simply measuring frequency, and by taking a ratio of natural constants, $e/\hbar$. Other applications to be discussed are SQUIDs used for measuring the tiny magnetic fields of the heart and brain, and qubits envisaged for quantum computing. The equivalent circuit of an actual Josephson junction is drawn in Figure 15.16, where the resistance and capacitance can be designed to match the application. For example a parallel resistance in the form of a thin resistive film deposited across the junction will provide additional damping.

15.8 The Josephson voltage standard

A small DC voltage V, applied across a Josephson junction produces an oscillating supercurrent where the conversion factor from voltage to fre-

---

[16]See *Quantum Liquids* by A. J. Leggett, published by Oxford University Press (2006).

quency, $2e/h$, given by eqn. 15.69 is $4.856\,10^{14}\,\mathrm{HzV}^{-1}$. For example an applied voltage of a tenth of a millivolt produces current oscillations at 48.56 GHz, a frequency at which microwave technology is mature. Then all that is required to give a reproducible voltage standard is a Josephson junction and a measurement of frequency. How this measurement is made is explained next.

An additional small AC voltage is applied so that the total voltage across the junction is $V + v\cos\omega t$ with $v \ll V$. Then using eqn. 15.68 the time-dependence of the phase across the junction is

$$
\begin{aligned}
\phi(t) &= \phi_0 + \int_0^t (2e/\hbar)[V + v\cos\omega t]\mathrm{d}t \\
&= \phi_0 + \omega_{\mathrm{J}}t + [2ev/(\hbar\omega)]\sin\omega t, \quad (15.70)
\end{aligned}
$$

where $\omega_{\mathrm{J}} = 2eV/\hbar$. Thus the supercurrent through the Josephson junction is

$$
I = I_0 \sin\{\phi_0 + \omega_{\mathrm{J}}t + [2ev/\hbar\omega]\sin\omega t\}. \quad (15.71)
$$

Now expanding the right-hand side in Bessel functions gives

$$
I = I_0 \sum_{i=-\infty}^{\infty} (-1)^n J_n\left(\frac{2ev}{\hbar\omega}\right)\sin[(\omega_{\mathrm{J}} - n\omega)t + \phi_0]. \quad (15.72)
$$

The important conclusion to draw from this equation is that there is a DC component to the current

$$
I_0 (-1)^n J_n\left(\frac{2ev}{\hbar\omega}\right)\sin\phi_0 \quad (15.73)
$$

whenever the Josephson frequency is an integral multiple of the applied frequency:

$$
n\omega = \omega_{\mathrm{J}} = 2eV/\hbar. \quad (15.74)
$$

This condition provides a calibration for voltages in terms of a ratio of fundamental constants and frequency only; the frequency lies in the microwave range and hence can be measured with high precision. Figure 15.17 shows the response of a Josephson junction exposed to a microwave source. In the upper panel the response of the Josephson junction for light damping shows constant voltage risers occuring whenever the condition given in eqn. 15.74 is fulfilled: these are called *Shapiro steps*. The overall variation of the DC current follows $I = V/R$, where R is the junction resistance. More important, for producing voltage standards, is the response under heavy damping. This is produced by making the time constant of the junction, RC, (where C is the junction capacitance) short compared to the microwave period. The response under heavy damping is shown in the lower panel of Figure 15.17 with *Shapiro spikes* where the current rises through zero at voltages satisfying the quantization condition given in eqn. 15.74. At low voltages away from this condition the time averaged Cooper pair current ranges into negative values to cancel the current due to electrons excited across the gap.

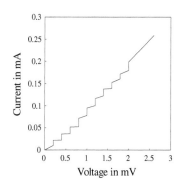

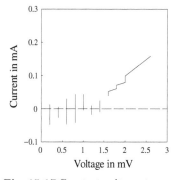

Fig. 15.17 Constant voltage steps seen when a Josephson junction is exposed to microwaves: in the upper/lower panel with weak/heavy damping.

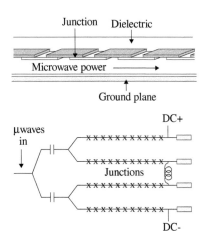

Fig. 15.18 Josephson voltage source. In the upper panel a segment of the microwave stripline with the junctions mounted on it. In the lower panel are shown the parallel microwave feeds to the junctions (crosses) and the microwave termination resistances (boxes). The DC current source supplies $\sim 1\mu$A in series to each junction. Adapted from Figure 5, C. A. Hamilton: *Reviews of Scientific Instruments* 71, 3611 (2000).

Using typically 70–90 GHz microwave sources the quantized voltage developed across a single junction is a fraction of a millivolt. From which it follows that practical voltage references, of for example one volt, must be built from thousands of junctions in series. In addition the junctions must share a common current bias. This becomes feasible using heavily damped junctions operating at Shapiro spikes. Conveniently the operational current ranges of all the Shapiro spikes do overlap. In practice the bias currents employed are in the microamp range. The SI unit of voltage is currently implemented in this way with a conversion factor derived from eqn. 15.74:

$$\text{voltage} = \text{microwave frequency} / 483597.9\,\text{GHz}. \qquad (15.75)$$

The key feature ensuring reproducibility is that the voltage depends only on the ratio e/h and a frequency measurement: it is thus independent of the choice of superconductor or the fabrication process. Figure 15.18 is a sketch of the layout of a typical Josephson voltage standard. The upper panel shows a section of a string of junctions with the structure forming a low impedance stripline along which the microwaves travel. The lower panel shows the layout of the series array of Josephson junctions, marked by crosses along the striplines. Microwaves are coupled in from the left and are absorbed at the right by resistive terminations in order to eliminate reflections. Feeding the microwaves on four striplines in parallel reduces the variation of the microwave power along the chain of junctions. The DC supply provides a common current bias of order 1 µA. Junctions are around 10 µm by 15 µm, often in the form of Nb/AlO$_x$/Nb sandwiches. Agreement between Josephson voltage standards located at different laboratories across the world is better than parts per billion.

15.9 SQUIDs

Superconducting quantum interference devices or SQUIDs are rings of superconductor broken by one or more Josephson junctions. Here the emphasis will be on DC SQUIDs, having just two weak junctions around the ring, as shown in Figure 15.19. The action of the SQUID is analysed first before considering applications.

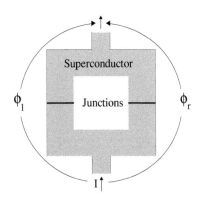

Fig. 15.19 A DC SQUID with two Josephson junctions.

The magnetic flux threading a SQUID is maintained by a circulating current flowing on the superconductor surface. Applying eqn. 15.51 and integrating round the closed loop gives

$$0 = \frac{2e}{\hbar} \oint \mathbf{A} \cdot d\mathbf{r} + \phi_\text{l} - \phi_\text{r}, \qquad (15.76)$$

where ϕ_l and ϕ_r are the phase changes along the two arms and $\mathbf{A}$ is the magnetic vector field. But

$$\oint \mathbf{A} \cdot d\mathbf{r} = \int \mathbf{B} \cdot d\mathbf{S} = \Phi \qquad (15.77)$$

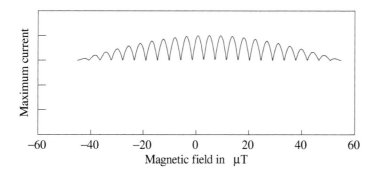

Fig. 15.20 Maximum current through a SQUID as a function of the applied magnetic field. Adapted from Figure 7a in R. C. Jaklevic, J. Lambe, J. E. Mercereau and A. H. Silver, *Physical Review* 140, A1628 (1965).

where Φ is the flux threading the loop. Then combining the last two equations

$$\phi_{\mathrm{r}} - \phi_{\mathrm{l}} = 2e\Phi/\hbar. \qquad (15.78)$$

Taking the junctions to be identical, and taking the phase change across each to be ϕ

$$\phi_{\mathrm{r}} = \phi + e\Phi/\hbar, \quad \phi_{\mathrm{l}} = \phi - e\Phi/\hbar. \qquad (15.79)$$

Thus the current flowing *through* the SQUID is

$$I = I_0[\sin(\phi + e\Phi/\hbar) + \sin(\phi - e\Phi/\hbar)] = 2I_0 \sin\phi \cos(e\Phi/\hbar). \qquad (15.80)$$

This produces the equivalent of a Young's two slit interference pattern when the flux, Φ, through the SQUID changes. In practice the modulation pattern looks like that in Figure 15.20 with an envelope that results from the individual junctions having finite width; in optical terms the envelope is the single slit diffraction pattern. There are current maxima whenever the flux threading the SQUID

$$\Phi = n\Phi_0, \qquad (15.81)$$

with n an integer. The material of SQUIDs are often $\mathrm{Al/AlO}_x/\mathrm{Al}$ or $\mathrm{Nb/AlO}_x/\mathrm{Nb}$ junctions. An insulated coupling coil of say 50 turns is fabricated on the SQUID *washer* which transfers the signal from the input coil, all cooled by liquid Helium. By measuring the current developed through the ring a flux change through the aperture of as small as $10^{-5}\Phi_0$ can be detected, with a response faster than kiloherz speed of the brain. This makes it possible to detect synchronous brain signals from as few as 10,000 synapses. The technique is called Magnetoencephalography (MEG).

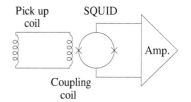

Fig. 15.21 A DC SQUID magnetometer circuit.

15.10   Qubits

The advantages of computing with quantum bits, *qubits*, rather than classical binary bits have been outlined in Chapter 10. A qubit could in principle be any two level quantum system, with its possible states

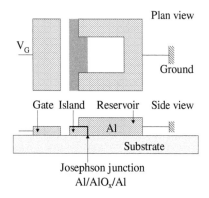

Fig. 15.22 Transmon qubit, incorporating a Cooper pair box.

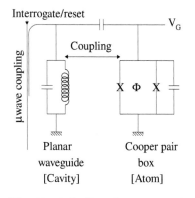

Fig. 15.23 Outline of transmon qubit circuit. Crosses mark the Josephson junctions and Φ represents the tuning flux.

defined by the surface of the Bloch sphere shown in Appendix D. The classical 0/1 state is replaced by a state capable of representing any linear superposition of the two classical states. The role of currents of electrons in controlling and interrogating bits in a classical computer is taken over by photons. Some progress towards producing practical elements of a quantum computer has been made using superconducting qubits held at mK temperatures; the interrogation and manipulation being performed using microwave photons of $\sim 10\,\text{GHz}$. One qubit architecture that shows significant potential will be discussed: the *transmon qubit*.[17] The basic elements of a transmon are shown in Figure 15.22. An $Al/AlO_x/Al$ Josephson junction separates a larger grounded reservoir from an isolated island capacitively coupled to a gate electrode at a potential V_g. Together the island and reservoir form a DC SQUID with two weak links. The state of the condensate of Cooper pairs on the island is the qubit. This arrangement is known as a *Cooper pair box* (CPB). Figure 15.23 is a sketch of the type of circuit using planar waveguides to interrogate and reset the qubit. A waveguide alongside the CPB forms an LC resonant circuit, which bathes the CPB with microwaves. The CPB and waveguide are the respective equivalents of an atom and a cavity enclosing it. The term Circuit QED (CQED) has been coined to describe this new combination. The other waveguide elements shown in the figure are used to interrogate the state of the resonant circuit, and hence the state of CPB qubit. This makes the qubit available for computation.

15.11　Topology and Dehmelt's question

Dehmelt asked why the quantization in superconductivity remained so well protected that voltage standards could be made of diverse and relatively impure materials. Looking back we see that flux quantization follows from the condensate phase being unique and from Stokes' Law requiring the phase change of the condensate round a closed loop be equal to the magnetic flux through the loop. Then the flux within can only be an integer times Φ_0 either through a hole in the superconductor or a vortex. There is no continuous symmetry involved of the sort that leads to conservation and quantization of angular momentum of atomic orbitals. Because quantization effects seen with superconductors are topological they are not destroyed by impurities or lack of symmetry. This parallels the topological origin of quantization in the Aharonov–Bohm effect and the integral quantum Hall effect described in Chapter 17.

Although the precision of quantization of circulation in He-II is less

[17]See for example C. R. Rigetti et al., *Physical Review B*86 100506(R) (2012); R. Barends et al., *Physical Review Letters* 111, 080502 (2013); J. Kelly et al., *Nature* 519, 66 (2015); J. Majer et al., Nature 449, 443 (2007); J. Koch et al., *Physical Review A*76 042319 (2007).

well established it is reasonable to accept that here too the origin of quantization is topological.

15.12   Further reading

The second edition of *Introduction to Superconductivity* by M. Tinkham, published by Dover Publications (2004) is a standard advanced text on classical superconductivity, with material on ^{3}He and some on HTSCs.

The second edition of *Introduction to Superconductivity* by A. C. Rose–Innes and E. H. Rhoderick, published by Pergamon Press (1978) has a clear and helpful approach that has not aged.

Superconductivity, Superfluids and Condensates by J. F. Annett published by Oxford University Press (2004) fleshes out the Landau and Ginzburg analysis of superfluids and superconductors. It has sections on high temperature superconductivity and ^{3}He.

'Superconducting Circuits and Quantum Information' by J. Q. You and F. Neri in *Physics Today* 58, 42 (2005) and 'Superconducting Quantum Bits' by J. Clarke and F. K. Wilhelm, *Nature* 453, 1031 provide useful short introductions to the field. There are articles by D-Wave Systems, for example M. W. Johnson *et al.* in *Superconducting Science and Technology* 23 065004 (2010).

Exercises

(15.1) A superconducting lead film is deposited on a 2 cm long rectangular insulating rod leaving the ends open. The circumference of the path in the lead film is 2 mm. This circuit in lead has a self-inductance 10^{-13} H. It is held at a temperature below the critical temperature of lead. The London penetration depth in lead is 50 nm. Suppose the current established in the lead loop decays by at most 2 per cent in 7 hours. Calculate the lower limit to the resistivity of superconducting lead. Based the experimental results of D. J. Quinn and W. B. Ittner, *Journal of Applied Physics* 33, 748 (1962).

(15.2) The temperature dependence of the critical field for some superconductors takes the approximate form

$$H_c(T) = H(0)[1 - (T/T_c)^2].$$

Work out the corresponding temperature depen-

dence of the Gibbs free energy and the entropy change between states.

(15.3) Assuming the temperature dependences of the last exercise, show that the change in heat capacity between the two phases of aluminium in zero applied field at $T = T_c$ is $4B_c^2(0)/[\mu_0 T_c^2]$. Compare the value you find with the measured value for aluminium presented in Figure 15.8. Take T_c to be 1.18 K and $B_c(0)$ to be 10^{-2} T. The density of aluminium is 2700 kg m^{-3} and its atomic number is 27.

(15.4) A magnetic field B_0 is applied parallel (in the z-direction) to a thin superconducting slab of thickness $2t$ in the x-direction. Show that the field inside the superconductor has the form

$$B(x) = B_0 \cosh[x/\lambda]/\cosh[t/\lambda],$$

where λ is the penetration depth and x is measured from midway across the slab.

(15.5) Mercury has a critical temperature of $4\,\mathrm{K}$. The reflection coefficient for infrared radiation of frequencies above $4\,10^{11}\,\mathrm{Hz}$ is similar for mercury in the superconducting and normal states. Below this frequency the reflection from the superconducting phase is ten times larger than in the normal phase. Explain this change.

(15.6) Show that the gauge transformations

$$\mathbf{A} \rightarrow \mathbf{A} - (\hbar/e^*)\nabla\xi$$
$$\psi \rightarrow \psi \exp(i\xi)$$

leave unchanged the magnetic field and the superconducting current given by eqn. 15.42.

(15.7) Use the results of Deaver and Fairbank's experiment to deduce the area of cross-section of the cylinder of tin.

(15.8) Show that there must be an upper limit to possible critical field values given by

$$B \simeq k_{\mathrm{B}}T_{\mathrm{c}}/\mu_{\mathrm{B}}, \tag{15.82}$$

where μ_{B} is the Bohr magneton.

(15.9) Infer the maximum supercurrent velocity in tin by following the argument made for obtaining the maximum superfluid velocity in He-II. Take the value of k_{F} to be $1.2\,10^{10}\,\mathrm{m}^{-1}$, and $3.72\,\mathrm{K}$ to be the critical temperature.

Gaseous Bose–Einstein condensates

16

16.1 Introduction

In 1995 the nearest approach to Einstein's condensate of bosons with vanishing interactions was achieved. Minute clouds of alkali metal atoms held in an otherwise hard vacuum within an magneto-optical trap were cooled below a critical temperature $T_c \sim 1\,\mu\mathrm{K}$.[1] Bosons are required so that the atoms involved must contain an even number of their fermion components: that is of neutrons plus protons plus electrons. The clouds are very dilute gases, around a million times less dense than air, containing $\sim 10^8$ atoms. Their states are indicated deep in the grey region of instability in figure 3.8. The instability is due to three body scatterings that form complexes that deplete the condensate. Two body scatters on the other hand lead to thermal equilibrium. It turns out that for condensates containing the alkali metals of interest $^{87}\mathrm{Rb}$ and $^{23}\mathrm{Na}$ the rate of two-body scatters is a hundred times the three-body scattering rate: these condensates can come to thermal equilibrium and be studied for up to minutes before depletion. At these ultra-low temperatures two-body interactions are limited to the s-wave scattering discussed in Section 7.9. The expression relating the potential U_0, to the scattering length a_s, is given in eqn. 7.72

$$U_0 = \frac{4\pi\hbar^2 n a_s}{m} \qquad (16.1)$$

where m is the atom's mass. Figure 7.11 shows that when a_s is negative the atoms are pulled together so that the condensate volume collapses, and depletion processes dominate. With a positive scattering length the atoms are held apart: this is the case for $^{87}\mathrm{Rb}$ which has $a_s = +98a_0$ ($+5.2\,\mathrm{nm}$); and for $^{23}\mathrm{Na}$ which has $a_s = +19a_0$ ($+1\,\mathrm{nm}$). These scattering lengths are of order 50 times the Bohr radius so that the two body scattering cross-sections are boosted by a factor of 2500; this boost accounts for the essential rapid progress to thermal equilibrium. The combination of a weakly repulsive interatomic s-wave potential and zero point motion stabilize the condensates of $^{87}\mathrm{Rb}$ and $^{23}\mathrm{Na}$ once they are formed in a trap.

The cooling techniques parallel those described in Chapter 12 for cooling ions. Cooling is mainly by the action of laser beams with as a final

[1] The leaders of the teams making the discovery, E. A. Cornell and C. E. Wieman of NIST and the University of Colorado and W. Ketterle of MIT, received the 2001 Nobel Prize in Physics.

Quantum 20/20: Fundamentals, Entanglement, Gauge Fields, Condensates and Topology.
Ian R. Kenyon. © 2020. Published in 2020 by Oxford University Press.
DOI: 10.1093/oso/9780198808350.001.0001

step the controlled evaporation of the hottest atoms; magnetic fields or again laser beams are used to trap the clouds of atoms. The teams mentioned used ^{87}Rb and ^{23}Na which have strong transitions from the ground state, $S_{1/2} \to P_{3/2}$, in the near infrared and visible respectively. This is convenient for cooling and imaging the state of the gas sample with readily available lasers. It is equally useful that alkali metals have large magnetic moments, making them easy to manipulate with magnetic fields. These species continue to be widely used.

The overall wavefunction of the condensate, Ψ, is the repeated product of one common wavefunction, $\psi(\mathbf{r})$, describing the motion of each and every atom in the condensate:

$$\Psi(\mathbf{r}_1, \mathbf{r}_2, \cdots, \mathbf{r}_n, t) = \prod_{i=1}^{i=n} [\psi(\mathbf{r}_i, t)]. \qquad (16.2)$$

Methods of trapping and cooling are described first in Section 16.2. The dilute gaseous condensates are fragile so that imaging, described in Section 16.3 is all done optically. The discussion of evidence for the existence of dilute gaseous Bose–Einstein condensates, BECs, and their properties is contained in Section 16.4.

The Gross–Pitaevskii equation, GPE, presented in Appendix J provides a simple framework for the analysis of BEC properties. It is a Schrödinger equation for the atoms of the condensate which incorporates a local potential due to the trap and a second effective potential due to the weak interaction of an atom with surrounding atoms. This latter is the potential already deduced and presented in eqn. 7.72 for low energy s-wave interactions between atoms.

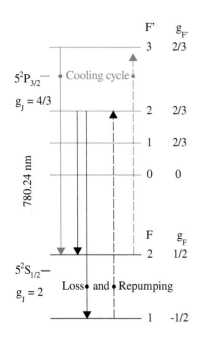

Fig. 16.1 Laser cooling and repumping cycles used for ^{87}Rb. Repumping is required when in the cooling cycle atoms enter the F' = 2 level rather than the F' = 3 level and decay to the F = 1 level. After repumping to the F' = 2 level atoms mostly decay to the F = 2 level thus re-entering the cooling cycle.

The technique of neutron scattering that was used to study excitations in He-II would simply blow the condensate away. Instead *Bragg spectroscopy* was invented in which an atom absorbs a photon from one laser beam while a second counterpropagating laser beam stimulates emission. By tuning the lasers the experimenter can input a range of energy-momentum combinations to the condensate in order to determine which provoke actual excitations of the condensate. This technique for obtaining the dispersion relation is detailed in Section 16.5.

Recalling the behaviour of He-II and superconductors, we expect vortex formation in dilute gaseous condensates. This topic is treated in Section 16.6.

A final section is used to present and analyse condensates achieved with cold dilute *fermion* gases. Condensation can happen when pairs of atoms form diatomic molecules or when they become weakly bound in structures analogous to Cooper pairs. Simply by sweeping an applied magnetic field through the value 832 G (gauss) a degenerate Fermi gas of ^{6}Li atoms can be tuned from a molecular BEC to a superconducting like condensate. This intriguing phase change is known as the BEC/BCS crossover.

16.2   Making the BEC

First some parameters of interest are deduced to give a feeling for what is involved. As mentioned, the emphasis will be on the commonly used alkali metal bosonic isotopes ^{87}Rb and ^{23}N, usually at densities in the range 10^{13}–10^{15} cm^{-3} in clouds consisting of 10^4–10^8 atoms inside an otherwise evacuated volume of cm dimensions. The gases are trapped in an harmonic potential well provided by magnetic fields or a laser beam. Chapter 3 showed that Bose–Einstein condensation occurs when the quantum extent of the individual atoms, that is their de Broglie wavelength grows large enough that they overlap and become indistinguishable. The expression for the critical temperature at which condensation occurs, eqn. 3.54, is repeated here

$$k_\mathrm{B}T_\mathrm{c} = 0.94\hbar\omega_0[N]^{1/3}, \tag{16.3}$$

where there are N atoms in an harmonic well with oscillation angular frequency ω_0 along all three axes. Taking the example of 10^6 atoms of ^{87}Rb in a well with angular frequency $1000\,\mathrm{rad\,s^{-1}}$ we get a critical temperature of $0.7\,\mu$K. At this temperature the atomic velocity is $1.2\,\mathrm{cm\,s^{-1}}$. The de Broglie wavelength at T_c (and also the atomic separation) is given by eqn. 3.48

$$\lambda_\mathrm{deB} = (2\pi\hbar^2/mk_\mathrm{B}T)^{1/2}, \tag{16.4}$$

which is $290\,$nm for the example stated. This condensate has a density 10^{-5} of atmospheric air. The classical excursion corresponding to zero point motion in an harmonic well at the angular frequency ω_0 is given by eqn. 2.32:

$$a_\mathrm{zpf} = \sqrt{\hbar/(2m\omega_0)}, \tag{16.5}$$

which gives $0.60\,\mu$m in the case considered. Mutual repulsion spreads the atoms out by factor $\sim\!10$ as detailed in Appendix J.

Atoms are produced by heating a wire covered in a suitable salt at $\sim\!500\,$K.[2] It follows that a reduction in temperature by a factor 10^9 is required to give condensation. Trapping and laser cooling reach $100\,\mu$K, *optical molasses* of laser beams cools to $10\,\mu$K and finally, evaporation of the hottest atoms carries the temperature below $\sim\!1\,\mu$K.

Figure 16.1 shows the energy levels that are of importance in cooling and probing Rubidium atoms.[3] Figure 16.2 is a sketch of apparatus for cooling a cloud of alkali metal bosonic atoms. Gas from a heated filament flows through a narrow metre-long channel to the trap. A laser

[2]The source could equally be an ampoule of gas.

[3]The ^{23}Na level structure mimics that shown, with a change in the principal quantum number of the electron: $5^2\mathrm{P}_{3/2}$ is replaced by $3^2\mathrm{P}_{3/2}$ and $5^2\mathrm{S}_{1/2}$ by $3^2\mathrm{S}_{1/2}$.

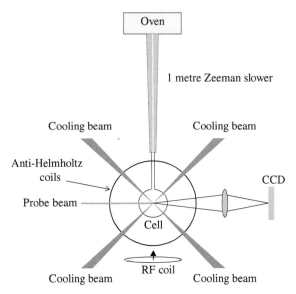

Fig. 16.2 Sketch of cooling apparatus with the cloud of atoms at the centre of the evacuated cm sized cell. The Zeeman slower is on a reduced scale to fit in one figure. The third laser pair point in/out of the paper.

beam red detuned with respect to the $S_{1/2} \to P_{3/2}$ atomic transition points upstream along the channel. As a result the atoms see the on-coming photons Doppler shifted back into resonance. The atoms absorb these photons, then rapidly de-excite, in the process emitting photons in random directions: there is consequently a net loss in the forward momentum of the atoms. However the effect of the deceleration of the atoms is that the resonance condition is lost. In order to maintain the cooling action a magnetic field is applied that gets weaker along the channel in step with the atom's slowing down. The Zeeman shift in the frequency of the atomic transition is tuned to maintain resonance along the length of the channel. The atoms emerge from this *Zeeman slower* at around 1 K.

Next the atoms enter an evacuated glass cell and are confined there within a *magneto-optical trap* (MOT). The magnetic component is pro-vided by a pair of coils wound in the anti-Helmholtz configuration and centred on the cell.[4] The upper panel of figure 16.3 shows the magnetic field lines in an MOT, the atomic energy levels (broken lines) and the counterpropagating circularly polarized laser beams (σ_+) used to con-fine the gas. Capture in the trap relies on the variation of the Zeeman shift across the trap. We take the quantization axis to point *right-ward*. At the centre the two laser beams, red detuned with respect to

---

[4]The coils are wound in opposition and placed about one radius apart. The magnetic field is a quadrupole viewed in any plane containing the axis and is exactly zero at the midpoint along the axis.

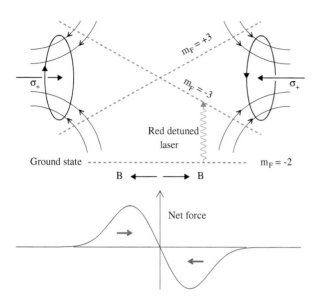

Fig. 16.3 The function of a magneto-optical trap. The field lines carry arrowheads. Only the relevant magnetic substates are indicated. σ_+ indicates incoming right circularly polarized photons.

the $F = 2 \to F' = 3$ transition, are absorbed equally by the atoms. However on the right the incoming right circularly polarized photons, $(J = 1, m_J = -1)$ from the right, are preferentially absorbed: for example the atoms in the $(F = 2, m_F = -2)$ state are excited into the $(F' = 3, m_F = -3)$ state:[5]

$$(2, -2) + (1, -1) \to (3, -3). \tag{16.6}$$

This leads to an effective force on the gas acting toward the trap centre. A similar effect is obtained on the left. The distribution of restoring force is displayed in the lower panel of figure 16.3. At this stage there are typically 10^9 atoms in a volume of dimensions of order $10 \, \mu\text{m}$. Then the trap is turned off and cooling initiated using laser beams red detuned from the $F = 2 \to F' = 3$ transition. Six beams are directed at the trap centre along the $+x$, $-x$, $+y$, $-y$, $+z$ and $-z$ directions. The effective force now opposes atoms travelling ouward in any direction; its viscous nature led to it being described as *optical molasses*. The limit of cooling appears to be determined by the balance between Doppler cooling and recoil heating, given in eqn. 12.19. Using the lifetime of 26 ns of the 5^2P state suggests a limiting temperature of $150 \, \mu\text{K}$. In fact cooling continues well below this limit but it is inadequate to reach T_c.

This sub-Doppler cooling comes about because the counterpropagating laser beams form standing waves across the gas. As a result the ground state magnetic substates oscillate in energy as they travel: for half a wavelength one has the higher energy, for the next half wavelength the other has higher energy. An atom can absorb a photon at an

[5] There is a subtle point here. The left-going photons with σ_+ polarization are in the (1,-1) state so that the m_F value of an atom drops by unity when it absorbs one. Right-going photons with the same σ_+ polarization are in the (1,+1) state so that m_F increases when an atom absorbs one.

[6]This is known as Sisyphus cooling by analogy with the torment of Sisyphus in a classical fable: he was condemned to roll a stone up an incline, but on reaching the top he found the stone inevitably fell back to the starting point.

energy peak; decay to the state of opposite magnetic quantum number and lower energy; then climb the potential hill and once more absorb a photon. This extends the effective lifetime by the duration of the hill climb.[6] There is however an inescapable lower limit to the energy that can be reached in laser cooling. This is the *recoil energy* of an atom when it emits a photon and reverts to its ground state:

$$E_r = p^2/2m = h^2/(2\lambda^2 m) \tag{16.7}$$

where p is the photon momentum, m is the atomic mass and the transition wavelength is λ. The temperature attainable after cooling would be $T_r = E_r/k_B$. In fact the acceleration of atoms in the periodic potential worsens this limit to $\sim 25 T_r$. With laser wavelength of 780 nm this gives 5 μK.

In order to retain and compress the cloud magnetic trapping can be used. Coils additional to the anti-Helmholtz coils are needed to remove the region of zero magnetic field around the centre; otherwise atoms atoms would spin flip there without any energy penalty, and escape the cooling mechanism. This improved configuration is called a *Ioffe–Pritchard trap*. The resulting field has a spatial distribution:

$$B \approx B_0 + B_z z^2 + B_r(x^2 + y^2), \tag{16.8}$$

where the coordinates are measured from the trap centre, z being the coordinate along the anti-Helmholtz coils' axis. The energy of the atoms in the field is

$$E = g_F m_F \mu_B B, \tag{16.9}$$

where g_F is the Landé factor whose value for the various states is displayed in figure 16.1. Atoms in states with positive magnetic energy are trapped and undergo harmonic motion around the trap centre. These states with positive magnetic energy are called *low field seeking*.

In practice the lowest temperatures reached after Sisyphus cooling are a few μK, and a supplementary means of cooling is needed to achieve condensation. This final step is to evaporate the most energetic atoms and then allow the remainder to come to a new thermal equilibrium. In this way temperatures below 0.1 μK can be attained so that the ^{87}Rb cloud condenses. In detail the sequence starts with the atoms being pumped into the $(F = 1, m_F = -1)$ state. Referring to figure 16.1 we see that these atoms have $g_F = -1/2$ making their magnetic energy, given by eqn. 16.9 positive. Hence they are low field seeking atoms and the three-dimensional magnetic trap captures them.

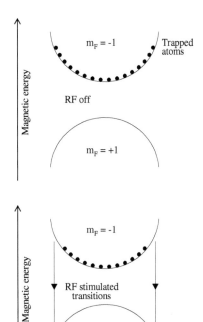

Fig. 16.4 Radio frequency evaporation of atoms from a magnetic trap.

At this point radio frequency radiation is used to flip the spins of the most energetic atoms so that they become high field seekers and leave the trap as shown in Figure 16.4. The energy involved in the transition from the $m_F = -1$ state to the $m_F = +1$ antitrapped state is just $\mu_B B$. Recall that the field is largest at the trap circumference and hence the difference in energy between the spin states is largest there also. Initially the radio frequency is set just high enough to flip the atoms in the outer layer of the trap. Suppose the field there is 3 gauss, then the frequency required is $\mu_B B/h$ that is 4.2 MHz. After the rapid re-thermalization the frequency is lowered to strip off another layer and the cycle repeated until the desired temperature is attained. This final cooling step is only possible because re-thermalization by two body scattering is sufficiently rapid in ^{87}Rb and ^{23}Na. Typically there remain 10^6 atoms at a density of 10^{14} cm^{-3} and the total time taken for cooling is $\sim$10 s.

16.3 Imaging

First the cloud of atoms is released from the trap by turning the magnetic field off. The cloud of atoms expands ballistically and after a delay the image of the cloud is recorded. Its image's size is determined by the expansion rate and so gives information about the velocity spectrum of the atoms at the moment of release. In the case of a pure condensate the atoms are expected to share a common velocity, whereas before condensation they have a thermal distribution.

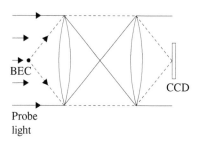

Measurements of the clouds are necessarily all optical because the cloud of atoms is so fragile. Two techniques will be mentioned. The upper panel in Figure 16.5 shows the arrangement for absorptive imaging which destroys the cloud while recording it. A laser beam resonant with a strong transition from the ground state of the atoms projects an image of the cloud onto a CCD, of the sort found in digital cameras. The transmitted intensity distribution $I(x,y)$ contains information about the cloud's density distribution in space $n(x,y,z)$:

$$I(x,y) = I_0 \exp\left[-\int n(x,y,z)\sigma \mathrm{d}z\right] \qquad (16.10)$$

where x,y are orthogonal coordinates on the plane of the CCD, and z is the coordinate along the laser beam. I_0 is the intensity incident on the cloud; σ is the absorption cross-section for the laser light on a condensate atom, which is $2.9\,10^{-9}$ cm^2 for the transition, $F = 2, m_F = 2 \rightarrow F' = 3, m_F = 3$, used in the case of rubidium. The directly measurable quantity is the column density through the cloud at x,y:

$$n_c(x,y) = \mathrm{OD}(x,y)/\sigma, \qquad (16.11)$$

where the optical density OD is $-\ln[(I - I_B)/(I_R - I_B)]$ with I_R and I_B being respectively the intensities when the cloud is absent, and when both cloud and probe beam are absent. This measurement of the optical density eliminates the effect of scattered light and of electronic noise in the CCD. If the potential well and cloud have a simple symmetry (e.g. spherically symmetric), then the three dimensional form of the cloud,

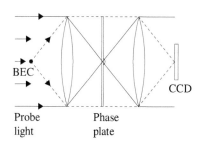

Fig. 16.5 Optical layout for destructive imaging (upper panel) and phase contrast imaging (lower panel) of the BEC. The cone of light scattered by the BEC follows the broken lines.

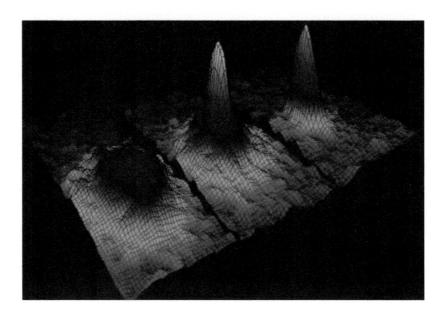

Fig. 16.6 ^{87}Rb clouds at temperatures well above, around and well below the critical temperature. Figure 13 in E. Cornell, *Journal of Research of the National Institute of Standards and Technology* 101, 419 (1996). Courtesy Professor Cornell and NIST.

$n(x, y, z)$ can be inferred using the analyses presented in Appendix J.

In one non-destructive form of imaging the laser probing the cloud is far off resonance from any atomic transition, making the photon-atom interaction dispersive rather than absorptive. The atoms do not recoil in dispersive imaging which gives the huge advantage that a sequence of images can be taken of a single cloud as it evolves. A disadvantage is that data from dispersive imaging is harder to interpret quantatatively: it would, otherwise, displace destructive imaging entirely.

The optical layout is shown in the lower panel of Figure 16.5. A lens brings the unscattered beam to a focus on a small boss located at the centre of a transparent sheet. This produces a phase delay of $\pi/2$ relative to light scattered by the cloud which almost all misses the boss. As before a CCD is placed at the image plane of the cloud. At this surface the light amplitude is the sum of the unscattered light $E_0 \cos(\omega t)$ and that scattered by the cloud

$$E_0 \sin(\omega t + \varepsilon) = E_0 \sin(\omega t) + \varepsilon E_0 \cos(\omega t), \qquad (16.12)$$

where ε is the small phase change due to travel through the cloud. The average intensity is thus

$$
E_0^2 \int_0^{2\pi/\omega} [\sin \omega t + (1 + \varepsilon) \cos \omega t]^2 \, \mathrm{d}t \Big/ \int_0^{2\pi/\omega} \mathrm{d}t
$$
$$
= \quad E_0^2 [1 + (1 + \varepsilon)^2]/2 \approx E_0^2 (1 + \varepsilon) \qquad (16.13)
$$

making it possible to extract the cloud's contribution. If there were no phase shift the intensity would be

$$E_0^2 \int_0^{2\pi/\omega} [\, 2\cos\omega t - \varepsilon\sin\omega t\,]^2 \, \mathrm{d}t \Big/ \int_0^{2\pi/\omega} \mathrm{d}t = E_0^2(2 + \varepsilon^2/2) \quad (16.14)$$

in which the cloud's effect is swamped. The technique is called *phase contrast imaging*.

A spectacular display of condensation is shown in Figure 16.6. The left-hand image was recorded for a cloud at a temperature well above T_c and shows only a broad Gaussian shape indicating a thermal distribution of momenta. The central image was recorded after cooling near T_c, and the right-hand image at a temperature well below T_c: this last shows the characteristic very narrow peak that identifies the presence of a condensate of atoms in a single momentum state.

16.4   Condensate properties

This section relates results from experiments that established the existence and properties of condensates in dilute boson gases, principally using the alkali metals ^{87}Rb and ^{23}Na cooled to microkelvin temperatures. Figure 16.7 shows the absorptive images of ^{23}Na clouds of volume 10^{-8} cm^3 during evaporation to a BEC: what is shown therefore is a sequence of images of identically prepared clouds. The geometric mean of the frequencies of oscillation of atoms in the trap in the three coordinate directions was 415 Hz. After evaporation to one of the radio frequencies indicated in the figure the trap was switched off and the cloud allowed to expand for 6 ms and imaged. The expanding clouds are seen to develop a sharp spike as the evaporation frequency falls through 0.7 MHz. This spike demonstrates that before release from the trap a fraction of the the atoms in the cloud form a condensate whose relative motion is zero point motion. The image seen is thus *bimodal* with a broad component from atoms remaining in thermal equilibrium and a narrow component from the condensate. A fit is made to each image with a combination of a thermal Maxwell–Boltzmann distribution and a Gaussian for the sharp peak due to the condensate, in thermal equilibrium. In the case considered the fitting procedure gave a condensation temperature of $2.0\pm0.5\,\mu$K. Then eqn. 16.3 was used to determine the expected number of atoms in the condensate at that temperature for an oscillation frequency of 415 Hz. This value is $1.4\,10^6$, which agrees within errors with the number, $0.7\,10^6$, derived directly from fitting the image. This makes the condensate volume $\sim$(37µm)^3.

Figure 16.8 shows values of the condensate fraction found from fitting a bimodal distribution to the images of ^{87}Rb clouds for a range temperatures following cooling. An estimate of the transition temperature,

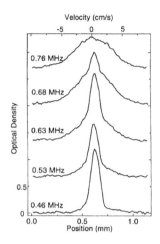

Fig. 16.7 Images of ^{23}Na clouds after evaporative cooling to the final radio frequencies indicated. Images are displaced for clarity. Figure 3 in K. B. Davis, M.-O. Mewes, M. R. Andrews, N. J. Van Druten, D. S. Durfee, D. M. Kurn and W. Ketterle, *Physical Review Letters* **75**, 3969 (1995). Courtesy Professor Ketterle and The Amercan Physical Society.

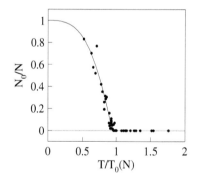

Fig. 16.8 Fraction of condensate in an ^{87}Rb cloud as a function of temperature. Adapted from Figure 1 in J. R. Ensher, D. S. Jin, M. R. Matthews, C. E. Wieman and E. A. Cornell, *Physical Review Letters* **77**, 4984 (1996).

T_0 was inferred from eqn. 16.3 where N is the number of atoms in the cloud and ω_0 the oscillation angular frequency of atoms in the magnetic potential well. It is clear that there is a threshold, very close to T_0, below which the condensate grows at the expense of the thermal component. The full line drawn in Figure 16.8 is a fit to the expected dependence from eqn. 3.55

$$N_0/N = 1 - (T/T_c)^3 \qquad (16.15)$$

and yields $T_c = 0.945 T_0$. Below a temperature $T_c/2$ the residual thermal component is insufficient in number to allow a reliable determination of the temperature.

The key demonstration of the long range coherence in dilute gas condensates was provided by the MIT group[7] using a condensate of $5\,10^6$ ^{23}Na atoms following cooling, trapping and evaporation. An argon laser provided a beam at 514 nm, blue shifted from the sodium D-line at 589 nm. This beam was directed to form a sheet across the midplane of the magnetic trap. The laser induced AC Stark shift of atomic states is given by eqn. 12.11, Δ being the transition frequency minus the laser frequency. A blue shifted laser will raise the energy of the ground state atoms and repels them, in this way separating the cloud into two independent condensates 40 μm apart. The magnetic trap and the argon laser were switched off simultaneously and after 40 ms flight time the atoms were pumped into the $F = 2$ state. Then after a 10 μs delay an absorptive image of the cloud was obtained. In Figure 16.9 two images are displayed showing the typical interference fringes with spacings of 15–20 μm. If the sources are a distance d apart and the delay before imaging is t then at any point on the interference pattern the relative velocity of the the two clouds must be $v = d/t$ and their relative momentum

$$p = mv = md/t. \qquad (16.16)$$

where m is the mass of a ^{23}Na atom. In turn this implies that the wavelength of the interference pattern seen is

$$\lambda = h/p = ht/(md). \qquad (16.17)$$

The relative phase between the two condensates changes randomly from image to image: as a result the fringes move en bloc but the spacing remains the same. It is clear that there is a high degree of coherence between the condensates because the fringes extend over all the overlap region.

Hänsch and colleagues[8] demonstrated the degree of coherence across a ^{87}Rb BEC by releasing two separated segments of a cloud and recording them after they had fallen some distance under gravity. In essence this

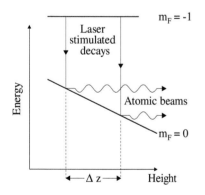

Fig. 16.9 Interference between laser and evaporatively cooled clouds of sodium atoms after 40 ms time of flight. Images are 1.1 mm horizontally by 0.5 mm vertically. Figure 2 from M. R. Andrews, C. G. Townsend, H.-J. Miesner, D. S. Durfee, D. M. Kuhn and W. Ketterle, *Science* 275, 637 (1997). Courtesy Professor Ketterle. Reprinted with permission from The American Association for the Advancement of Science.

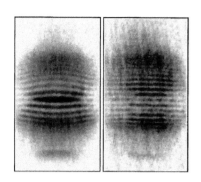

Fig. 16.10 Production of two separate coherent beams of atoms from a single condensate. The $m_F = -1$ trapped state is the central portion of that shown in Figure J.1.

[7]M. R. Andrews, C. G. Townsend, H.-J. Miesner, D. S. Durfee, D. M. Kuhn and W. Ketterle, *Science* 275, 637 (1997).
[8]I. Bloch, T. W. Haensch and T. Esslinger, *Nature* 403, 168 (2000).

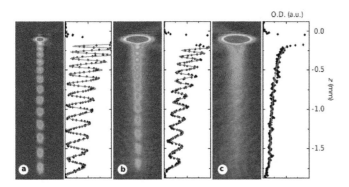

Fig. 16.11 Interference patterns observed between atomic beams outcoupled from ^{87}Rb atomic clouds in free fall. On the left at a temperature well below T_{c}; on the centre at a temperature close to T_{c}; and on the right, well above T_{c}. Figure 1 in I. Bloch, T. W. Haensch and T. Esslinger, *Nature* 403, 166 (2000). Courtesy Professor Haensch and Springer Verlag.

is a two slit interference experiment. Absorption of two weak radio frequency beams with slightly differing frequencies, ω and ω', transfered atoms from a trapped state $(F = 1, m_F = -1)$ into an untrapped state $(F = 1, m_F = 0)$ as indicated in Figure 16.10. The variation in energy of the untrapped state, relative to that of the trapped state with height, is due to gravity displacing the minimum of the trap potential from that of the magnetic field vertically by 12 μm. As a result the radio frequency beams release the atoms at heights differing by Δz, where

$$mg\Delta z = \hbar(\omega - \omega'),\qquad(16.18)$$

m being the atomic mass of a ^{87}Rb atom. Figure 16.11 shows the interference patterns observed in the image intensity of the outcoupled beams as the fall together. The vertical separation of of the release locations is 465 nm: on the left at a temperature well below T_{c}; in the centre at a temperature close to T_{c}; and on the right at a temperature well above T_{c}. The visibility of the interference patterns is defined by

$$V = (I_{\max} - I_{\min})/(I_{\max} + I_{\min})\qquad(16.19)$$

where $I_{\max}(I_{\min})$ is the maximum (minimum) intensity. The visibility would be 1.0 for full coherence and zero if the beams of atoms were incoherent. This quantity is plotted versus Δz in Figure 16.12 for two cases: at 290 μK well above, and at 250 μK well below the critical temperature. The visibility for the thermal state above T_{c} is a Gaussian in Δz whose width matches the de Broglie wavelength. Below T_{c} a corresponding Gaussian sits on a constant plateau in visibility of around 0.75: this latter feature demonstrates that the Bose–Einstein condensate possesses long range coherence. The parallel case of He-II appeared in Figure 14.9.

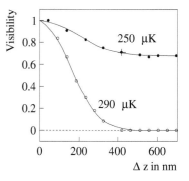

Fig. 16.12 Spatial correlation of a trapped Bose gas. The visibility for temperatures above and below T_{c}. The curves are Gaussian fits plus a long range component below T_{c}. Adapted from Figure 4 in I. Bloch, T. W. Haensch and T. Esslinger, *Nature* 403, 166 (2000). Courtesy Professor Haensch.

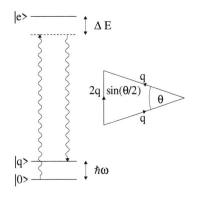

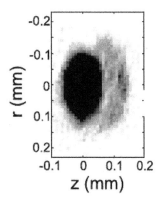

Fig. 16.13 Bragg excitation of condensate from rest $|0\rangle$ to a state of motion $|q\rangle$.

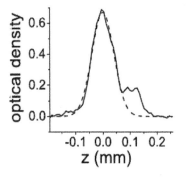

Fig. 16.14 Bragg displaced (right) and parent condensate clouds (left) in the upper panel; below a section through this image. r/z is the radial/axial coordinate. Adapted from Figure 1, *Physical Review Letters* **88**, 120407 (2002). Courtesy Professor Davidson and The American Physical Society.

16.5　Bragg spectroscopy of dispersion relations

Gaseous BECs are too delicate for their excitations to be investigated in the same way as He-II, that is using neutron scattering. Instead a less violent yet more precise technique was developed, albeit limited in its application to gases. In this *Bragg spectroscopy* two laser beams illuminate the the gas cloud from opposite directions. This is shown in Figure 16.13. Their energies are offset from a transition between the ground state $|0\rangle$ and an excited state $|e\rangle$. Atoms in the ground state in the condensate can absorb photons from one laser and be stimulated to emit by the other laser, returning them to their ground state, but now having collective motion $|q\rangle$. The lasers are tuned to slightly different frequencies ω and $\omega + \Delta\omega$ so the process resembles Raman scattering. Thus the energy given to a condensate atom is $\hbar\Delta\omega$ and typically has a value $\sim 10^{-10}\hbar\omega$. If θ is the angle between the laser beams then the momentum received by the atom is $\Delta q = 2q \sin \theta/2$, where q can be taken as the momentum of a photon from either laser because their energies differ so little.

At the Weizmann Institute[9] ^{87}Rb atoms prepared in the $F = 2, m_F = 2$ ground state were cooled to produce a condensate of 10^5 atoms with a thermal fraction of less than 0.05. Bragg scattering was carried out with the laser frequencies detuned 6.5 GHz below the $F = 2 \rightarrow F' = 3$ transition frequency. The measurement sequence started by pulsing the Bragg lasers on simultaneously, and at the end of the pulse the cloud was allowed to expand for 38 ms, after which an absorption image was recorded. Figure 16.14 shows the shadows of the undisturbed condensate and the excited atoms in profile and projection. The dashed line is a Gaussian fit to the undisturbed condensate cloud. The column density of the weaker right hand spot relative to the total gives the probability that excitation at the selected energy $\hbar\omega$ and momentum $\hbar q$ would occur. This reveals the essence of Bragg spectroscopy: the energy and momentum can be separately tuned through the choices of the frequency difference and the angle between the lasers' beams. Figure 16.15 shows the observed excitation spectrum: the inset is an enlargement of the low wave number region. The solid line is the prediction using the Bogoliubov approximation described in Appendix J. Free particle behaviour is shown by the broken line. We see that the dispersion relation of excitations in the condensate is phonon-like (energy proportional to momentum) at low energies and from the slope the speed of sound was determined to be 2 ± 0.1 mm s^{-1}. The measured dispersion relation converges to the quadratic form at high energies indicating the onset of free particle behaviour.

[9] J. Steinhauer, R. Ozeri, N. Katz and N. Davidson, *Physical Review Letters* **88**, 120407 (2002).

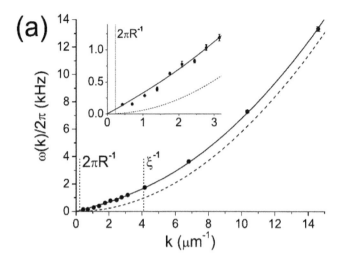

Fig. 16.15 The measured excitation spectrum of a trapped Bose–Einstein ^{87}Rb condensate. The solid line is the Bogoliubov spectrum with $gn(r = 0)/\hbar$ equal to 1.91 kHz. The dashed line is the free-particle spectrum. The inset shows the linear phonon regime. ξ is the healing length given by eqn. J.6. Adapted from Figure 3 in J. Steinhauer, R. Ozeri, N. Katz and N. Davidson, *Physical Review Letters* 88, 120407 (2002). Courtesy Professor Davidson and The American Physical Society.

16.6   Vortices

From the argument made in the case of He-II in Section 14.10 it follows that vortex formation is also to be expected in a dilute gaseous BEC. Around any closed path in a condensate the change in phase, ϕ of the condensate has to be an integral multiple n, of 2π because the wavefunction is single valued at any point. The superfluid velocity is connected to the rate of change in phase through eqn. 14.13

$$\mathbf{v}_s = (\hbar/m)\nabla\phi, \tag{16.20}$$

where m is the mass of an atom in the condensate. Hence the circulation round a closed path

$$\oint \mathbf{v}_s \cdot d\mathbf{r} = nh/m. \tag{16.21}$$

Each vortex contributes h/m to the circulation. Close to the axis of a vortex the gas would need to travel faster than the speed of sound: there is therefore a core of thermal gas entrained around the vortex axis. The radius of this normal core, r_c is at least the healing length, given by eqn. J.6, and in the (Thomas–Fermi) limit when the interaction energy dominates the kinetic energy of the condensate $r_c \sim 1.94\xi$.

The quantum number n is another topological quantum number, a *winding number*, independent of any symmetry of the path. In this case it counts the number of vortices, regions where there is a hole in the condensate, enclosed by the path. The quantization of circulation in He-II and of enclosed flux in superconductors share this topological origin. They are not reliant on a symmetry in the manner of quantization of angular momentum.

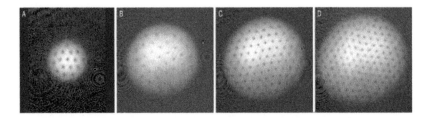

Fig. 16.16 Observation of clouds with around (A) 16, (B) 32, (C) 80, and (D) 130 vortices. They crystallize in a triangular Abrokosov lattice. In (D) the cloud diameter after ballistic expansion was 1 mm. Figure 1 in J. R. Abo–Shaeer, C. Raman, J. M. Vogels and W. Ketterle, *Science* 292, 476 (2001). Courtesy Professor Ketterle. Reprinted with permission from The American Association for the Advancement of Science.

Figure 16.16 is made up of a set of images of vortex patterns produced by stirring a ~30 μm radius cloud of $5\,10^7$ ^{23}Na atoms, of which ≥ 90 per cent are in the condensate. Rotation was effected by stirring for 40 ms with a laser of wavelength 532 nm blue shifted with respect to the atomic transition, and then allowing 500 ms for the motion to equilibrate. Finally, the gas was released from the trap and after 35 ms of ballistic expansion an absorption image recorded. These images show arrays of vortices, arranged with the same triangular symmetry as He-II vortex arrays. These are known as *Abrikosov* lattices, which have the property of distributing the orbital angular momentum as smoothly as possible. As with vortices in He-II their areal density is $2\omega m/h$. On the coarse scale the behaviour mimics that of a rigid body. The vortices persist for about 40 s, growing weaker and less numerous with time, but retaining an Abrikosov lattice array. The calculated healing length was 0.2 μm implying that the cores had radius 0.4 μm. Measurements made by K. E. Wilson and colleagues with megapixel CCDs[10] confirm this inference.

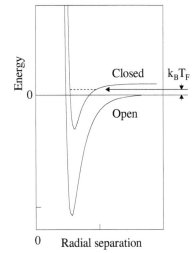

Fig. 16.17 Feshbach resonance. Energies of the open and closed channels versus the atomic separation.

16.7 BEC to BCS crossover

Here the topic is the condensates produced in dilute gases such as ^{6}Li at ultracold temperatures. ^{6}Li atoms, having three electrons, are fermions and have to be manipulated to form pairs, thus making bosons that can condense. The surprise is that besides there being a gaseous condensate of ^{6}Li$_2$ molecular bosons, condensation also occurs in which pairs of ^{6}Li atoms form the analogue of Cooper pairs. It only requires a change in applied magnetic field to get from one condensate to the other, thanks to what is called a *Feshbach resonance*. This transition between condensates is called the *BEC/BCS crossover*, which is now explored.

[10]K. E. Wilson, Z. L. Newman, J. D. Lowney and B. P. Anderson, *Physical Review* A91, 023621 (2015).

An ultracold fermion gas is degenerate with all states filled up to the Fermi level. In the case of ^{6}Li the number density employed, n, is typically $2.5\,10^{12}\,\mathrm{cm}^{-3}$. Then the Fermi level parameters are, reusing eqn. 5.5

$$k_\mathrm{F} = [3\pi^2 n]^{1/3} = 4.1\,\mu\mathrm{m}^{-1}, \tag{16.22}$$

$$T_\mathrm{F} = [\hbar^2 k_\mathrm{F}^2/2m]/k_\mathrm{B} = 1.5\,\mu\mathrm{K}. \tag{16.23}$$

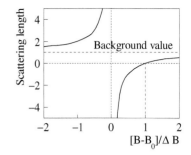

Figures 16.17 and 16.18 illustrate the behaviour of a dilute ultracold ^{6}Li fermion gas under the influence an applied magnetic field around 832 G (using the units Gauss equal to 10^{-4} T). In Figure 16.17 the variation of the energy of a pair of ^{6}Li atoms is shown as a function of their distance apart. They approach in the *entrance/open channel* with an energy $k_\mathrm{B}T_\mathrm{F}$. The main requirement for having a Feshbach resonance is that there should be a *closed channel* with different quantum numbers having a bound state close to threshold. Then if the incoming energy matches the energy of the bound state in the closed channel there is a Feshbach resonance manifested by a high cross-section for scattering from the open to the closed channel. In the case of ^{6}Li an applied magnetic field is used to tune the relative energies of the open and closed channels, making use of the difference in their Zeeman shifts. Both channels are s-wave: the open channel has electron spin 0, positive parity and corresponding molecular notation $^1\Sigma_g$; the closed channel has spin 1, negative parity and notation $^3\Sigma_u$. The relevant state in the closed channel is a vibrational level bound by $1.07\,\mu\mathrm{eV}$.

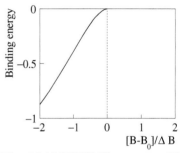

Fig. 16.18 BEC/BCS crossover. The scattering length is normalized to the background value. ΔB is the resonance width.

The interaction between the atoms can be parametrized in terms of the scattering length[11]:

$$a_s(B) = a_\mathrm{bg}[1 - \Delta B/(B - B_0)], \tag{16.24}$$

[11] See Section 7.9.

where a_bg is the background value of the scattering length well away from resonance, the magnetic field value at resonance is B_0 and ΔB is the resonance width. The upper panel of Figure 16.18 shows how the ^{6}Li scattering length varies with the applied field around the Feshbach resonance at 832 G. Below resonance the scattering length is positive, while just above it the scattering length is negative: the values diverge to infinity at resonance, reflecting the growing strength of the interaction.

With magnetic fields below 832 G pairs of ^{6}Li atoms bind to form a boson. Close to resonance the scattering length becomes powers of ten larger than the Bohr radius so that the atoms are quite widely separated: such a structure is known as a *dimer*. The lower panel in Figure 16.18 shows how the binding energy varies with the applied field; very close to resonance it has a quadratic dependence $E = \hbar^2/(ma_s^2)$. Once the boson dimers come into existence they form a Bose–Einstein condensate.

Above resonance the scattering length is negative with the interaction strength being generally inadequate to give bound dimers. However

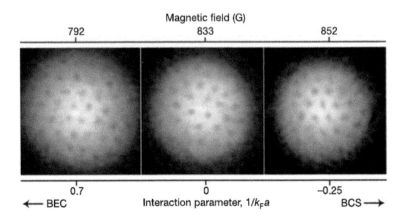

Fig. 16.19 Optimized vortex lattices at the BEC/BCS crossover. After creation at 812 G, the field was ramped in turn to indicated values and the cloud held for 50 ms, followed by ballistic expansion and imaging. The fields of view are 780 μm squares. Figure 3 in M. W. Zweierlein, J. A. Abo–Shaeer, A. Schirotzek, C. H. Schunk and W. Ketterle, *Nature* 435, 1047 (2005). Courtesy Professor Ketterle and Springer Verlag.

experiments like that of the MIT group[12] have discovered that condensation occurs in ^{6}Li both above as well as below the Feshbach resonance. For each observation a cloud of $\sim 10^6$ ^{6}Li atoms was cooled into a degenerate state (of fermions) in a 50:50 mix of the two hyperfine state ($m = \pm 1/2$). A collision between atoms one with $m = +1/2$, the other with $m = -1/2$, constitutes the entry channel. The clouds were trapped in a volume 110 μm axially and 45μm radially, with chemical potential 200 nK, and $1/k_F \approx 0.3$ μm. Each cloud was rotated about its long axis to produce the vortices shown in Figure 16.19. Evidently there is superflow around vortices on either side of resonance. The presence of superflow is evidence for the existence of a condensate above resonance. The interpretation made is that the atoms just above the Fermi sea pair up forming Cooper pairs: the members of a pair have equal and opposite momenta in the rest frame of the condensate and equal and opposite spins. At the Feshbach resonance there is then a BEC/BCS crossover between two different varieties of condensate.

16.8   Further reading

When Atoms behave as Waves: Bose–Einstein Condensation and the Atomic Laser, the Nobel Lecture by Wolfgang Ketterle, Laureate 2001, on the Nobel website is an engaging account of the discovery of Bose–Einstein condensation, and puts over the physics clearly. The lectures by his fellow Laureates, Cornell and Wieman are also highly recommended.

---

[12] M. W. Zweierlein, J. A. Abo–Shaeer, A. Schirotzek, C. H. Schunk and W. Ketterle, *Nature* 435, 1047 (2005).

The second edition of *Bose–Einstein Condensation in Dilute Gases* by C. J. Pethick and H. Smith, published by Cambridge University Press is an essential source for extensive coverage of gaseous BECs.

Bose–Einstein condensates and cold atoms are used in interferometry, time standards and gravimetry: see 'Bose–Einstein Condensates in microgravity' *Applied Physics B* 84 663 (2006); 'Atom interferometer Measurement of the Newtonian Constant of Gravity' by J. B. Fixler, G. T. Foster, J. M. McGuirk and M. A. Kasevich, *Science* 315, 74 (2007).

Exercises

(16.1) Show that the loss of atoms from a dilute boson gas cloud at microkelvin temperatures due to three-body collisions is six times greater for a thermal cloud than for a condensate.

(16.2) What is the critical temperature for a cloud of 10^6 atoms confined in a harmonic potential with angular frequency 10^3 rad s^{-1}? What fraction of the atoms are in the thermal cloud at $0.5\,\mu\text{K}$?

(16.3) In a dilute boson gas at microkelvin temperatures a weak interaction can cause scatters

$$1 + 2 \rightleftharpoons 3 + 4.$$

The direct and inverse reactions have the same matrix element M. Then the forward rate is

$$W_f = |M|^2 n_1 n_2 (n_3 + 1)(n_4 + 1).$$

What do the component parts of the terms (n_3+1) and (n_4+1) represent? Show that in thermal equilibrium

$$\frac{n_1 n_2}{(n_1 + 1)(n_2 + 1)} = \frac{n_3 n_4}{(n_3 + 1)(n_4 + 1)}.$$

Evidently in this case n and the fraction $n/(n+1)$ are functions of the temperature alone. Use this fact to establish the Bose–Einstein distribution.

(16.4) Consider a condensate of N atoms confined in spherically symmetric harmonic trap for which the angular frequency of oscillation is ω_0. The mass of the atoms is m and the scattering length is a_s. Show that the radius of the condensate in the Thomas–Fermi limit is

$$R_\text{TF} = [(15 N a_\text{s} \hbar^2)/(m^2 \omega_0^2)]^{1/5}.$$

Hence prove eqn. J.10.

(16.5) From the result of the last exercise prove that

$$\mu = [\hbar\omega_0/2][15 N a_\text{s}/a_\text{ho}]^{2/5}.$$

(16.6) What is the speed of sound in a ^{87}Rb condensate with density 10^{14}cm^{-3} and scattering length 100 times the Bohr radius?

(16.7) A spheroidal condensate contains a single vortex along its long axis. The density in the transverse plane is of the Thomas–Fermi shape: $n(r) = n_0(1 - r^2/R_\text{TF}^2)$. Calculate the total angular momentum of a unit thickness slice perpendicular to the vortex axis. What limits does this result impose on vortex formation?

(16.8) Estimate the length of a Zeeman slower for ^{87}Rb atoms leaving their source with velocity around $300\,\text{m s}^{-1}$. The half life of the 5P excited state is 27 ns, the atomic mass is $1.44\ 10^{-25}$ kg, and the wavelength of the laser 780 nm.

(16.9) A spherically symmetric cloud of 10^9 ^{23}Na atoms at a density 10^{16} m^{-3} at $20\,\mu\text{K}$ is deposited in a spherically symmetric trap. What should the trapping angular frequency be in order that there is minimum loss?

(16.10) 10^7 ^{87}Rb atoms are trapped in a CO_2 laser beam of 10.6 µm wavelength. The scattering rate of photons per atom is $7\ 10^{-3}$ Hz. Calculate the energy gain per collision and the heating per second.

Quantum Hall effects

17

17.1 Introduction

When a magnetic field is applied transverse to a conductor, as in the upper panel of Figure 17.1, an electric field E_y develops transverse to both the current i_x and applied field B. This is the Hall effect introduced in Section 5.2.3. Rearranging eqn. 5.19 the classical expression for the Hall resistance is, using the notation of the figure,

$$\rho_{yx} = E_y/i_x = B/(ne), \qquad (17.1)$$

where n is the electron density. In addition the longitudinal resistance, $\rho_{xx} = E_x/i_x$, would be independent of the magnetic field. Dramatically different behaviour was observed by von Klitzing when he measured the Hall effect for MOSFET and HEMT devices at high magnetic fields and low temperatures. As explained in Section 6.4 the current in such semiconductor heterostructures is carried by a two-dimensional electron gas, *2DEG*, of thickness $\sim 10\,\text{nm}$. This gas is indicated in the upper panel of the figure by broken lines. The footprint of a typical six terminal Hall bar of length $\sim 1\,\text{cm}$ is shown in the lower panel of the figure. The connections are shown for measuring the longitudinal voltage V_x, the Hall voltage V_y and the longitudinal current i_x. The Hall voltage is measured with the necessary high precision using very high input impedance equipment. Figure 17.2 shows the variation of the Hall resistance and of the longitudinal resistance of a GaAs/AlGaAs heterojunction held at $40\,\text{mK}$ as the applied magnetic field is swept up to $15\,\text{T}$. It differs significantly from the classical prediction. The plateaux in the Hall resistance occur where its inverse, the *Hall conductance*, is an exact integral multiple of e^2/h. Coincident with the plateaux, the longitudinal resistance is zero. Experiments with other 2DEG in Si/SiO_2 heterostructures and in graphene (two-dimensional sheets of carbon) all give precisely the same conductance quantization independent of the geometry of the device, or the (weak) level of impurities. The measured values of e^2/h are consistent to one part in a billion: ρ_{xx} is vanishingly small over the ranges of magnetic field giving plateaux in ρ_{xy}. This behaviour, whose discovery earned von Klitzing the Nobel prize in Physics in 1985, is known as the *Integer Quantum Hall Effect* (IQHE). Two things are very surprising about these results: the stability of the quantization of the Hall conductance over a range of magnetic field, and the dissipationless longitudinal current flow over the same range. The discovery and explanation of the IQHE showed it to be a first example of a *topological, rather than*

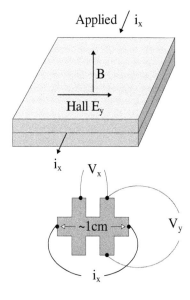

Fig. 17.1 Quantum Hall bar containing the 2DEG indicated by the broken line. The lower image is a plan view of a six terminal device.

Quantum 20/20: Fundamentals, Entanglement, Gauge Fields, Condensates and Topology.
Ian R. Kenyon. © 2020. Published in 2020 by Oxford University Press.
DOI: 10.1093/oso/9780198808350.001.0001

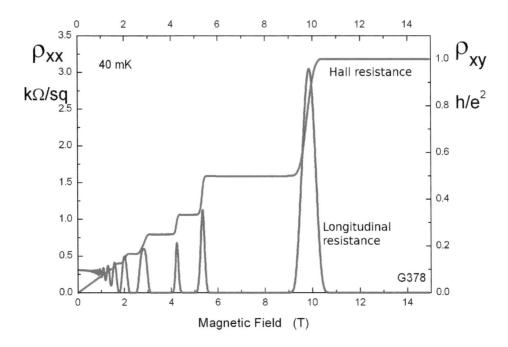

Fig. 17.2 Quantum Hall effect recorded by D. R. Leadley, T. Foxon, P. Gee and R. Nicholas in 1987. GaAs/GaAlAs heterojunction at 40 mK and in fields up to 15 T. Electron mobility 58 000 cm^{2}V^{-1}s^{-1}. Courtesy Professor Leadley (Warwick University). This is a particularly simple plot to interpret. There can be complications, from the presence of several valleys in the conduction band with a corresponding increase in the number of plateaux; also where the alternative spin alignments double the number of plateaux. In this chapter these complications are avoided in order to concentrate on fundamentals.

symmetry based quantization process. Other examples have followed including the Quantum Spin Hall effect, also discussed in this chapter.

In Section 17.2 basic conductivity formulae are presented. Then in Section 17.3 the quantum states of a 2DEG in a strong magnetic field, the *Landau levels*, are determined. The observed quantized conductance plateaux occur whenever all Landau levels up to a given level are full of electrons and higher levels empty: there are energy states between Landau levels, which are localized round impurities and therefore cannot contribute any current. Thus between Landau levels there is a *mobility gap* rather than an energy gap. In Section 17.4 the question of bulk versus edge flow is addressed. The robustness of the IQHE is discussed in Section 17.5. Thouless' analysis which identified the topological nature of the IQHE quantization is given in Section 17.6. The relationship of the quantum of conductivity to the fine structure constant is discussed in Section 17.6.1

In 1982 Tsui, Störmer and Gossard using higher magnetic fields and lower temperatures discovered further plateaux in the Hall conductance at values $(e^2/h)/m$ with m being an odd integer. This is the *frac-*

tional quantum Hall effect (FQHE). The explanation in terms of compound fermions as the active charge carriers was given by Laughlin and Halperin: this analysis is sketched in Section 17.7.

In certain quantum well devices edge flow can be established without any applied magnetic field. The best studied material has the quantum well structure CdTe/HgTe/CdTe. There is flow of spin along the device edges without any corresponding flow of charge. This example of a different type of topological quantization is discussed in Section 17.8.

17.2   Conductance relations

Using the coordinate frame of Figure 17.1 the relations between the electric field in Vm^{-1} and the current density in Am^{-1} for a 2DEG are

$$\begin{bmatrix} E_x \\ E_y \end{bmatrix} = \begin{bmatrix} \rho_{xx} & \rho_{xy} \\ \rho_{yx} & \rho_{yy} \end{bmatrix} \begin{bmatrix} i_x \\ i_y = 0 \end{bmatrix}$$

$$\begin{bmatrix} i_x \\ i_y = 0 \end{bmatrix} = \begin{bmatrix} \sigma_{xx} & \sigma_{xy} \\ \sigma_{yx} & \sigma_{yy} \end{bmatrix} \begin{bmatrix} E_x \\ E_y \end{bmatrix}$$

where ρ is a resistivity and σ a conductivity. Note that Hall current is null because the Hall voltage is measured with a high input resistance device. The force acting on an electron of velocity $\mathbf{v}$ in a magnetic field $\mathbf{B}$ is $-e\mathbf{v} \wedge \mathbf{B}$ so that $\rho_{yx} = -\rho_{xy}$ and $\sigma_{yx} = -\sigma_{xy}$. In an isotropic medium $\rho_{yy} = \rho_{xx}$ and $\sigma_{yy} = \sigma_{xx}$. Then these equations become

$$\begin{align} E_x &= \rho_{xx} i_x, & (17.2)\\ E_y &= -\rho_{xy} i_x, & (17.3)\\ i_x &= \sigma_{xx} E_x + \sigma_{xy} E_y & (17.4)\\ 0 &= -\sigma_{xy} E_x + \sigma_{xx} E_y. & (17.5) \end{align}$$

In the plateaux regions $E_x = 0$, that is to say there is dissipationless longitudinal current flow. Hence the above equations collapse to

$$\begin{align} \rho_{xx} &= \sigma_{xx} = 0, & (17.6)\\ i_x &= \sigma_{xy} E_y, & (17.7)\\ \rho_{xy} &= -1/\sigma_{xy} & (17.8) \end{align}$$

Surprisingly, the longitudinal resistance and the longitudinal conductance can simultaneously vanish.[1] Note also that in two dimensions the resistance, R and resistivity, ρ have the same dimensions, Ohms: $R = \rho \frac{length}{width}$.

[1] This is thanks to the interlocking of electrical parameters in the two dimensions by the action of the magnetic field.

17.3   Landau levels

Electrons in a 2DEG under a transverse magnetic field B in the z-direction occupy quantized states akin to classical cyclotron orbits. The energy levels are known as *Landau levels*. In the case of a typical GaAs/AlGAs Hall bar in a $10\,\mathrm{T}$ field at $0.1\,\mathrm{K}$ the energies to consider are these: thermal $k_{\mathrm{B}}T = 8.6\,\mu\mathrm{eV}$; Zeeman energy $g\mu_{\mathrm{B}}B = 0.25\,\mathrm{meV}$ where the Landé factor is 0.44; cyclotron energy $\hbar eB/m^* = 24\,\mathrm{meV}$ taking the electron effective mass m^* in a MOSFET channel to be $0.07m$. Hence the energy of the cyclotron motion dominates and we ignore the others. In particular there is negligible thermal excitation from one Landau level to the next. There are choices to be made for the electron basis states in the analysis: the important states for the IQHE extend along the full length of the Hall bar, namely the x-direction, and as we shall see this makes them narrow in the y-direction. The appropriate choice of the magnetic vector potential is this

$$\mathbf{A} = -By\mathbf{x} \tag{17.9}$$

$\mathbf{x}$ being a unit vector along the x-axis. As before the y-axis is in the direction of the Hall voltage and the x-axis is along the direction of the longitudinal current. Using eqn. 1.78, the Hamiltonian (energy) of the electron has this operator form

$$\hat{E} = (\hat{\mathbf{p}} + e\hat{\mathbf{A}})^2/2m^*$$
$$= [-\hbar^2\frac{\partial^2}{\partial x^2} - \hbar^2\frac{\partial^2}{\partial y^2} + 2ie\hbar By\frac{\partial}{\partial x} + e^2B^2y^2]/(2m^*). \tag{17.10}$$

A general solution to Schrödinger's equation with this expression for the energy is

$$\Psi(x, y) = \psi(y)\exp(ikx), \tag{17.11}$$

where $k = 2\pi j/L$, with L being the bar length and j is an integer. The step size in k is

$$\Delta k = 2\pi/L. \tag{17.12}$$

Entering $\Psi(x, y)$ in Schrödinger's equation gives

$$[-\frac{\hbar^2}{2m^*}\frac{d^2}{dy^2} + \frac{1}{2}m^*\omega_{\mathrm{c}}^2(y - y_0)^2]\psi(y) = E\psi(y) \tag{17.13}$$

where ω_{c} is the cyclotron angular frequency eB/m^* and

$$y_0 = \frac{\hbar k}{eB} \tag{17.14}$$

is called the *guiding centre* of the wavefunction. Evidently the motion across the bar is simple harmonic motion around the guiding centre and has energy

$$E_n = (n + 1/2)\hbar\omega_{\mathrm{c}}, \tag{17.15}$$

with n being an integer that labels the *Landau levels*. The spacing between the eigenstates in y is obtained using eqns. 17.12 and 17.14

$$\Delta y = \frac{\hbar}{eB}\Delta k = \frac{h}{eBL}, \tag{17.16}$$

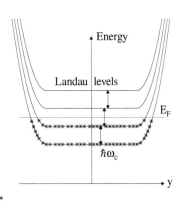

Fig. 17.3 Landau levels across the width of the Hall bar ignoring the Hall voltage. E_{F} marks the Fermi level. The stars mark the occupied electron states.

which is around 0.04 pm in the case of a bar of 1 cm length under 10 T. This gives a degeneracy for a bar of width W equal to $W/\Delta y$ or $eBLW/h$. Thus the degeneracy, the number of distinct quantum states corresponding to a given Landau level, per unit surface area of the 2DEG, is

$$D = eB/h, \qquad (17.17)$$

which is $2.42 \, 10^{11} \, \mathrm{cm}^{-2}$ for GaAs/AlGaAs in a 10 T field.[2] A quantity called the magnetic length, l_B is defined

$$l_B = \sqrt{\hbar/eB}, \qquad (17.18)$$

and $2\pi l_B^2$ is the area per eigenstate whatever its shape. At 10 T l_B is 8.1 nm. The *filling fraction*, ν, is defined to be the density of electrons divided by the density of available electron quantum states

$$\nu = \frac{n_e}{D}. \qquad (17.19)$$

When ν is an integer j, the first j Landau levels are full.

17.4   Current flow

Figure 17.3 shows how the energies of the Landau levels vary across the width of a Hall bar and their populations: the energies rise as the edge of the bar, and the vacuum state, is approached. When the magnetic field is reduced the degeneracy of the Landau levels given in eqn. 17.17 falls, and the Fermi level must rise. When the Fermi level reaches the flat central region of the next, the $(j+1)$th Landau level, there is a pulse in the longitudinal conductivity visible in Figure 17.1 and in synchrony the Hall conductance σ_{xy} rises from je^2/h to a new plateau $(j+1)e^2/h$.

Figure 17.4 shows the quantum and semi-classical views of the current flow in edge states. In the classical view an electron on recoiling from the edge follows a new cyclotron orbit so that it continues in the same sense along the edge. Contact with impurities does not deflect the sense of progress along an edge, nor does a kink in the edge. Scattering from one edge to another is very improbable because of the nanometre scale of the magnetic length. The eigenstates of the quantum analysis in the preceding section are narrow channels parallel to the bar length. Those near the edge where the population of the Landau levels reaches the Fermi level can be looked on as edge channels for current, one per occupied Landau level. The details of how the Hall voltage is distributed across the width of the Hall bar has been investigated by E. Ahlswede, P. Weitz, J. Weis, K. von Klitzing and K. Eberle (*Physica* B298, 562 (2001)).

We can infer the Hall conductivity taking account of the whole width of the Hall bar. Figure 17.5 shows an individual Landau level with a uniform tilt to indicate the effect of the Hall voltage. We use the

[2]Note that each electron eigenstate occupies an area exactly equal to that containing one flux quantum, $\Phi_0 = h/e$, which we can attribute to the common action of the uncertainty principle.

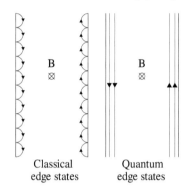

Classical edge states   Quantum edge states

Fig. 17.4 Classical and quantum pictures of the edge flow.

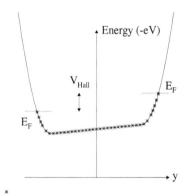

Fig. 17.5 Landau levels across the width of the Hall bar taking account of the Hall voltage. E_F marks the Fermi level. The stars show the electron population.

standard, perfectly valid approach, of considering the Hall voltage to drive the longitudinal current. Then the electron group velocity in the x direction along the Hall bar is

$$v_x = \frac{\partial[\text{energy}]}{\partial[\text{x} - \text{momentum}]} = \frac{-e}{\hbar}\frac{\partial V}{\partial k}, \qquad (17.20)$$

where V is the Hall voltage. Electrons throughout the bulk are available to carry longitudinal current. Averaging over all the parallel states across the Hall bar the longitudinal current density is

$$i_x = -e \int [dk/2\pi] v_x(k)/W, \qquad (17.21)$$

where W is the bar's width and $dk/2\pi$ is the density of the essentially one dimensional eigenstates extending along x, given by eqn. B.8. Then

$$i_x = \frac{e^2}{h}\frac{V}{W} = \frac{e^2}{h}E_y, \qquad (17.22)$$

where E_y is the Hall field.[3] Whence the conductivity per occupied Landau level is e^2/h.

This correctly reproduces the values of quantized conductance in Figure 17.2 but does not explain their persistence over a finite variation of magnetic field. In a 2DEG in a real Hall bar there is a continuum of energy levels due to lattice imperfections and impurities. The key question is why σ_{xy} and σ_{xx} should remain fixed at values corresponding to an exactly full Landau level, during considerable reduction in the magnetic field value. Why doesn't the overflow of electrons from a Landau level contribute to the conductivity?

17.5   The stability of IQHE

The answer is provided by Figure 17.6, which shows the density of states at any given location in the Hall bar. The Landau levels account for the sequence of narrow ranges of states indicated in gray. The broader distributions are those of states in which electrons are trapped by impurities and other defects in the crystal structure. These states are local and as a result do not contribute to conduction: they provide a reservoir of levels that will be filled as the Fermi level advances between the sharply defined Landau levels. There is thus no *energy gap* between Landau levels but instead a *mobility gap* and it is this that guarantees the stability of the IQHE. Impurities are essential; if they are gradually removed the conductance plateaux get progressively narrower.

The conditions for successful observation of the IQHE are now summarized. First the temperature must be much less than energy separation of successive Landau levels: this requires high fields and low temperatures. Next the interval between scatters should be much longer than

[3]The same result can also be obtained by using $i_x = eDv_x$ where v_x is the electron velocity, $D = eB/h$, and we take the average Hall field $E_y = v_x B$.

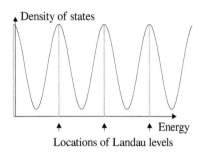

Locations of Landau levels

Fig. 17.6 Density of states at any point along the Hall bar.

the cyclotron period which requires crystals of high purity. Thirdly the mean free path λ must be longer than the magnetic length, otherwise scattering can disrupt the Landau levels: usually the third requirement is well satisfied with $\lambda \sim$ microns. There is a balance to be struck between having high enough purity to give a long scattering length and having enough impurities to maintain a mobility gap.

17.6 Topological quantization

Thouless and colleagues revealed the underlying topological nature of the IQHE.[4] For this and other work Thouless received a Nobel Prize in Physics in 2016. The Hall conductance was shown to be a function on the two-dimensional space formed by the magnetic fluxes ϕ_j and ϕ_v through the longitudinal current and transverse voltage circuits shown in Figure 17.7. This restricts possible values of the Hall conductance to integral multiples of e^2/h: the integer counts both the number of vortices in the electron wavefunction across this space and the number of occupied Landau levels. The reader may wish to jump this intricate but critical proof and proceed direct to eqn. 17.39.

Figure 17.7 shows a Hall bar with connections for the longitudinal current and transverse voltage. Magnetic flux through the current and voltage loops, ϕ_j and ϕ_v respectively is produced by superconducting solenoids so that local to the circuits there is no magnetic field.[5] We start from the Kubo expression K.19 for the Hall conductance

$$\sigma_{jv} = i\hbar \sum_{n \neq 0} \left[\frac{\langle \psi_0 | J_v | \psi_n \rangle \langle \psi_n | J_j | \psi_o \rangle - \langle \psi_0 | J_j | \psi_n \rangle \langle \psi_n | J_v | \psi_o \rangle}{(E_n - E_0)^2} \right], \quad (17.23)$$

with J_v and J_j being the currents in the Hall direction and in the applied current direction respectively. This formula exposes the symmetry between the current and voltage loops. E_n is the energy of an excited quantum state ψ_n, E_0 that of the ground state ψ_0.

We consider the effect of adding a flux ϕ in one loop which perturbs the energy by

$$w = -J\phi, \quad (17.24)$$

so that there is a new ground state, given to first order in w by eqn. E.12,

$$|\psi_0\rangle' = |\psi_0\rangle + \sum_{n \neq 0} \left[\frac{\langle \psi_n | w | \psi_0 \rangle}{E_0 - E_n} |\psi_n\rangle \right]. \quad (17.25)$$

For the loop in question

[5]This arrangement simplifies the analysis in the same way as in the Aharonov–Bohm effect in Section 13.4.

[4]D. J. Thouless, M. Kohmoto, M. P. Nightingale and M. de Nijs, *Physical Review Letters* 49, 405 (1982).

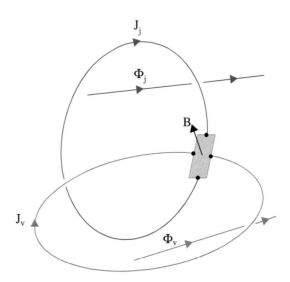

Fig. 17.7 Hall bar and connections with flux through the current and voltage loops due to current flowing in superconducting solenoids.

$$\frac{\partial|\psi_0\rangle}{\partial\phi} = -\sum_{n\neq0}\left[\frac{\langle\psi_n|J|\psi_0\rangle}{E_0 - E_n}|\psi_n\rangle\right]. \tag{17.26}$$

Making this substitution for each current in the equation for the Hall conductivity gives the expectation of a commutator in the ground state

$$\sigma_{jv} = i\hbar\left\langle\left[\frac{\partial\psi_0}{\partial\phi_v}, \frac{\partial\psi_0}{\partial\phi_j}\right]\right\rangle, \tag{17.27}$$

for particular values of the fluxes ϕ_j and ϕ_v. After changing ϕ_v or ϕ_j from 0 to h/e, the phase returns to its initial value. At this juncture it helps to picture ϕ_v and ϕ_j as orthogonal coordinates in (ϕ_v,ϕ_j) space. The (ϕ_v,ϕ_j) space is toroidal because the points $\phi_j = 0$ and h/e are equivalent and hence coincide; equally $\phi_v = 0$ and h/e coincide. Kubo's conductance should be insensitive to the boundary conditions imposed by particular values of ϕ_v and ϕ_j. Thus it is justified to average over the whole of the toroidal surface.[6] Then the expectation value of the conductance is

[6]Q. Niu and D. J. Thouless, *Physical Review B*35, 2188 (1987) and Q. Niu, D. J. Thouless and Y.-S. Wu, *Physical Review B*31, 3372 (1985).

$$\sigma_{jv} = \int\int i\hbar\left\langle\left[\frac{\partial\psi_0}{\partial\phi_v}, \frac{\partial\psi_0}{\partial\phi_j}\right]\right\rangle \mathrm{d}\phi_v\,\mathrm{d}\phi_j / \int\int \mathrm{d}\phi_v\,\mathrm{d}\phi_j, \tag{17.28}$$

where all four integrals run over the range 0 to h/e. Then we have

$$\sigma_{jv} = (e/h)^2 i\hbar \int\int\left\langle\left[\frac{\partial\psi_0}{\partial\phi_v}, \frac{\partial\psi_0}{\partial\phi_j}\right]\right\rangle \mathrm{d}\phi_v\,\mathrm{d}\phi_j. \tag{17.29}$$

This integral is more readily interpreted when it is appreciated that it is an example of Berry's phase with the integrand being the curl of *Berry's connection* **S**. Explicitly the connection[7] has components

$$S_j = \frac{1}{2}\left[\left\langle\frac{\partial\psi_0}{\partial\phi_j}|\psi_0\right\rangle - \left\langle\psi_0|\frac{\partial\psi_0}{\partial\phi_j}\right\rangle\right],$$

$$S_v = \frac{1}{2}\left[\left\langle\frac{\partial\psi_0}{\partial\phi_v}|\psi_0\right\rangle - \left\langle\psi_0|\frac{\partial\psi_0}{\partial\phi_v}\right\rangle\right]. \tag{17.30}$$

The component of the curl of **S** normal to the ϕ_j, ϕ_v surface is

$$\nabla_\phi \wedge \mathbf{S} = \frac{\partial}{\partial\phi_v}S_j - \frac{\partial}{\partial\phi_j}S_v$$

$$= \left\langle\left[\frac{\partial\psi_0}{\partial\phi_j}, \frac{\partial\psi_0}{\partial\phi_v}\right]\right\rangle. \tag{17.31}$$

Thus as anticipated eqn. 17.29 becomes[8]

$$\sigma_{xy} = \frac{-ie^2}{2\pi h}\int_T \nabla_\phi \wedge \mathbf{S}\, d\phi_v\, d\phi_j, \tag{17.32}$$

where the T signifies that the integral is over the torus.

We now examine the behaviour of the wavefunction ψ_0 bearing in mind that it can have singularities. We can apply a gauge transformation

$$\psi \to \psi\exp[i\eta(\phi_v, \phi_j)] \tag{17.33}$$

which will not affect any observable. Berry's connection will also change[9]

$$\mathbf{S} \to \mathbf{S} + i\nabla_\phi\eta(\phi_v, \phi_j), \tag{17.34}$$

but this does not affect the conductivity. It is here finally that the topological core of the IQHE emerges. It is not in general possible to have a smoothly varying phase η over the whole toroidal (ϕ_v,ϕ_j) space and avoid discontinuities. For example,[10] suppose there are two regions, 1 and 2, between which the mismatch at the boundary is

$$\psi_2 = \exp[i\eta_d(\phi_v, \phi_j)]\psi_1 \tag{17.35}$$

with η_d being a smooth function of ϕ_v and ϕ_j. This topology carries over to Berry's connection

$$\mathbf{S}_2(\phi_v, \phi_j) = \mathbf{S}_1(\phi_v, \phi_j) + i\nabla_\phi\eta_d(\phi_v, \phi_j) \tag{17.36}$$

We now apply Stokes' theorem to the surface integral of $\nabla_\phi \wedge \mathbf{S}$ over these separate regions of the (ϕ_j, ϕ_v) space and add them. Because together these regions cover the whole of the toroidal space the line integral is along their mutual boundary and is traversed in opposite senses for the two regions. The result of applying Stokes' theorem is a line integral

[8]Mathematicians recognize that the integral

$$\frac{i}{2\pi}\int_T \nabla_\phi \wedge \mathbf{S}\, d\phi_j\, d\phi_v$$

is a topological invariant, the *first Chern class* of a U(1) principal fibre bundle on a torus. This first Chern number is an integer. We proceed more slowly.

[9]In detail

$$\mathbf{S} \to \mathbf{S} + i\frac{\partial\eta}{\partial\phi_j}\hat{\phi}_j + i\frac{\partial\eta}{\partial\phi_v}\hat{\phi}_v$$

where $\hat{\phi}_j$ and $\hat{\phi}_v$ are unit vectors along the ϕ_j and ϕ_v directions respectively.

[10]The rigorous treatment given by M. Kohmoto considers more general possibilities: see *Annals of Physics* 160, 343 (1985).

---

[7]The reader should refer back to Section 13.5 for details about Berry's analysis.

of the change in Berry's connection at the boundary between the two regions

$$\int_T \nabla_\phi \wedge \mathbf{S} \, \mathrm{d}\phi_v \, \mathrm{d}\phi_j = -i \int_{\partial\phi} \nabla_\phi \eta_d \cdot \mathbf{d}\phi, \qquad (17.37)$$

where the subscript $\partial\phi$ signifies the integral is along the boundary. Then the only possibility for a match at the boundary is if

$$-i \int_{\partial\phi} \nabla_\phi \eta_d \cdot \mathbf{d}\phi = 2\pi i n, \qquad (17.38)$$

where n is an integer making the phase difference η_d equal to $2n\pi$. Whatever division is made between the two segments of the (ϕ_v, ϕ_j) space this must hold. It is clearly a topological constraint: a property of the whole surface rather than a property of any particular part of it. n is a topological invariant equivalent to Berry's phase in eqn. 13.35. Its physical significance is the number of full Landau levels and topologically it is the net number of vortices in the wavefunction over the (ϕ_j, ϕ_v) surface. Finally, we get the Hall conductance by substituting for the integral in eqn. 17.32

$$\sigma_{xy} = ne^2/h, \qquad (17.39)$$

showing that the IQHE is the result of topological quantization rather than some symmetry. Because the topology of the (ϕ_v, ϕ_j) space determines the quantization of the Hall conductance the quantization is robust to parts in 10^9 against changes in the choice of geometry of the Hall device or the semiconductors from which it is constructed, and against impurities. The results here parallel those found in Chapter 6: a single quantum mode, whether a Landau level in a quantum Hall device or a mode travelling through a quantum point contact, gives conductance e^2/h.

17.6.1   Resistance standards

[11]This precludes any useful comparison between the values of e^2/h, or of the fine structure constant, determined via the IQHE and via the $g - 2$ experiment described in Section 12.7.

Measurements on individual Hall bars give conductances agreeing to parts in 10^9. However standard resistances are only reliable[11] to one part in 10^7! As a palliative measure a new resistance scale has been introduced based on the IQHE with the quantum Hall resistance being defined to be $R_K = 25\ 812.807\ 449$ Ohms.

17.7   The fractional quantum Hall effect

By raising the magnetic field on a GaAs/AlGaAs Hall bar, and making measurements at lower temperatures Tsui, Störmer and Gossard discovered a plateau in σ_{xy} at a filling factor of 1/3. Figure 17.8 shows data taken later by the same group which exhibits several example of plateaux in σ_{xy} and the accompanying vanishing of σ_{xx}, all at fractional fillings. The study of the *fractional quantum Hall effect (FQHE)* required purer samples with mobilities $\sim 10^{6-7} \mathrm{cm}^2 \mathrm{V}^{-1} \mathrm{s}^{-1}$, high fields up to 30 T, and

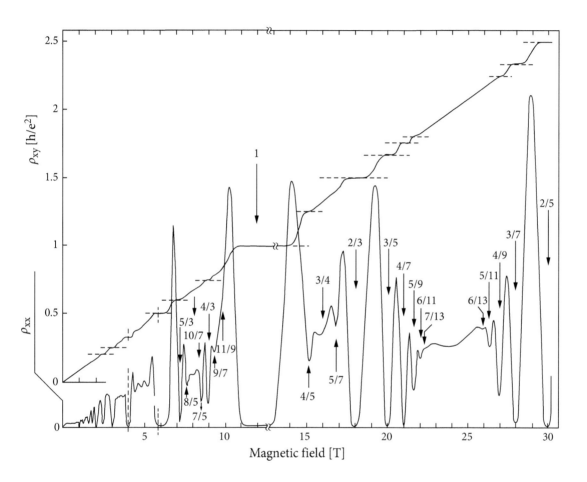

Fig. 17.8 Fractional quantum Hall effect observed with a GaAs/AlGaAs heterostructure. Adapted from Figure 1 in R. Willett, J. P. Eisenstein, H. L. Störmer, D. C. Tsui, A. C. Gossard and J. H. English, *Physical Review Letters* 59, 1776 (1987). Courtesy Professor Störmer and The American Physical Society.

temperatures 10–100 mK. With very rare exceptions the plateaux in σ_{xy} have been found to occur at fractional fillings which have an odd denominator: 1/3, 1/5, 4/7, 2/5, 5/3....

The physical process that leads to the FQHE is the interaction between electrons. This interaction could be adequately approximated by a background mean field in the interpretation of the IQHE effect. However, at the lower temperatures and higher magnetic fields that characterize the FQHE the interaction between the electrons leads to the appearance of quasi-particles made up of electrons and associated magnetic flux. Laughlin showed that these features would account for quantization at fractional fillings.

Laughlin proposed a trial wavefunction for the electrons which produces the FQHE with filling factors of $1/(2p+1)$, p being an integer. Each electron is to be associated with $2p$ flux quanta (h/e) turning it into a quasi-particle. These quasi-particles respond to the residual field remaining after their associated flux quanta are subtracted. Such quasi-particles are *composite fermions* which themselves undergo an integer quantum Hall effect in the residual field. Calculations are made using the *symmetric gauge* in which the magnetic vector potential

$$\mathbf{A} = \mathbf{B} \wedge \mathbf{r}/2 \tag{17.40}$$

where $\mathbf{r}$ is the distance from a reference point in the plane of the 2DEG. A new notation is introduced

$$z = x + iy, \tag{17.41}$$

where x and y are orthogonal components of $\mathbf{r}$ in the plane of the electron gas. Note that in this section z is not, as might be anticipated, the distance from the plane along the field direction. Laughlin's single particle wavefunction has the form

$$\chi_p(z) = C(z/l_B)^p \exp\left[-\frac{|z|^2}{4l_B^2}\right], \tag{17.42}$$

which has an angular momentum component $p\hbar$ perpendicular to the 2DEG; C is a normalizing constant. This wavefunction forms an annulus of mean radius $\sqrt{2p}\,l_B$ and its radial spread about the mean is l_B. Laughlin's wavefunction of all the electrons in the 2DEG has the form

$$\phi_p = \prod_{i<j}(z_i - z_j)^{2p+1} \prod_k \exp\left[-\frac{|z_k|^2}{4l_B^2}\right]. \tag{17.43}$$

The term $(z_i - z_j)^{2p+1}$ imposes the requirement that two electrons cannot occupy the same state, a requirement that can be looked on as the result of Coulomb repulsion. The odd power of $(z_i - z_j)$ in the first term anti-symmetrizes the wavefunction under the interchange of electrons. An even power would give symmetrization and a fractional quantum

Hall effect with an even denominator. The probability distribution corresponding to Laughlin's wavefunction is

$$|\phi_p|^2 = \exp\left[2(2p+1)\sum_{i,j}\ln|z_i - z_j| - \sum_k |z_k|^2/2l_B^2\right]. \qquad (17.44)$$

We can write this as $|\phi_p|^2 = \exp(-H)$ where H has the same mathematical form as the Hamiltonian of a two-dimensional classical planar plasma. The first term in the exponent is equivalent to

$$V = -q^2 \sum \ln|z_i - z_j|, \qquad (17.45)$$

the potential arising from the mutual interactions of carriers with charge q in two dimensions. The second term is equivalent to

$$V' = \frac{\pi}{2}\rho q^2 \sum |z_k|^2, \qquad (17.46)$$

the potential due to the neutralizing background for a carrier density ρ. When V and V' are equated to the two terms in eqn. 17.44 this gives

$$q^2 = 2(2p+1) \quad \rho = 1/[2\pi(2p+1)l_B^2]. \qquad (17.47)$$

Such a plasma spreads out uniformly to minimize its energy. Thus Laughlin's wavefunction describes a two-dimensional fluid of composite fermions. Variational calculations have been carried out with large numbers of electrons; in these cases Laughlin's wavefunction is found to be very close to the exact solution.

The interpretation of the FQHE as the IQHE for composite fermions in an effectively weaker magnetic field leads logically to a hierarchy of states. The first composites can become the input for forming further composites that absorb more flux quanta; these in turn can undergo the IQHE in the weaker residual magnetic field, and so on. Such observed quantum Hall fractions as 5/3 are accounted for in this way.

At sufficiently low filling factors, that is at very high fields and low electron density the electron-electron interaction will dominate and it is expected that the electrons lock into a *Wigner crystal*, different in kind from Laughlin's fluid. The transition to the crystalline form is expected to occur around a filling factor of 1/7; and in fact no FQHE effects have been observed at lower filling factors than this.

17.8   Quantum spin Hall effect

The applied magnetic field is essential to the IQHE and FQHE: it guarantees that electron flow along each edge is in one sense only: this is known as *chiral flow*. Such behaviour violates the principle of *time reversal invariance*, TRI, namely that reversing the flow of time converts

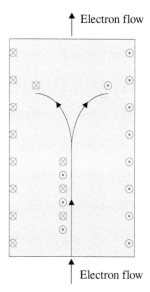

Fig. 17.9 Spin-dependent scattering of electrons from spinless scatterers. Spin-up(down) electrons are shown as circles with a dot(cross).

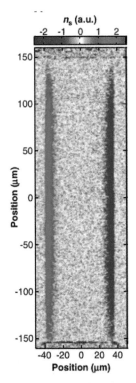

n_s (a.u.)

Fig. 17.10 Results of Kerr rotation spectroscopy on a n-GaAs film carrying current with no magnetic field. Adapted from Figure 2 in *Science* **306**, 1910 (2004). Courtesy Professor Gossard. Reprinted with permission of The American Association for the Advancement of Science.

one physical process into another actual physical process. In the case of the quantum Hall flow time reversal would give the opposite of what is observed. This violation is intrinsic to electromagnetism because the magnetic force $-e\mathbf{v} \wedge \mathbf{B}$ reverses under time reversal. Take away the magnetic field and chiral edge flow would seem unlikely. However Hirsch noticed that in the absence of any applied magnetic field the two edges of a Hall bar would be preferentially populated by electrons with spin-up at one edge and spin-down at the other: this is the *spin Hall effect*. The mechanism responsible is simple scattering from spinless impurities, for which the scattering potential is

$$V = V_c(r) + V_s(r)\boldsymbol{\sigma} \cdot \mathbf{L}. \qquad (17.48)$$

Here $\boldsymbol{\sigma}$ is the electron spin and $\mathbf{L}$ its orbital angular momentum around the scatterer; V_c is the radial potential and V_s its radial gradient. In the equation the second term changes sign under time reversal so its effect mimics that of an imposed magnetic field, as illustrated in Figure 17.9. Gossard and colleagues[12] studied this effect in a 2 μm film of n-GaAs grown on a 2 μm AlGaAs substrate. The probe used to detect the electrons' spin alignment was Kerr rotation spectroscopy. Light from a polarized laser beam with spot diameter 1.1 μm was incident normally on the film and the beam was raster scanned over the film. The returning light suffers a rotation of its polarization axis dependent on whether the direction of the spins of the electrons in the film are up toward the beam or downward away from it. Figure 17.10 shows the distribution of the observed rotation of the laser polarization projected onto a plan view of the film. The regions of intense colour near both edges indicate strong rotation signals, and hence a concentration of spin polarized electrons. In this colour-coded plot the electron polarization is seen to be oppositely directed at the two edges.

What are the implications of these results for heterostructures containing a 2DEG in the absence of any applied magnetic field? Following from Hirsch's observation it is anticipated that one dimensional edge currents will flow, this time carrying spin. Spin currents were first observed in a thin layer of HgTe sandwiched between layers of CdTe so as to form a quantum well where the 2DEG can form. In such specially designed heterostructures there will be a pair of channels at each edge of a Hall bar: one with spin-up electrons travelling in one sense, the other with spin-down electrons travelling in the opposite sense. Consequently the spin flow is continuous in one sense around the periphery of the Hall bar. This is the *quantum spin Hall effect*, QSHE. Figure 17.11 contrasts the edge flow in the QSHE with that in the IQHE. In the QSHE the edge currents are known as *helical* to emphasize that there is spin flow as well as charge flow.

[12]Y. K. Koto, R. C. Myers, A. C. Gossard and D. D. Awschalom, *Science* **306**, 1910 (2004).

17.8.1 TRI and Kramer pairs

Here the analysis is restricted to currents in 2DEGs in the absence of any applied magnetic field. Hence time reversal invariance holds. The time reversal operator acting on an electron spin state, using the notation of Appendix D, is

$$T = -i\sigma_y K = \begin{bmatrix} 0 & -1 \\ 1 & 0 \end{bmatrix} K, \qquad (17.49)$$

where the matrix acts on the spin-components on the Bloch sphere, and K produces complex conjugation. Then

$$T^2 = \begin{bmatrix} -1 & 0 \\ 0 & -1 \end{bmatrix}. \qquad (17.50)$$

Suppose we test the simple proposition that after time reversal the only change to an electron wavefunction, ψ is a phase change χ,

$$T\psi = \exp[i\chi]\psi, \qquad (17.51)$$

then

$$T^2\psi = T\exp[i\chi]\psi = \exp[-i\chi]T\psi = \psi. \qquad (17.52)$$

This flatly contradicts eqn. 17.50 so that the time-reversed state must be distinct from the original electron state. These pairs of states are known as *Kramer pairs*. Time reversal invariance therefore requires that the channels at the edges of a Hall bar come in Kramer pairs. Figure 17.12 shows the dispersion relations where Kramer pairs of edge channels cross between conduction and valence bands. The Fermi level can intersect the Kramer pair wherever it lies in the gap between the bulk bands. In this way conduction in the edge channels is guaranteed *topologically*. Such a state would exhibit the quantum spin Hall effect with topologically protected pairs of edge states. A material which is a *bulk insulator* and an *edge conductor* is known as a *topological insulator*. If one of the Kramer pair forms a loop hanging from the conduction band and the other a loop up from the valence band without the loops crossing the material remains a *trivial insulator*. Notice that the velocity of travel given by eqn. 5.60

$$v = \mathrm{d}E/\mathrm{d}[\hbar k] \qquad (17.53)$$

is opposite for the two spin channels at a given edge.

At first sight it would appear that back scattering between the two opposite spin states at one edge should mix them and destroy any coherence. However, back scattering with spin reversal has two contributions of equal amplitude which, thanks to time reversal invariance, cancel exactly. One contribution to back scattering with spin reversal has the electron spin rotating 180° in one sense, while the other contribution has the electron spin rotating 180° in the opposite sense. These processes are time-reversed images of each other and will have equal strengths. Crucially rotation of the electron undergoing the first scattering will differ by 360° from the rotation resulting from the alternative scattering.

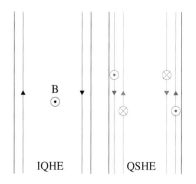

Fig. 17.11 Edge channels in IQHE and QSHE. Spin-up(down) electrons are indicated by crosses(dots) in circles. The B field points toward the reader.

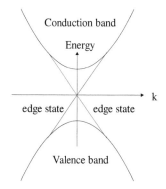

Fig. 17.12 Dispersion relations showing Kramer pairs. Wherever the Fermi level lies between the valence and conduction bands it must cut the pair of edge states. This makes the device a topological insulator which exhibits the QSHE.

This has deep significance. Appendix D shows that when a spinor is rotated 360° around its axis it returns to its initial state but with the phase changed by π. As a result the two contributions to back scattering with spin-reversal have equal and *opposite* amplitudes and will cancel precisely.

In the QSHE there are no Landau levels and hence the topology is distinct from that of the IQHE. Instead there are just the two possibilities either that illustrated by Figure 17.12 or non-crossing edge states. This makes the topological classification Z_2, a two-dimensional group of real integers. $Z = 1$ specifies the topological insulator and $Z = 0$ the trivial insulator. Three-dimensional topological insulators are also known: the volume is insulating while the surface conducts. References are given at the end of the chapter for information on the developing research field of three-dimensional topological insulators.

17.9   (HgCd)Te quantum well Hall bars

The first observation of the QSHE was made by Molenkamp and colleagues using a heterostructure in which a HgTe layer several nm thick was sandwiched between CdTe layers. The difference between the energy bands of HgTe and CdTe are such that electrons see the HgTe layer as a square potential well. A 2DEG is captured in the quantum states of this electrostatic potential well. The bulk energies of the eigenstates in the well are illustrated in Figure 17.13 for two different HgTe thicknesses, one smaller and one larger than the critical thickness $d_c \sim 6.5$ nm.

For future reference remember that these diagrams are energy diagrams at a slice through the Hall bar: the plane of the 2DEG runs perpendicular to these diagrams.

In the case that the thickness $d > d_c$ the energy levels in the quantum well are inverted with the valence state H1 above the conductance state E1. This has the consequence that at the edges of the Hall bar these levels form a Kramer pair that cross between the conduction and valence bands: this is shown in Figure 17.12. The material is therefore a topological insulator. When the thickness $d < d_c$ the effect of the CdTe overrides the HgTe ordering: the valence state H1 now lies below the conductance state E1. Correspondingly no edge states cross the band gap and the material remains a trivial insulator.

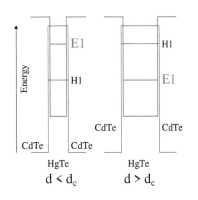

Fig. 17.13 Reversal of the level order of the conduction (red) and valence (black) bound states in broad HgTe quantum wells embedded in CdTe.

Figure 17.14 shows the longitudinal four terminal resistance $R_{14,23}$ measured by Molenkamp and colleagues on (Hg,Cd)Te quantum wells of well thickness 5.5 nm and 12 nm. $R_{14,23}$ is the resistance when the current is measured between terminals 1 and 4, while the voltage is measured between terminals 2 and 3 located on the same side of the

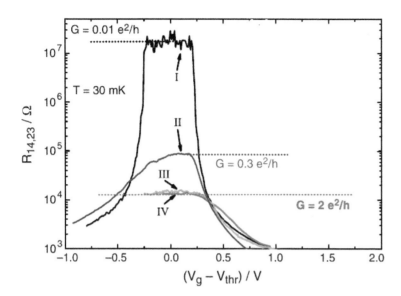

Fig. 17.14 Longitudinal four terminal resistance of CdTe/HgTe/CdTe quantum well heterostructures at 30 mK; HgTe thickness 5.5 and 12 nm. The individual plots are discussed in the text. Figure 4 in M. König, S. Wiedmann, C. Brüne, A. Roth, H. Buhmann, L. W. Molenkamp, X.-L. Qi and S.-C. Zhang, *Science* 318, 766 (2007). Courtesy Professor Molenkamp. Reprinted with permission from The American Association for the Advancement of Science.

bar. The temperature was maintained at 30 mK, the electron mobility being $\sim 1.5\,10^5$ cm^{2}V^{-1}s^{-1}. In the Figure $R_{14,23}$ is plotted against the gate voltage applied to control the electron numbers and hence the Fermi level. Over the voltage range ~ -0.3 to $+0.3$ V the Fermi level lies in the bandgap and consequently the conductance is confined to the edge states. At other bias voltages the Fermi level intersects either the conduction band or the valence band and the bar reverts to being a good conductor. Consider the resistance data labelled I. For this device the width of the HgTe quantum well is less than d_c and no edge states cross the gap between the valence and conduction bands. Consequently, over the voltage range ~ -0.3 to $+0.3$ V the resistance climbs to of order 10 MΩ. The data labelled III and IV refer to devices having $d = 12$ nm, making them topological insulators. For these devices, over the voltage range ~ -0.3 to $+0.3$ V, current can be carried by the edge channels. The corresponding conductance observed for these two devices agrees well with the value, $2e^2/h$, expected when Kramer pairs at each edge provide a one dimensional channel with conductance e^2/h. Device II had larger dimensions which exceeded the coherence length of the edge states and its conductance was therefore weaker.

A parallel study was made by the same experimental team using a six terminal device having $d > d_c$. The results are shown in Figure 17.15. Here also resistance maxima are seen where conduction is limited to the edge states formed by a Kramer pair. In the figures, $R_{ij,kl}$ specifies the resistance where the current is measured between terminals i and j, while the voltage is measured between terminals k and l,

$$R_{ij,kl} = V_{kl}/I_{ij}. \tag{17.54}$$

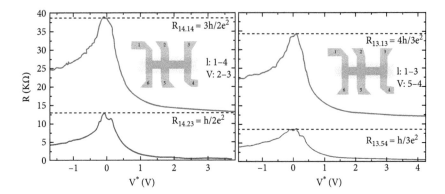

Fig. 17.15 QSHE observed with a six terminal resistance CdTe/HgTe/CdTe quantum well heterostructure at 30 mK. Figure 3 in A. Roth, C. Brüne, H. Buhmann, L. W. Molenkamp, J. Maciejko, X.-L. Qi and S.-C. Zhang, *Science* 325, 294 (2009). Courtesy Professor Molenkamp. Reprinted with permission from The American Association for the Advancement of Science.

The inserts in the figures show the labelling of the terminals. Referring to Section 6.6 the conductance of a one dimensional channel is

$$G = T(e^2/h) \qquad (17.55)$$

where T is the transmission coefficient. We write the transmission coefficient for an edge channel in the spin Hall regime T_{mn}^s: while that for the quantum Hall regime we write T_{mn}. The only quantum Hall transmission that is non-zero is the forward transmission between adjacent terminals $T_{i,i+1} = 1$. Allowing for the two spin channels the non-zero spin Hall transmission coefficients are $T_{i,i+1}^s = T_{i+1,i}^s = 1$. On this basis the prediction for $R_{14,23}$ is readily computed. With obvious notation

$$
\begin{aligned}
i_{12} &= (e^2/h)V_{12} \text{ and because} \\
i_{12} &= i_{23} = i_{34}; \text{ and } V_{12} = V_{23} = V_{34} \\
\text{we have } i_{12} &= (e^2/h)V_{23}.
\end{aligned}
$$

Also i_{16} is identical to i_{12} so that we have finally

$$i_{14} = i_{12} + i_{16} = (2e^2/h)V_{23} \text{ and } R_{14,23} = h/2e^2. \qquad (17.56)$$

The predictions for $R_{13,54}$, $R_{14,14}$ and $R_{13,13}$ all agree with the observed values: $h/3e^2$, $3h/2e^2$ and $4h/3e^2$ respectively. These observations provide a clear demonstration of the QSHE.

17.10   Further reading

The Quantum Hall Effect by D. Yoshioka, published by Springer (2002) provides a clear, but extended account of the IQHE and the FQHE.

'Topological Insulators' by M. Z. Hasan and C. L. Kane, *Reviews of Modern Physics* 82, 3045 (2010) gives an excellent introduction to two- and three-dimensional topological insulators.

'Topological Quantum Matter' by F. D. Haldane, *Reviews of Modern Physics* 89, 40502 (2017) tackles the more formal aspects of the subject in his Nobel lecture.

The second edition of *Advanced Solid State Physics* by P. Phillips, published by Cambridge University Press has an up-to-date and readable chapter on the IQHE and other topological states.

Exercises

(17.1) Show that the quantum spin Hall resistance $R_{13,54}$ for a six terminal HgTe/(Hg,Cd)Te bar is $h/3e^2$.

(17.2) Show that the quantum spin Hall resistance $R_{14,14}$ for a four terminal HgTe/(Hg,Cd)Te bar is $3h/4e^2$.

(17.3) Show that

$$\sigma_{xx} = \rho_{xx}/[\rho_{xx}^2 + \rho_{xy}^2]$$

(17.4) A GaAs/AlGaAs Hall bar has 10^{12} electrons per cm^2. What is the filling factor at a field of 10 T? What are the values of σ_{xx} and σ_{xy}?

(17.5) Show that the peak value of χ_p in eqn. 17.42 is at $r = \sqrt{2p}l_B$.

(17.6) What effect does time reversal have on position, momentum, magnetic moment and kinetic energy?

(17.7) Show that the eigenvalue of the angular momentum component perpendicular to the 2DEG for the wavefunction in eqn.17.42 is $p\hbar$.

(17.8) Show that the eigenvalue of the angular momentum component perpendicular to the 2DEG of the wavefunction given in eqn. 17.43 is $N(N-1)(2p+1)\hbar/2$, where N is the number of composite fermions.

Particle physics I

18

18.1 Prologue

The standard model of particle physics is the outcome of reductionism, that is to say searching for the minimum number of postulates from which to explain as many properties of matter as possible. Ideally it would become possible to deduce the totality of all physical effects. However this aim is illusory because the sheer number of particles in any macroscopic object makes such a calculation infeasible. What happens instead, as previous chapters have revealed, is that simple phenomena emerge at the macroscopic scale, for example, superconductivity. These phenomena are analysed by introducing appropriate pseudoparticles and interactions. The success of physical theory has come from pushing reductionism as far as possible while at the same time developing models for emergent phenomena. It is striking that concepts and techniques developed to analyse emerging phenomena have made our understanding of particle physics more fruitful. Witness the use of spontaneous symmetry breaking, taken over from superconductivity, to resolve the subtleties of the electroweak interaction.

18.2 Introduction to the particles and forces

Elementary particle physics is the study of the particles from which all else is made, the study including primarily their interactions. The interactions of elementary particles naturally conserve energy, momentum and charge. At present the known elementary *material particles* are all fermions, with intrinsic angular momentum (spin) $\hbar/2$ in common with the electron. Each has an antiparticle having the same mass and the opposite charge and magnetic moment. On bringing a particle-antiparticle pair together at rest they annihilate to give an energy release equal to the sum of their masses. The elementary fermions fall into two distinct categories. In the first category are the leptons, which feel the weak force responsible for power generation in the sun and, if they are charged, the electromagnetic force also. In the second category are the quarks, which feel both these forces and in addition the strong force that binds quarks permanently into either *baryons* made of three quarks or into *mesons* made from a quark-antiquark pair. The proton (two up quarks and a down quark) and neutron (one up quark and two down

Quantum 20/20: Fundamentals, Entanglement, Gauge Fields, Condensates and Topology.
Ian R. Kenyon. © 2020. Published in 2020 by Oxford University Press.
DOI: 10.1093/oso/9780198808350.001.0001

Table 18.1 The three quark and lepton families, with their masses times c^2 and charges. All are fermions with spin $\hbar/2$.

generations	1	2	3	Q/e
quarks	up 2.2 MeV	charm 1.3 GeV	top 173 GeV	+2/3
	down 4.5 MeV	strange 100 MeV	beauty 4.2 GeV	-1/3
leptons	ν_e $<\sim 1\,\mathrm{eV}$	ν_μ $<\sim 1\,\mathrm{eV}$	ν_τ $<\sim 1\,\mathrm{eV}$	0
	electron 0.511 MeV	μ-lepton 106 MeV	τ-lepton 1.8 GeV	-1

[1] Consider identical systems made from n fermions: interchanging two systems will multiply the overall wavefunction by a factor $(-1)^n$.

1 GeV is 1000 MeV or 10^9 eV.

quarks) are the lightest baryons. Baryons being made of three fermions are thus always fermions and the mesons being made from two fermions are always bosons.[1] Given their quark content the baryons and mesons interact strongly and are collectively known as *hadrons*. There is a residual component of the strong force, just as the Van der Waals force is the residual electromagnetic interaction between neutral atoms. Protons and neutrons are bound into nuclei by this residual strong interaction. Collisions between electrons and protons at momentum transfers of up to $\sim 300\,\mathrm{GeVc}^{-1}$ have not revealed any substructure in quarks or electrons, and by extension in any leptons. Applying Heisenberg's uncertainty relation they are thus smaller than 10^{-18} m. We shall treat them as point-like. Free quarks are never observed directly, only as the bound constituents of hadrons. The ingenious ways their properties have been determined are described here.

Table 18.1 shows the quarks and leptons, together with their masses and charges. They come in pairs/doublets which mimic the spin states of electrons and so they are called *isospin doublets*: (u,d), (c,s), (t,b); (ν_e,e), (ν_μ,μ), (ν_τ,τ). Absorption or emission of photons or gluons (the carriers of the strong force) leaves a u-quark as a u-quark and so on: the weak interaction induces transitions. The particles also fall into three families of increasing mass. A first family is composed of u, d, e and ν_e (the electron neutrino) and their antiparticles: these particles are apparently all that are required to account for chemical, biological and cosmological phenomena. There are nonetheless two other families that mimic the first. They are dramatically heavier and therefore only produced in high energy collisions: we don't know why this replication occurs.

In addition to the material particles there are the *field particles*, the *carriers* of the forces between them. These are *vector bosons*, that is bosons with unit intrinsic angular momentum (spin) $\hbar$. One is the well known photon, massless and carrying the electromagnetic force. Massless *gluons* carry the strong force and the W$^\pm$-bosons and neutral Z-boson, both having roughly 90 times the proton mass, carry the weak force.

Finally, we come to the Higgs boson: the only elementary particle with spin zero, that is to say a scalar particle. It was first observed in 2012 and has a mass $125\,\text{GeV}c^{-2}$. Its existence is central to the understanding of the unified quantum theory of the electromagnetic and weak forces. Crucially it is the interaction of other elementary particles with the Higgs field that is responsible for the generation of their masses.

The action of field particles is illustrated by the *Feynman diagram* in Figure 18.1. A photon is exchanged between a pair of electrons and transfers energy and momentum between them. Material particles are represented by *straight full lines*; photons by, as here, *wavy lines*. The other field particles appear in later diagrams: the W-bosons and Z-bosons (the carriers of the weak force) also represented by *wavy lines*; the gluons (the fields particles of the strong force) by *curly* lines. Lastly, Higgs particles are represented by *broken straight lines*. Time flows rightward on the diagram. Fermion lines carry a *forward pointing arrowhead*, their antiparticles a *backward pointing arrowhead*. Bosons lines have *no arrowheads*. The permitted types of intersection points, the *vertices*, are discussed later when each interaction is discussed in detail.

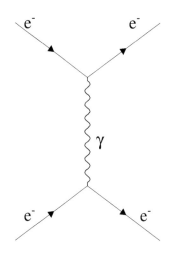

Fig. 18.1 Feynman diagram for the exchange of a photon between a pair of electrons.

There is a basic difficulty in conserving energy and momentum simultaneously in this exchange as in other exchanges. To see why, consider what happens in the rest frame of the emitting electron. The conservation of (energy/momentum) equation at the vertex where the photon is emitted is

$$(m_ec^2, 0) = (E_e, p) + (E_\gamma, -p), \tag{18.1}$$

when conservation of momentum is applied. The energy balance is then

$$m_ec^2 = \sqrt{m_e^2c^4 + p^2c^2} + pc, \tag{18.2}$$

which is patently incorrect. When the photon is reabsorbed into the other electron conservation of energy and momentum obtains once more. Exchange is only possible thanks to the uncertainty principle: an imbalance of energy (ΔE) can endure for a short period (Δt) provided that $\Delta E\Delta t \sim \hbar$. This restriction imposes a limit to the range of the force

$$\Delta R \sim c\Delta t \sim \hbar c/\Delta E. \tag{18.3}$$

When the particle exchanged has a mass M then the range of force has an absolute limit of $\hbar/Mc$. This argues that the electromagnetic force carried by a photon has infinite range because the photon is massless; which agrees with all we know about electromagnetism. The weak force is carried by the W-boson of mass $80.4\,\text{GeV}c^{-2}$ and Z-boson of mass

91.2 GeVc^{-2} so it is short range $\sim 10^{-17}$ m, less than the nuclear size. Finally, the strong force is carried by the massless *gluons* and by the same argument would be expected to have long range. This conclusion conflicts with the fact that nuclear sizes are limited to $\sim 10^{-15}$ m. The resolution of the conflict will emerge when the quantum gauge theory of the strong force is presented later.

Chapter 7 showed how to determine the rate of processes using Fermi's golden rule. A comparison is made here of decays via the three forces taking examples where the energy release is similar. With this selection the phase space available to the decay products is similar in each case. It is then solely the strength of the interaction that determines the relative decay rates. The stronger the force the shorter the mean lifetime, t_m, as we now see:

$$\text{weak}\quad \mu^- \;\rightarrow\; e^- + \bar{\nu}_e + \nu_\mu \quad 105\,\text{MeV}$$
$$t_\mathrm{m} \;=\; 2.2\,10^{-6}\,\text{s}$$
$$\tag{18.4}$$

$$\text{EM}\quad \pi^0 \;\rightarrow\; \gamma + \gamma \qquad 125\,\text{MeV}$$
$$t_\mathrm{m} \;=\; 0.8\,10^{-10}\,\text{s}$$
$$\tag{18.5}$$

$$\text{strong}\quad \Delta^{++} \;\rightarrow\; p + \pi^+ \qquad 154\,\text{MeV}$$
$$t_\mathrm{m} \;=\; 10^{-23}\,\text{s}. \tag{18.6}$$

In the third, strong reaction, baryon number is conserved: the Δ^{++} and p are both baryons. Bosons of all sorts are not conserved, as for example in the second reaction, where a spin zero π^0-meson decays to two photons and in the third reaction a π-meson is created.

Besides the π^+-meson ($u\bar{d}$) and the Δ^{++} (uuu) baryon there are thousands of hadrons made from all possible combinations of u-, d-, s-, c- and b-quarks. We shall limit consideration to the lightest hadrons in which the quarks are in relative s-states of motion, that is with no relative orbital angular momentum. Higher mass hadrons decay in a cascade to these lightest hadrons, the baryons ending as protons or neutrons, the lightest mesons decaying to leptons. The exception is the top quark whose mass, 173 GeV, is so large that it decays immediately via

$$t \rightarrow b + e^+ + \nu_e \tag{18.7}$$

before it can be trapped in a hadron with other quarks. Table 18.2 lists the conservation laws obeyed by the three forces. For example, in the reaction 18.7, ν_e is a lepton and e^+ an antilepton; hence the net number of electron leptons starts as zero and remains zero.

Table 18.2 The quantum conservation laws.

	Weak	EM	Strong
Quark flavour	NO	YES	YES
Lepton flavour	NO	YES	YES
Net quark, lepton, baryon number	YES	YES	YES
Colour, charge	YES	YES	YES

18.3   What follows

In the remainder of this chapter calculations in QED using Feynman diagrams are sketched, and the basic properties of the weak interaction are described. Chapter 19 covers the quantum gauge theories of the strong force and of electroweak unification.

A first section below presents the *natural units* used in particle physics. An example of an early particle discovery is the next topic: the 1936 discovery of the positron (the antiparticle of the electron). By contrast at the end of Chapter 19 the components of a generic modern detector are described. Next the calculation of simple QED processes using Feynman diagrams is performed. Then the renormalization procedure is described by which unwanted infinities are eliminated from predictions.

After this attention switches to the weak interaction and its general features including the breakdown of parity and charge conjugation invariance. The chapter concludes with sections discussing oscillation: those between neutrino species, as well as particle-antiparticle oscillations of neutral mesons. One example using the entanglement of beauty particles to observe CP violation is presented in Appendix I.

The relativistic equation for the electron, the *Dirac equation* and its spinor solutions are covered in Appendix G. Reference to this is made as appropriate in the text.

18.4   Natural units

Particle physicists use units appropriate to studies in which the distances of interest are $\sim 10^{-15}$ m, and the energies ~ 1 GeV (thousand million electron volts). This also has the effect of eliminating the numerous cs and $\hbar$s that would otherwise festoon the equations. The simple rules are:

- Set $\hbar = 6.582 \, 10^{-25}$ GeV s to unity;
- Set $\hbar c = 1.973 \, 10^{-16}$ GeV m to unity;
- Use GeV as the energy unit.

Then the unit of time is $6.582 \, 10^{-25}$ s, the unit of distance is $0.1973 \, 10^{-15}$ m (10^{-15} m is one femtometre) and of area (for cross sections) $0.389 \, 10^{-31}$ m^2 (0.389 millibarns, one barn being 10^{-28} m^2). Thus if a calculation of a cross-section in these new units gives X then the cross-section is 0.389 X millibarns.

When expressing interaction strengths set $\varepsilon_0 = \mu_0 = 1$, which is the Heaviside–Lorentz choice. As a result the fine structure constant is now $e^2/4\pi$ which being a pure number still has the value $1/137$.

18.5 An early discovery

Figure 18.2 of a cloud chamber established the existence of the positron, the antiparticle of the electron, with the equal mass and opposite charge to the electron.[2] The photo is of a 14 cm diameter 1 cm deep chamber containing saturated water vapour. A charged particle traversing the chamber ionizes the atoms in its path and these ions become condensation centres. Consequently the droplets formed follow the path of the particle. A 1.5 T magnetic field, B, acts along the line of sight toward the viewer. This permits a determination of the momentum p, from the path curvature r, thus: $p = qBr$, for a particle of charge q. The observed track crosses the 6 mm lead plate lying across the centre of the chamber. Hence the parent particle must have travelled upward because the track's radius of curvature is less above the plate than below. With the given field orientation the charge of the particle generating the track must be positive. As to the particle species, electrons travel similar distances without loss of energy in the vapour while protons have short heavy tracks. The track is therefore that of a positron, the first identified antiparticle.

Decades of research with accelerators have produced ever more energetic particles, while detectors of growing size and sophistication have been essential tools used in unravelling what we know about elementary particles. This culminated with the discovery of the Higgs boson in 2012 at the Large Hadron Collider, a 27 km roughly circular accelerator. Detectors tens of metres in size are located at the crossing points where the counter-rotating 6.5 TeV proton beams are steered into collision. We shall work our way to this key moment in these two chapters.

We continue from here on in using natural units.

18.6 Field equations

Some simple inferences about exchanges can be made by considering the relativistic equivalent of Schrödinger's equation for relativistic massive

2C. D. Anderson, *Physical Review* **43**, 491 (1933).

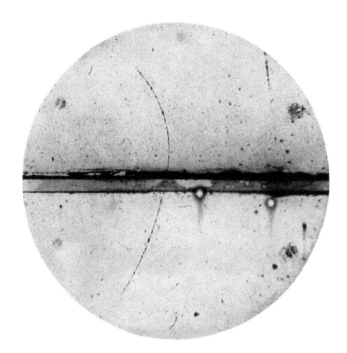

Fig. 18.2 Photograph from a cloud chamber of a 63 MeV positron travelling upward, entering a 6 mm lead plate and emerging with 23 MeV. Figure 1 in C. D. Anderson, *Physical Review* 43, 491 (1933). Courtesy The American Physical Society.

spinless particles:

$$(E^2 - p^2 - m^2)\phi = 0, \tag{18.8}$$

where ϕ is the particle wavefunction. Replacing observables by operators gives the Klein–Gordon equation introduced in Section 1.16

$$-\partial^2\phi/\partial t^2 + \nabla^2\phi = m^2\phi. \tag{18.9}$$

Taking the static limit

$$\nabla^2\phi = m^2\phi. \tag{18.10}$$

Assuming spherical symmetry this becomes

$$\frac{1}{r^2}\frac{\partial}{\partial r}\left(r^2\frac{\partial\phi}{\partial r}\right) = m^2\phi, \tag{18.11}$$

which has the solution known as the *Yukawa potential*

$$\phi(r) = \frac{g_0}{4\pi r}\exp(-r/R), \tag{18.12}$$

where $R = 1/m$ specifies the range, while g_0, as yet undetermined, specifies the strength of the interaction. In the case of the Coulomb potential $e/[4\pi r]$ the mass of the photon exchanged is zero and the range is infinite. Yukawa argued from the nuclear size of 1 fm that the carrier of the strong force had mass $1\,\text{fm}^{-1}$ or 0.197 GeV. This is not far from the mass of the charged π-meson triplet ($\pi^{\pm}$ at 0.1396 GeV and π^0 at 0.1350 GeV). These spinless particles are the lightest and the most copiously produced hadrons in high energy collisions. π-meson exchange is a fair first approximation to the residual strong force, mentioned above, between nucleons (proton and neutron) holding them in nuclei.

It is useful at this point to connect these results with relativistic field theory. The Lagrangian density of a scalar field ϕ with particles of mass m is

$$
\begin{aligned}
L &= T - V = \text{kinetic energy} - \text{potential energy} \\
&= [\dot{\phi}]^2/2 - [\nabla\phi]^2/2 - m^2\phi^2/2,
\end{aligned}
\tag{18.13}
$$

where $\dot{\phi}$ is $\partial\phi/\partial t$. The equation of motion for the field is then obtained by applying Lagrange's equation,[3] which minimizes the action:

$$
\frac{\mathrm{d}}{\mathrm{d}t}\frac{\partial L}{\partial\dot{\phi}} + \nabla\cdot\frac{\partial L}{\partial(\nabla\phi)} - \frac{\partial L}{\partial\phi} = 0.
\tag{18.14}
$$

The result is

$$
\frac{\partial^2\phi}{\partial t^2} - \nabla^2\phi + m^2\phi = 0,
\tag{18.15}
$$

which reproduces eqn. 18.9. A gauge invariant version of eqn. 18.13 will be made use of when discussing the mechanism discovered by Brout, Englert and Higgs that enabled the theories of the electromagnetic force and the weak force to be unified by Glashow, Weinberg and Salam. This topic is covered in Chapter 19.

[3] See for example 3rd edition of *Classical Mechanics* by H. Goldstein, C. P. Poole and J. L. Safko, published by Pearson International (2002).

18.7 $e^- + e^+ \rightarrow \mu^- + \mu^+$

This purely leptonic process will serve to introduce the use of Feynman diagrams in calculating the matrix elements used with Fermi's golden rule to obtain a reaction rate. Initially the rate will be calculated assuming the leptons are spinless, that is scalar particles. After this a simple ansatz will be used to obtain the cross-section when the leptons are treated more correctly as fermions.

In high energy particle physics experiments the particle energies eclipse the masses so that the masses can be neglected. A particle's energy and momentum are equal in this *ultrarelativisic regime*, $E = p$. Important experiments have taken place at electron-positron colliders in which the incoming electron and positron have equal and opposite momentum. The laboratory frame is then conveniently the frame in which the centre of mass of the particles is at rest. Figure 18.3 shows the simplified kinematics: the arrowheads show the direction of the particle momentum in this diagram. The centre of mass (*cm*) energy squared, s is given by

$$
s = \left(\sum_i E_i\right)^2 - \left(\sum_i \mathbf{p}_i\right)^2.
\tag{18.16}
$$

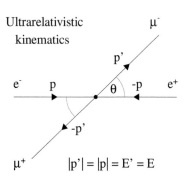

Ultrarelativistic kinematics

$|p'| = |p| = E' = E$

Fig. 18.3 Two-particle to two-particle scattering in the centre of mass frame for the ultrarelativistic regime.

The sum is over either all the incoming particles or equally over all those outgoing, so that s is $4E^2$ for the ultrarelativistic example displayed, where $p_i = E_i = E$. The cross-section derived from eqn. 7.32 is

$$\sigma = |M|^2 \rho_{\mathrm{f}}/F, \qquad (18.17)$$

where M is the matrix element for the process in question. F is the flux per unit area of the incident particles. The wavefunction normalization used here has

$$\int_V \phi^* \phi \mathrm{d}V = 2EV \qquad (18.18)$$

which is Lorentz invariant: a boost alters $E \rightarrow E\gamma$ and the reference volume $V \rightarrow V/\gamma$. By convention the flux factor picks up the product of the incoming energies making it also Lorentz invariant.[4] In the ultra-relativistic regime $F = 2s$. The density of final states is correspondingly Lorentz invariant. Here we express it explicitly in the centre of mass frame: the density of states accessible when scattering into a solid angle $\mathrm{d}\Omega$, with the chosen wavefunction normalization, is

$$\rho_{\mathrm{f}} = \mathrm{d}\Omega/(32\pi^2). \qquad (18.19)$$

Inserting these values in eqn 18.17 gives

$$\frac{\mathrm{d}\sigma}{\mathrm{d}\Omega} = |M|^2/(64\pi^2 s). \qquad (18.20)$$

At this point Feynman's universally used diagrammatic method of cal-culating matrix elements is first introduced. The details of the diagrams will depend on whether the interaction is, as here, electromagnetic or weak or strong.

Feynman diagrams joining the initial to the final state particles are made up of internal lines and vertices. In Figure 18.4 a complete lowest-order Feynman diagram is drawn for the reaction of interest. At the first vertex the e^+ and e^- annihilate to give a photon: this latter converts later to a $\mu^+\mu^-$ pair. We now evaluate the matrix element from this diagram, which is the product of a *vertex* factor for each vertex (here two) and a *propagator* for each (here one) internal line. The propagator for the the internal line is tackled first, initially assuming that the particle involved is a scalar. We go back to eqn. 18.8 and rewrite it for the case that a source Q emits a scalar,

$$(E^2 - \mathbf{p}^2 - m^2)\phi = Q. \qquad (18.21)$$

This can be recast as

$$\phi = Q/(E^2 - \mathbf{p}^2 - m^2) = Q/(q^2 - m^2), \qquad (18.22)$$

where q^2 is the *four-momentum transfer squared*

$$q^2 = E^2 - \mathbf{p}^2, \qquad (18.23)$$

[4]This is the usual Lorentz factor $\gamma = 1/(1 - v^2/c^2)$. For more of the messy details on the flux factor and density of states see *Elementary Particle Theory* by A. D. Martin and T. D. Spearman, published by North-Holland and Elsevier (1970).

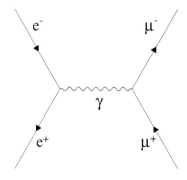

Fig. 18.4 Feynman diagram for $e^- + e^+ \rightarrow \mu^- + \mu^+$ via an intermediate vir-tual photon.

which in the case of a free particle is the rest mass squared. From eqn. 18.22 we infer that propagator is

$$X = \frac{1}{q^2 - m^2}. \tag{18.24}$$

Energy/momentum is not conserved in exchanges so that the denominator does not equal zero as it would be for a free particle. The exchanged particle is *virtual* since it has a limited life and is said to be *off the mass shell* meaning that $E^2 - p^2 = q^2 \neq m^2$.

In fact the internal particle is a photon, so the mass is zero in the propagator, and there is an additional multiplicative factor to take account of its polarization:

$$-ig_{\mu\nu}/q^2. \tag{18.25}$$

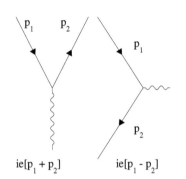

$ie[p_1 + p_2]$ $ie[p_1 - p_2]$

Fig. 18.5 Vertices for QED Feynman diagrams.

Here $g_{\mu\nu}$ is the diagonal time-space metric of special relativity:

$$\begin{bmatrix} 1 & 0 & 0 & 0 \\ 0 & -1 & 0 & 0 \\ 0 & 0 & -1 & 0 \\ 0 & 0 & 0 & -1 \end{bmatrix}$$

Figure 18.5 shows examples of vertices in Feynman diagrams together with the multiplicative factors they contribute to a matrix element. Those for the initial and final vertices in $e^-(1) + e^+(2) \rightarrow \mu^-(3) + \mu^+(4)$ are

$$ie(p_1 - p_2)_\mu \quad \text{and} \quad ie(p_3 - p_4)_\nu, \tag{18.26}$$

where p_1 etc. are all four vectors. Taking the product of these two vertex factors and of the propagator, the resulting matrix element is

$$-iM = ie(p_1 - p_2)_\mu(-ig_{\mu\nu}/q^2)ie(p_3 - p_4)_\nu, \tag{18.27}$$

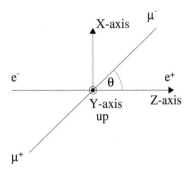

Fig. 18.6 $e^- + e^+ \rightarrow \mu^- + \mu^+$. Coordinates for spin analysis.

Recall that sums are to be taken over repeated subscripts, so the result is a Lorentz invariant quantity. In the present case $q^2 = (p_1 + p_2)^2 = s$ the centre of mass energy squared. Implementing the metric gives

$$M = -\frac{e^2}{s}(p_1 - p_2)_\mu(p_3 - p_4)_\mu. \tag{18.28}$$

At high energy the term in energy (dimensional subscript 0) vanishes leaving the spatial parts

$$M = \frac{e^2}{s}(\mathbf{p_1} - \mathbf{p_2}) \cdot (\mathbf{p_3} - \mathbf{p_4}) = e^2 \cos\theta, \tag{18.29}$$

where θ is the scattering angle between the incoming e^- and outgoing μ^-. Inserting this result in eqn. 18.20 gives

$$\frac{d\sigma}{d\Omega} = \frac{e^4}{64\pi^2 s}\cos^2\theta = \frac{\alpha^2}{4s}\cos^2\theta, \tag{18.30}$$

where α is the fine structure constant.

In this process we can also infer the cross-section when lepton spin is taken into account. Use is made of the fact that at high enough energy a lepton's spin is antiparallel to its direction of travel while an antilepton's spin is parallel; this makes them equivalent to right and left circularly polarized photons respectively. Therefore, in Figure 18.6, the electric vector is transverse to the electron's direction of travel, along Ox or Oy with equal probability. If it is along Oy it can be transferred directly to the μ^- with unit probability. If it lies along Ox then there is a mismatch between the electron and μ^- axes of θ. Malus' law from optics gives the probability of the transfer to be $\cos^2\theta$. The polarization contributions are incoherent and add, whence we estimate:

$$\frac{\mathrm{d}\sigma}{\mathrm{d}\Omega} = \frac{\alpha^2}{4s}[1 + \cos^2\theta]. \tag{18.31}$$

This duplicates the result of a full calculation in which spinors are used to describe the leptons (see Appendix G). Integrating over the polar angle gives a total cross-section for leptons with spin at high energy

$$\sigma = \int \frac{\mathrm{d}\sigma}{\mathrm{d}\Omega} 2\pi \sin\theta\,\mathrm{d}\theta = \frac{4\pi\alpha^2}{3s}. \tag{18.32}$$

Some key features of general application can be noted:

- Each vertex gives a factor α so incorporating the interaction strength.
- The propagator factor $1/s$ means that the cross-section falls rapidly with increasing energy.
- Factors in θ are determined by the spins involved.

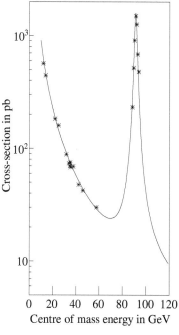

Fig. 18.7 Cross-section for $e^- + e^+ \to \mu^- + \mu^+$. Data taken from W. Bartel et al., *Zeitschrift fur Physik* C28, 507 (1985) ; G. Abbiendi et al., *European Physical Journal* C19, 587 (2001); M. Miura et al., *Physical Review* D57, 5345 (1998).

18.8 Z^0-boson production

Figure 18.7 shows the way the actual measured cross-section for $e^- + e^+ \to \mu^- + \mu^+$ varies as the cm energy increases to over $100\,\mathrm{GeV}$. At first, the cross-section falls smoothly with increasing energy, as predicted by eqn. 18.32. The fall is interrupted by a sharp peak in cross-section at $91.2\,\mathrm{GeV}$ of full width at half height $2.5\,\mathrm{GeV}$. This feature is a resonance due to the production and decay of the Z-boson:

$$e^+ + e^- \to Z^0 \to \mu^+ + \mu^-. \tag{18.33}$$

The corresponding Feynman diagram is drawn in Figure 18.8. As mentioned earlier the Z-boson and $W^\pm$-bosons are the carriers of the weak force. Evidently the Z-boson rest mass M_Z is $91.2\,\mathrm{GeV}$. The width of the resonant peak Γ_Z ($2.5\,\mathrm{GeV}$) is the uncertainty in the resonance energy: then using Heisenberg's uncertainty principle the lifetime of the Z-boson is $1/\Gamma_Z$, or around 10^{-25} s.

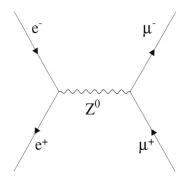

Fig. 18.8 Feynman diagram for $e^- + e^+ \to \mu^- + \mu^+$ via an intermediate virtual Z^0.

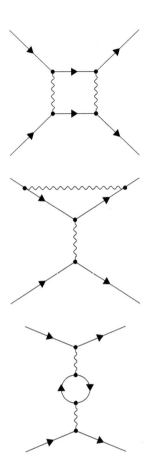

Fig. 18.9 Second-order Feynman diagrams for $e^- + e^- \rightarrow e^- + e^-$.

The full matrix element for the reaction must therefore contain the sum of contributions from diagrams with the photon and Z-boson intermediate states shown in Figures 18.4 and 18.8, respectively. The propagator for the Z-boson is $1/[q^2 - M_Z^2]$ where q^2 is the four momentum squared carried by the Z-boson. Consequently, the Z-boson contribution to cross-sections is suppressed at low energies by a factor q^4/M_Z^4 relative to the photon contribution, but becomes of comparable importance once q^2 approaches M_Z^2.

The take-home message is that the weak processes at low energies are weak relative to electromagnetic processes due to the large masses of the weak vector bosons that appear on the internal lines of Feynman diagrams. At energies approaching the rest mass energy of the weak vector bosons the weak coupling strength approaches that of the electromagnetic interaction.

18.9   Higher order processes and renormalization

The matrix element for a given process is the sum of contributions from all possible Feynman diagrams. Figure 18.9 shows some examples of *second order* Feynman diagrams for the reaction $e^- + e^- \rightarrow e^- + e^-$. The two additional vertices bring an additional multiplicative factor α in the matrix element. The contributions from individual second-order diagrams to the cross-section are smaller by a factor α^2. Processes of even higher order are correspondingly suppressed. Consequently sufficiently accurate predictions can generally be made by summing the contributions of Feynman diagrams up to a few orders. In Chapter 12 it was pointed out, that over 10,000 diagrams were necessary to predict $g - 2$ for the electron to parts in 10^{10}. Fortunately there are computer codes that generate complex diagrams and their amplitides on command.

A problem arises with loops like that in the third panel: the amplitude becomes infinite and this led to the introduction of a technique called *renormalization* to render such infinities innocuous. The momentum, k, circulating around a closed loop as opposed to the momentum entering and leaving the loop q, is unbounded. As a result the contribution to the matrix element from this diagram contains a factor that diverges when integrated over this circulating momentum:

$$\int_{q=0}^{\Lambda} \mathrm{d}^4 k / k^2 \rightarrow \infty \text{ as } \Lambda \rightarrow \infty, \tag{18.34}$$

because the volume element $\mathrm{d}^4 k$ grows faster than k^2. The outcome of such divergent contributions is that in any matrix element calculated at first order e^2 is replaced by:

$$\lim_{\Lambda \rightarrow \infty} e^2 \left[1 - \frac{\alpha}{3\pi} \ln \frac{\Lambda^2}{m^2} \right] C(q^2) \tag{18.35}$$

where $C(q^2)$ is calculable. At the two limits

$$\text{for } q^2 = 0 \quad C(q^2) = 1, \tag{18.36}$$

$$\text{for } q^2 \gg m^2 \quad C(q^2) = \left[1 + \frac{\alpha}{3\pi} \ln \frac{|q^2|}{m^2} \right]. \tag{18.37}$$

Therefore, the charge measured in Millikan's oil drop experiment, where $q^2 = 0$, is

$$e_R^2 = \lim_{\Lambda \to \infty} e^2 \left[1 - \frac{\alpha}{3\pi} \ln \left(\frac{\Lambda^2}{m^2} \right) \right]. \tag{18.38}$$

This is the key result: a finite measurable quantity has absorbed the divergent contribution to the charge. Then provided we replace the charge e everywhere by the charge e_R we will have rendered the divergent term innocuous. This procedure is known as renormalization. This is not the end of the story because there are diagrams with multiple loops that generate infinite amplitudes at higher order. It turns out that only a single correction to the charge is needed to make finite all the matrix elements using diagrams up to a particular order.

A similar procedure is needed to absorb infinities of a second type. In this case it is the mass of the electron that is renormalized. Quantum theories like QED in which the infinities in matrix elements can be absorbed by a finite number of measurable quantities (here the mass and charge of the electron) are said to be *renormalizable*. Crucially this property seems to be confined to quantum gauge theories: other theories generate growing numbers of renormalizing terms as the order of the Feynman diagrams increases and are thus rendered non-predictive. Quantum gauge theories have been discovered for the weak and strong processes of particle interactions. In the case of the electromagnetic and weak interactions the addition of internal lines and vertices leads to amplitudes of diminishing importance so that the sum of these contributions converges rapidly.

The result is that the fine structure constant (coupling strength) $\alpha = e^2/4\pi$ varies with q^2 after renormalization, for $q^2 \gg m^2$ like

$$\alpha(q^2) = \alpha(\mu^2) \left[1 - \frac{\alpha(\mu^2)}{3\pi} \ln \frac{|q^2|}{\mu^2} \right]^{-1}. \tag{18.39}$$

where μ^2 is a reference four-momentum squared. The q^2 dependence is inherited from that in eqn. 18.37. α becomes a *running* coupling constant; it varies with q^2, being $1/137$ statically, $1/126$ at the vector boson mass and $1/100$ at 10^6 GeV. We can interpret this behaviour using Figure 18.10. This shows the vacuum around a charge where, as described in Section 8.2.1, electron-positron pairs are constantly coming into existence giving rise to the zero point fluctuations in the electromagnetic field. The electron from such a pair is repelled and the positron is attracted by the negative source charge, thus shielding this charge. By

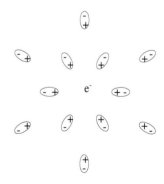

Fig. 18.10 Shielding of charge by vacuum fluctuations.

virtue of the uncertainty principle we can understand that as the q^2 exchanged by the probe charge increases the probe will approach the source charge more closely and the shielding becomes less effective. Hence the effective charge increases and ultimately in the limit of ultra-high q^2 the full source charge would be exposed.

18.10   Neutrinos

Pauli deduced the existence of neutrinos from the apparent violation of the principle of conservation of energy in nuclear β-decay. If the only decay products were the daughter nucleus and the electon, these would travel back to back with equal and opposite momentum in the rest frame of the parent nucleus. In fact the electron kinetic energy, as shown for example in Figure 7.6, has a continuous spectrum up to a value close to, and at that time indistinguishable from $[M_{\text{parent}} - M_{\text{daughter}} - m_{\text{electron}}]$. Pauli suggested that a third particle was emitted: an unknown weakly interacting, neutral, massless particle, which has come to be called the electron neutrino. This particle was detected by Reines and Cowan in 1959[5] in inverse β-decay initiated by the copious antineutrinos from a nuclear reactor

$$\overline{\nu}_e + \text{p} \rightarrow \text{n} + \text{e}^+. \tag{18.40}$$

Detection of the process involved the observation of three photons. The positron annihilated on an electron to give a characteristic pair of photons each of energy 0.511 MeV, and the neutron was captured on cadmium, the resultant nucleus emitting a third photon microseconds later.

Measurements on the production and decay of the Z-boson at the Large Electron-Positron collider LEP, at CERN made it possible to count the number of light neutrino species. The total decay width of the Z-boson seen in Figure 18.11 is the sum of contributions from decay to quark-antiquark pairs, charged lepton-antilepton pairs and neutrino-antineutrino pairs:

$$Z^0 \rightarrow \nu_x + \overline{\nu}_x. \tag{18.41}$$

The rate of decay to any detectable particles is measurable and all rates are calculable in the standard model. Using this information the predictions of the Z-boson resonance shape for 2, 3 and 4 light neutrino species are shown in the figure.[6] This fixes the number of light neutrino species to be three: these are the ν_e, ν_μ and ν_τ of the standard model. The ν_μ and ν_τ have also been detected experimentally.[7] As noted earlier the neutrinos form doublets with the charged leptons (e^-, ν_e), (μ^-, ν_μ) and (τ^-, ν_τ). It is the sum of each leptonic species that is found to be conserved: for example $n(\text{e}^-) + n(\nu_e) - n(\text{e}^+) - n(\overline{\nu}_e)$.

[6]The more channels that are open the shorter the lifetime, and hence, using Heisenberg's uncertainty principle, the more uncertain the energy/mass.

[5]F Reines and C. Cowan, *Physical Review* 113, 273 (1959).
[7]G. Danby et al., *Physical Review Letters* 9, 36 (1962); Kodama et al., *Physics Letters* B504, 218 (2001).

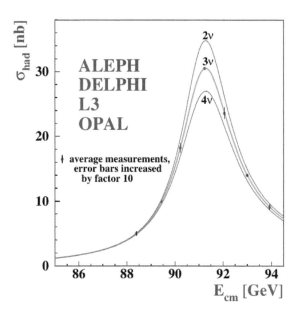

Fig. 18.11 Measured width of Z^0-boson compared to predictions for 2, 3 and 4 light neutrino species. Figure 1.13 in S. Schael et al., *Physics Reports* 427, 257 (2006). 1 nb is 10^{-9} barns. Courtesy Professor Grunewald and Elsevier.

Attempts to measure the electron neutrino mass have involved measuring carefully the upper limit of the electron energy spectrum emitted in the β-decay

$$^3\text{H} \rightarrow {}^3\text{He} + e^- + \overline{\nu}_e \tag{18.42}$$

If the neutrino is massive then this cut-off would be less than the available energy calculated assuming a massless neutrino. The null results[8] lead to an upper limit of 2.0 eV for the electron neutrino mass. The fits of cosmological models to astronomical data[9] lead to a limit for the sum of the masses of the three species of 0.23 eV. Evidently the neutrinos are around a million times lighter than the next lightest elementary particle, the electron, but are not exactly massless, which was what Pauli had proposed.

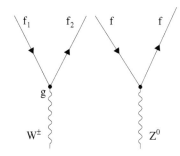

Fig. 18.12 Weak interaction vertices.

[8] V. N. Aseev et al., *Physical Review* D84, 112003 (2011).

[9] Planck Collaboration (P. A. R. Ade et al.), *Astronomy and Astrophysics* 571, A16 (2014).

18.11   The weak interaction

Use will be made from time to time of results presented and discussed in Appendix G on the Dirac equation. This is the relativistic equivalent of Schrödinger's equation and incorporates the spin degree of freedom.

Nuclear β-decay was used in Chapter 7 to illustrate a decay rate calculation. Instead of photon exchange, weak interactions involve W-boson exchange or Z-boson exchange. These field particles have respective masses 80.4 and 91.2 GeV which implies, according to the Yukawa argument, a force with range 10^{-17} m, less than the size of a nucleus. Figure 18.12 shows the interaction vertices that are elements of the Feynman diagrams for weak interactions. In the case of W-boson exchange the fermions f_1 and f_2 are either both leptons or both quarks with charges differing by e. The coupling strength is written g. We shall see later that $g = e \sin\theta_W = 0.4808e$, where θ_W is the *Weinberg angle*. With Z-boson exchange the fermions at the vertex are identical but the coupling strength varies with the fermion species. The purely leptonic decays of these vector bosons are compared here

$$
\begin{aligned}
W^+ &\to\ e^+ + \nu_e \ \text{ or } \mu^+ + \nu_\mu \text{ or } \tau^+ + \nu_\tau, \\
Z^0 &\to\ e^+ + e^- \ \text{ or } \mu^+ + \mu^- \text{ or } \tau^+ + \tau^- \text{ or } \nu_x + \overline{\nu}_x,
\end{aligned}
$$

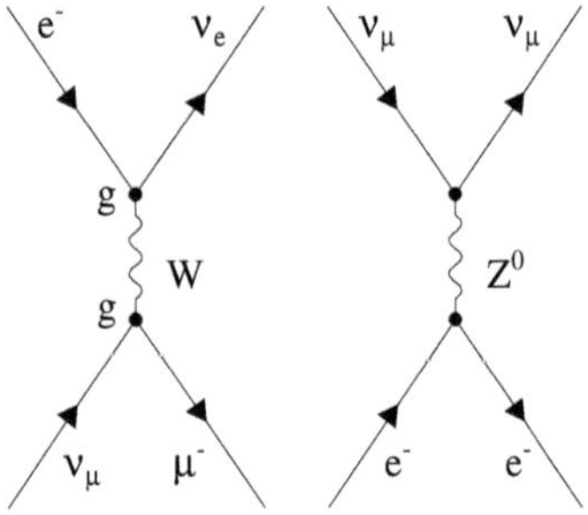

with ν_x specifying any neutrino species. When the vector boson decays into a quark-antiquark pair the particles actually observed are hadrons (π-mesons, etc.).

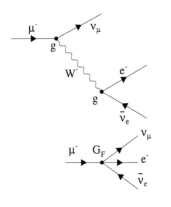

Fig. 18.13 First-order Feynman diagrams for the weak interaction.

Figure 18.13 shows the lowest order contributions to the reactions

$$
\begin{aligned}
e^- + \nu_\mu &\to\ \nu_e + \mu^-, \\
\nu_\mu + e^- &\to\ \nu_\mu + e^-.
\end{aligned}
$$

In calculations of matrix elements the appropriate propagator is the W-boson or Z-boson propagator given by eqn. 18.24

$$ 1/[M^2 - q^2], \tag{18.43} $$

where M is the boson mass and q^2 the four momentum transfer squared it carries. The upper panel in Figure 18.14 has the same particles and exchanges as the left-hand panel in Figure 18.13, but rearranged as a μ-lepton decay.

In low-energy processes such as nuclear β-decay the propagator collapses to $1/M^2$. This simplification is behind the success of Fermi's approximation that the interaction is the point contact interaction drawn in Figure 18.14. It follows that the Fermi coupling constant used in Chapter 7 is related to the weak interaction coupling constant g,

$$ G_F = g^2/[4\sqrt{2}M_W^2] = 1.1664 \, 10^{-5} \, \text{GeV}^{-2}. \tag{18.44} $$

The neutrino scattering cross-sections are the smallest for any observed particle because they only feel the weak force. Using the standard model the cross-sections for neutrino interactions on leptons and quarks are of order $G_F^2 s$, where s is the centre of mass energy squared. At accelerator energies of order $1\,\text{GeV}$ this gives cross-sections of order 0.04 picobarns per nucleon. Thus the mean path length before interaction in lead would be around $4 \, 10^4 \, \text{m}$.

Fig. 18.14 Weak vertices of μ^- lepton decay and the equivalent Fermi point contact interaction.

18.12 Parity, CP and CPT

It seems obvious that if you take any physical process and reflect it through some reference point as origin so that $\mathbf{r} \to -\mathbf{r}$ then the result should be another physically realizable process. This is the *parity trans-formation*. The idea holds true for strong and electromagnetic processes but in weak processes it fails: *parity* is said to be violated. The crucial test came in an experiment measuring the angular distribution of electrons from the decay

$$^{60}\text{Co} \rightarrow {}^{60}\text{Ni} + e^- + \overline{\nu}_\text{e}, \qquad (18.45)$$

in which the cobalt nucleus was polarized by an applied magnetic field, cobalt being ferromagnetic. The distribution in polar angle θ, relative to the nuclear spin, was found to be peaked *backward* with respect to the spin

$$I(\theta) = 1 - \beta \cos\theta, \qquad (18.46)$$

where β is the electron velocity (over c). A parity transformation would leave the magnet current and nuclear polarization unchanged, but would reverse the electron direction. The outcome of the parity transformation would be as shown in the lower panel of Figure 18.15 with an angular decay distribution

$$I(\theta) = 1 - \beta \cos(180° + \theta) = 1 + \beta \cos\theta. \qquad (18.47)$$

This distribution is peaked *forward* with respect to the nuclear spin in contradiction to the *observed* behaviour. This is direct evidence of parity non-conservation in weak interactions: the Nobel prize in Physics for 1957 was awarded to Lee and Yang for predicting this outcome.

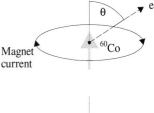

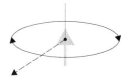

Fig. 18.15 ^{60}Co decay and parity transformed decay.

    A quantum number *parity* can be defined for an elementary particle; either positive if the parity transformation leaves the internal wavefunction unaffected or negative if this wavefunction reverses sign. The parity transformation also affects the motion of particles in space: a state of orbital angular momentum $L\hbar$ has a parity $(-1)^L$. Parity is a multiplicative quantum number so that the overall parity of two particles is the product of their individual intrinsic parities, and the parity of their relative motion. The classical electromagnetic field A is a vector, with spin parity $J^P = 1^-$ making the photon a vector particle. Equally the Z-boson and W-bosons are vector particles. Appendix G shows that a fermion and its antiparticle have opposite parity. For definitenes, the fermions are assigned positive parity. Thus the lowest mass mesons, including the π-mesons and K-mesons which are a quark-antiquark pair in a relative s-state of motion have spin-parity 0^-.

    Beside the parity transformation P there are two related discrete transformations: time reversal T that involves reversing a process in time, and charge conjugation C which reverses the signs of electric charges. Applying either to electromagnetic or strong processes always results in valid processes, but not in the case of the weak interaction. Neutral mesons can be assigned definite C-parities so the notation for labelling them is extended to J^{PC}. Photons can convert to an e^+e^- pair and so have negative C-parity. C-parities are multiplicative like parities:

Table **18.3** The CPT conservation laws

	Weak	EM	Strong
CPT	YES	YES	YES
C, P and T	NO	YES	YES

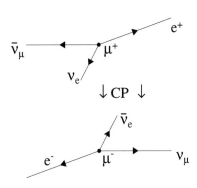

Fig. 18.16 $\mu^+ \to e^+ + \overline{\nu}_\mu + \nu_e$ and the CP transformation to $\mu^- \to e^- + \overline{\nu}_e + \nu_\mu$.

[10] Even the experiments used to study the stable antiparticles (antiprotons and positrons) are difficult because anti-particles annihilate on matter.

π^0-mesons decay to pairs of photons so their C-parity is positive, making their state 0^{-+}. When the compound transformation CP is applied to weak processes a physically realizable process is obtained. A good example is shown in Figure 18.16 where the process considered is the decay

$$\mu^+ \to e^+ + \overline{\nu}_\mu + \nu_e. \tag{18.48}$$

As a result CP is apparently a good symmetry though separately C and P are not. We shall find that there is, for example, a violation of CP at the level of 0.1 per cent in K^0-meson decays. This, and like violations of CP conservation, have deep implications in cosmology.

There is powerful theorem which states that the product CPT gives a realizable physical process, a theorem based only on Lorentz invariance and the requirement that interactions are local. One consequence of this CPT theorem is that a particle and its antiparticle have identical masses, as well as equal and opposite charges and magnetic moments. All experiments support the CPT theorem. For example, the measured masses of the proton and antiproton agree to better than parts in 10^8, and their Landé g-factors agree to a similar precision.[10] The most precise test is provided by the K^0-meson and $\overline{K^0}$-meson masses which agree to better than parts in 10^{18}. Table 18.3 summarizes the CPT conservation laws.

18.13   Handedness of the weak interaction

Appendix G introduces the fully relativistic Dirac equation for the electron. This became the starting point for the development of the quantum field theory of electromagnetic interactions of fermions known as quantum electrodynamics, whose properties we have met. Matrix elements that take the place of the spinless electromagnetic matrix element, given in eqn. 18.27, become quite involved when explicit account is taken of spins. We shall concentrate on the most significant feature of interest in a matrix element involving W-boson exchange. This contains a factor $(1 - \gamma_5)$, explained in Appendix G, and is the origin of parity non-conservation. Fermions selected by the operator $P_L = (1 - \gamma_5)/2$ are those of left-handed *chirality* while $P_R = (1 + \gamma_5)/2$ selects those of right-handed chirality. In the ultrarelativistic limit P_L chooses fermions with spin aligned opposite to their direction of travel and antiparticles with spin aligned along this direction. These spin aligned states are known as states of definite helicity, $+1$ for alignment parallel to the motion and -1 for alignment opposite to the direction of motion. In the non-relativistic regime the chiral states become a superposition of the helicity states. The action of the factor $1-\gamma_5$ in the weak interaction via W-boson exchange limits interaction to the fermions satisfying $P_{\mathrm{L}}\psi = \psi$ and antiparticles of the opposite chirality $P_R\psi = \psi$. That means that W-boson only couples to purely left-handed particles and purely right-handed antiparticles in the ultrarelativistic limit. In the non-relativistic regime we can always turn a helicity $+1$ state into a helicity -1 state by moving to a frame in which its direction of travel is reversed. Chiral states, however, are Lorentz invariant, which is essential if the matrix element for the weak interaction is to be Lorentz invariant. An exercise involving decays of the $\pi^\pm$-meson is designed to bring these features into focus.

18.14 Neutrino oscillations

Experiments measuring the neutrino flux from the sun, and from reactor and accelerator sources have shown that neutrinos oscillate from one species to other neutrino species. This oscillation comes about because the *weak eigenstates* in which neutrinos are created in a weak process are not the *mass eigenstates* that propagate freely in space, but rather linear superpositions of them. The unitary matrix connecting the neutrino weak and mass eigenstates is called the Pontecorvo–Maki–Nakagawa–Sakata (PMNS) matrix.

The first evidence for neutrino oscillations came from experiments to measure the neutrino flux from the sun. This flux can be reliably predicted from two inputs. First the nuclear reaction at work in the sun generating neutrinos

$$4\,^1\mathrm{H} \;\rightarrow\; {}^4\mathrm{He} + 2\mathrm{e}^+ + 2\nu_{\mathrm{e}} \tag{18.49}$$

releases $24.7\,\mathrm{MeV}$ in energy. Second the resulting energy flux (flow per unit time) from the sun is well measured. The rate of the reaction is then the sun's energy flux divided by $24.7\,\mathrm{MeV}$, and twice this quantity is the neutrino flux. Numbers are spectacular, the neutrino flux from the sun at the earth's surface is $6.5\,10^{14}\,\mathrm{m}^{-2}\mathrm{s}^{-1}$. Neutrinos were detected via the reaction

$$\nu_{\mathrm{e}} + e^- \rightarrow \nu_{\mathrm{e}} + e^-. \tag{18.50}$$

in the Super-Kamiokande detector. This vessel, containing 40 tonnes of pure water, is located $1000\,\mathrm{m}$ below a mountain in Japan. The overburden of rock reduces the background from charged cosmic rays to an acceptable level. An electron emerging from the above reaction occuring within the vessel is travelling faster than the local speed of light (in water) and produces a shock wave of light, Cerenkov light, as shown in Figure 18.17. This is analogous to the sound produced by a plane breaking the sound barrier. The Cerenkov light travels at a fixed angle θ with

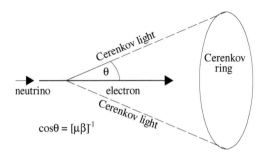

$$\cos\theta = [\mu\beta]^1$$

Fig. 18.17 Emission of Cerenkov radiation from an electron in water.

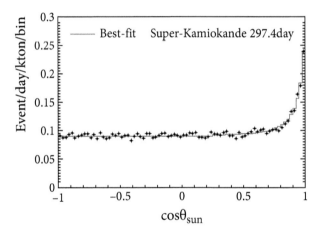

Fig. 18.18 Angular distribution of neutrinos detected with respect to the sun's instantaneous direction. Figure 2 in Y. Fukuda et al., (Super-Kamiokande Collaboration), *Physical Review Letters* 81, 1158 (1998). Courtesy Professor T. Kajita, for the Super-Kamiokande Collaboration, and The American Physical Society.

respect to the flight path of the electron and by extension to that of the incident neutrino: $\cos\theta = 1/\mu\beta$ where μ is the refractive index of water and β is the electron velocity in units of c. In water with $\beta \approx 1.0$, θ is 43°. This hollow cone of light is detected by photomultipliers that line the inside of the containing vessel. Any neutrino that interacts therefore produces a Cerenkov ring of responding photomultipliers. Figure 18.18 shows the angular distribution of the neutrinos detected relative to the instantaneous orientation of the sun. A signal can clearly be associated with the sun as origin. It was found that only 46 per cent of the neutrino flux expected from the sun had been detected. Davis, who pioneered neutrino detection from the sun, and Koshiba, the leader of the Super-Kamiokande collaboration, shared the 2002 Nobel prize in physics for their daring experiments.

Subsequently the total neutrino flux from the sun was detected at the

Sudbury Neutrino Observatory (SNO) in Canada. The detector used was a 1000 tonne vessel filled with heavy water, which was viewed by photomultipliers lining its inner surface. The reactions possible, denoting the deuteron as d, are

$$\nu_e + d \rightarrow p + p + \mathbf{e^-}, \tag{18.51}$$

$$\nu_x + e^- \rightarrow \nu_x + \mathbf{e^-} \tag{18.52}$$

$$\nu_x + d \rightarrow \nu_x + p + \mathbf{n}. \tag{18.53}$$

The detected particle in each reaction is written in bold. ν_x signifies that all three neutrino species produce this interaction. Neutrons from the third reaction are captured on a deuteron to give ^{3}H. Subsequently ^{3}H decays releasing a 6.25 MeV gamma ray, which in turn produces an electromagnetic shower with associated Cerenkov light. The total count of all neutrino species agrees well with the expected flux from the sun. This pins down the cause of the loss of electron neutrinos to neutrino oscillations from one species to another. Next we investigate how to analyse and interpret neutrino oscillations.

A. B. MacDonald, the leader of the SNO group, shared the 2015 Nobel prize in physics with T. Kajita of the Super-Kamiokande collaboration which discovered that neutrinos produced by cosmic rays entering the earth's atmosphere oscillate from ν_μ to ν_e.

If the neutrino eigenstates produced in weak processes are not the same as the mass eigenstates then an electron neutrino will be a superposition of the mass eigenstates. The wavefunctions of the mass eigenstates develop with time in this way in their rest frame

$$\psi = \psi_0 \exp(imt), \tag{18.54}$$

where m is the rest mass. As time passes the phase relation between the mass eigenstates changes and this alters an electron neutrino into some superposition of electron, muon and tau neutrinos. Evidently if neutrinos were massless then species would propagate without change. For simplicity we consider the case that just two species, ν_e and ν_μ, are produced by weak interactions, while ν_1 and ν_2 are the two mass eigenstates. Then in terms of the amplitudes (wavefunctions)

$$\begin{aligned} \nu_e &= \nu_1 \cos\theta + \nu_2 \sin\theta, \\ \nu_\mu &= \nu_2 \cos\theta - \nu_1 \sin\theta, \end{aligned} \tag{18.55}$$

where θ is the mixing angle. Applying these equations to an initially pure ν_e beam at time $t = 0$ gives

$$\begin{aligned} \nu_e(0) &= \nu_1(0)\cos\theta + \nu_2(0)\sin\theta \\ 0 &= \nu_2(0)\cos\theta - \nu_1(0)\sin\theta, \end{aligned} \tag{18.56}$$

whence

$$\nu_1(0) = \nu_e(0)\cos\theta \text{ and } \nu_2(0) = \nu_e(0)\sin\theta. \tag{18.57}$$

Each eigenstate evolves as it propagates, and with x being the distance travelled in time t:

$$\nu_i(t, x) = \nu(0, 0)\exp[i(-E_i t + p_i x)], \tag{18.58}$$

where i is 1 or 2, E_i is the energy and p_i the momentum. The momenta, p, must be identical in order that the eigenstates are in phase at time zero. Hence we can ignore the common factor $\exp[ipx]$. After time t the initially pure state has evolved to

$$
\begin{aligned}
\nu_e(t) &= \nu_1(t)\cos\theta + \nu_2(t)\sin\theta \\
&= \nu_1(0)\exp(-iE_1 t)\cos\theta + \nu_2(0)\exp(-iE_2 t)\sin\theta \\
&= \nu_e(0)[\cos^2\theta \exp(-iE_1 t) + \sin^2\theta \exp(-iE_2 t)]. \quad (18.59)
\end{aligned}
$$

Then the probability of the initial ν_e remaining as a ν_e is

$$
\begin{aligned}
I(t) &= \cos^4\theta + \sin^4\theta + 2\sin^2\theta\cos^2\theta\cos[(E_1 - E_2)t] \\
&= 1 - \sin^2(2\theta)\sin^2[\Delta m^2 t/4E], \quad (18.60)
\end{aligned}
$$

where we have used eqn. 1.76

$$
E_1 - E_2 \approx \frac{m_1^2}{2p} - \frac{m_2^2}{2p} \approx \frac{m_1^2 - m_2^2}{2E} = \frac{\Delta m^2}{2E} \quad (18.61)
$$

in the ultrarelativistic limit appropriate for neutrinos at MeV energies, with $\Delta m^2 = m_1^2 - m_2^2$. The probability for the detection of a ν_μ is just $1 - I(t)$. The distance travelled by the neutrinos during one cycle of oscillation, while $\Delta m^2 t/4E$ changes by π, is

$$
L = 2.476 \left[\frac{E \text{ in GeV}}{\Delta m^2 \text{ in eV}^2} \right] \text{ km.} \quad (18.62)
$$

Disappearance experiments observe the loss of the initially pure species and *appearance experiments* observe the other species. The neutrino sources used are reactors, cosmic rays, the sun and targets exposed to accelerators. Figure 18.19 shows the difference in mass squared and the lepton flavour content of the mass eigenstates deduced from fitting all available oscillation data. The normal hierarchy with the predominantly τ mass eigenstate higher in mass is shown. It is expected that the data will soon be sufficiently precise to distinguish this from the inverted hierarchy with the predominantly τ mass eigenstate lower than the other two.

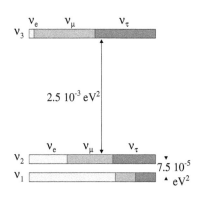

Fig. 18.19 Neutrino mixing parameters: ν_1, ν_2 and ν_3 are the mass eigenstates that propagate freely in vacuum.

Now that the values of Δm^2 are quite well known it is possible when designing an experiment with an accelerator source of neutrinos to tune the energy content and locate the detector far enough from the source so that it lies at a peak or trough in the cycle of oscillations. The PMNS matrix can be fully expressed using three mixing angles and a phase. The values of the mixing angles determined thus far are:

$$
\begin{aligned}
\sin^2\theta_{12} &= 0.270 - 0.344, \\
\sin^2\theta_{23} &= 0.385 - 0.644, \\
\sin^2\theta_{13} &= 0.02 - 0.03.
\end{aligned}
$$

The phase is important in determining the degree of CP violation in the mixing process, but currently less well determined. This topic will be developed more fully in connection with quark mixing.

18.15   Quark flavour oscillations

Quark flavour oscillations are caused by another mismatch of eigenstates. This time between the quark eigenstates associated with the strong and weak interactions. Quarks are bound in hadrons so that when quark flavour oscillations occur what are observed are the oscillations between hadron species. Charge is conserved so that the only examples of oscillations are provided by neutral mesons that undergo weak decays: K^0 mesons ($d\bar{s}$), D^0-mesons ($\bar{u}c$), B^0_d mesons ($\bar{b}d$) and B^0_s mesons ($\bar{b}s$), plus their antiparticles.

The first example seen was that of neutral K-meson oscillations. Neutral K-mesons are produced in eigenstates of the strong interaction: K^0 ($d\bar{s}$) and $\overline{K^0}$ ($s\bar{d}$). For example,

$$\pi^- + p \rightarrow K^0(d\bar{s}) + \Lambda^0(uds), \quad (18.63)$$

where the K^0-meson has mass 0.4976 GeV and the Λ^0-baryon has mass 1.1157 GeV. The K^0-mesons decay via the weak interactions in two distinct eigenstates: K^0_L with a relatively long lifetime $5.12\,10^{-8}$ s and K^0_S with a relatively short lifetime $0.896\,10^{-10}$ s.[11] These decay modes are

$$\begin{aligned} K^0_S &\rightarrow \pi^+ + \pi^-, \\ K^0_L &\rightarrow \pi^+ + \pi^0 + \pi^-. \end{aligned} \quad (18.64)$$

In both cases the energy released is sufficiently small that the π-mesons are in s-states of relative motion (with parity +). It follows that the parity is the product of the π-meson intrinsic parities, and that the two decay modes have opposite parities: positive for K^0_S and negative for K^0_L. K^0_S and K^0_L are to high precision the eigenstates of CP:

$$\begin{aligned} K^0_S &= \frac{K^0 + \overline{K^0}}{\sqrt{2}} \quad [CP+] \\ K^0_L &= \frac{K^0 - \overline{K^0}}{\sqrt{2}} \quad [CP-]. \end{aligned} \quad (18.65)$$

Equally

$$\begin{aligned} K^0 &= \frac{K^0_S + K^0_L}{\sqrt{2}}, \\ \overline{K^0} &= \frac{K^0_S - K^0_L}{\sqrt{2}}. \end{aligned} \quad (18.66)$$

The overall effect of the mismatch between the strong and weak eigenstates is that mesons in a beam of initially pure K^0-mesons will oscillate into $\overline{K}^0$-mesons. A significant difference from neutrino oscillations is that the K-mesons are decaying rather than simply changing their nature. In order to follow the oscillation some decay channel must be chosen that

[11] In the absence of weak interactions the K^0-meson and $\overline{K^0}$-meson would propagate as mass/strong eigenstates without change.

The existence of these opposite parity decay modes of what were apparently the same neutral particle was initially puzzling. It led Martin Block to suggest that parity could be violated. Lee and Yang discovered the formal theory of a parity violating weak force and their predictions were confirmed initially by Mme. Wu's observation of the parity violation in ^{60}Co decay described in Section 18.12.

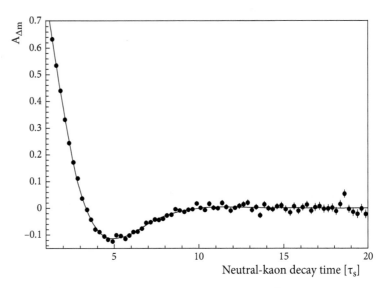

Fig. 18.20 $A_{\Delta m}$ as a function of time in units of the K_S^0 lifetime. Figure 2 in A. Angelopoulos *et al.* (CPLEAR Collaboration), *Physics Letters* B444, 38 (1998). Courtesy Professor J. Fry and Elsevier Science BV.

is only accessible to K^0 or $\overline{K}^0$. The most useful channels are these

$$K^0(d\bar{s}) \rightarrow \pi^-(d\bar{u}) + \nu_e + e^+,$$

where $\bar{s} \rightarrow \bar{u} + \nu_e + e^+;$ (18.67)

$$\overline{K^0}(s\bar{d}) \rightarrow \pi^+(u\bar{d}) + \overline{\nu}_e + e^-,$$

where $s \rightarrow u + \overline{\nu}_e + e^-.$

It is the presence of the u ($\bar{u}$) in the π^+ (π^-) that assures us the parent was a $\overline{K}^0$ (K^0). Following a similar analysis to that performed for neutrino oscillations we get the survival intensity for an initially pure K^0-meson beam

$$|\langle K^0|K^0(t)\rangle|^2 = \frac{1}{4}\{\exp(-\Gamma_S t) + \exp(-\Gamma_L t)$$
$$+ 2\exp[-(\Gamma_S + \Gamma_L)t/2]\cos(\Delta m t)\}, \quad (18.68)$$

where Γ_S (Γ_L) is the K_S^0 (K_L^0) decay rate and Δm is the K_L^0–K_S^0 mass difference. In the experimental determination of Δm by the CPLEAR Collaboration[12] care was taken to eliminate biases due to the variation of detection efficiency between channels: beams of either pure K^0 or pure $\overline{K}^0$ mesons were generated and two overall rates were constructed. These were the rate, R of the decays labelled to be in the same strangeness state as the parent neutral K-meson and the rate, R' of decays in the opposite strangeness state. Then a ratio is defined

$$A_{\Delta m} = (R - R')/(R + R'), \quad (18.69)$$

[12]R. Adler et al., *Physics Letters* B444, 38 (1998).

whose time-dependence can be inferred from eqn. 18.68

$$A_{\Delta m} = \frac{2\exp[-(\Gamma_L + \Gamma_S)t/2]\cos(\Delta mt)}{\exp(-\Gamma_L t) + \exp(-\Gamma_S t)}. \tag{18.70}$$

The observed ratio is plotted as a function of time in units of the K_S^0 lifetime in Figure 18.20. The fitted curve is that predicted from eqn. 18.70 for a mass difference between the K_L^0 and K_S^0 states of $3.50\,\mu\text{eV}$.

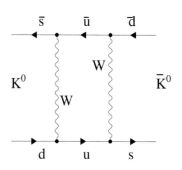

Fig. 18.21 Feynman diagram with double W-boson exchange contributing to $K^0 \rightleftharpoons \overline{K}^0$.

Underlying the neutral K-meson oscillations is the unequal coupling of the W-boson to the quark states produced in strong interactions. These couplings are summarized by a unitary three-by-three matrix like the PMNS matrix. This is the Cabbibo–Kobayashi–Maskawa (CKM) matrix. It relates the weak isospin partners of the u, c and t quarks; namely d′, s′ and b′ to the eigenstates d, s and b produced in the strong interactions:

$$\begin{bmatrix} d' \\ s' \\ b' \end{bmatrix} = \begin{bmatrix} V_{ud} & V_{us} & V_{ub} \\ V_{cd} & V_{cs} & V_{cb} \\ V_{td} & V_{ts} & V_{tb} \end{bmatrix} \begin{bmatrix} d \\ s \\ b \end{bmatrix}. \tag{18.71}$$

The numerical values of the components are given approximately by

$$\begin{bmatrix} V_{ud} & V_{us} & V_{ub} \\ V_{cd} & V_{cs} & V_{cb} \\ V_{td} & V_{ts} & V_{tb} \end{bmatrix} \approx \begin{bmatrix} 1 - \lambda^2/2 & \lambda & A\lambda^3(\rho - i\eta) \\ -\lambda & 1 - \lambda^2/2 & A\lambda^2 \\ A\lambda^3(1 - \rho - i\eta) & -A\lambda^2 & 1 \end{bmatrix}. \tag{18.72}$$

Overall fits to the data require $\lambda = 0.22506 \pm 0005$, $A = 0.811 \pm 0.026$, $\rho = 0.124 \pm 0.018$ and $\eta = 0.356 \pm 0.011$. The phase appearing and parametrized by η and ρ will be returned to shortly. Figure 18.21 exhibits one of the Feynman graphs by which the process

$$K^0 \rightleftharpoons \overline{K}^0 \tag{18.73}$$

may occur, showing how it relies on the presence of the off diagonal elements in the CKM matrix.

Fitch and Cronin discovered rare examples of $K_S^0 \rightarrow 3\pi$ decays at the level of $3.5\,10^{-7}$, and of $K_L^0 \rightarrow 2\pi$ at the level of $2\,10^{-3}$ showing that CP invariance was violated: for this discovery they were awarded the 1980 Nobel prize in Physics. The CPLEAR collaboration[13] studied this CP violating amplitude by producing K^0 and $\overline{K}^0$ from proton-antiproton annihilations at rest (via the strong reactions)

$$\begin{aligned} p\overline{p} &\rightarrow K^0(d\overline{s}) + K^-(s\overline{u}) + \pi^+(u\overline{d}) \\ p\overline{p} &\rightarrow \overline{K}^0(s\overline{d}) + K^+(u\overline{s}) + \pi^-(d\overline{u}). \end{aligned}$$

The sign of the charged K-meson unambiguously labels the nature of the accompanying neutral K-meson. Figure 18.22 shows the decay rates

[13]R. Adler et al., *Physics Letters* B363, 243 (1995).

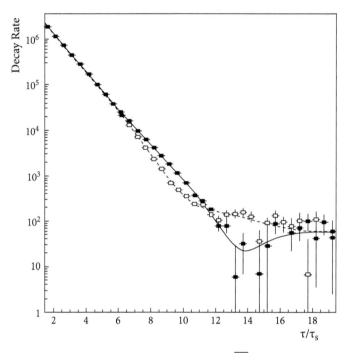

Fig. 18.22 Comparison of the decays of K^0 □ and $\overline{K^0}$ ● to $\pi^+ + \pi^-$ as functions of time in units of the K_S^0 lifetime. Figure 1 in R. Adler et al., (CPLEAR Collaboration), *Physics Letters* B363, 243 (1995). Courtesy Professor Fry and Elsevier Science BV.

of K^0 and $\overline{K^0}$ into $\pi^+ + \pi^-$ as a function of time in units of the K_S^0 lifetime. The effect of CP violation (*CPV*) is small but unambiguous. The amplitude required in the matrix element to produce the fits shown is only 0.1 per cent of the CP conserving amplitude.

CP violation is traceable to the phase in the CKM matrix. Now an n by n unitary matrix is necessary to describe transitions between n generations, and its full description requires $n(n-1)/2$ real parameters and $(n-1)(n-2)/2$ complex phases. At a time when only the first two generations of quarks were known Kobayashi and Maskawa noticed that if there were only two generations there would be no complex phase β and no chance of CP violation. Violation only becomes feasible with three generations.[14] They were awarded half the 2008 Nobel prize in Physics for this insight.

[14] Equally this is the case for the PMNS matrix that parametrizes neutrino mixing.

CP violation is an essential ingredient required for the explanation of how the current matter-filled universe developed from an initial state containing identical amounts of matter and antimatter. The matter and antimatter content has almost all annihilated to photons and they now constitute the cosmic microwave background. Comparing the density of this annihilation radiation to the matter density from astronomical observations shows that only 10^{-10} of the initial matter is present in the universe today. Sakharov deduced that a matter-antimatter imbalance could develop provided both CP and baryon number conservation are violated, and that the universe has a period when it is out of thermal equilibrium. This looks promising as an explanation of the observed imbalance. However, on the basis of the CP violation observed in K^0, B^0_d and B^0_s decays a survival rate for matter of only 10^{-17} is predicted. CP violation in the neutrino sector is yet to be measured precisely: if it is important it will still require a mechanism to transfer its effects to the quark (hadron) sector.

18.16   Further reading

Modern Particle Physics by Mark Thomson, published by Cambridge University Press (2013) is a comprehensive reference work covering modern developments.

An Introduction to the Standard Model of Particle Physics by W. N. Cottingham and D. A. Greenwood published by Cambridge University Press (1995) is dated but provides a clear, theoretically oriented introduction.

Detectors for Particle Radiation by K. Kleinknecht published by Cambridge University Press (1988) gives a brief account of the basic techniques used in detectors.

Exercises

(18.1) Show that the reactions producing the ρ and ω mesons

$$K^-(\bar{u}s) + p(uud) \rightarrow \Lambda^0(uds) + \rho,$$
$$K^- + p \rightarrow \Lambda^0 + \omega,$$

have equal cross-sections.

$$\rho \equiv (u\bar{u} + d\bar{d})/\sqrt{2},$$
$$\omega \equiv (u\bar{u} - d\bar{d})/\sqrt{2}, \quad (18.74)$$

with masses 0.770 and 0.783 GeV respectively.

(18.2) The proton spin wavefunction can be written in terms of the quarks as

$$\sqrt{\frac{2}{3}}\chi_u(1,1)\chi_d(1/2,-1/2) - \sqrt{\frac{1}{3}}\chi_u(1,0)\chi_d(1/2,1/2).$$

The quark state $\chi(s, m_s)$ indicates total spin s and component m_s. Take the magnetic moment of the

quarks to be $(q/2m)\sigma$ where q is the charge, m the mass and σ the spin. Assume the u-quark and d-quark have identical masses. Show that this implies the magnetic moments of the proton and neutron are in the ratio $\mu_n/\mu_p = -2/3$. The observed ratio is -0.685.

(18.3) Which of the following are valid decays of the baryon Λ^0 (sdu) with mass 1.115 GeV?

$$\Lambda \rightarrow p(0.939) + \pi^0(0.135),$$
$$\Lambda \rightarrow p(0.939) + \pi^-(0.140),$$
$$\Lambda \rightarrow p(0.939) + \pi^-(0.140) + \pi^0(0.135)$$
$$\Lambda \rightarrow \pi^+(0.140) + \pi^-(0.140),$$

(18.4) A π-meson of mass 0.139 GeV has a momentum 0.1 GeV. What is its velocity and energy?

(18.5) Show that the process $\gamma \rightarrow e^- + e^+$ requires some other particle in order to conserve energy and momentum. In the case of a 0.5 GeV gamma what is the four momentum imbalance?

(18.6) In a reaction $A + B \rightarrow C + D$ the four momentum transfers of interest are $s = (a + b)^2$ the centre of mass energy, $t = (a - c)^2$ and $u = (a - d)^2$ where a, b, c and d are the four momenta of the respective particles. Show that

$$s + t + u = M_A^2 + M_B^2 + M_C^2 + M_D^2.$$

Recall that $a^2 = M_A^2$ etc.

(18.7) Consider the decays

$$
\begin{aligned}
\pi^+ &\rightarrow \mu^+ + \nu_\mu, \\
\pi^+ &\rightarrow e^+ + \nu_e,
\end{aligned}
\tag{18.75}
$$

where the masses of the particles involved are $M_\mu = 0.1057$ GeV and $M_\pi = 0.1396$ GeV. First show that the momentum of the μ-lepton in the rest frame of the π-meson is

$$p_\mu = (M_\pi^2 - M_\mu^2)/2M_\mu,$$

and its velocity $p_\mu/\sqrt{p_\mu^2 + M_\mu^2}$. Hence calculate numerically the velocities of the μ-lepton and the electron in the rest frame of the π-meson. These reactions have rates in the ratio $10^4{:}1$ despite the very much larger phase space available to the second decay. Explain how the requirement of left handedness in weak decays mediated by the W-boson suppresses the positron decay. Use eqn. G.25.

(18.8) Using a 1 GeV μ-neutrino beam from an accelerator complex what is the appropriate distance at which the detector should be located to maximize the appearance of τ-neutrinos?

(18.9) In the reaction

$$\nu_\mu + e^- \rightarrow \mu^- + \nu_e$$

show that the maximum deviation of the μ-lepton from the flight path of the incident neutrino is $\sqrt{2m_e/E_\nu}$ where E_ν is the laboratory energy of the neutrino. Take E_ν to be much larger than the masses involved.

(18.10) Using results given in Appendix G show that

$$i\gamma_0\gamma_1\gamma_2\gamma_3 = \gamma_5.$$

Particle physics II

19.1   Introduction

In this chapter the success of extending quantum gauge theory beyond
QED to explain the features of the strong and weak forces is described.
This is the core of the standard model of particle physics. The first topic
discussed is the quantum gauge theory of the strong interaction, *quantum
chromodynamics* (QCD). The SU(3) colour symmetry underlying QCD
and the associated *gluons* that carry the strong force are introduced in
a first section. Then drawing on the analogy with QED the expected
behaviour of strong interactions of low and high four-momentum transfer squared (q^2) is deduced. At low q^2 the strong coupling constant is
so large that all quarks are bound in hadrons: by contrast they become
free at asymptotically large values of q^2. The structure of the proton is
then discussed stressing the dependence on the four momentum transfer
from the probe.

The second topic is the unification of the description of the weak
and electromagnetic interactions in a single quantum gauge theory. The
symmetry underlying the weak interaction is $SU(2)_L$ where L signifies
that the W-boson only interacts with left-handed fermions. The unified
quantum gauge theory of the *electroweak* sector $U(1) \otimes SU(2)_L$ would be
expected to have massless field particles, whereas the actual W-bosons
and Z-boson are massive. The way that the *Brout–Englert–Higgs* (BEH)
mechanism resolves this contradiction is described. It required the real-
ization that a scalar field should exist with its quantum being a massive
scalar boson called the Higgs boson. Discovery experiments at CERN
at the $Sp\overline{p}S$ collider, LEP and LHC are described. These validated the
electroweak gauge theory through the observation of the field particles,
the $W^{\pm}$-bosons and the Z^0-boson, in 1983 and of the Higgs boson in 2012.

Finally, some puzzles and mysteries connected with the standard model
are discussed briefly.

19.2   The strong interaction and QCD

The quarks inside the lowest mass baryons are in an s-state of motion,
that is to say they have no relative orbital angular momentum. Certain
of such low mass baryons have spin $3\hbar/2$, so that the quark spins must
be aligned parallel. Two examples are Ω^- (1670 MeV), containing three

Quantum 20/20: Fundamentals, Entanglement, Gauge Fields, Condensates and Topology.
Ian R. Kenyon. © 2020. Published in 2020 by Oxford University Press.
DOI: 10.1093/oso/9780198808350.001.0001

s-quarks, while the $\Delta^{++}(1236)$ contains three u-quarks. It was a puzzle that interchange of a pair of s-quarks in the Ω^- or of u-quarks in Δ^{++} leaves their wavefunctions unchanged. Such a situation violates the Pauli principle. Greenberg deduced that there must be a further symmetry and an associated quantum number: under this symmetry quark interchanges lead to a reversal of the wavefunction, restoring overall antisymmetry under interchange of identical fermions. This new symmetry has to be a threefold symmetry in order to antisymmetrize all interchanges among three quarks. Including this *colour symmetry*, the wavefunction of the quarks making up a hadron becomes

$$\psi = \psi_{\text{spatial}}\psi_{\text{spin}}\psi_{\text{colour}}. \tag{19.1}$$

The colour symmetry is SU(3), a unitary symmetry group whose members transform complex vectors in a three-dimensional *colour* space so that their norms remain unchanged.[1] If the quarks are fundamental particles it makes sense that their properties match those of the fundamental 3 representation[2] of the group, so that the colour label of a quark is either *red* r, *green* g or '*blue* b. There is an inequivalent fundamental representation $\bar{3}$ that accommodates the antiquarks whose colour labels are $\bar{r}$, $\bar{g}$ or $\bar{b}$. Leptons, on the other hand, do not feel the strong force so they are singlets as far as colour is concerned.

A useful check of these ideas was to measure the hadron to lepton production ratio, R, following e^+e^- annihilations

$$R = \frac{\sigma(e^+e^- \to \text{hadrons})}{\sigma(e^+e^- \to \mu^+\mu^-)} = \frac{\sum \sigma(e^+e^- \to q\bar{q})}{\sigma(e^+e^- \to \mu^+\mu^-)} \tag{19.2}$$

where σ signifies the cross-section, with the sum taken over all the active quark species. Measurements were made at the Deutsches Elektronen Synchratron laboratory, Hamburg (DESY) on the PETRA e^+e^- collider.[3] At the 40 GeV centre of mass energy the $u\bar{u}$, $d\bar{d}$, $c\bar{c}$, $s\bar{s}$ and $b\bar{b}$ quark pairs can be produced, but not $t\bar{t}$ because the t-quark has a mass of 173.0±0.4 GeV. The coupling is purely electromagnetic so that cross-sections are proportional to the charge squared of the particles produced. Summing over the five active quark flavours and the N_c colours the prediction is[4]

$$R = \frac{N_c[4/9 + 1/9 + 4/9 + 1/9 + 1/9]}{1} = 1.22\,N_c. \tag{19.3}$$

The measured value is 3.9 ±0.3 so that three colours are required.

The product of the 3 representation (quark) with an $\bar{3}$ representation (antiquark) decomposes in SU(3) into an octet and a singlet representation:

$$\bar{3} \otimes 3 = 1 \oplus 8. \tag{19.4}$$

The singlet is the colourless state of the mesons ($r\bar{r}$ + $g\bar{g}$ + $b\bar{b}$). The octet of states matches the colour content of the field particles, the *gluons*, that carry the strong force between quarks. Their colour content

[1] After a transformation from $a\hat{r} + b\hat{g} + c\hat{b}$ to $A\hat{r} + B\hat{g} + C\hat{b}$, $|A|^2 + |B|^2 + |C|^2$ equals $|a|^2 + |b|^2 + |c|^2$. The S stands for special, meaning that the determinant of the transformations is unity. This excludes an inversion of a coordinate, equivalent in three dimensions to a parity transformation.

[2] When any operation of the group is applied to a member of a representation then the result lies in the same representation.

[3] H. J. Behrends et al., Physics Letters 144B, 297 (1984).

[4] See Table 18.1 for quark and lepton charges.

includes the other combinations such as r$\bar{\text{g}}$ and g$\bar{\text{b}}$. The quantum gauge theory of the strong force, quantum chromodynamics (QCD), was developed by Gross, Wilczek and Politzer and they shared the 2004 Nobel prize in Physics for this work. By analogy with QED, transformations under the group SU(3) replace the complex one-dimensional gauge transformations of electromagnetism, with their simpler group U(1). These eight massless vector bosons (gluons) replace the single photon. The gluons are neutral and do not feel the electroweak force. Fundamental particles that interact strongly, that is gluons, quarks and antiquarks, are known collectively as *partons*.

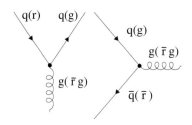

The vertices for the QCD model of strong interactions are drawn in Figure 19.1. In addition to the q$\bar{\text{q}}$g vertex there are ggg and gggg vertices setting the theory apart from QED: a gluon carries colour charge so it interacts with other gluons, whereas photons, not carrying electric charge, do not interact directly with other photons. This is a consequence of SU(3) being a non-Abelian group, which means that the effect of successive transformations depends on their ordering.

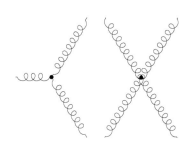

Fig. 19.1 QCD vertices with colour labels in brackets.

Like QED, QCD is a renormalizable quantum gauge theory, making it an effective tool for calculations. The strong coupling constant, α_{s}, which expresses the q$\bar{\text{q}}$g vertex coupling strength is renormalized thus

$$\alpha_{\text{s}}(q^2) = \alpha_{\text{s}}(\mu^2)\Big/[1 + \beta_{\text{s}}\alpha_{\text{s}}(\mu^2)\ln(|q^2|/\mu^2)] \qquad (19.5)$$

where q^2 is the four-momentum transfer squared, and μ^2 is some reference value. This result can be compared to the renormalization of the fine structure constant

$$\alpha(q^2) = \alpha(\mu^2)\Big/[1 - \frac{\alpha(\mu^2)}{3\pi}\ln(|q^2|/\mu^2)], \qquad (19.6)$$

from eqn. 18.39. In the QCD expression β_s is the sum of contributions from the internal loops drawn in Figure 19.2

$$\beta_{\text{s}} = (11 - 2N_{\text{F}}/3)/(4\pi). \qquad (19.7)$$

N_{F} is the number of quark flavours active at q^2. The gluon loop gives the factor 11, which is larger than $2N_{\text{F}}/3$; thus $\beta_{\text{s}}\alpha_{\text{s}}$ is positive. This gives a *positive sign* to the coefficient of the logarithmic dependence on the four-momentum transfer squared $|q^2|$ in eqn. 19.5. This is in contrast to the *negative sign* in QED, $-\alpha/3\pi$. As a consequence the asymptotic $|q^2|$ dependence inverts that for QED:

$$\alpha_{\text{s}}(q^2) \to 0 \ \text{ as } \ q^2 \to \infty, \qquad (19.8)$$

while as q^2 falls $\alpha_{\text{s}}(q^2)$ eventually saturates at unity resulting in the bound hadrons. Any larger value would lead to the probability of reactions exceeding unity, breaking the so-called *unitarity bound*. The

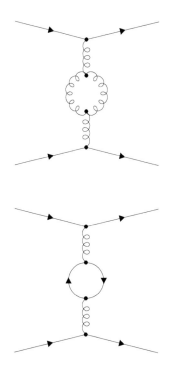

Fig. 19.2 QCD internal loops.

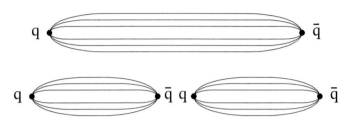

Fig. 19.3 Gluon field lines showing how they snap at an extension of ~1 fm.

self-interaction of the gluons causes the colour force to grow with increasing distance (lower q^2). Schematically the gluon field lines connecting a quark and antiquark in a meson dispose themselves as shown in the upper panel of Figure 19.3. Their mutual attraction draws them together, which is different from how the field lines between two electric charges diverge to fill all space. If the quark and antiquark move apart the field energy increases linearly with the separation. Eventually the point is reached where a lower energy state can be reached by snapping the lines and forming a new quark and new antiquark as shown in the lower panel. Both of the quark-antiquark pairs are bound in mesons. An effective potential of the form

$$V(r) = -\frac{4\alpha_s}{3r} + \lambda r, \qquad (19.9)$$

has been used to calculate the energies of the bound states of *quarkonia*, which are mesons made up from heavy quark-antiquark partners $c\bar{c}$ and $b\bar{b}$. The relative masses of the heavy quarkonia can be predicted with some accuracy using coefficients $4\alpha_s \sim 0.2$ and $\lambda \sim 0.16$. The potential terminates at some separation where it becomes energetically favourable to produce a quark-antiquark pair. This is around 1 fermi and is the effective range of the strong force.

It is the growth of the QCD coupling at low energies that undoes the expectation of having an infinite range of force for massless gluon exchange.

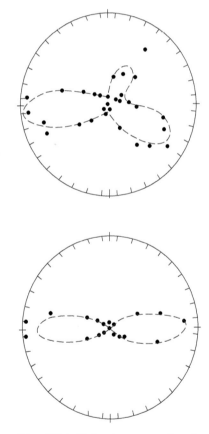

Some indication as to why it is that only $q\bar{q}$ and qqq systems are bound can be obtained from the group structure of SU(3). The force between quarks is proportional to the product of the colour charges just as the force between charged particles is proportional to the product of their charges. Now the product of colour charge depends on the representation to which the quarks belong. The only cases where the force is attractive are

Fig. 19.4 Hadronic energy flow in e^+e^- annihilations with and without a gluon being radiated. Figure 3 in D. P. Barber et al., *Physical Review Letters* 43, 830 (1979). Courtesy Professor Chen and The American Physical Society.

$$q\bar{q} \text{ in singlet,}$$
$$qq \text{ in } \bar{3},$$
$$\bar{q}\bar{q} \text{ in } 3. \qquad (19.10)$$

With higher representations the forces are repulsive. The quark combination $q\bar{q}$ in a colour singlet can be bound into mesons. The other

possibility is to take a q in the triplet representation and a qq in the an-
titriplet representation to form qqq in the singlet representation. Both
steps give binding and thence comes the final colourless baryon.

The first evidence for the existence of gluons was obtained at DESY.
The observations were made at the PETRA e^+e^- collider having a centre
of mass energy of 30 GeV. The energy flow in the hadrons from the e^+e^-
annihilations was observed to have the two distinct patterns shown in
Figure 19.4. These patterns are interpreted as arising from the reactions
at the quark level

$$e^+ + e^- \rightarrow q + \bar{q}, \tag{19.11}$$
$$e^+ + e^- \rightarrow q + \bar{q} + g, \tag{19.12}$$

with corresponding Feynmann diagrams shown in Figure 19.5. The lower
panels of Figures 19.4 and 19.5 refer to the reaction in eqn. 19.11 with
an emerging quark and antiquark. The upper panels refer to the reaction in eqn. 19.12 where a gluon is radiated by a quark (or antiquark).

In interactions at energies large compared to 1 GeV, the baryon mass
scale, α_s becomes small and perturbation theory can be applied to the
primary interaction in e^+e^-, proton-proton or proton-antiproton collisions. Such interactions are known as *hard processes*. An emerging quark
can radiate gluons and a gluon can materialize as a quark antiquark pair.
A cascade occurs with the mean energy per particle steadily falling at
each step. Only the early steps in which α_s remains small can be predicted by perturbative QCD. The later steps culminate in the dressing
of partons to produce the hadrons observed by the experimenter. Each
of the high energy partons emerging from primary process gives rise to a
group of hadrons which is more collimated the higher the energy of the
initiating parton. Such a structure is called a *jet*. This structure can be
recognized in Figure 19.4

How emerging partons convert into hadrons is not easily followed in
detail.[5] There are semi-empirical models guided by QCD for each stage
which are tuned to reproduce the experimentally observed overall hadron
distributions. Simulated data sets are produced by following a cascade,
making at each step a random choice from the corresponding statistical
distribution. This is called a Monte Carlo simulation. Computer power
permits rapid accumulation of large simulated data sets, which can be
compared with experimental data. For example, this approach will generate the backgrounds expected from standard model processes when
searching for new processes.

To summarize: quarks behave as if nearly free when the four momentum transfer squared of processes is large compared to 1 GeV^2; quarks
are bound in hadrons by processes in which the four-momentum transfer
squared is small on this scale. In a word, there is asymptotic freedom
at high q^2 and infrared slavery at low q^2.

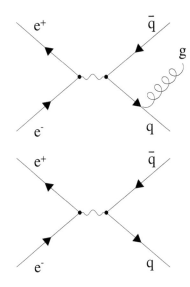

Fig. 19.5 Initial hard process in e^+e^-
annihilation to a quark-antiquark pair.
In the upper diagram the quark radiates a hard gluon.

[5]There is a mismatch between the
colour octet gluons or colour triplet
quarks/antiquarks emerging from any
hard collision and the final observed
colour singlet hadrons. Thus the dressing has to involve exchange of colour
through the medium of low energy partons; something that is not calculable
because α_s is correspondingly large.

19.3   Proton structure

A first point to absorb is that the sum of two u-quark masses and one d-quark mass given in Table 18.1 falls well short of the proton mass. The masses entered there have been deduced from the electroweak behaviour of the quarks[6]. Evidently the greater part of the proton mass is due to the energy of the internal motion and interaction of the quarks and the gluons. The effective mass of the quarks, for example, $0.938\,\mathrm{GeV}/3$ in a proton, is called the *constituent mass*.

In the view taken so far a proton has quark content (uud). These three

[6]Coming from coupling to the Higgs.

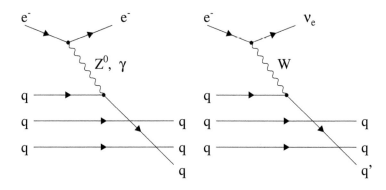

Fig. 19.6 Lowest-order Feynman diagrams contributing to deep inelastic scattering. The charge exchange process, in which the emitted neutrino is not detected, will not be discussed further.

valence quarks are bound by gluon exchange. Each gluon involved can convert to a colour mismatched quark-antiquark pair and these quarks can radiate gluons in turn. More refined information on proton structure has been obtained by using electroweak probes, $e^{\pm}$ and neutrinos, which are themselves structureless (at least down to $10^{-18}\,\mathrm{m}$). The important first-order Feynman diagram involved in the inclusive process

$$e^- + p \to e^- + \text{hadrons} \tag{19.13}$$

is shown in the left-hand panel of Figure 19.6. When the four momentum transfer from the electron is larger than a few GeV/c this is known as *deep inelastic scattering*. Afterwards the disrupted proton fragments into hadrons while the electron escapes as a free particle. Figure 19.7 compares the measured dependence on four-momentum transfer squared q^2 of elastic scattering, in which the proton remains intact, and of deep inelastic scattering. Each cross-section has been divided by the Mott scattering cross-section, that is the cross-section expected if both electron and proton are pointlike particles. We can interpret these results in the Born approximation of Section 7.7. The elastic scattering amplitude obtained is just the Fourier transform of the target shape, with contributions from all parts of the target being coherent. Pursuing this line of argument the flat distribution in q^2 of deep inelastic scattering is the

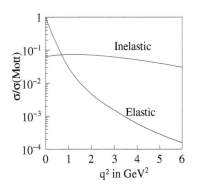

Fig. 19.7 Elastic and deep inelastic scattering cross-sections. Adapted from M. Breidenach, J. I. Friedman, H. W. Kendall, E. M. Bloom, D. H. Coward, H. deStaebler, J. Drees, L. M. Mo and R. E. Taylor, *Physical Review Letters* **23**, 935 (1969).

Fourier transform of a point target. In other words, the electron is scattering off *pointlike* constituents in the proton, namely the quarks. This discovery won the 1990 Nobel prize in Physics for Friedmann, Kendall and Taylor.

The definitive experiments to measure proton structure were performed on the HERA 2 km diameter collider at DESY. 27.5 GeV electrons (or positrons) collided head-on with 920 GeV protons, giving a cm energy of 300 GeV and four-momentum transfers squared of up to 10^5 GeV^2. This explored the proton at a physical scale of $1/300$ in natural units, that is 10^{-18} m and no signs of quark or electron substructure were observed. Hence the focus here is on the measurement of the distribution of the parton momenta in the proton. The fraction

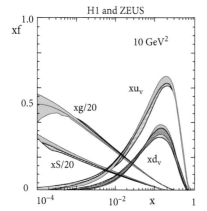
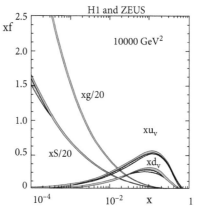

Fig. 19.8 Parton distributions in the proton at four-momentum transfer squared of 10 GeV^2 and 10 000 GeV^2. $xu_v(d_v)$ refers to the up(down)-valence quarks, xg to the gluons, and xS to the sea quarks generated from *gluon splitting* into quark-antiquark pairs. Note that sea and gluon distributions are scaled down by a factor 20. Figures 56 and 57 in H. A. Abramowicz et al. (H1 and ZEUS Collaborations), *European Physical Journal C*75, 580 (2015). Courtesy Professor Paul Newman and Springer Link.

of momentum, x, carried by the parton (quark or antiquark) struck in the primary scattering step, can be inferred as follows[7]. Immediately after the primary collision the parton *four-momentum* is $xP + q$ where P is the proton *four-momentum* and q that of the photon emitted by the electron and absorbed by the parton. Taking the quark mass to be m we have

$$m^2 = (xP + q)^2 = x^2P^2 + q^2 + 2xP \cdot q. \qquad (19.14)$$

The quark mass squared before the collision is $(xP)^2$. At the high energies involved we may neglect the mass m so that $x^2P^2 = m^2 = 0$. Using these equalities in the previous equation the fraction of the proton momentum carried by the parton is

$$x = -q^2/2P \cdot q. \qquad (19.15)$$

Measurement of the electron momentum alone fixes q^2 and $P.q$. This makes the determination of x both simple and independent of the details of the debris of the proton. Figure 19.8 shows the distributions of the fraction of proton momentum carried by the parton species at two different four-momentum transfers. In this plot $xf\mathrm{d}x$ is the fraction of the proton momentum carried by partons with x values between x and

[7]A simplified view is taken, neglecting the very much smaller momentum components transverse to the proton's momentum in the laboratory.

$x + \mathrm{d}x$. The label xg refers to gluons, xS refers to the sea of quarks and antiquarks created from gluon splitting. xu_v and xd_v refer to the initial up and down valence quarks. The valence quark contributions give a peak at around x equal to $1/6$; the integrated contribution of this peak over all x dominates at low q^2. The parton distributions evolve as q^2 increases thanks to the increasing number of generations of radiated gluons and the *sea quarks and antiquarks* from gluon splitting. This fuels the migration to very low x-values seen in the right-hand panel of Figure 19.8. This situation is indicated in the cartoon Figure 19.9. The

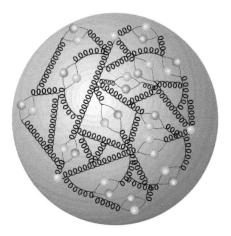

Fig. 19.9 Cartoon of the proton structure as seen at large four-momentum transfers. There are a growing number of gluons and quark-antiquark pairs. Credits: DESY (Deutsches Elektronen-Synchrotron Laboratory, Hamburg).

[8]This information was vital when predicting the backgrounds to Higgs production at LHC.

evolution of the parton distributions with q^2 above $\sim 1\,\mathrm{GeV}^2$ is predicted reliably[8] within the framework of QCD, taking account of the evolution of the coupling constant α_s with q^2.

19.4   Electroweak unification

The weak interaction via W-boson exchange couples the members of the weak isospin quark doublets (u,d′), (c,s′) and (t,b′): it also couples the lepton doublets (ν_e, e^-), (ν_μ, μ^-) and (ν_τ, τ^-). Following the same line of argument that led from colour triplets to QCD we argue that these doublets lie in the simplest, the fundamental, representation of another Lie symmetry group $SU(2)_L$. This is the special unitary group in two dimensions, whose elements rotate the complex amplitudes of the quark and lepton weak isospin states over the surface of the Bloch sphere, described in Appendix D. Then, for example, ν_e would lie at one pole and the electron at the other pole. The subscript L signifies that the coupling is to left-handed fermions, picked out by the $(1 - \gamma_5)$ factor in the W-boson coupling to fermions[9]. The right-handed states lack this coupling and are then singlets of $SU(2)_L$. For example, the left-handed components of u and d′ have weak isospin $t^L = 1/2$, with components t_3^L equal to $+1/2$ and $-1/2$ respectively. By contrast, the right-handed components have $t^L = t_3^L = 0$.

[9]See Appendix G.

The overall symmetry needed to describe the electroweak sector is thus the product $SU(2)_L \otimes U(1)$. As we shall see $U(1)$ is not precisely the symmetry group of electromagnetism. For this reason the additional quantum number for $U(1)$ is chosen not to be the particle charge, Q, but the hypercharge y, where

$$Q = t_3^L + y/2. \tag{19.16}$$

Left-handed quarks have $y = 1/3$, left-handed leptons have $y = -1$ and right-handed quarks and leptons have $y = 2Q$. A quantum gauge theory of the electroweak sector was built using this product symmetry by Weinberg, Salam and Glashow.[10] Now, by extension, the argument made in Section 13.3 for the masslessness of the photon, implies that all gauge field bosons should be massless. However, the actual W-boson and Z-boson exchanged are massive.

[10]They shared the 1979 Nobel prize in Physics for this work.

The resolution of this contradiction was discovered by Brout, Englert and Higgs.[11] They perceived that the symmetry underlying the electroweak interaction was spontaneously broken, so that the vacuum state of the universe did not possess the full symmetry obeyed by the Hamiltonian. This situation parallels that met earlier in the examples of He-II and superconductivity where it is the phase choice of a boson condensate which breaks the symmetry. In the electroweak case the vacuum itself becomes a condensate with equivalent vacua differing solely in the choice of phase. Transitions between these equivalent vacua cost no energy and hence the corresponding scalar quanta are massless. At first sight this looks to make the difficulty worse because there are now unobserved massless scalar bosons to add to the massless vector field particles!

[11]Higgs and Englert shared the 2013 Nobel prize in Physics given for this achievment, Brout having died earlier. See F. Englert and R. Brout, *Physical Review Letters* 13, 321 (1964). P. W. Higgs, *Physics Letters* 12, 132 (1964); *Physical Review Letters* 13, 508 (1964).

The approach due to Brout, Englert and Higgs, which rescues the theory, follows the implications further of having scalar fields appropriate to the symmetry $SU(2)_L \otimes U(1)$. The choice made is to have two complex scalar fields in a weak isospin doublet:

$$\phi = \begin{bmatrix} \phi_1 \\ \phi_2 \end{bmatrix} \qquad \phi^\dagger = \begin{bmatrix} \phi_1^* & \phi_2^* \end{bmatrix}, \tag{19.17}$$

where ϕ_1 has positive charge e and ϕ_2 is neutral. The most general expression for the Lagrangian density including only these fields and their first derivatives is (compare with eqn. 18.13)

$$L = T - V = D_\mu \phi^\dagger D_\mu \phi - [\mu^2 \phi^\dagger \phi + \lambda (\phi^\dagger \phi)^2]. \tag{19.18}$$

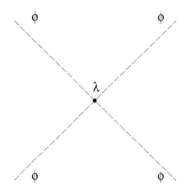

Fig. 19.10 Scalar self interaction.

Here D_μ is the covariant derivative needed to include the influence of the gauge fields, like the covariant derivative of QED given in eqn. 13.13. This guarantees invariance under changes of the phase of ϕ, that is to say invariant under transformations of the overall $SU(2)_L \otimes U(1)$ electroweak symmetry. λ is a dimensionless coupling constant for the self interaction energy corresponding to Figure 19.10, and is therefore positive. Inclusion of higher order terms in $\phi^\dagger \phi$ would lead to the theory

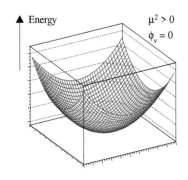

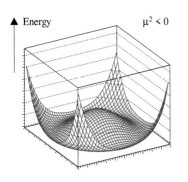

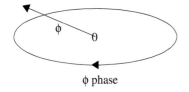

Fig. 19.11 Alternative vacua for scalar fields. The equivalent Higgs vacua lie at points along the minimum in the trough. The two surfaces touch at $\phi = 0$.

[13]See Appendix H for details of the evaluation. Terms in the Higgs interaction like hhh, hhhh, hBB, hWW, hhBB and hhWW are omitted. These terms correspond to valid Feynman vertices like that in Figure 19.10 for hhhh.

becoming non-renormalizable.

In the lowest energy state, the vacuum, the field ϕ_v will be constant, with the potential V at its lowest value. There are two very different possible outcomes depending on the sign of μ^2. If $\mu^2 > 0$, the vacuum field $\phi_v = 0$. If $\mu^2 < 0$ the vacuum field we get by minimizing V with respect to the variation of $\phi^\dagger \phi$ is

$$\phi_v^\dagger \phi_v = -\mu^2/(2\lambda) \equiv v^2/2, \qquad (19.19)$$

which is non-zero. Here we have the vacuum field recognized by Brout, Englert and Higgs as the actual vacuum field in nature.[12] Both possibilities for the field energy in the vacuum are displayed in Figure 19.11. In the case discovered by BEH the value of ϕ for the vacuum will be located somewhere around the lowest level of the circular trough. The symmetry breaking behaviour is just this: although eqn. 19.18 is invariant under rotations of the phase of weak isospin, the vacuum field ϕ_v has a definite phase, indicated here by the arrow in the lower panel in Figure 19.11.

The choice of the direction of the isospin axis is quite arbitrary locally and is chosen to make the neutral states real; the vacuum and excited states being respectively

$$\phi_v = \frac{1}{\sqrt{2}} \begin{bmatrix} 0 \\ v \end{bmatrix}, \text{ and } \phi = \frac{1}{\sqrt{2}} \begin{bmatrix} 0 \\ v + h(x) \end{bmatrix}, \qquad (19.20)$$

which have $t_3^L = -1/2$ and $y = 1$. $h(x)$ is the excitation from the vacuum, which depends on the location in space-time, x, whereas v is the universal vacuum. Real (radial) excursions in Figure 19.11 about the minimum will require energy, and the quanta of these excitations are the *Higgs bosons*.

From now on the development of electroweak field theory can follow the path taken for QED. The covariant derivative is more complicated, reflecting the nature of the product symmetry group $SU(2)_L \otimes U(1)$:

$$D_\mu = \partial_\mu + igt_i^L W_{i\mu} + i(g'/2)yB_\mu, \qquad (19.21)$$

where g and g' are the respective coupling constants for weak isospin and hypercharge. There are three weak isospin gauge fields W_i and one hypercharge gauge field B, W_3 and B being neutral. t_i^L and y act on the Higgs. Substituting the covariant derivative from eqn. 19.21 in eqn. 19.18 and keeping terms of interest:[13]

$$\begin{aligned} L &= (1/8)g^2v^2[W_1^2 + W_2^2] + (1/8)v^2[gW_3 - g'B]^2 \\ &\quad - \lambda v^2 h^2. \end{aligned} \qquad (19.22)$$

[12]This behaviour is different from that of the electromagnetic field whose lowest energy (vacuum) state has a mean value of zero, with non-zero *fluctuations*. This *Higgs vacuum* is non-zero at all points in space-time.

We see immediately that each gauge field appears quadratically multiplied by a coefficient, which is the (mass squared)/2 of the particle associated with the field. Each mass squared is proportional to v^2 where v is the Higgs vacuum field. The acquisition of mass in this way by the gauge bosons is known as the *BEH mechanism*. The essential point is that the formalism is gauge invariant, and yet delivers the quadratic terms that give the gauge bosons their mass.

Note that eqn. 19.18 lacks a factor 1/2 multiplying the mass squared compared to eqn. 18.13. Then the choice of $(v+h)/\sqrt{2}$ rather than the more consistent $v + h$ for the Higgs fields restores this factor. See the second edition of *Quarks, Leptons and Gauge Fields* by K. Huang, published by World Scientific (1992). The factors used in this chapter are now standard.

The actual neutral fields observed are linear superpositions of W_3 and B such that one of them, the photon, is massless. This is achieved by taking

$$Z = W_3 \cos\theta_W - B\sin\theta_W, \qquad (19.23)$$
$$\gamma = W_3 \sin\theta_W + B\cos\theta_W, \text{ photon} \qquad (19.24)$$

where θ_W is the *Weinberg angle*

$$\sin\theta_W = \frac{g'}{[g^2 + (g')^2]^{1/2}}. \qquad (19.25)$$

Then the components of the Lagrangian density[14] become more recognizable

$$L = (1/8)g^2v^2[(W^+)^2 + (W^-)^2 + Z^2/\cos^2\theta_W] - \lambda v^2 h^2. \qquad (19.26)$$

The masses of the weak vector bosons can now be extracted from the coefficients of the fields in this equation:

$$M(W) = gv/2 \text{ and } M(Z) = M(W)/\cos\theta_W. \qquad (19.27)$$

Put in words, three components of the scalar field ϕ have morphed via the BEH mechanism into the three longitudinal spin components of the massive vector bosons, $W^\pm$ and Z^0. The remaining component has become the Higgs boson. Now that the photon is identified and because its coupling to charge is well known we can use eqn. 19.24 to obtain

$$e = g\sin\theta_W \qquad (19.28)$$

linking the weak and electromagnetic couplings. Then, after some manipulation, the Fermi constant used in Section 7.8 is found to be

$$G_F = g^2/(4\sqrt{2}M_W^2). \qquad (19.29)$$

The vacuum expectation value of the Higgs field from eqn. 19.19

$$v = \frac{\mu}{\sqrt{\lambda}} = \frac{2M_W}{g} = 246 \text{ GeV} \qquad (19.30)$$

giving the electroweak energy scale. Finally the Higgs mass appearing in eqn. 19.26 is $\sqrt{2\lambda v^2}$, in which λ is unknown. Thus the standard model predicts the existence of the Higgs and its couplings but *not* its mass. However, the expectation was that its mass should be of order v. The photon remains massless and therefore has just two circular polarization states with spin notation (s = 1, $m_s = \pm 1$). This analysis parallels the (non-relativistic) Ginzberg–Landau analysis of superconductivity in Section 15.4, with ϕ being the equivalent of their order parameter.

[14]Using the weak isospin relations for $t^L = 1$ states:

$$W^\pm = (W_1 \mp iW_2)/\sqrt{2},$$
$$W_1^2 + W_2^2 = (W^+)^2 + (W^-)^2.$$

19.5   The discovery experiments

In the late 1970s the production of the electroweak vector bosons presented a difficulty. If the value of the Fermi coupling constant is inserted into eqn. 19.29 the masses of the vector bosons are predicted to be around 100 GeV. At that time the proton beam from the CERN SPS accelerator had an energy (E_p) of 260 GeV. The accelerator is an evacuated tube in a roughly circular closed ring with circumference 7 km. RF cavities around the ring accelerate the protons to full energy, after which they coast at that energy; dipole and quadrupole magnets steer and focus the proton beam. The protons come in pencil-like bunches, ~ 10 cm long and $\sim 10\mu$m in diameter; ~ 100 such bunches circulate equally spaced around the ring at close to the speed of light. A pulsed electric field would be used to extract the protons. When one of these protons strikes a nucleon of mass m the cm energy is $\sqrt{2mE_p}$: this energy is only 23 GeV, inadequate to create a vector boson.

In a bold stroke CERN converted the SPS accelerator into a collider with protons travelling in one sense around the evacuated ring and antiprotons in the other sense in the same ring. In collision this yields a proton-antiproton cm energy of $2\times 260 = 520$ GeV: or on average $520/6$ $= 87$ GeV in annihilations of quark on antiquark. All this energy is available for creating new particles. Taking account of the fluctuations in energy around the mean, this is adequate for generating a vector boson. The techniques for collecting enough antiprotons and inserting them into the accelerator's phase space earned Simon van der Meer half the 1985 Nobel prize in Physics. He shared it with Carlo Rubbia, who initiated the programme and led the UA1 experimental team that discovered the W-boson and Z-boson. Before proceeding, some remarks are necessary about the generic features of modern detectors at particle colliders.

The colliding beams travel in an evacuated beam pipe and are brought to collision at designated collision points. A detector is centred on a collision point and consists of a number of components. Wrapped immediately around the beam pipe are tracking chambers that record the tracks of the charged particles emerging from the interaction and leaving the beam pipe. Modern tracking chambers are made of multiple thin layers of silicon similar to those in CMOS digital cameras. A uniform, usually axial magnetic field is applied over the tracker region so that track curvature can be used to determine particle momenta. If the track radius of curvature is r, the momentum is $p = Be/r$ in a plane perpendicular to the field B Tesla. The silicon layers form successive closed cylinders around the interaction point spaced at radii that optimize the determination of track curvature and direction at origin.

Multiple layers of sampling calorimeters surrounding the tracker measure the energy of emerging particles. In an inner electromagnetic calorimeter lead sheets cause the photons and electrons to interact and produce electromagnetic showers of electrons, photons and positrons. Between the lead layers are sheets of scintillator giving light proportional to the number of particles in the shower. With enough layers the total light is proportional to the energy deposited. Photomultipliers detect the light and convert the signal to electrical pulses that can be processsed to reconstruct the energies. The electromagnetic shower development is more rapid in a material with high atomic number (charge), hence the preference for lead. Measurement of the energy of hadrons such as protons, neutrons and charged mesons requires a deeper outer calorimeter of instrumented steel and scintillator to contain the hadronic showers. Finally, outer layers of *muon chambers* identify the muons (μ-leptons). As regards electromagnetic interactions muons behave like 210 times heavier electrons, which reduces their rate of energy loss through the calorimeters comparably. Hence muons alone emerge from the calorimeters and leave ionization signals in the outer muon detector layers. Figure 19.12 summarizes the essential features of such detectors which contribute to particle identification.

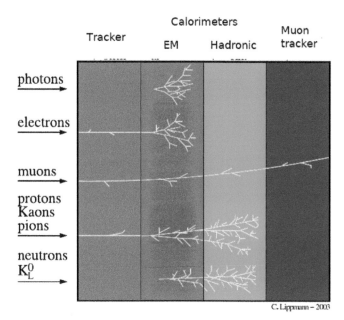

Fig. 19.12 The principles of measurement and identification of particles from interactions at high energies. Adapted from Figure 1 in C. Lippmann, *Nuclear instruments and methods A*666, 148 (2012). Courtesy Dr. Lippmann and Elsevier.

Returning to the discoveries, the production and leptonic decay of a W⁻-boson occured in around one in every 10^7 proton-antiproton interactions

$$\begin{aligned} \mathrm{p} + \overline{\mathrm{p}} &\rightarrow \mathrm{W}^- + \text{hadrons}, \\ \mathrm{W}^- &\rightarrow \mathrm{e}^- + \overline{\nu}_\mathrm{e}, \end{aligned} \qquad (19.31)$$

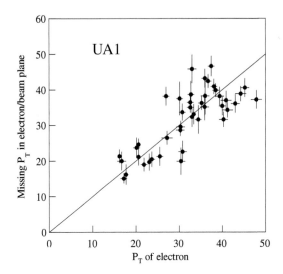

Fig. 19.13 Missing transverse momentum versus the electron transverse momentum. Adapted from Figure 3b in G. Arnison et al., *Physics Letters* 129B, 273 (1983).

where the underlying production process is

$$d + \bar{u} \rightarrow W^-, \tag{19.32}$$

with equivalent processes involving W^+-bosons. In such events an isolated track identified as that of an electron having a momentum component of $\sim$30–50 GeV transverse to the beam axis was detected. What was unanticipated, but critical to proving the existence of the W-boson, was that the neutrino from the W-boson decay was signalled by an imbalance in the momentum of the particles detected in the plane transverse to the beams. In Figure 19.13 this missing transverse momentum is seen to balance the transverse momentum of the electron. Assigning this imbalance to a neutrino allowed a determination of the W-boson mass of 80.9± 3.5 GeV from the events observed. Further information could be extracted from the angular distribution of the electrons. The angular distribution shown in Figure 19.14 is that of the $e^\pm$ from the decay of the $W^\pm$ with respect to the proton/antiproton direction in the centre of mass frame. It is in good agreement with the superposed $[1 + \cos\theta]^2$ shape required for the decay of a vector boson ($J^P = 1^-$). Examples of the ten times less frequent production of the Z-boson decaying via

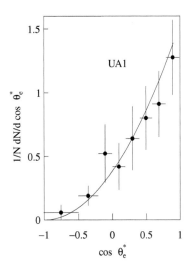

Fig. 19.14 The angular distribution of the $e^\pm$ from the decay of the $W^\pm$ with respect to the proton/antiproton direction in the centre of mass frame. Adapted from Figure 6a in G. Arnison et al., *Europhysics Letters* 1, 327 (1986). The prediction for the production and decay of a vector boson is superposed.

$$
\begin{aligned}
Z^0 &\rightarrow e^+ + e^-, \text{ or} \\
Z^0 &\rightarrow \mu^+ + \mu^-,
\end{aligned}
\tag{19.33}
$$

observed by the UA1 Collaboration (ibid) gave a Z-boson mass of 95.6± 3.5 GeV. The current accepted values of the W-boson and Z-boson masses are 80.379± 0.012 GeV and 91.188± 0.002 GeV, respectively.

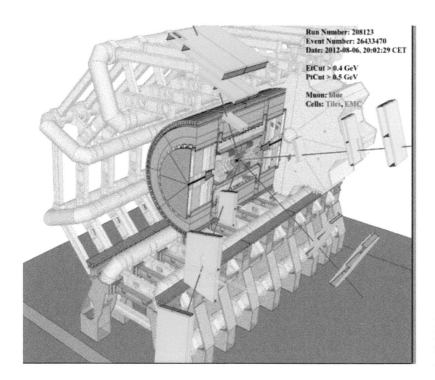

Run Number: 208123
Event Number: 26433470
Date: 2012-08-06, 20:02:29 CET

EtCut > 0.4 GeV
PtCut > 0.5 GeV

Muon: blue
Cells: Tiles, EMC

Fig. 19.15 Four μ-lepton decay of a Higgs boson observed in the ATLAS detector at the LHC. Courtesy CERN and ATLAS Collaboration.

The Large Electron Positron Collider (LEP) of 27 km circumference was subsequently built at CERN to explore the electroweak sector with great precision. Electron and positron beams were accelerated and then collided head on. The centre of mass energy range was tunable up to 209 GeV. By tuning across the mass of the Z-boson it was straightforward to explore the properties of the Z-boson and test an array of predictions of the standard model with exquisite precision. In particular, the determination of the number of light neutrino species at LEP was recounted in Chapter 18.

CERN built the next collider the Large Hadron Collider (LHC), to explore the electroweak energy scale given by eqn. 19.30 and discover whether the Higgs boson existed. The LHC is a proton-proton collider with currently 6.5 TeV (6500 GeV) beams. Taking account of the sharing of energy between gluons and quarks the average parton-parton centre of mass collision energy is ~2 TeV, adequate for the purpose. The LHC occupies the 27 km long tunnel dug originally for LEP.

As the beam energy increases so the demands on detector technology increase. The length and spread of the shower of particles generated in matter by an incident particle increases. Thus the size of calorimeters increases. The magnetic field and measurement precision must improve to match the reduction in track curvature (straighter tracks). This requires the detector granularity to increase correspondingly. The two huge detectors built for the Higgs search, ATLAS 46 m long and of 20 m diameter, and CMS, are located at collision points at diametrically opposite points around the LHC ring.

We look at the successful first detection of the Higgs decaying in the two channels

$$H \to l_1^+ + l_1^- + l_2^+ + l_2^-. \qquad (19.34)$$

$$H \to \gamma + \gamma, \qquad (19.35)$$

where l_1 and l_2 can specify either an electron or a muon. The discovery was made by the ATLAS and CMS collaborations in 2012. Figure 19.15 shows a cutaway view of the ATLAS detector with an event assigned to Higgs production and its decay by the channel given in eqn. 19.34. The tracks are reconstructions from the electronic signals recorded by the detector of particles travelling out from the head-on collision between the 6.5 TeV counterpropagating protons. Among the tracks are four muons, two positive and two negative, that are the decay products of a Higgs boson. They are recognizable as muons because they are

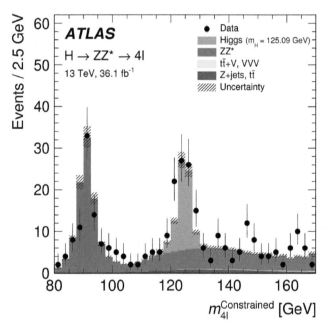

Fig. 19.16 The effective mass distribution for the four lepton channels from the ATLAS Collaboration. Data is given as points with statistical errors. The histogram shows the prediction of the standard model including known backgrounds. These include the left hand peak due to the rare four lepton decay of the Z^0-boson. Figure 3 in Aaboud et al., https://arxiv.org/abs/1712:02304 (also *JHEP* 03, 095 (2018)). Courtesy CERN and ATLAS Collaboration.

seen to travel right through the detector from the interaction and leave ionization signals in the outer muon detector layer. For clarity, only the muon detector elements struck by particles are displayed: these are the rectangular boxes, apparently floating free.

It is straightforward to calculate the mass of the parent of these four muons. If $(E_i, \mathbf{p}_i)$ is the four-momentum of the ith daughter then we can form a Lorentz invariant quantity that is unchanged in the decay

$$\left(\sum_i E_i\right)^2 - \left(\sum_i \mathbf{p}_i\right)^2 = E_{\text{parent}}^2 - p_{\text{parent}}^2 \qquad (19.36)$$

which is precisely the squared mass of the parent. If the four leptons are not the daughters of a particular particle but random combinations of leptons then the distribution of the effective masses obtained in this way would be a smooth distribution with no particular shape. If however the leptons are from the decay of a definite species there would be an accumulation in the effective mass distribution around the parent mass. The effective mass distributions for the Higgs decay channels given in eqns. 19.34 and 19.35 are plotted in Figures 19.16 and 19.17, respectively, using the ATLAS data collected at 13 GeV centre of mass energy. The signal peaks at 125 GeV in both channels were seen independently by the CMS experiment making it unequivocal that these peaks are due to decays of the searched-for Higgs boson. These, and other anticipated

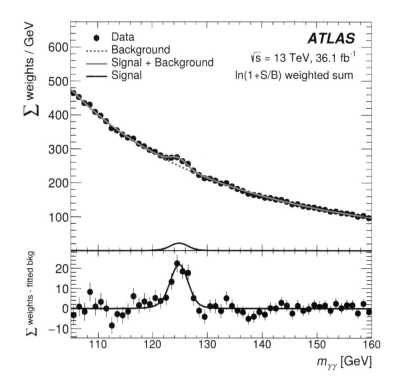

Fig. **19.17** The effective mass distribution for the two photon channel from the ATLAS Collaboration. The lower plot shows the data points with statistical error bars after subtracting a smooth background. Figure in ATLAS Collaboration, e-print arXiv:1806.00242. Courtesy CERN and ATLAS Collaboration.

decay modes, are all observed at rates consistent with the predictions of the standard model. From the analysis of the angular distribution of the outgoing particles in several decay channels it is concluded that the

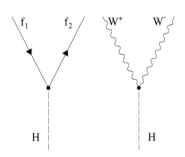

Fig. 19.18 Single Higgs couplings.

Higgs boson has spin-parity 0^+.

An important question that has to be addressed is how the Higgs field interacts with the fermions. We start by attempting to write an expression for the fermion masses. Applying blindly the formula that emerged for the W- and Z-bosons we get

$$\lambda \psi^\dagger \psi = \lambda [\psi_L^\dagger \psi_L + \psi_L^\dagger \psi_R + \psi_R^\dagger \psi_L + \psi_R^\dagger \psi_R]/\sqrt{2}, \qquad (19.37)$$

where the subscripts identify the handedness of the fermion. Now, $\psi_L^\dagger \psi_L = \psi^\dagger P_R P_L \psi = 0$, and equally $\psi_R^\dagger \psi_R = 0$, so that

$$\lambda \psi^\dagger \psi = \lambda [\psi_L^\dagger \psi_R + \psi_R^\dagger \psi_L]/\sqrt{2}. \qquad (19.38)$$

Any expression for a mass should be an $SU(2)_L$ scalar. However ψ_L transforms as a $SU(2)_L$ doublet while ψ_R transforms as a singlet: so the expression written above fails the test. A suitable expression is obtained by including the Higgs particle ϕ in the formula: $\lambda \psi^\dagger \phi \psi$. This is a scalar and hence viable as a mass. Labelling the fermion involved as f we can expand this and find that there are two terms

$$
\begin{aligned}
\lambda \psi^\dagger \phi \psi &= \lambda \left([0, f]_L \begin{bmatrix} 0 \\ v+h \end{bmatrix} f_R + f_R^\dagger [0, v+h] \begin{bmatrix} 0 \\ f \end{bmatrix}_L \right)/\sqrt{2} \\
&= \lambda [v+h][f_L^\dagger f_R + f_R^\dagger f_L]/\sqrt{2} \\
&= \lambda v f^\dagger f/\sqrt{2} + \lambda h f^\dagger f/\sqrt{2}. \qquad (19.39)
\end{aligned}
$$

The first of these terms is the energy of the interaction of the fermion with the vacuum Higgs field and gives mass to the fermion. The second term is the energy of the interaction of the fermion with the Higgs boson, which relates to Feynman diagram vertex shown in Figure 19.18. It follows that the Higgs-fermion coupling should be proportional to the fermion mass. Equivalently the Higgs decay rate to a fermion-antifermion pair should be proportional to the fermion mass squared. Figure 19.19 shows the experimental test made by the ATLAS and CMS Collaborations. The data points agree within the experimental precision with the linear prediction of the standard model.

19.6   Puzzles with the standard model

Intriguing puzzles remain following the successful explanation of particle properties and their weak, electromagnetic and strong interactions using quantum gauge theories. The notable items are

(1) The calculation of the Higgs mass gives a divergent result. This could be cured if *supersymmetric particles* exist. These supersymmetric particles, if detected, could also account for the dark matter in the universe. Each normal particle would have a supersymmetric partner with spin offset by $\hbar/2$: for example the W-boson having a spin $\hbar/2$ partner the Wino. These supersymmetric particles

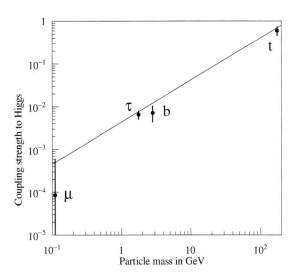

Fig. 19.19 Correlation of the Higgs couplings with the particle mass. The prediction from the standard model for the Higgs coupling gives the line drawn. Adapted from Figure 19, ATLAS and CMS Collaborations, *Journal of High Energy Physics* 8, 045 (2016).

would only interact either gravitationally or through some new ultraweak interaction with normal matter. Divergences in the Higgs mass calculation due to Feynman diagrams including only normal particles would be exactly cancelled by Feynman diagrams containing supersymmetric particles. Direct detection of supersymmetric particles will have to rely on observing the recoil of a nucleus when struck by a supersymmetric particle, at the same time ruling out normal processes. Thus far experiments have failed to detect any supersymmetric signal.

(2) Cosmological arguments indicate that 75 per cent of the energy in the universe is so-called *dark energy*. Unlike normal matter or radiation the pressure and density of dark energy have opposite signs leading to continual unstoppable expansion of space-time. No conventional particle interpretation seems likely to work.

(3) Calculation of the energy of the vacuum due to the various fields gives a quantity about 10^{50} times larger than the value cosmological measurements produce. This is an astonishingly large discrepancy.

(4) The matter-antimatter imbalance in the universe is not explained in the standard model. We have seen that there is a discrepancy of a factor 10^7 between what can be inferred from the CP violation observed in neutral K-meson and B-meson decays and the actual fraction of residual matter.

(5) The standard model has no explanation for the lepton and quark

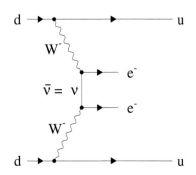

Fig. 19.20 Neutrinoless double β-decay.

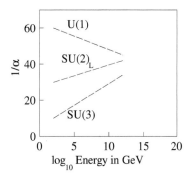

Fig. 19.21 Extrapolation of coupling constants according to the standard model.

mass spectrum: masses range from a fraction of an eV for neutrinos to 172 GeV for the top quark. In all there are more than 20 constants needed to specify the standard model and their origins are not explained. One point often made is that varying each of these constants ever so slightly appears to upset some step in the chain of physical processes that lead to our existence.

(6) The nature of the neutrino is not at all settled. Here, the view has been taken that neutrinos are fermions and that an antineutrino is quite distinct from a neutrino. However there is another possibility that cannot be ruled out by any of the experiments performed so far. Majorana suggested that neutrinos could be their own antiparticles. What would clinch his proposal is the observation of neutrinoless double β-decay of heavy nuclei. In the usual double β-decay two successive decays of a neutron to a proton are each accompanied by a neutrino and an electron. If the neutrino were a Majorana particle then the neutrino emitted in one β-decay could be absorbed in the second β-decay, as shown in Figure 19.20. Detection of this neutrinoless double β-decay requires that both electrons are detected and their energies measured. If the pair are from a neutrinoless double β-decay their energies would be unique, each equal to half the total available energy. The nuclei of interest are extremely long lived, in the case of ^{130}Te the lifetime is $8\,10^{20}$ years. Hence the search must be made using tons of ultrapure material and a period of observation extending over many years.

(7) Figure 19.21 shows the predicted energy dependence of the coupling constants of the U(1), SU(2)$_L$ and SU(3) forces in the standard model.[15] After a very long extrapolation they appear to converge at energies around 10^{15} GeV. The implication is that there may be an underlying overall symmetry. A possibility is SU(5) giving *grand unification*. One implication would be that the proton can decay, by for example p$\rightarrow$ e$^+$ + π^0. Searches in Super-Kamiokande have put a limit of greater than 10^{34} years but this does not exhaust the lifetime range possible in grand unified theories. The universe is rather youthful on this scale, being only 10^{10} years old.

(8) General relativity is a classical gauge theory that is completely successful in explaining gravitional effects on the macroscopic scale. However at a certain energy scale the gravitational self energy becomes equal to the rest mass energy:

$$G_N M^2 / \lambda = M, \tag{19.40}$$

G_N being the gravitational constant and λ the Compton wavelength which in natural units is M^{-1}. This critical mass, called the *Planck mass*, is thus $(G_N)^{-1/2}$. In natural units G_N is $6.67\,10^{-11}$ making the Planck mass equal to $1.22\,10^{19}$ GeV. At this energy

[15]See for example U.Amaldi et al., *Physical Review D*36, 1385 (1987).

scale quantum effects must surely come into play. How general relativity can be reconciled with quantum mechanics is however unknown despite decades of theoretical effort. This is the key question facing theoretical physicists.

19.7 Further reading

Modern Particle Physics by Mark Thomson, published by Cambridge University Press (2013) is a comprehensive reference work covering modern developments.

Quantum Field Theory and the Standard Model, by M. D. Schwartz, Cambridge University Press (2016) probes more deeply into the theoretical underpinnings with broad coverage.

Reality is Not What it Seems: The Journey to Quantum Gravity, by C. Rovelli, published by Penguin Random House (2017) gives a lively account of the field by a current practitioner in quantum gravity research.

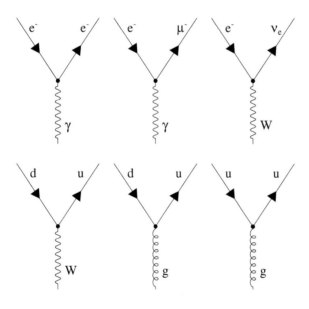

Fig. 19.22 Vertices quiz.

Exercises

(19.1) In Figure 19.22 which are the valid vertices?

(19.2) Equation 19.5 can be rewritten as

$$\alpha_s(q^2) = [\beta_s \ln(|q^2|/\Lambda^2)]^{-1}.$$

Deduce the expression for Λ. If Λ has value $0.2\,\text{GeV}$ calculate α_s at q^2 of $10^6\,\text{GeV}^2$.

(19.3) Calculate the cross-section for

$$e^+ + e^- \to \mu^+ + \mu^-$$

at centre of mass energies 2 and $40\,\text{GeV}$. What is the corresponding cross-section for

$$e^+ + e^- \to \tau^+ + \tau^- \ ?$$

(19.4) In the decay $H \to \gamma + \gamma$ show that the polarization state is $LL' + RR'$ for an even parity Higgs and $LL' - RR'$ for an odd parity Higgs. L indicates that the photon is left-handedly polarized, R that it is right-handedly polarized.

(19.5) An electron of $200\,\text{GeV}$ energy is scattered inelastically from a proton and emerges with energy $160\,\text{GeV}$ travelling at an angle $1°$ from its incident direction. Show that the four-momentum transfer squared is given by

$$q^2 = -2EE'(1 - \cos\theta)$$

where E is the incident electron energy, E' the scattered electron energy and θ the scattering angle. Calculate the four momentum transfer squared. What is the x-value of the target parton? What is the probability that the struck parton is a quark rather than an antiquark?

(19.6) What fraction of the Z-boson is B and what fraction is W_3?

(19.7) How much weaker is the Higgs to electron coupling expected to be than the Higgs to μ-lepton coupling? Is the determination going to be difficult?

(19.8) What is the force between two quarks implied by the potential in eqn. 19.9?

(19.9) Show that the Higgs cannot decay to three π-mesons each with spin parity 0^-.

(19.10) In the standard model the partial width of the Z-boson from a neutrino final state $\nu_x + \bar{\nu}_x$ is $[g^2 M_Z]/[96\pi \cos^2\theta_W]$. Calculate the total width due to all neutrino decays. Use the fact that $e^2/4\pi$ is $1/126$ at the energy of the Z-boson mass.

Appendix: Vector calculus

In spherical polar coordinates $\hat{r}$, $\hat{\theta}$ and $\hat{\phi}$ are unit vectors pointing along the radial, polar angle and azimuthal directions respectively. The vector differentials are given below.

$$\nabla\psi = \frac{\partial\psi}{\partial r}\hat{r} + \frac{1}{r}\frac{\partial\psi}{\partial\theta}\hat{\theta} + \frac{1}{r\sin\theta}\frac{\partial\psi}{\partial\phi}\hat{\phi}. \tag{A.1}$$

$$\nabla\cdot\mathbf{A} = \frac{\partial A_r}{\partial r} + \frac{2}{r}A_r + \frac{1}{r}\frac{\partial A_\theta}{\partial\theta} + \frac{\cot\theta}{r}A_\theta + \frac{1}{r\sin\theta}\frac{\partial A_\phi}{\partial\phi}, \tag{A.2}$$

$$\nabla^2\psi = \frac{\partial^2\psi}{\partial r^2} + \frac{2}{r}\frac{\partial\psi}{\partial r} + \frac{1}{r^2}\frac{\partial^2\psi}{\partial\theta^2} + \frac{\cot\theta}{r^2}\frac{\partial\psi}{\partial\theta} + \frac{1}{r^2\sin^2\theta}\frac{\partial^2\psi}{\partial\phi^2}. \tag{A.3}$$

$$(\nabla\wedge\mathbf{A})_r = \frac{1}{r\sin\theta}\left[\frac{\partial(A_\phi\sin\theta)}{\partial\theta} - \frac{\partial A_\theta}{\partial\phi}\right], \tag{A.4}$$

$$(\nabla\wedge\mathbf{A})_\theta = \frac{1}{r\sin\theta}\frac{\partial A_r}{\partial\phi} - \frac{1}{r}\frac{\partial(rA_\phi)}{\partial r}, \tag{A.5}$$

$$(\nabla\wedge\mathbf{A})_\phi = \frac{1}{r}\frac{\partial(rA_\theta)}{\partial r} - \frac{1}{r}\frac{\partial A_r}{\partial\theta}. \tag{A.6}$$

In cylindrical polar coordinates $\hat{\rho}$, $\hat{\phi}$ and $\hat{z}$ are unit vectors pointing radially in the transverse plane, azimuthally and in the axial direction respectively. The vector differentials are given below.

$$\nabla\psi = \frac{\partial\psi}{\partial\rho}\hat{\rho} + \frac{1}{\rho}\frac{\partial\psi}{\partial\phi}\hat{\phi} + \frac{\partial\psi}{\partial z}\hat{z}. \tag{A.7}$$

$$\nabla\cdot\mathbf{A} = \frac{A_\rho}{\rho} + \frac{\partial A_\rho}{\partial\rho} + \frac{1}{\rho}\frac{\partial A_\phi}{\partial\phi} + \frac{\partial A_z}{\partial z}, \tag{A.8}$$

$$\nabla^2\psi = \frac{1}{\rho}\frac{\partial\psi}{\partial\rho} + \frac{\partial^2\psi}{\partial\rho^2} + \frac{1}{\rho^2}\frac{\partial^2\psi}{\partial\phi^2} + \frac{\partial^2\psi}{\partial z^2}. \tag{A.9}$$

$$(\nabla\wedge\mathbf{A})_r = \frac{1}{\rho}\frac{\partial A_z}{\partial\phi} - \frac{\partial A_\phi}{\partial z}, \tag{A.10}$$

$$(\nabla\wedge\mathbf{A})_\phi = \frac{\partial A_\rho}{\partial z} - \frac{\partial A_z}{\partial\rho}, \tag{A.11}$$

$$(\nabla\wedge\mathbf{A})_z = \frac{A_\phi}{\rho} + \frac{\partial A_\phi}{\partial\rho} - \frac{1}{\rho}\frac{\partial A_\rho}{\partial\phi}. \tag{A.12}$$

Also we have:

$$\nabla \wedge \nabla \psi = 0, \tag{A.13}$$

$$\nabla \cdot (\nabla \wedge \mathbf{A}) = 0, \tag{A.14}$$

$$\nabla \wedge (\nabla \wedge \mathbf{A}) = \nabla(\nabla \cdot \mathbf{A}) - \nabla^2 \mathbf{A}. \tag{A.15}$$

Again:

$$\nabla \cdot (\psi \mathbf{A}) = \psi(\nabla \cdot \mathbf{A}) + \mathbf{A} \cdot \nabla \psi, \tag{A.16}$$

$$\nabla \wedge (\psi \mathbf{A}) = \psi \nabla \wedge \mathbf{A} - \mathbf{A} \wedge \nabla \psi, \tag{A.17}$$

$$\nabla \cdot (\mathbf{A} \wedge \mathbf{B}) = \mathbf{B} \cdot (\nabla \wedge \mathbf{A}) - \mathbf{A} \cdot (\nabla \wedge \mathbf{B}), \tag{A.18}$$

$$\nabla \wedge (\mathbf{A} \wedge \mathbf{B}) = \mathbf{A}\nabla \cdot \mathbf{B} - \mathbf{B}\nabla \cdot \mathbf{A} + (\mathbf{B} \cdot \nabla)\mathbf{A}$$
$$- (\mathbf{A} \cdot \nabla)\mathbf{B}. \tag{A.19}$$

Finally, the integral equations due to Gauss and Stokes are

$$\int_{\text{shell}} \mathbf{A} \cdot d\mathbf{S} = \int_{\text{volume}} (\nabla \cdot \mathbf{A})dV, \tag{A.20}$$

$$\oint_{\text{circuit}} \mathbf{A} \cdot d\ell = \int_{\text{surface}} (\nabla \wedge \mathbf{A}) \cdot d\mathbf{S}. \tag{A.21}$$

Use the thumbs-up right-hand rule to orient the positive surface normal and the circuit direction. Then the thumb points along the normal and the circuit follows the fingers from palm to nail.

Appendix: Mode densities

B

Mode densities are defined as the number of distinguishable quantum states in a given setting, which might be for photons or electrons in vacuum or in matter. The method is to count the number of waves of the form $\exp(i\mathbf{k} \cdot \mathbf{r})$ that can fit in the n-dimensional space, $\mathbf{k}$ being the wave numbers and $\mathbf{r}$ the coordinates in n-dimensions. In a single dimensional enclosure length L we must have

$$\exp(ikx) = exp[ik(x + L)], \qquad (\text{B.1})$$

so that $k = 2n\pi/L$ with n being an integer. Counting the allowed wave numbers below a maximum of k gives

$$N = \int_0^k \mathrm{d}k/[2\pi/L] = kL/(2\pi), \qquad (\text{B.2})$$

and the mode density is

$$g(k) = \mathrm{d}N/\mathrm{d}k = L/(2\pi). \qquad (\text{B.3})$$

For two dimensional motion over an area A with $(k_x^2 + k_y^2) < k^2$

$$\begin{aligned} N &= \int\int \{\mathrm{d}k_x/[2\pi/L]\}\{\mathrm{d}k_y/[2\pi/L]\} \\ &= L^2(\pi k^2)/[4\pi^2] = Ak^2/(4\pi). \end{aligned} \qquad (\text{B.4})$$

Then the mode density for two-dimensional motion is

$$g(k) = \mathrm{d}N/\mathrm{d}k = Ak/(2\pi). \qquad (\text{B.5})$$

For three dimensional motion within a volume V with $(k_x^2 + k_y^2 + k_z^2) < k^2$

$$\begin{aligned} N &= \int\int\int \{\mathrm{d}k_x/[2\pi/L]\}\{\mathrm{d}k_y/[2\pi/L]\}\{\mathrm{d}k_z/[2\pi/L]\} = \\ &= L^3(4\pi k^3/3)/(8\pi^3) = Vk^3/(6\pi^2). \end{aligned} \qquad (\text{B.6})$$

The corresponding mode density for three-dimensional motion is

$$g(k) = \mathrm{d}N/\mathrm{d}k = Vk^2/(2\pi^2). \qquad (\text{B.7})$$

These results are valid in general for areas and volumes of any shape. The mode density will be multiplied by a factor $(2S+1)$ where the particles involved have spin S and all the magnetic substates are accessible. In the case of photons there are only two polarizations, and the mode density is multiplied by two. Collecting the densities of states per spin orientation:

$$1 - \text{dimensional line length } L \qquad g(k)\,\mathrm{d}k = L\,\mathrm{d}k/(2\pi), \qquad \text{(B.8)}$$
$$2 - \text{dimensional area } A \qquad g(k)\,\mathrm{d}k = A\,k\,\mathrm{d}k/(2\pi), \qquad \text{(B.9)}$$
$$3 - \text{dimensional volume } V \qquad g(k)\,\mathrm{d}k = V\,k^2\,\mathrm{d}k/(2\pi^2). \text{(B.10)}$$

Alternative expressions for the density of states in terms energy (E), angular frequency (ω) and frequency (f) are also useful. In order to make the conversion we need the *dispersion relation* appropriate to the situation.

For photons in vacuum and also, for most practical purposes, in air the dispersion relation is

$$E = h\,f = \hbar\,\omega = \hbar\,c\,k,$$

while in a medium of refractive index μ we have

$$E = h\,f = \hbar\,\omega = \hbar\,(c/\mu)\,k.$$

The equalities

$$g_f(f)\,\mathrm{d}f = g_E(E)\,\mathrm{d}E = g_\omega(\omega)\,\mathrm{d}\omega = g_k(k)\,\mathrm{d}k,$$

are essential when making the conversions. For example, photons in free space have a density of states per polarization/spin state in angular frequency for *three-dimensional* motion

$$g_\omega(\omega) = g_k(k)(\mathrm{d}k/\mathrm{d}\omega) = V\,k^2/(2\pi^2 c) = V\,\omega^2/(2\pi^2 c^3), \qquad \text{(B.11)}$$

and in frequency

$$g_f(f) = g_\omega(\omega)(\mathrm{d}\omega/\mathrm{d}f) = 4\pi\,V\,f^2/c^3. \qquad \text{(B.12)}$$

The dispersion relation for non-relativistic particles of mass m is the familiar result

$$k = p/\hbar = \sqrt{2mE/\hbar^2}, \text{ and} \qquad \text{(B.13)}$$
$$\mathrm{d}k/\mathrm{d}E = \sqrt{m/(2E\hbar^2)}. \qquad \text{(B.14)}$$

Thus, in the non-relativistic approximation, the density of states in *three dimensions*, per spin orientation, is

$$g_E(E) = g_k(k)(\mathrm{d}k/\mathrm{d}E) = V\,(2m/\hbar^2)^{3/2}\,E^{1/2}/(4\pi^2). \qquad \text{(B.15)}$$

If the particles have spin S and are unpolarized the result should be multiplied by a further factor $(2S+1)$. A useful simplification can be obtained for non-relativistic particles: first replace k in eqn. B.6 by $\sqrt{2mE/\hbar^2}$,

$$N = V\,(2mE/\hbar^2)^{3/2}/(6\pi^2);$$

then divide eqn B.15 by this result to give

$$g_E(E) = \frac{3N}{2E}. \tag{B.16}$$

Again in the non-relativistic approximation, the density of states in *two dimensions*, per spin orientation, is

$$g_E(E) = Am/(2\pi\hbar^2) = N/E. \tag{B.17}$$

Finally, in the non-relativistic approximation, the density of states in *one dimension*, per spin orientation, is

$$g_E(E) = L\sqrt{m/2E}/(2\pi\hbar) = N/(2E). \tag{B.18}$$

B.1 Harmonic wells

Harmonic potential wells are used to store atoms at low temperatures and the density of states must be recalculated to take account of the confinement. The potential can be anisotropic:

$$U(\mathbf{r}) = (K_x x^2 + K_y y^2 + K_z z^2)/2,$$

where K_x, K_y and K_z are the force constants for displacements from the origin. If the atoms have mass m, then the angular frequencies of their oscillations along the three axes are $\omega_x = \sqrt{K_x/m}$ and so on. Then

$$U(\mathbf{r}) = (m/2)(\omega_x^2 x^2 + \omega_y^2 y^2 + \omega_z^2 z^2). \tag{B.19}$$

The energy of an atom in such a potential is

$$E = E_x + E_y + E_z + E_0, \tag{B.20}$$

where $E_i = n_i \hbar \omega_i$ and n_i is a non-negative integer, and E_0 is the zero-point energy $\hbar(\omega_x + \omega_y + \omega_z)/2$. The states with energies less than some value ϵ of $E - E_0$ all lie between the origin of the $[E_x, E_y, E_z]$ space and the plane defined by

$$\epsilon = E_x + E_y + E_z.$$

Each distinguishable quantum state takes up a volume $(\hbar\omega_x)(\hbar\omega_y)(\hbar\omega_z)$ in this space. Hence the number of states with excitation less than ϵ is

$$\begin{aligned} G(\epsilon) &= \int_0^\epsilon dE_x \int_0^{\epsilon - E_x} dE_y \int_0^{\epsilon - E_x - E_y} dE_z/(\hbar^3 \omega_x \omega_y \omega_z) \\ &= \epsilon^3/(6\hbar^3 \omega_x \omega_y \omega_z), \end{aligned}$$

and the corresponding density of states with excitation energy around ϵ is

$$g(\epsilon) = dG/d\epsilon = \epsilon^2/(2\hbar^3 \omega_x \omega_y \omega_z). \tag{B.21}$$

If the potential is isotropic this reduces to

$$g(\epsilon) = \epsilon^2/(2\hbar^3 \omega^3) = \epsilon^2/(2\epsilon_0^3), \tag{B.22}$$

where ϵ_0 is the quantum of mode energy. These state densities must be multiplied by $2S + 1$ for particles of spin S when all their magnetic substates are accessible.

Appendix: Density matrix

Quantum mechanics predicts the expectation value, that is the mean of measurements made on an *ensemble* of identically prepared states. If the relevant state is simply a pure state of a quantum system its wavefunction can always be expanded as a linear superposition of orthonormal eigenstates $|n\rangle$

$$\sum_n p_n |n\rangle, \tag{C.1}$$

where p_n is the amplitude of the eigenstate $|n\rangle$ in the superposition, so that $\sum_n p_n^* p_n = 1$. Then the expectation value of an observable is given by

$$\langle \hat{A} \rangle = \sum_{m,n} p_m^* p_n \langle m | \hat{A} | n \rangle. \tag{C.2}$$

A less simple case is of an ensemble consisting of a mix of orthonormal states. Then the expectation value of the observable is

$$\langle \hat{A} \rangle = \sum_n P_n \langle n | \hat{A} | n \rangle, \tag{C.3}$$

where P_n is the probability of the state $|n\rangle$ in the mixture, all the P_ns are real and positive and $\sum_n P_n = 1$.

The more complex situation is where the states in the mix, $|\psi_i\rangle$ with probabilities P_i, are not orthogonal, but still normalized. The P_is are still real and positive and $\sum_i P_i = 1$. An example might be a mix of electrons, half with spin along the x-axis, half with spin along the y-axis. In 1927 von Neumann invented the *density matrix* formalism to handle mixed or pure states. The density matrix elements are defined in terms of the orthonormal set $|\{n\}\rangle$ by

$$\rho_{nm} = \sum_i P_i \langle n | \psi_i \rangle \langle \psi_i | m \rangle. \tag{C.4}$$

Then the expectation of the value of the observable A for the mixed state, introduced here, is

$$\begin{aligned} \langle \hat{A} \rangle &= \sum_i P_i \langle \psi_i | \hat{A} | \psi_i \rangle \\ &= \sum_i P_i \sum_{m,n} \langle \psi_i | m \rangle \langle m | \hat{A} | n \rangle \langle n | \psi_i \rangle, \end{aligned} \tag{C.5}$$

where the closure relation, eqn. 1.61, has been used twice. Then

$$\langle \hat{A} \rangle = \sum_{m,n} \rho_{nm} A_{mn}, \tag{C.6}$$

where $A_{mn} = \langle m | \hat{A} | n \rangle$. When $|\psi\rangle$ is a pure state, that is a linear superposition of eigenstates, $\rho_{nm} = p_m^* p_n$: while, if the state is a mixture of orthogonal states, $\rho_{mn} = P_m \delta_{mn}$. We can show from eqn. C.4 that the density matrix is hermitian:

$$\begin{aligned} \rho_{mn}^* &= \sum_i P_i \langle m | \psi_i \rangle^* \langle \psi_i | n \rangle^* \\ &= \sum_i P_i \langle \psi_i | m \rangle \langle n | \psi_i \rangle = \rho_{nm}. \end{aligned} \tag{C.7}$$

It also has unit trace, where the trace is the sum of the diagonal elements of a matrix, something independent of the choice of the set of eigenstates,

$$\sum_m \rho_{mm} = \sum_m \sum_i P_i \langle m | \psi_i \rangle \langle \psi_i | m \rangle = \sum_i P_i \langle \psi_i | \psi_i \rangle = 1. \tag{C.8}$$

If there is partial coherence the values of the corresponding off-diagonal elements ρ_{mn} depart from zero and their presence will be signalled by the effects associated with coherence. For example, interference fringes appear with low visibility if there is weak coherence between the sources. Systems can lose coherence, as will happen when atoms excited by a laser into Rabi oscillations then undergo collisions that randomly disturb their phases.

Corresponding to the density matrix there is a *density operator*

$$\hat{\rho} = \sum_i P_i |\psi_i\rangle\langle\psi_i| = \sum_{i,m,n} P_i |n\rangle\langle n|\psi_i\rangle\langle\psi_i|m\rangle\langle m| = \sum_{m,n} |n\rangle\langle m|\rho_{nm}. \tag{C.9}$$

This is unlike all operators met previously, because it is defined afresh for each mixture of states considered. Using eqn. C.6

$$\begin{aligned} \langle \hat{A} \rangle &= \sum_{m,n} \rho_{nm} A_{mn} = \sum_{m,n} \langle n|\hat{\rho}|m\rangle\langle m|\hat{A}|n\rangle \\ &= \sum_n \langle n|\hat{\rho}\hat{A}|n\rangle. \end{aligned} \tag{C.10}$$

An alternative way of presenting this result is

$$\langle \hat{A} \rangle = \text{Trace}[\hat{\rho}\hat{A}]. \tag{C.11}$$

This result shows that with a suitable choice of the eigestates the calculation of $\langle \hat{A} \rangle$ can be considerably simplified. Note that $\text{Trace}(AB) = \text{Trace}(BA)$, even if $BA \neq AB$. In a pure eigenstate $|m\rangle$ we have

$$\hat{\rho} = |m\rangle\langle m|, \qquad \hat{\rho}^2 = |m\rangle\langle m|m\rangle\langle m| = \hat{\rho} \qquad \text{and} \qquad \text{Trace}(\hat{\rho}^2) = 1, \tag{C.12}$$

which provides a useful test of the purity of a state. For an mixed state

$$\hat{\rho}^2 \neq \hat{\rho}, \text{and Trace}[\hat{\rho}^2] < 1. \tag{C.13}$$

Appendix: The Bloch sphere, spin-1/2, and qubits

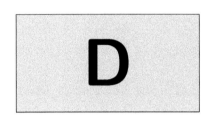

The Bloch sphere shown in Figure D.1 provides a simple way to display the pure or mixed states of a quantum system having two orthonormal eigenstates $|0\rangle$ and $|1\rangle$. The states could be the ground and excited states of an atom, the vertical and horizontal polarization states of a photon, the spin-up and spin-down states of an electron or the corresponding up and down weak isospin states of the quarks and leptons. Any other pure states will be a superposition

$$|\psi\rangle = c_0|0> +c_1|1 >= \begin{bmatrix} c_0 \\ c_1 \end{bmatrix} \qquad (D.1)$$

with $|c_0|^2 + |c_1|^2 = 1$, and it can be displayed as a point on the surface of the Bloch sphere with angular coordinates θ and ϕ such that

$$|\psi\rangle = \cos\left(\frac{\theta}{2}\right)|0\rangle + e^{i\phi}\sin\left(\frac{\theta}{2}\right)|1\rangle = \begin{bmatrix} \cos(\theta/2) \\ e^{i\phi}\sin(\theta/2) \end{bmatrix}, \qquad (D.2)$$

with $0 \leq \theta \leq \pi$, and $0 \leq \phi < 2\pi$. The states at the poles with $\theta = 0$ and π are respectively $|0\rangle$ and $|1\rangle$. The components along the Cartesian coordinates are

$$\begin{aligned} \psi_x &= \sin\theta\cos\phi = 2\cos(\theta/2)\sin(\theta/2)\,\mathrm{Re}[\exp(i\phi)] \\ &= 2\,\mathrm{Re}[c_0 c_1^*], \qquad (D.3) \\ \psi_y &= \sin\theta\sin\phi = 2\cos(\theta/2)\sin(\theta/2)\,\mathrm{Im}[\exp(i\phi)] \\ &= -2\,\mathrm{Im}[c_0 c_1^*], \qquad (D.4) \\ \psi_z &= \cos^2(\theta/2) - \sin^2(\theta/2) = |c_0|^2 - |c_1|^2. \qquad (D.5) \end{aligned}$$

For example a state on the equator of the Bloch sphere at an azimuth ϕ has a state vector

$$|\psi\rangle = [|0\rangle] + e^{i\phi}|1\rangle]/\sqrt{2} = \begin{bmatrix} 1/\sqrt{2} \\ e^{i\phi}/\sqrt{2}) \end{bmatrix}, \qquad (D.6)$$

Pure states such as $|\psi\rangle$ are the *qubits* in quantum computation. Bits in classical computers can only take discrete values: either zero corresponding to $|0\rangle$ or unity corresponding to $|1\rangle$. Qubits contain additional

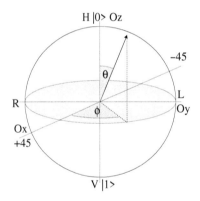

Fig. D.1 Bloch sphere. The relative orientations are shown for the horizontal and vertical plane polarized, $\pm 45°$ plane polarized, right, and left circularly polarized photon states.

phase information and can be manipulated in ways not possible using classical bits.

In the case of an atom the ground state $|g\rangle = |0\rangle$ and the excited state $|e\rangle = |1\rangle$ can carry the qubits. Photons can carry qubits in a pair of orthogonally polarized states. These can be the familiar states of linear, or circular polarization. Representing photon states on the Bloch sphere for the case that horizontal and vertical polarization are the base states, we have

$$|H\rangle = |0\rangle = \begin{bmatrix} 1 \\ 0 \end{bmatrix}; \quad |V\rangle = |1\rangle = \begin{bmatrix} 0 \\ 1 \end{bmatrix}; \tag{D.7}$$

also

$$|L\rangle = [|H\rangle + i|V\rangle]/\sqrt{2}; \quad |R\rangle = [|H\rangle - i|V\rangle]/\sqrt{2}; \tag{D.8}$$

and

$$|+\rangle = [|H\rangle + |V\rangle]/\sqrt{2}; \quad |-\rangle = [|H\rangle - |V\rangle]/\sqrt{2}. \tag{D.9}$$

L, R indicate left/right circularly polarization; $+/-$ indicate polarization at $\pm 45°$ to the horizontal. $|L\rangle$, $|R\rangle$, $|+\rangle$ and $|-\rangle$ all lie in the equatorial plane.

Equally electron spin states can be represented by points on the Bloch sphere. Electrons with their spin aligned along the spatial $\pm x$, $\pm y$, $+z$ and $-z$-axes are represented on the Bloch sphere by

$$\frac{1}{\sqrt{2}}\begin{bmatrix} 1 \\ \pm 1 \end{bmatrix}; \quad \frac{1}{\sqrt{2}}\begin{bmatrix} 1 \\ \pm i \end{bmatrix}; \quad \begin{bmatrix} 1 \\ 0 \end{bmatrix}; \quad \begin{bmatrix} 0 \\ 1 \end{bmatrix}, \tag{D.10}$$

respectively. These states are equivalent to the photon states $|+/-\rangle$, $|L/R\rangle$ and $|H/V\rangle$ respectively. They and their equivalents are the eigenstates of the respective Pauli matrices:

$$\sigma_x = \begin{bmatrix} 0 & 1 \\ 1 & 0 \end{bmatrix}; \quad \sigma_y = \begin{bmatrix} 0 & -i \\ i & 0 \end{bmatrix}; \quad \sigma_z = \begin{bmatrix} 1 & 0 \\ 0 & -1 \end{bmatrix}; \tag{D.11}$$

with eigenvalues ± 1 for the spin along the $\pm$ axes.

Referring to electrons, the *spin* operators S_x etc. are constructed from the Pauli matrices σ_x etc:

$$S_x = \sigma_x \hbar/2. \tag{D.12}$$

These matrices satisfy the usual commutation relations for angular momentum components:

$$[S_x, S_y] = i\hbar S_z \text{ and equivalently } [\sigma_x, \sigma_y] = 2i\sigma_z, \tag{D.13}$$

with similar equations obtained by cycling x, y and z. In terms of these components the total *spin* operator is

$$S^2 = \sigma^2(\hbar/2)^2 = [\sigma_x^2 + \sigma_y^2 + \sigma_z^2](\hbar/2)^2 = \begin{bmatrix} 1 & 0 \\ 0 & 1 \end{bmatrix}[3\hbar^2/4]. \tag{D.14}$$

Operating with S^2 on each of the above eigenstates gives, as expected, $j(j+1)\hbar^2$ with j equal to $1/2$.

Notice that a full rotation through 2π of an electron in ordinary space rotates its internal wavefunction, *a spinor*, by π in the *spin space* defined by the surface of the Bloch sphere. Consequently the electron wavefunction changes by a factor -1^1; it requires a rotation of 4π in ordinary space to restore the wavefunction to its original value. Equally the rate of electron spin precession in a magnetic field is twice the rate of phase precession.

[1]This phase change of π under a 2π rotation was first verified for neutrons, also spin 1/2 particles, by S. A. Werner, R. Colella, A. W. Overhauser and C. F. Eagan: Physical Review Letters 35, 1053 (1975).

The operations describing transitions between the two base states, whether they are electron spin states, photon polarization states or atomic states, are

$$\sigma_- = (\sigma_x - i\sigma_y)/2 = \begin{bmatrix} 0 & 0 \\ 1 & 0 \end{bmatrix},$$

$$\sigma_+ = (\sigma_x + i\sigma_y)/2 = \begin{bmatrix} 0 & 1 \\ 0 & 0 \end{bmatrix}. \qquad (D.15)$$

Thus

$$\sigma_- \begin{bmatrix} 1 \\ 0 \end{bmatrix} = \begin{bmatrix} 0 \\ 1 \end{bmatrix}, \quad \sigma_+ \begin{bmatrix} 0 \\ 1 \end{bmatrix} = \begin{bmatrix} 1 \\ 0 \end{bmatrix}. \qquad (D.16)$$

σ_+ and σ_- are respectively known as *raising* and *lowering* operators. Note that unphysical processes give a null result,

$$\sigma_- \begin{bmatrix} 0 \\ 1 \end{bmatrix} = \sigma_+ \begin{bmatrix} 1 \\ 0 \end{bmatrix} = \text{null}. \qquad (D.17)$$

The counting operator is

$$n = \sigma_+\sigma_- = (1 + \sigma_z)/2 = \begin{bmatrix} 1 & 0 \\ 0 & 0 \end{bmatrix}. \qquad (D.18)$$

A rotation of α around, for example, the y-axis has the form

$$\begin{aligned} \exp(-i\alpha\,\sigma_y/2) &= I - i\alpha\,\sigma_y/2 + [-i\alpha\,\sigma_y/2]^2/2! + [-i\alpha\,\sigma_y/2]^3/3! + ... \\ &= I - i\alpha\,\sigma_y/2 - [\alpha/2]^2\,I/2 + i[\alpha/2]^3\,\sigma_y/6 + ... \\ &= I\cos(\alpha/2) - i\sigma_y\sin(\alpha/2). \qquad (D.19) \end{aligned}$$

Generalizing this for a rotation α about a unit vector $\mathbf{s}$ gives

$$\exp(-i\alpha\mathbf{s}\cdot\boldsymbol{\sigma}/2) = I\cos(\alpha/2) - i(\mathbf{s}\cdot\boldsymbol{\sigma})\sin(\alpha/2). \qquad (D.20)$$

Another useful relation involving Pauli matrices is this:

$$(\mathbf{b}\cdot\boldsymbol{\sigma})(\mathbf{c}\cdot\boldsymbol{\sigma}) = (\mathbf{b}\cdot\mathbf{c})I + i\boldsymbol{\sigma}\cdot(\mathbf{b}\wedge\mathbf{c}). \qquad (D.21)$$

All the two state systems used to carry qubits can be analysed and manipulated with the mathematical formulae given here, but the corresponding physical manipulations require different conditions and apparatus depending on the objects involved.

A state which is an incoherent mix of pure states is located within the volume of the Bloch sphere. All states, whether within or on the Bloch sphere, can be fully described using a density matrix ρ. In the case of a pure state

$$|\psi\rangle = c_0|0\rangle + c_1|1\rangle,\tag{D.22}$$

where $|c_0|^2 + |c_1|^2 = 1$, the density matrix is

$$\rho = \begin{bmatrix} |c_0|^2 & c_0 c_1^* \\ c_1 c_0^* & |c_1|^2 \end{bmatrix}.\tag{D.23}$$

The components of the state vector on the Bloch sphere are then

$$\begin{aligned}
\psi_x &= \rho_{01} + \rho_{10} = 2\mathrm{Re}[\rho_{01}], & \text{(D.24)}\\
\psi_y &= i[\rho_{01} - \rho_{10}] = -2\mathrm{Im}[\rho_{01}], & \text{(D.25)}\\
\psi_z &= \rho_{00} - \rho_{11}, & \text{(D.26)}
\end{aligned}$$

consistent with eqns. D.3, D.4 and D.5. The off-diagonal terms are known as *coherences*. In the case of a mixed state located at the centre of the Bloch sphere $\rho_{00} = \rho_{11} = 0.5$, with $\rho_{01} = \rho_{10} = 0$, which is an equal mix of $|0\rangle$ and $|1\rangle$: the coherences are zero so they are completely incoherent. Pure states are located on the Bloch sphere surface and always have $\rho^2 = \rho$. Whenever the carrier of a qubit interacts with its surroundings it will generally enter a mixed state or undergo a phase change.

There is a similarity between the transformations in a two-dimensional space with complex dimensions, and rotations in a three-dimensional space with real dimensions. The transformations in two space that leave the complex length unchanged and have unit determinant form a group SU(2), while the rotations in real three space form a group SO(3). The *fundamental representation* of SU(2) corresponds to particles with spin one half.[2] Half-integer spins can be combined vectorially to give any whole-integer or a half-integer spin (times $\hbar$) and thus generate all the representations of SU(2). On the other hand SO(3) only has representations with orbital angular momentum equal to a whole integer times $\hbar$. SU(2) is technically a double-covering of the group SO(3). Rotating a body with integral spin through 2π in three space reproduces its initial state. By contrast we have observed that rotating an electron through 2π changes its phase by π; thus it requires a rotation through 4π to reproduce the electron's initial state. This indicates the importance of the electron's quantum state being referenced to an underlying complex two-dimensional space, rather than the usual three-dimensional space. An electron's intrinsic angular momentum is decoupled from the idea, or existence, of some corresponding rotating structure in normal three-dimensional space. This implies a similar decoupling for the spins of all the leptons and quarks.

[2] The $2j+1$ members of any spin j multiplet make up a representation of the group. Then the group of operations transforms any member of such a representation into all members of that representation and into nothing else.

Appendix: Perturbation theory

E

Here we investigate the effect of a small time independent alteration (*perturbation*) in the Hamiltonian from a form for which the wavefunctions are known or readily determined. Take the complete Hamiltonian to be

$$H = H_0 + \varepsilon H', \tag{E.1}$$

where $\varepsilon H' \ll H_0$ and the dimensionless parameter ε is unity. Then it helps to picture the perturbation being turned on by increasing ε from zero to unity, and this is the way we proceed, making any term in ε^n smaller than any term in ε^{n-1} for all positive integers n. Let $\{\psi_n^{(0)}\}$ be the set of wavefunctions for the unperturbed Hamiltonian H_0, with the corresponding energy eigenvalues $\{E_n^{(0)}\}$. Then let those for the full Hamiltonian H be

$$\psi_n = \sum_m c_m \psi_m^{(0)}. \tag{E.2}$$

The coefficients c_m can be expanded as a *perturbation series* of terms with values decreasing by a factor ε at each step:

$$c_m = c_m^{(0)} + \varepsilon c_m^{(1)} + \varepsilon^2 c_m^{(2)} + ..., \tag{E.3}$$

and the corresponding corrections to the energies has a similar expansion

$$E_n = E_n^{(0)} + \varepsilon E_n^{(1)} + \varepsilon^2 E_n^{(2)} + ... \tag{E.4}$$

The unperturbed values are by definition $E_n^{(0)}$. Consider how a state $\psi_n^{(0)}$ is modified by the perturbation. We have in the absence of perturbation

$$c_m^{(0)} = 0 \text{ for } m \neq n, \text{ and } c_n^{(0)} = 1. \tag{E.5}$$

The first order corrections can be obtained by equating coefficients of ε in the equation

$$H\psi_n = E_n\psi_n. \tag{E.6}$$

This gives

$$H' \sum_m \psi_m^{(0)} c_m^{(0)} + H_0 \sum_m \psi_m^{(0)} c_m^{(1)} = E_n^{(0)} \sum_j \psi_j^{(0)} c_j^{(1)} + E_n^{(1)} \sum_j \psi_j^{(0)} c_j^{(0)}. \tag{E.7}$$

Now taking the product of eqn. E.7 with $\psi_n^{(0)*}$ and integrating over space gives:

$$\sum_m H'_{nm} c_m^{(0)} + E_n^{(1)} c_n^{(1)} = E_n^{(0)} c_n^{(1)} + E_n^{(1)} c_n^{(0)}, \quad (E.8)$$

where the orthogonality of the wavefunctions $\psi_j^{(0)}$ to one another has been made use of. The shorthand used here is

$$H'_{mn} = \int \psi_m^{(0)*} H' \psi_n^{(0)} \mathrm{d}\mathbf{r}. \quad (E.9)$$

$c_n^{(1)}$ and $c_m^{(0)}$ are both zero so this collapses to

$$E_n^{(1)} = H'_{nn} \quad (E.10)$$

giving H'_{nn} as the energy change of state n. Taking the product of eqn. E.7 with $\psi_k^{(0)*}$, where $k \neq n$, and repeating the integration over space gives

$$\sum_m H'_{km} c_m^{(0)} + E_k^{(0)} c_k^{(1)} = E_n^{(0)} c_k^{(1)}. \quad (E.11)$$

Now using $c_n^{(0)} = 1$ this collapses to

$$c_k^{(1)} = \frac{H'_{kn}}{E_n^{(0)} - E_k^{(0)}}, \quad \text{for } k \neq n. \quad (E.12)$$

The second order corrections in ε^2 are

$$c_k^{(2)} = \sum_{j \neq n} \frac{H'_{kj} H'_{jn}}{\left(E_n^{(0)} - E_j^{(0)}\right)\left(E_n^{(0)} - E_k^{(0)}\right)} - \frac{H'_{nn} H'_{kn}}{\left(E_n^{(0)} - E_k^{(0)}\right)^2} \quad (E.13)$$

for $k \neq n$ and

$$c_n^{(2)} = -\frac{1}{2} \sum_{j \neq n} \frac{|H'_{jn}|^2}{\left(E_n^{(0)} - E_j^{(0)}\right)^2}. \quad (E.14)$$

The additional correction to the energy to second order in ε is

$$E_n^{(2)} = \sum_{j \neq n} \frac{|H'_{jn}|^2}{E_n^{(0)} - E_j^{(0)}}. \quad (E.15)$$

Further corrections are of order ε^3. The perturbation expansion is only applicable if the perturbation is smaller than the separation between the individual energy levels, that is

$$H'_{jn} \ll |E_n^{(0)} - E_j^{(0)}| \quad (E.16)$$

for all j.

Appendix: RSA encryption

F.1 RSA encryption

Suppose that Alice wishes to receive messages encrypted so that only she and the sender Bob can read them. First she selects two different large primes p and q, then takes the products $n = pq$ and $t = (p-1)(q-1)$. She selects a *public key*, e, $1 < e < t$ with e not equal to p or q. Finally she calculates a *private key*, d, such that $ed = 1 + jt$, j being a positive integer. Thus

$$ed = 1 \bmod t. \tag{F.1}$$

Alice makes n and e known to all the world.

Bob first converts his alphameric message into long digital strings, each with an even number of digits, m, keeping $m < n$. Each such number M is encrypted using n and the public key e,

$$K = M^e \bmod n, \tag{F.2}$$

and K is sent to Alice. Alice simply calculates

$$K^d \bmod n, \tag{F.3}$$

which we can show is precisely Bob's message M.

The decryption produces

$$K^d \bmod n = M^{ed} \bmod n. \tag{F.4}$$

However, from eqn. F.1 we have for some integer j, $ed = 1 + j(p-1)(q-1)$, so that

$$M^{ed} \bmod n = M^{ed} \bmod p = M[M^{j(q-1)}]^{p-1} \bmod p. \tag{F.5}$$

Now we must invoke *Fermat's little theorem*. This states that if p is not a factor of some number a then

$$a^{p-1} = 1 \bmod p. \tag{F.6}$$

Using this result eqn. F.5 is reduced to

$$M^{ed} \bmod p = M \bmod p. \tag{F.7}$$

Similarly

$$M^{ed} \bmod q = M \bmod q. \tag{F.8}$$

Together the last two equations require that

$$M^{ed} \bmod (pq) = M \bmod (pq), \qquad (F.9)$$

and eqn. F.4 collapses to

$$K^d \bmod n = M \bmod n. \qquad (F.10)$$

Now M was selected to be less than n so that Alice's procedure has retrieved M, Bob's message, intact.

Appendix: The Dirac equation

A critical step in quantum physics, made in 1929 by Dirac, was the discovery of the relativistic wave equation for an electron. This provided the basis from which the quantum field theory of electromagnetism, QED, and the use of Feynman diagrams as calculation tools could be developed. The equation is linear in the coordinates, and using the natural units explained in Chapter 18,

$$\left[i\gamma_\mu \frac{\partial}{\partial x_\mu} - m \right] \psi = 0. \tag{G.1}$$

Here the subscript values run from 0 to 3 corresponding to the time and space dimensions. Einstein's convention is also used. When a subscript is repeated it is summed over: thus $A_\mu B_\mu$ is $A_0 B_0 - A_x B_x - A_y B_y - A_z B_z$. The equation describes particles with spin-1/2 (in natural units) having a Lande g-factor exactly 2.0.

The function ψ is a *spinor* having dimension 4. These are *spin* dimensions as opposed to spatial dimensions. Crudely speaking one dimension is required to accommodate a spin-up electron, one for a spin-down electron, one for a spin-up positron and one for a spin-down positron. The gamma matrices are space-time vectors each with a four-dimensional spin structure. Explicitly

$$\gamma_0 = \begin{bmatrix} I & 0 \\ 0 & -I \end{bmatrix}, \ \gamma_i = \begin{bmatrix} 0 & \sigma_i \\ -\sigma_i & 0 \end{bmatrix}. \tag{G.2}$$

In addition we define

$$\gamma_5 = \begin{bmatrix} 0 & I \\ I & 0 \end{bmatrix}. \tag{G.3}$$

where σ_i and I are the two-dimensional spinor matrices described in Appendix D. The γ matrices anticommute

$$\gamma_\mu \gamma_\rho + \gamma_\rho \gamma_\mu = \{\gamma_\mu, \gamma_\rho\} = 2g_{\mu\rho}, \tag{G.4}$$

where $g_{\mu\rho}$ is the space-time metric tensor given in eqn. 1.71.

The solutions of the Dirac equation are four-component spinors, those with positive energy E, with momentum $\mathbf{p}$ and with spin component $s\hbar/2$ are

$$u(\mathbf{p}, s) = \sqrt{E+m} \begin{bmatrix} \phi_s \\ \frac{\boldsymbol{\sigma} \cdot \mathbf{p}}{E+m} \phi_s \end{bmatrix}, \tag{G.5}$$

where ϕ_s are a set of three two-dimensional spinors:

$$\phi_+ = \begin{bmatrix} 1 \\ 0 \end{bmatrix}, \quad \phi_0 = \begin{bmatrix} 0 \\ 0 \end{bmatrix}, \quad \phi_- = \begin{bmatrix} 0 \\ 1 \end{bmatrix}. \tag{G.6}$$

When these two-component spinors are acted on by the σ operators we get

$$
\begin{aligned}
\boldsymbol{\sigma} \cdot \mathbf{p}\phi_+ &= p_x \sigma_x \phi_+ + p_y \sigma_y \phi_+ + p_z \sigma_z \phi_+ \\
&= p_x \begin{bmatrix} 0 & 1 \\ 1 & 0 \end{bmatrix} \begin{bmatrix} 1 \\ 0 \end{bmatrix} + p_y \begin{bmatrix} 0 & -i \\ i & 0 \end{bmatrix} \begin{bmatrix} 1 \\ 0 \end{bmatrix} + p_z \begin{bmatrix} 1 & 0 \\ 0 & -1 \end{bmatrix} \begin{bmatrix} 1 \\ 0 \end{bmatrix} \\
&= p_x \begin{bmatrix} 0 \\ 1 \end{bmatrix} + p_y \begin{bmatrix} 0 \\ i \end{bmatrix} + p_z \begin{bmatrix} 1 \\ 0 \end{bmatrix} \\
&= \begin{bmatrix} p_z \\ p_x + ip_y \end{bmatrix},
\end{aligned}
\tag{G.7}
$$

and

$$\boldsymbol{\sigma} \cdot \mathbf{p}\phi_- = \begin{bmatrix} p_x - ip_y \\ -p_z \end{bmatrix}. \tag{G.8}$$

For an electron at rest

$$u(0, s) = \sqrt{2m} \begin{bmatrix} \phi_s \\ \phi_0 \end{bmatrix}. \tag{G.9}$$

The corresponding complementary row spinors are

$$\overline{u} = [\overline{u}_1, \overline{u}_2, \overline{u}_3, \overline{u}_4] = u^\dagger \gamma_0, \tag{G.10}$$

where as usual $u^\dagger$ is the complex conjugate transpose of u, that is its Hermitian adjoint. The spinors for spin-up and spin-down particles are orthogonal

$$\overline{u}(\mathbf{p}, 1/2)u(\mathbf{p}, -1/2) = 0. \tag{G.11}$$

With these definitions the normalization is

$$\overline{u}(\mathbf{p}, s)u(\mathbf{p}, s) = 2m. \tag{G.12}$$

The power of Dirac's formalism is that the spinors also describe the positrons as well as electrons. Indeed Dirac's work preceded the discovery of the positron by four years so that initially the antiparticle aspect appeared hard to justify. Positrons have spinors of the form

$$v(\mathbf{p}, s) = \sqrt{E + m} \begin{bmatrix} \frac{\boldsymbol{\sigma} \cdot \mathbf{p}}{E+m}\chi_s \\ \chi_s \end{bmatrix}, \tag{G.13}$$

where

$$\chi_+ = \begin{bmatrix} 0 \\ 1 \end{bmatrix}, \quad \chi_0 = \begin{bmatrix} 0 \\ 0 \end{bmatrix}, \quad \chi_- = \begin{bmatrix} 1 \\ 0 \end{bmatrix}, \tag{G.14}$$

inverting the electron spinors ϕ_+ and ϕ_-. The corresponding row spinors are

$$\overline{v} = v^\dagger \gamma_0. \tag{G.15}$$

For a positron at rest we have

$$v(0, s) = \sqrt{2m} \begin{bmatrix} \chi_0 \\ \chi_s \end{bmatrix}. \qquad \text{(G.16)}$$

Not only electrons and positrons obey the Dirac equation but also any structureless spin-1/2 fermion; that is to say all the leptons and quarks in the absence of interactions. The eigenstates presented above provide the basis for calculation even in the presence of interactions.

In high energy experiments the energies far exceed the particle masses and in this limit the weak interaction acts only on left-handed leptons and quarks which have the spin vector pointing oppositely to the direction of motion: and their right-handed antiparticles. All neutrinos have masses less than $1\,\text{eV}$ so the approximation is adequate at and above the energies met in nuclear reactions. In order to handle this behaviour formally we introduce operators to select a particular handedness.

The operator that picks out the left-handed component of a spinor is

$$P_\text{L} = \frac{1}{2} \begin{bmatrix} I & -I \\ -I & I \end{bmatrix} = \frac{1}{2}(1 - \gamma_5). \qquad \text{(G.17)}$$

For example,

$$P_\text{L} \begin{bmatrix} w_1 \\ w_2 \end{bmatrix} = \begin{bmatrix} w \\ -w \end{bmatrix}, \qquad \text{(G.18)}$$

where $w = (w_1 - w_2)/2$. Inserting this into the Dirac equation for massless particles gives

$$\frac{\partial}{\partial t} \begin{bmatrix} w \\ -w \end{bmatrix} = \boldsymbol{\sigma} \cdot \nabla \begin{bmatrix} w \\ -w \end{bmatrix}, \qquad \text{(G.19)}$$

so that only a two-component spinor is required. For such a particles with four-momentum $(E, \mathbf{p})$, $E = p$ and then

$$Ew = -\boldsymbol{\sigma} \cdot \mathbf{p}w. \qquad \text{(G.20)}$$

The spin is therefore opposite to the momentum: the particle is said to have *negative helicity*. The corresponding operator for picking out right-handed states is

$$P_\text{R} = \frac{1}{2} \begin{bmatrix} I & I \\ I & I \end{bmatrix} = \frac{1}{2}(1 + \gamma_5). \qquad \text{(G.21)}$$

Then

$$P_\text{R} \begin{bmatrix} w_1 \\ w_2 \end{bmatrix} = \begin{bmatrix} w' \\ w' \end{bmatrix}, \qquad \text{(G.22)}$$

where $w' = (w_1 + w_2)/2$.

When spin-1/2 fermions are not in the ultrarelativistic regime their spin is no longer forced to lie antiparallel to the momentum or parallel in the case of antiparticles. The handedness operators P_L and P_R pick out what are called *chiral* states. It is with the left-handed chiral component of particles and right-handed chiral component of antiparticles alone that the W-boson couples (as in the ultrarelativistic limit). On the other hand states with spin aligned along the direction of motion or against it are called states of definite *helicity*. Chirality is a Lorentz invariant property, but helicity is not. Consider an electron in one frame of reference having helicity $\hbar/2$: purely by changing to a reference frame in which its motion is reversed the helicity is reversed.

Using eqn. G.5 we can write the positive and negative helicity states of an electron travelling along the z-axis as

$$
u_\uparrow = N \begin{bmatrix} 1 \\ 0 \\ \kappa \\ 0 \end{bmatrix} \quad u_\downarrow = N \begin{bmatrix} 0 \\ 1 \\ 0 \\ -\kappa \end{bmatrix} \tag{G.23}
$$

Here $N = \sqrt{E+m}$ and $\kappa = p/(E+m) \approx \beta$ where β is the particle velocity in units c. The left-handed chiral component of $u_\uparrow$ is thus

$$
\begin{aligned}
P_L u_\uparrow &= \frac{1}{2} \begin{bmatrix} 1 & 0 & -1 & 0 \\ 0 & 1 & 0 & -1 \\ -1 & 0 & 1 & 0 \\ 0 & -1 & 0 & 1 \end{bmatrix} \begin{bmatrix} 1 \\ 0 \\ \kappa \\ 0 \end{bmatrix} \\
&= \frac{1}{2} \begin{bmatrix} 1-\kappa \\ 0 \\ -1+\kappa \\ 0 \end{bmatrix} .
\end{aligned} \tag{G.24}
$$

Hence the fraction of left-handedness in the positive helicity fermion is

$$
\frac{1}{2}(1-\kappa) \approx \frac{1}{2}(1-\beta). \tag{G.25}
$$

This diminishes as the particle's energy increases, so that in the ultrarelativistic limit the helicity up state is purely right-handed chiral. Correspondingly the helicity down state becomes pure left-handed chiral.

Under the parity transformation spatial dimensions reverse so that eqn. G.1 becomes

$$
\left[i\gamma_0 \frac{\partial}{\partial t} + i\boldsymbol{\gamma} \cdot \nabla - m \right] P\psi = 0, \tag{G.26}
$$

where P is the parity operator acting on the spinor. Multiplying this on the left by γ_0, and then using $\gamma_0 \boldsymbol{\gamma} = -\boldsymbol{\gamma}\gamma_0$, gives

$$
\left[i\gamma_0 \frac{\partial}{\partial t} - i\boldsymbol{\gamma} \cdot \nabla - m \right] \gamma_0 P\psi = 0. \tag{G.27}
$$

This has recovered the original Dirac equation provided that

$$\gamma_0 P = I. \tag{G.28}$$

Now $P^2 = I$, so we have

$$P = \pm\gamma_0. \tag{G.29}$$

Choosing $P = \gamma_0$ gives the fermion positive parity and the anti-fermion negative parity. If $P = -\gamma_0$ is chosen the reverse is true. What we can say for sure is that fermion and anti-fermion have *opposite* parity.

Appendix: Higgs Lagrangian

Expanding the covariant derivative in eqn. 19.21 gives

$$
\begin{aligned}
D_\mu\phi &= \partial_\mu\phi + igt_i^L\phi W_{i\mu} + i(g'/2)y\phi B_\mu \\
&= \partial_\mu\phi + ig\left(\frac{1}{2}\begin{bmatrix} 0 & 1 \\ 1 & 0 \end{bmatrix}\phi W_{1\mu} + \frac{1}{2}\begin{bmatrix} 0 & -i \\ i & 0 \end{bmatrix}\phi W_{2\mu}\right) \\
&\quad + \left(ig\frac{1}{2}\begin{bmatrix} 1 & 0 \\ 0 & -1 \end{bmatrix}\phi W_{3\mu} + \frac{ig'}{2}y\phi B\right) \\
&= \partial_\mu\phi + ig\left(\frac{1}{2}\begin{bmatrix} (v+h)/\sqrt{2} \\ 0 \end{bmatrix}W_{1\mu} - \frac{i}{2}\begin{bmatrix} (v+h)/\sqrt{2} \\ 0 \end{bmatrix}W_{2\mu}\right) \\
&\quad + \left(-\frac{ig}{2}\phi W_{3\mu} + \frac{ig'}{2}\phi B\right).
\end{aligned} \tag{H.1}
$$

The isospin operators used here are $t_i^L = \frac{1}{2}\tau_i$, where τ_i has the same formal structure as the spin operator σ_i given in Appendix D. Also

$$
\begin{aligned}
D_\mu\phi^\dagger &= \partial_\mu\phi^\dagger - ig\left(\frac{1}{2}\begin{bmatrix} (v+h)/\sqrt{2} \\ 0 \end{bmatrix}W_{1\mu}^* + \frac{i}{2}\begin{bmatrix} (v+h)/\sqrt{2} \\ 0 \end{bmatrix}W_{2\mu}^*\right) \\
&\quad + \left(+\frac{ig}{2}\phi W_{3\mu}^* - \frac{ig'}{2}\phi B^*\right).
\end{aligned} \tag{H.2}
$$

Then the kinetic term in eqn. 19.18 is

$$
\begin{aligned}
T &= D_\mu\phi^\dagger D_\mu\phi \\
&= (\partial_\mu\phi)^2 + g^2[W_1^2 + W_2^2]\frac{(v+h)^2}{8} \\
&\quad + [gW_3 - g'B]^2\frac{(v+h)^2}{8}.
\end{aligned} \tag{H.3}
$$

The potential term to be subtracted is

$$
\begin{aligned}
V &= \mu^2\phi^2 + \lambda\phi^4 = \mu^2(v+h)^2/2 + \lambda(v+h)^4/4 \\
&= \mu^2\left(\frac{v^2}{2} + vh + \frac{h^2}{2}\right) + \frac{\lambda}{4}(v^4 + 4v^3h + 6v^2h^2 + 4vh^3 + h^4)
\end{aligned}
$$

Now using $\mu^2 = -v^2\lambda$ we get

$$
V = -v^4\lambda/4 + \lambda v^2 h^2 + O(h^3) = \text{Higgs vacuum value} + \lambda v^2 h^2 + O(h^3). \tag{H.4}
$$

Inserting T and V in eqn. 19.18 produces eqn. 19.22.

Appendix: Entangled beauty and CP violation

Oscillations between the beauty meson $B^0(d\bar{b})$ and its antiparticle, both of mass 5.280 GeV, provided another system that has been studied to detect and measure CP violation. Electron-positron colliders had been built at the accelerator laboratories, KEK in Japan and SLAC in the USA, to study the physics of beauty particles. Tuned at a centre of mass energy of 10.58 GeV the e^+e^- annihilations produced the $\Upsilon(4S)$ vector meson with quark content $(b\bar{b})$ exclusively. The reactions possible are then limited to

$$
\begin{aligned}
e^- + e^+ \rightarrow \Upsilon(4S)(b\bar{b}) \quad &\rightarrow \quad B^0(d\bar{b}) + \overline{B^0}(\bar{d}b)\ 50\%, \\
&\rightarrow \quad B^+(u\bar{b}) + B^-(\bar{u}b)\ 50\%.
\end{aligned}
$$

There is 20 MeV of the centre of mass energy remaining after creating the beauty mesons, not enough to create any further particles. In order to give these mesons useful flight lengths before they decay the electron and positron beam energies are made unequal, 9 GeV and 3.1 GeV respectively, so that the centre of mass is moving relative to the laboratory with a $\beta\gamma$ of 0.56. The beauty mesons acquire a momentum of about $0.56 \times 5.280 = 3$ GeV. The CP and mass eigenstates analogous to K_S^0 and K_L^0 that propagate freely are labelled B_L and B_H. Both have so many decay modes because of their large mass that their lifetimes are effectively equal. The significant parameter is their mass difference ($\Delta m = 0.335$ meV), which, as in the case of K^0-mesons, determines the frequency of oscillation between B_H and B_L, and between B^0 and $\overline{B^0}$: this is 0.51 ps^{-1}. With the B^0 lifetime being 1.52 ps there are 1.52×0.51 or 0.775 cycles of oscillation in the decay time, about twice that for neutral K-mesons.

Figure I.1 shows the reconstruction of an event in the BaBar detector at the SLAC collider as seen looking along the beams' axis. All the decay products are charged and their tracks registered by the detector. A uniform magnetic field is applied parallel to the beam axis; that is perpendicular to the figure. From the track curvature the particle momentum can then be determined.

What is especially important is that the $\overline{B^0}$ and B^0 mesons are created in an *entangled* state: their identity can only be revealed when one or

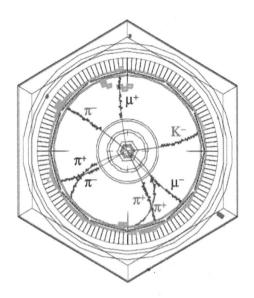

Fig. I.1 Fully reconstructed event at BaBar. The black dots are the hits recorded in the gaseous tracking chamber. The green splashes indicate energy deposition in the calorimeters around the tracker. The blue dots are hits in the muon chambers outside the calorimeters to which only the μ leptons can penetrate. Courtesy the BaBar Collaboration and Chris Hawkes.

other decays. We can work out their Bell state by utilizing the conservation of C, P and CP in the production and decay of the $\Upsilon(4S)$, which are respectively electromagnetic and strong interaction processes. The parent $\Upsilon(4S)$ has the same quantum numbers as the photon by virtue of the electromagnetic production process which conserves C-parity and parity: $J^{PC} = 1^{--}$ and CP+. These must also be the overall quantum numbers of the pair of beauty mesons because they are produced by a strong interaction process. Both neutral B-mesons have $J^P = 0^-$, hence in order to conserve angular momentum in the production process they must emerge with unit relative orbital angular momentum: L=1. This orbital motion has parity $((-1)^L = -)$ and hence the parity of the $\overline{B^0}B^0$ system is negative like the parent $\Upsilon(4S)$. The product of their C-parities is positive which makes the product CP negative. Therefore, the entangled state of the neutral B-mesons has to be antisymmetric under their interchange so that their overall state has CP+ like the parent $\Upsilon(4S)$. In symbols:

$$|\Upsilon(4S)\rangle \rightarrow \frac{1}{\sqrt{2}}[|B^0_1\rangle|\overline{B^0}_2\rangle - |\overline{B^0}_1\rangle|B^0_2\rangle]. \qquad (I.1)$$

The subscripts added here specify the order in which they decay.[1] A unique approach is then possible for observing CP violation. Modes of decay are selected for study that are identical for particle and antiparticle parents. The so-called *golden modes* are in the f_- channel

$$B^0 \text{ or } \overline{B^0} \rightarrow K^0_S + J/\psi \qquad (I.2)$$

and the corresponding f_+ channel involving a K^0_L. They are known as golden channels because it is easy to recognize and reconstruct the

[1]See J. Bernabeu and F. Martinez-Vidal, *Reviews of Modern Physics* 87, 165 (2015) for more details.

kinematics of common decays of the K^0_S, K^0_L and of the J/ψ. J/$\psi(c\bar{c})$ of mass 3.097 GeV decays via the simple channels

$$J/\psi \to \mu^+ + \mu^- \ \ \text{or} \ \ e^+ + e^-, \tag{I.3}$$

both with a branching ratio of 6 per cent. In the golden channels the

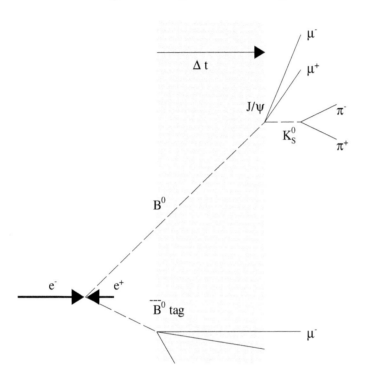

Fig. I.2 Production and decays of the entangled B-mesons. The interval Δt can be positive or negative. The corresponding path lengths are only of order 500 µm. Broken lines indicate neutrals only detected through their daughter particles.

B^0 and the K^0 have spin-parity $J^P = 0^-$ while J/ψ has $J^{PC} = 1^{--}$. In order to conserve angular momentum in the decay of the beauty meson the daughters, K^0 and the J/ψ, must have unit relative orbital angular momentum. Thus the final state f_- has CP given by

$$\text{CP}(f_-) = \text{CP}(K^0_S) \times \text{CP}(J/\psi) \times \text{CP}(L=1) = (+)(+)(-) = -,$$

and when a K^0_L appears $\text{CP}(f_+) = +$. This brings us to the essential point of the analysis. If CP is conserved in weak interactions then the rates of the processes $B^0 \to f_{CP}$ and $\overline{B^0} \to f_{CP}$ will be equal: any differences will signal CP violation. The source of CP violation could occur through interference between the channels

$$B^0 \to f_{CP}, \tag{I.4}$$
$$B^0 \to \overline{B^0} \to f_{CP}, \tag{I.5}$$

or the corresponding processes for $\overline{B^0}$.

The experimental procedure is sketched in Figure I.2. After some interval following their creation one neutral B-meson decays. If the decay includes a negative lepton then it *tags* its parent as $\overline{B^0}(\overline{db})$: decaying via $b \to l^- + c + \overline{\nu}$. A positive lepton signals the parent is a B^0. This means in the example shown that the surviving meson is a B^0 at time zero, which after a time t decays as a CP− state. Asymmetry parameters that would vanish if CP were conserved are defined and measured

$$A_{\pm} = \frac{R(\overline{B^0} \to f_{\pm}) - R(B^0 \to f_{\pm})}{R(B^0 \to f_{\pm}) + R(\overline{B^0} \to f_{\pm})}, \tag{I.6}$$

where R is the rate for a given reaction, and where the specified identity of the beauty meson is that indicated by the tag at time zero. The variations of the two asymmetries, $A_{\pm}$, are shown in Figure I.3 as functions of proper time interval between the *tag* and the decay. These asymmetries are seen to oscillate showing that a neutral B-meson oscillates between B^0 and $\overline{B^0}$. They are in antiphase as expected for states (f_+ and f_-) with opposite CP.

The CP violation can be analysed in an analogous manner to the neutral K-meson case. This gives

$$A_{\pm} = \pm \sin 2\beta \sin \Delta m t \tag{I.7}$$

where t is the proper time between the tag and the decay. $\sin 2\beta$ is a CP violating parameter. The experimental value is currently 0.691 ± 0.017. The connection with the CKM matrix, given in Chapter 18, is that $\tan \beta = \eta/(1 - \rho)$.

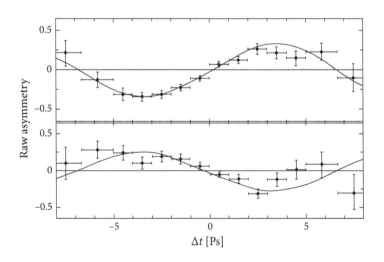

Fig. I.3 Raw asymmetry as a function of the proper time for the CP− J/ψK$_S^0$ channel in the upper panel, and for CP+ J/ψK$_L^0$ in the lower panel. Adapted from figure 2 in B. Aubert et al, *Physical Review Letters* 99, 171803 (2007). Courtesy The American Physical Society.

Appendix: The Gross–Pitaevskii equation

This appendix introduces a formalism for calculating the behaviour of a boson gas condensate at a temperature close to absolute zero in an externally applied potential well, in addition taking into account the weak interactions between the atoms. The interactions are weak enough and the temperature low enough so that only s-wave interactions are significant, and these can be treated by the Born approximation. We can take over the result from Chapter 7 showing that in such cases each atom feels an effective potential

$$U_0 = gn = 4\pi\hbar^2 na_s/m, \tag{J.1}$$

where a_s is the s-wave scattering length, m the atomic mass and n the number density of the atoms.

A gaseous condensate has an overall wavefunction given by eqn. 16.2

$$\Psi(\mathbf{r}_1, \mathbf{r}_2, ..., \mathbf{r}_n, t) = \prod_{i=1}^{i=n} [\psi(\mathbf{r}_i, t)]. \tag{J.2}$$

The motion of individual atoms described by $\psi(\mathbf{r}_i, t)$ satisfies a time-dependent Schrödinger equation

$$i\hbar \frac{\partial \psi}{\partial t} = \left[-\frac{\hbar^2 \nabla^2}{2m} + V(\mathbf{r}) + gn \right] \psi, \tag{J.3}$$

where $V(\mathbf{r})$ is the trap potential. This is known as the time-dependent *Gross–Pitaevskii equation*. It has an obvious parallel in the Ginzberg–Landau equation, a simlarly useful tool for handling the macroscopic properties of superconductors. In a time-independent potential the energy is conserved and we have

$$\psi(\mathbf{r}, t) = \psi(\mathbf{r}) \exp(-i\mu t/\hbar) \tag{J.4}$$

with μ being the chemical potential, that is the energy required to add one atom to the condensate. Then substituting for $\psi(\mathbf{r}, t)$ in eqn. J.3 gives the time-independent Gross–Pitaevskii equation

$$\mu\psi = \left[-\frac{\hbar^2 \nabla^2}{2m} + V(\mathbf{r}) + gn \right] \psi. \tag{J.5}$$

We can show that very rapid changes of $\psi(\mathbf{r})$ with position are not possible. If there is a sizeable change in the wavefunction, meaning of order unity, in a distance x there is an associated momentum $\sim \hbar/x$ and a kinetic energy $\sim [\hbar^2/(2mx^2)]$. This kinetic energy must not exceed the potential energy $\sim ng$. Then by equating these energies we can determine a shortest distance ξ, over which a significant change can occur in ψ:

$$\xi = \hbar/\sqrt{2mng} = 1/\sqrt{8\pi a_s n}, \tag{J.6}$$

which is called the *healing length*.

A situation met with for many gaseous condensates is that the interaction energy swamps the kinetic energy: the limit in which the kinetic energy may be neglected is called the *Thomas–Fermi limit*. The solution to the Gross–Pitaevskii equation in this Thomas–Fermi limit is

$$n = |\psi|^2 = [\mu - V(\mathbf{r})]/g. \tag{J.7}$$

In an harmonic well the potential energy at a radius r given by eqn. 2.24 is $m\omega_0^2 r^2/2$, where ω_0 is the classical angular frequency of motion. Then in the Thomas–Fermi limit

$$|\psi|^2 = (\mu - m\omega_0^2 r^2/2)/g. \tag{J.8}$$

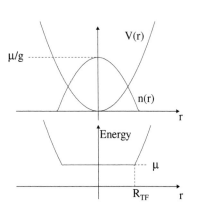

Figure J.1 shows the shapes of the potential $V(r)$, of the density $n(r)$ and of the total energy distribution. The radius indicated in the figure is the Thomas–Fermi radius outside which the wavefunction vanishes

$$R_{\mathrm{TF}} = \sqrt{2\mu/m\omega_0^2}. \tag{J.9}$$

In terms of a_s and the number of atoms in the well, N,

$$R_{\mathrm{TF}}/a_{\mathrm{ho}} = [15Na_s/a_{\mathrm{ho}}]^{1/5}, \tag{J.10}$$

Fig. J.1 The Thomas–Fermi distributions for an harmonic potential well.

where $a_{\mathrm{ho}} = \sqrt{\hbar/m\omega_0}$ gives the length scale appropriate to the well: see Section 2.3. This ratio is typically between 5 and 10, indicating that the condensate spreads well beyond the well due to the mutual repulsion between the atoms.

Figure J.2 shows the distribution along the long axis of cigar shaped condensate of ^{23}Na atoms measured by Hau and coworkers[1] This data neatly demonstrates the importance of interactions in spreading the condensate over a much larger volume than the trap.

J.1   Excitations and Bogoliubov theory

Dilute gaseous Bose-Einstein condensates are fragile so that excitations of experimental interest that do not disrupt them can be treated as

[1]L. V. Hau, B. D. Busch, C. Liu, Z. Dutton, M. M. Burns and J. A. Golovchenko, *Physical Review* A58 R54 (1998).

small perturbations. Working in this limit Bogoliubov expanded the wavefunction as

$$\psi = \psi_0 + \delta\psi, \tag{J.11}$$

where $\delta\psi$ is an excitation from a pure condensate ψ_0. Inserting this expansion into the time-dependent GPE, eqn. J.3 gives

$$-\frac{\hbar^2}{2m}\nabla^2\delta\psi + V\delta\psi + 2g|\psi|^2\delta\psi + g\psi^2\delta\psi^* = i\hbar\frac{\partial\delta\psi}{\partial t}. \tag{J.12}$$

Plane wave perturbations with energy $\hbar\omega$ and momentum $\hbar q$ have the form

$$\delta\psi = \exp[-i\mu t/\hbar]\left[u\exp[-i(\omega t + qr)] - v^*\exp[i(\omega t - qr)]\right]. \tag{J.13}$$

Inserting this in the previous equation and projecting out the coefficient of $\exp[-i\omega t]$ gives

$$-\frac{\hbar^2 q^2}{2m}u = Vu + 2ngu - gnv - \hbar\omega u - \mu u. \tag{J.14}$$

In the Thomas–Fermi limit $\mu = gn + V$, so that the previous equation simplifies to

$$\left[\frac{\hbar^2 q^2}{2m} + gn - \hbar\omega\right]u - gnv = 0. \tag{J.15}$$

Similarly projecting out the coefficient of $\exp[+i\omega t]$ we obtain

$$\left[\frac{\hbar^2 q^2}{2m} + gn + \hbar\omega\right]v - gnu = 0. \tag{J.16}$$

These last two coupled equations in u and v must satisfy a secular equation in order to be consistent,

$$\left[\frac{\hbar^2 q^2}{2m} + gn - \hbar\omega\right]\left[\frac{\hbar^2 q^2}{2m} + gn + \hbar\omega\right] - g^2 n^2 = 0, \tag{J.17}$$

which reduces to give the excitation energy

$$\varepsilon_q = \hbar\omega = \left[\frac{\hbar^2 q^2}{2m}\left(\frac{\hbar^2 q^2}{2m} + 2gn\right)\right]^{1/2}. \tag{J.18}$$

This *dispersion relation* reduces to simple forms at both low and high energy. At low energy the interaction dominates, $gn \gg \hbar^2 q^2/2m$, so that

$$\varepsilon_q = \hbar q\sqrt{gn/m}. \tag{J.19}$$

In this limit the motion is collective, the excitations being phonons. Their velocity is the speed of sound

$$v_s = \varepsilon_q/\hbar q = \sqrt{ng/m}. \tag{J.20}$$

At high excitation energies, $\hbar^2 q^2/2m \gg gn$, the dispersion relation collapses to

$$\varepsilon_q = \frac{\hbar^2 q^2}{2m}, \tag{J.21}$$

so that the excitations are particle-like.

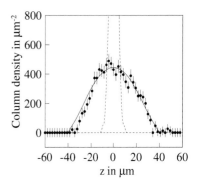

Fig. J.2 Density distribution of 80 000 sodium atoms in the trap of Hau et al. versus the axial coordinate. The experimental points are the column density of the cloud. The solid line gives the prediction for the mean-field theory of atom interactions: the dashed line that for a noninteracting gas in the same potential well. The profiles share the same normalization. The peak of the Gaussian lies near $5500\,\mu m^{-2}$. Adapted from Figure 3 in F. Dalfovo, S. Giorgini, L. P. Pitaevskii and S. Stringari, *Reviews of Modern Physics* 71, 463 (1999).

Appendix: Kubo's Hall conductance formula

Kubo derived[1] a widely used expression for the linear response of an observable to some perturbation that varies with time. Here the analysis is used to obtain an expression for the Hall conductance. The Hamiltonian (energy) is

$$H + v, \tag{K.1}$$

where v is the perturbation, small compared to H. In the interaction representation states transform like

$$|\psi(t)\rangle = U(t,0)\,|\psi_0\rangle, \tag{K.2}$$

where $|\psi_0\rangle$ is the state at time zero, and

$$U(t,0) = \exp[-(i/\hbar)\int_0^t v(t')\mathrm{d}t']. \tag{K.3}$$

Operators transform like

$$\mathrm{Op}(t) = \exp[iHt]/\hbar]\,\mathrm{Op}(0)\,\exp[-iHt/\hbar]. \tag{K.4}$$

Then a current in the voltage loop in the IQHE develops in this way

$$
\begin{aligned}
\langle J_v(t)\rangle &= \langle \psi(t)|J_v(t)|\psi(t)\rangle \\
&= \langle \psi_0|U^{-1}(t)J_v(t)U(t)|\psi_0\rangle \\
&= \langle \psi_0|J_v(t)|\psi_0\rangle + (i/\hbar)\langle \psi_0|\int_0^t [v(t'), J_v(t)]\mathrm{d}t'|\psi_0\rangle \\
&\quad + O(v^2),
\end{aligned} \tag{K.5}
$$

where the change in current is the term of interest.

$$\langle \delta J_v(t)\rangle = (i/\hbar)\int \langle \psi_0|[v(t'), J_v(t)]|\psi_0\rangle \mathrm{d}t'. \tag{K.6}$$

With the IQHE in mind the interaction is chosen to be between an applied electric field $E = E_0 \exp[-i\omega t]$ and the longitudinal current in its loop. Now

$$v(t) = -\mathbf{J}_j \cdot \mathbf{A}. \tag{K.7}$$

[1]R. Kubo, *Journal of the Physical Society of Japan* 12, 570 (1957).

Choosing a suitable gauge,

$$\mathbf{E} = -\partial \mathbf{A}/\partial t. \tag{K.8}$$

Then for the chosen perturbation

$$\mathbf{A} = \mathbf{E}/(i\omega), \tag{K.9}$$

and

$$v(t) = iJ_j E_0 \exp[-i\omega t]/\omega. \tag{K.10}$$

Substituting this in eqn. K.6 gives

$$
\begin{aligned}
\langle \delta J_v(t) \rangle &= \int_{-\infty}^{t} \langle \psi_0 | [J_v(t), J_j(t')] | \psi_0 \rangle E_0 \exp[-i\omega t'] \, \mathrm{d}t'/(\hbar\omega) \\
&= \frac{E_0}{\hbar\omega} \int_{-\infty}^{t} \langle \psi_0 | [J_v(0), J_j(t')] | \psi_0 \rangle \exp(-i\omega t') \mathrm{d}t'.
\end{aligned} \tag{K.11}
$$

The correlation only depends on $T = t - t'$ so that

$$
\begin{aligned}
\langle \delta J_v(t) \rangle &= \frac{E_o \exp(-i\omega t)}{\hbar\omega} \\
&\int_0^{\infty} \mathrm{d}T \exp(i\omega T) \langle \psi(0) | [J_v(0), J_j(T)] | \psi(0) \rangle.
\end{aligned} \tag{K.12}
$$

From this it follows that the conductance at angular frequency ω is

$$\sigma_{jv}(\omega) = \frac{1}{\hbar\omega} \int_0^{\infty} \langle \psi_0 | [J_v(0), J_j(T)] | \psi_0 \rangle \exp(i\omega T) \mathrm{d}T. \tag{K.13}$$

The unit operator is

$$I = \sum_n |\psi_n\rangle\langle\psi_n|, \tag{K.14}$$

$|\psi_n\rangle$ being a complete set of states. Inserting this into the last equation twice and rewriting T as t gives

$$
\begin{aligned}
\sigma_{jv} &= \int_0^{\infty} \sum_n \{ [\langle \psi_0 | J_v | \psi_n \rangle \langle \psi_n | J_j | \psi_0 \rangle \exp[i(E_n - E_0)t/\hbar] \\
&\quad - [\langle \psi_0 | J_j | \psi_n \rangle \langle \psi_n | J_v | \psi_0 \rangle \exp[i(E_0 - E_n)t/\hbar]\} \\
&\quad \exp(i\omega t)\mathrm{d}t/(\hbar\omega),
\end{aligned} \tag{K.15}
$$

where we have used

$$\langle \psi_n | J_j(t) | \psi_0 \rangle = \langle \psi_n | J_j(0) | \psi_0 \rangle \exp[i(E_n - E_0)t/\hbar]. \tag{K.16}$$

Carrying out the integration and putting $J_{v/j}(0) = J_{v/j}$ gives

$$\sigma_{jv} = \frac{-i}{\omega} \sum_{n \neq o} \left[\frac{\langle \psi_o | J_v | \psi_n \rangle \langle \psi_n | J_j | \psi_0 \rangle}{\hbar\omega + E_n - E_0} - \frac{\langle \psi_0 | J_j | \psi_n \rangle \langle \psi_n | J_v | \psi_0 \rangle}{\hbar\omega - E_n + E_0} \right]. \tag{K.17}$$

Then for the DC conductance the denominators become

$$(\hbar\omega \pm E_n \mp E_0)^{-1} = (\pm E_n \mp E_0)^{-1} - \hbar\omega(E_n - E_0)^{-2}. \qquad \text{(K.18)}$$

Note that the first term changes sign under the rotation $j \to v$, $v \to -j$ so it must vanish (also by gauge invariance). This leaves in the DC limit:

$$\sigma_{jv} = i\hbar \sum_{n \neq 0} \left[\frac{\langle \psi_0 | J_v | \psi_n \rangle \langle \psi_n | J_j | \psi_o \rangle - \langle \psi_0 | J_j | \psi_n \rangle \langle \psi_n | J_v | \psi_0 \rangle}{(E_n - E_0)^2} \right]. \qquad \text{(K.19)}$$

Index